EARTH SCIENCE

A Holistic Approach

EARTH SCIENCE

A Holistic Approach

DONALD J. CONTE
California University of Pennsylvania

DONALD J. THOMPSON
California University of Pennsylvania

LAWRENCE L. MOSES
California University of Pennsylvania

WCB **Wm. C. Brown Publishers**
Dubuque,Iowa•Melbourne,Australia•Oxford,England

Book Team

Project Editor *Lynne Meyers*
Developmental Editor *Tom Riley*
Production Editor *Daniel Rapp*
Designer *K. Wayne Harms*
Art Editor *Carla Goldhammer*
Photo Editor *Lori Hancock*
Permissions Coordinator *Gail I. Wheatley*
Art Processor *Joyce E. Watters*
Visuals/Design Developmental Consultant *Donna Slade*

Wm. C. Brown Publishers
A Division of Wm. C. Brown Communications, Inc.

Vice President and General Manager *Beverly Kolz*
Vice President, Publisher *Earl McPeek*
Vice President, Director of Sales and Marketing *Virginia S. Moffat*
National Sales Manager *Douglas J. DiNardo*
Marketing Manager *Amy Halloran*
Advertising Manager *Janelle Keeffer*
Director of Production *Colleen A. Yonda*
Publishing Services Manager *Karen J. Slaght*
Permissions/Records Manager *Connie Allendorf*

Wm. C. Brown Communications, Inc.

President and Chief Executive Officer *G. Franklin Lewis*
Corporate Senior Vice President, President of WCB Manufacturing *Roger Meyer*
Corporate Senior Vice President and Chief Financial Officer *Robert Chesterman*

Cover photo of Earth: NASA; cover photo of aurora borealis: © SPL/Photo
Researchers, Inc.

Composition by The Clarinda Company

Copyedited by Laura Beaudoin

The credits section for this book begins on page 431 and
is considered an extension of the copyright page.

Printed in the United States of America by Wm. C. Brown Communications, Inc.,
2460 Kerper Boulevard, Dubuque, IA 52001

10 9 8 7 6 5 4 3 2 1

DEDICATION

*J*ust as we write about interrelationships in this holistic text, we wish to dedicate this book to those whose relationships with us have made our lives whole and our planet truly a home.

To Donna, and to Michael, Melanie, and Christian Conte

To Ardith, and to Laura, Michael, and Matthew Thompson and Pamela Luketich

To Lisa and Leslie Moses

CONTENTS

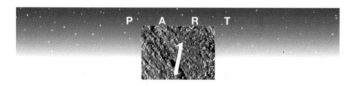

PART 1

INTRODUCTION TO A HOLISTIC SCIENCE

STUDY EARTH SCIENCE FOR OBVIOUS REASONS

PART 2

THE BIG PICTURE

THE UNIVERSE, MATTER, AND ENERGY

PART 3

THE INVISIBLE ENVELOPE

3

PRELUDE TO WEATHER

4

ORIGIN, STRUCTURE, AND CHARACTERISTICS OF THE ATMOSPHERE

5

UNDERSTANDING THE TROPOSPHERE

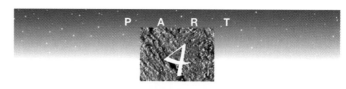

PART 4

GEOLOGICAL COMPOSITION AND PROCESSES

6

CLIMATE

7

EARTH MATERIALS

8

EXTERNAL FEATURES AND PROCESSES

9

EARTHQUAKES, GEOLOGIC STRUCTURES, AND EARTH'S INTERIOR

10

PLATE TECTONICS: A UNIFIED THEORY OF EARTH

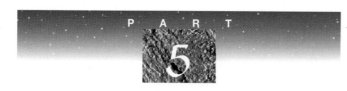

P A R T 5

OCEAN WATERS AND ENVIRONMENTS

11

OCEAN WATERS

12

MARINE AND NEAR-SHORE ENVIRONMENTS

PREFACE

he study of Earth is the study of a home. This book is about the nature of our residence, its origin, composition, and processes. To explain this complex planet, we aim to unify its many physical interrelationships in an approach we call holistic.

Earth science is traditionally subdivided into astronomy, meteorology, geology, and oceanography, and we have not ignored this sound practice. What is different about our approach is the integration we achieve within the limitations of an easy-to-read, relatively short exposition.

The overall structure of this approach can be thought of as a spiraling inward, from large, encompassing concepts to the specifics of the four earth science subdisciplines. Because many students are unfamiliar with the excitement of science, we begin with an overview that avoids describing a "scientific method" in sterile terms. Instead, we relate what scientists do to familiar experiences. Then, we explain the four earth science subdisciplines with specific examples. We believe that many nonscientists can find meaning in the explanation of a real event that affected real people, such as a hurricane or an earthquake. *Earth Science: A Holistic Approach* offers a very practical logic behind the scientific method and scientific inquiry because the book's focus is always on the planet as the personal home of its readers.

The intended audience of this book is the entry level student whose scientific background is limited and whose interest can be enhanced by a story-like, conversational style of narration. We want our readers to be comfortable as we lead them toward and through the specifics of earth science. Consequently, we define many terms within the context of a subject, rather than through strict, formal definitions—such definitions are available in the book's glossary.

The holistic integration that we present changes the typical categorization of earth science topics. For example, instead of grouping our explanations of coastal processes with oceanographic topics exclusively (a practice of others), we place our explanations with other surface processes, where they logically belong. We also incorporate descriptions of atmospheres on other planets into a discussion of Earth's atmosphere, tying our planet to the Solar System. Likewise, we treat the Sun, which is the source of the energy that drives our atmosphere, as a prelude to Earth's weather rather than as some distant, unrelated celestial object.

We summarize our holistic approach in two diagrams found early in the book. Figures 1.2 and 1.3 exemplify the interrela-tionships that we strive to convey. Such illustrations and the many cross references within this book enable a reader unfamiliar with earth science to imagine the relationships among the universe, the Solar System, and our planet, and the ties among the diverse earth features and processes. Although we would prefer our readers to follow the order of our chapters, which we believe builds rationally toward two large integrative chapters on resources and earth history, we do structure each part of the book so that readers may independently pursue any subdiscipline.

After an introductory chapter that lays the intellectual foundation for the holistic study of Earth, we begin the text proper in chapter 2 with astronomy. We explain the origin and fundamental composition of our universe, Solar System, and planet in a manner that leads to the other earth science subdisciplines. We return to develop some aspects of this topic further in chapter 14 in the context of Earth's history. Like other earth science texts, ours has an explanation of stars and stellar evolution in the astronomy section, but unlike traditional texts, we relate solar composition, structure, and processes directly to our planet.

The treatment of weather and climate, chapters 3 through 6, begins with a familiar experience as we narrate the passage of a hypothetical storm system as seen through the eyes of TV forecasters. The story of this dynamic weather system unfolds in the context of the televised forecasts that are familiar to the general public. We then explain the nature of Earth's atmosphere against the background of this storm system. Our explanations of weather culminate in chapter 6 with an integrative study of climate that contains numerous examples and simplified characteristics of weather patterns.

We explore the composition of the solid Earth by relating minerals and rocks in chapter 7 to discussions of matter in chapter 2. Our exposition of geology continues as we turn our readers' attention in chapter 8 to the processes that shape Earth's surface. By revealing the nature of earthquakes in chapter 9, we lay the groundwork for the holistic study of Earth systems in chapter 10, the study of plate tectonics. We treat this topic not only for the value of its integrative concepts, but also as a lesson in the scientific method. It is in chapter 10 that we explain the process of multiple working hypotheses by tracing the history of the theory in a story-like fashion.

When we turn our reader more specifically toward oceanography in chapter 11, we adopt an apparently traditional approach, but chapter 12 extends the chemistry, geology, and dynamic interrelationships of the oceans to the study of marine and near-shore environments in a way unique to this book. This approach enables us to show the complexity of ocean basins while relating oceans to surface processes, tectonics, and life.

Our book ends with two encompassing chapters. Chapter 13 ties the earth sciences together in the explanation of Earth's resources by integrating climate, tectonics, erosion, and sedimentation. The book culminates in a chapter on how our planet arrived at its current state by concentrating on the history of North America.

The book is well illustrated. Almost every major point is accompanied by a photograph, a line drawing, or a chart. We have also included two kinds of boxes on topics of interest. One variety of box, titled "Further Consideration," briefly explores related topics that are of interest, but are not essential to the understanding of principal concepts covered in the chapter. Often, these deal with new directions and initiatives in the earth sciences. The second type of box is one with an environmental flavor, called "Investigating the Environment," which shows how Earth's processes affect our lives.

To enhance the learning process, we include at the end of each chapter a list of significant points, a list of essential terms, review questions, and "challenges" to the students to explore some topics in greater depth. An *Instructor's Manual* with a *Test Item File* includes black and white line drawings suitable for conversion to transparencies. Classroom Testing Software available in DOS, Windows, and Macintosh formats, facilitates the instructor's evaluation of student progress. Full-color 35mm slides of photographs from the text are also available, as are full-color transparencies of key text diagrams. A *Student Study Guide,* with questions on each of the almost 500 illustrations and with practical earth science exercises for each chapter, will aid students in assimilating the details of earth science, provide a sound review of essential principles and terms, and give students laboratory-like experiences. The book also contains an extensive glossary and ten appendixes with matter germane to the study of earth science.

ACKNOWLEDGEMENTS

From inception to publication, this project had many helpers and benefactors. At California University of Pennsylvania, our colleague William A. Gustin provided us with valuable insights. An army of graduate students also performed a service by helping to compile the glossary, researching questions, and by running and fetching. We wish to thank William Belski, William Brotz, David Domen, Deborah Johnson, Kenneth Kretchun, Walter Thompson, Anthony Mauro, and Raymond Reed, all of whom demonstrated a professionalism in their diligence. The editors and staff at Wm. C. Brown Publishers kept the project on track. Without their constant help, suggestions, and constructive criticism, this book could never have been completed. We especially want to acknowledge Lynne Meyers: she was the beacon who guided us through some foggy times. Finally, we wish to thank our reviewers, whose names are on the next page.

LIST OF REVIEWERS

S. Roger Kirkpatrick, *Marietta College*
Alan L. Kafka, *Boston College*
Judy Ann Lowman, *Chaffey College*
John F. Looney, Jr., *University of Massachusetts–Boston*

John P. Szabo, *University of Akron*
Paul Nelson, *St. Louis Community College–Meramec*
Stephen Wareham, *Fullerton College*
Steve Leavitt, *Tree-Ring Lab, University of Arizona*

INTRODUCTION TO A HOLISTIC SCIENCE

Earth science is an encompassing discipline that incorporates astronomy, meteorology, geology, and oceanography into a holistic explanation of our planet. By clarifying the nature of Earth, this multidisciplinary approach provides us with information necessary for our well-being. We are earthbound beings, and this planet is our only home. After all, even astronauts must return from their space voyages. Learning about our home is a sensible practice.

AT THE END OF PART 1, YOU WILL BE ABLE TO

1. illustrate the holistic philosophy of earth science;
2. identify with the nature of scientific modeling;
3. distinguish between personal and scientific information;
4. justify the necessity of using scientific instruments;
5. distinguish between the disciplines of earth science;
6. recognize and cite evidence for the practicality of earth science studies.

Earth Prayer
Attributed by many to Chief Seattle

This Earth we know.
The Earth does not belong to us;
we belong to the Earth.
This we know.
All things are connected
like the blood which unites one family.
All things are connected.

1

STUDY EARTH SCIENCE FOR OBVIOUS REASONS

Basaltic lava fountaining from fissure.

Chapter Outline

INTRODUCTION

We don't get much of a choice when it comes to livable worlds: *Earth is all we have.* We can dream of future explorations and colonies on other planets, but the truth is that other worlds, even in our own Solar System, are either too far from Earth or too inhospitable for prolonged visits. So, our dependence upon this world is undeniably obvious, and understanding its composition, features, and processes makes good sense. Think of earth science as a way of learning about your home.

Pictures of Earth reveal a special world, one with vast liquid oceans surrounding land masses and a sphere of life-supporting gases and abundant water clouds (fig. 1.1). Supplied by energy from the sun and by gravity, the waters and gases of this predominantly blue world are in constant motion, shaping and changing Earth's surfaces. Occasionally, catastrophic events caused by either earthquakes or volcanic eruptions hint that below the planet's surfaces other processes also generate change. Earth science is an ongoing attempt to understand not only the agents of planetary change but also the resultant features, such as mountains, plains, lakes, ocean basins, river systems, glaciers, and weather patterns.

THE HOW AND WHAT OF EARTH SCIENCE

Generally, earth scientists are in the business of creating **models,** or conceptual representations of the real world. They constantly define our planet's processes, cycles, patterns, irregularities, composition, and structure to provide us with understandable representations. Ultimately, the work of earth scientists may lead to a holistic model of all that makes up our world. The encompassing nature of such a model would give a perspective from which humans could view all interrelationships among the world's component parts and functions. By tying together the origins, compositions, forms, and processes of our planet's features, earth scientists work to establish a unified theory of Earth, one that gives meaning to a complex planet and its universe.

Figure 1.1   Our home as viewed from space.

In science, a **theory** is a model based on fact, *not* an idle speculation. To form a unified theory of Earth, scientists must take into account natural forces, such as gravity, which govern the formation of earth features and the changes these features undergo. They must also rely on other disciplines, such as mathematics, chemistry, and physics. The variety and complexity of our planet's features and processes make the task of modeling a difficult one. As figure 1.2 shows, just one earth feature may represent many interrelationships. Great volcanoes like Mount Saint Helens in southern Washington or Mount Pinatubo in the Philippines are complexes of ash, rock, ice, and water that cast their influence over wide areas. Active volcanoes can affect the weather, cause devastating floods and mudflows, and blanket a region in ash. In turn, volcanoes result from unseen processes acting beneath the planet's surfaces. The lesson is that none of our planet's features or processes exists in a vacuum. Each either influences or is influenced by other features and processes. Holistic models, therefore, are attempts at clarifying the many interrelationships among Earth's processes and features.

SCIENTIFIC INFORMATION GATHERING

Each of us learns about the world through personal experiences, but the knowledge we attain is limited by our emotions, concentration, and previous experience. To avoid the limitations of personal involvement in learning about Earth, scientists work to eliminate, or at least reduce, subjectivity in their studies. Gathering information for science, therefore, is different from personal learning.

Human Perceptions and Conclusions

We know our surroundings through the perceptions of the senses. We see, hear, feel, taste, and smell the world around us, and we rely on these perceptions to furnish us with information about our environment. What we perceive through our senses, however, can be misleading. For example, if you have suffered from a fever, you know that the temperature you sense with a fever is different from that sensed by a healthy person. Similarly, a dry day may not appear to be as "hot" as a day of equal temperature but with higher humidity. The difference lies not in the actual heat of the air but in our perceptions. Because we rely on a mechanism such as perspiring to regulate body temperature during exercise and hot weather, our personal comfort—and, thus, our sense of temperature—depends on the capacity of the air around us to absorb moisture through evaporation, which is a cooling process (discussed in chapter 4).

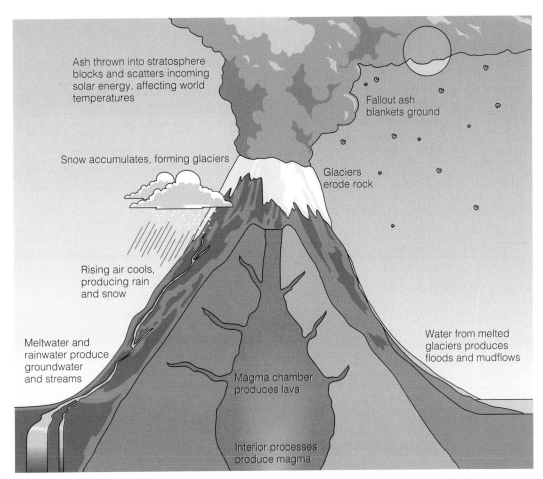

Figure 1.2 A volcano is a local complex of interrelated features and processes that may have a worldwide impact.

Scientific Observations and Conclusions

Because of human limitations, earth scientists cannot rely merely on the five senses to determine the physical nature of the world, that is, to explain Earth's composition, processes, and features. They know that much of our planet is invisible to us, or beyond our reach, and is odorless, colorless, and tasteless. They also realize that misinterpreting the information garnered by the senses is a common way of missing the truth. Although the senses don't lie to us personally, we often err when we assume what we sense is reality.

To avoid these mental traps, earth scientists try to recognize patterns in Earth's features and processes. Such patterns, revealed by the gathering of specimens and the use of instruments that record data without human variations, help scientists to create the models by which we understand Earth objectively.

CLASSIFICATION OF THE EARTH SCIENCES

As earth scientists have acquired knowledge about our home, they have established four inclusive classes of information to explain Earth as a planet: astronomy, meteorology, geology, and oceanography. Although these four categories are traditionally how earth science texts are divided, a unified theory must rely on seeing the interrelationships among the four disciplines. Thus, in their study of our planet, earth scientists borrow information from each of the four (fig. 1.3). In the study of the atmosphere, for example, earth scientists may draw on information about mountains and plains, ocean currents, and even solar events, such as increases in the number of sunspots.

This text treats these interrelationships as the foundation of earth studies, beginning with astronomy and the origins of the universe and Solar System. Because Earth is just one of countless bodies in the universe, a definition of its position and nature

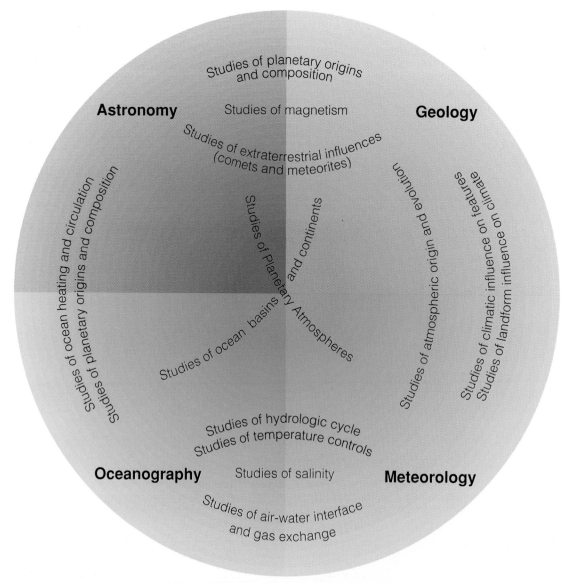

Figure 1.3 Earth scientists interrelate the four disciplines for a holistic modeling of our planet.

in time and space is important. The other three disciplines, meteorology, geology, and oceanography, follow in a sequence that makes necessary and frequent references to the preceding information. The composition of Earth's atmosphere as treated in this text, for example, appears in the context of planetary atmospheres; and changes along coastal zones, ordinarily covered in oceanography sections in the traditional approach, appear here in the context of other surface processes.

SCIENTIFIC RESEARCH THROUGH INSTRUMENTS

Because our senses cannot detect much of our world and because we often misinterpret exterior reality by misreading our sensory data, earth scientists use instruments specifically designed for registering what we cannot perceive through our five senses. An example of something invisible to us that requires instrumentation to detect is unseen waves of energy, some of which are harmful, that permeate our environment.

Instruments Detect the Electromagnetic Spectrum

One of the most obvious examples of studying the world invisible to the human eye is research into the nonvisible portions of the **electromagnetic spectrum.** The spectrum is a range of energy forms known as **radiation. Electromagnetic radiation** is energy in wave form that originates from every vibrating body in the universe (which is *every* body in the universe) and that travels across the near vacuum of space at the speed of light. Included within this general category are the forms of radiation familiar to many of us, at least in name: gamma rays, X rays, ultraviolet rays, light, infrared rays, microwaves, and radio waves. All of these forms are alike, except for their wavelengths, which vary from extremely short lengths for gamma and X rays to very long wavelengths for radio waves (fig. 1.4). Hotter bodies produce more short wavelengths than cooler bodies. Thus, the sun produces much more ultraviolet and visible light radiation than the much cooler Earth.

Instruments Find an Invisible Hole

Since the mid-1970s, scientists have become aware that solar **ultraviolet (UV) radiation,** which causes sunburn and skin cancer, has intensified in various high-latitude regions, particularly over Antarctica and latitudes above 50°N.

Further research into this invisible, but dangerous, radiation has revealed a link between its intensification and a "hole," or a decrease in abundance of **ozone** (O_3), a gas found at high altitudes in our atmosphere (fig. 1.5). Although ozone molecules form naturally in the stratosphere, they are also synthesized by humans at Earth's surface, primarily through the burning of fossil fuels and reactions with nitrogen compounds. This surface ozone is largely responsible for the photochemically produced smog that often shrouds the Los Angeles area. At Earth's surface,

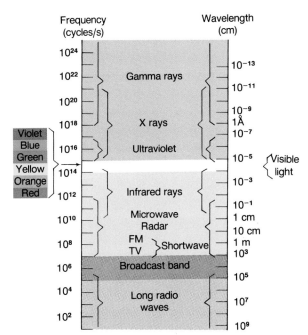

Figure 1.4 The electromagnetic spectrum. (Source: Alyn C. and Alison B. Duxbury, *An Introduction to the World's Oceans,* 3d ed. Copyright © 1991 Wm. C. Brown Communications, Inc., Dubuque, Iowa.)

this gas is a pollutant and a health hazard. Twenty-five kilometers (15.5 mi) above the surface, however, ozone acts like a protective filter.

In a photochemical reaction with the incoming solar radiation, ozone separates into O and O_2, absorbing the ultraviolet end of the spectrum in the process and reducing the amount of UV radiation (wavelengths between 0.2 and 0.3 µm) that reaches the surface. The discovery of the ozone "hole" at altitudes between 14 and 25 kilometers (9 and 15.5 mi) is typical of how earth scientists work to learn about Earth objectively and establish models that explain some of Earth's complex interrelationships, including those influenced by human activities.

The discovery of the ozone hole depended on the combined work of chemists, meteorologists, aerospace engineers, climatologists, and physicists. Now, many scientists have adopted a multidisciplinary approach to study the hole. Such an approach may lead to a model that can adequately explain the hole as a consequence of both natural and human processes (see box 4.2, "Ozone-depleting Compounds and Stratospheric Ice").

THE TRADITIONAL EARTH SCIENCE DISCIPLINES

Although meteorology, geology, and oceanography may seem to be the only relevant studies of Earth, the fourth earth science discipline, astronomy, provides us with very useful information. Its relevance, as well as the relevance of the other three sciences, is the product of the natural interrelationships that earth science attempts to reveal.

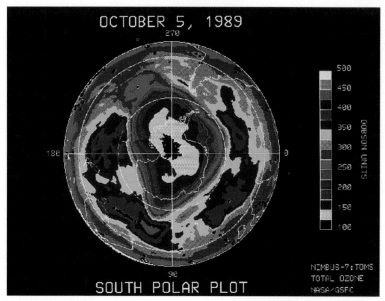

Figure 1.5   The ozone "hole" over the continent of Antarctica as mapped from space by TOMS (Total Ozone Mapping Spectrometer) in October 1989. Color intensity indicates ozone concentration in Dobson units, the standard measure.

Astronomy

There is a strong tradition among earth scientists to incorporate observations of bodies other than Earth in their studies of our planet. This is the study of **astronomy.** Earth does not exist in a void. It is part of a large Solar System that includes the sun, Earth's moon, and other planets. In addition, the Solar System is part of an even larger system of suns called the Milky Way Galaxy, and it, in turn, belongs to a cluster of sun systems. The connection between Earth and these systems is unavoidable. Electromagnetic radiation from many sources, as well as uncountable and unseen small particles, continuously bombard our planet (see box 2.3, "Will an Asteriod Hit You on the Head?"). Whenever we see the flash of a meteor, we know that collisions with comets and rocky bodies that cross Earth's orbit are possible. The Barringer Crater near Winslow, Arizona, is a good example of the devastating nature of such collisions (fig. 1.6). During the 1980s, the possibility of such impacts led some scientists to speculate that a very large impact may have dealt the last species of dinosaurs a death blow about 65 million years ago (chapter 14) and that periodic collisions may have influenced the evolution of life.

As large as our planet may seem to us when we traverse its continents and oceans, Earth is just a tiny part of the universe. This size relationship becomes apparent whenever we consider the distances to other bodies in the universe. For instance, after more than two decades of travel at 46,600 kilometers (28,900 mi) per hour, the *Pioneer 10* spacecraft is just a little more than 8 billion kilometers (5 billion mi) from Earth. If it continues uninterrupted in its journey, it will pass within 29 trillion kilometers (18 trillion mi) of the star called Ross 248 in 32,600 years. If those numbers seem large, you should realize that many objects lie millions of times farther from Earth than Ross 248 (see box 2.1, "Red Shifted Galaxies: The Farther They Are, the Faster They Go").

The objects of the universe beyond Earth and the composition of the basic constituents of matter are the topics of chapter 2.

Meteorology

We usually take for granted the everyday work of weather scientists. They tell us in a word or phrase that we should dress for heat, or cold, or rain, and we question their skills when they convey a faulty forecast. But the role these earth scientists play is more than that of fashion consultant.

Through one branch of atmospheric study, known as **meteorology,** we have learned about potential global catastrophes that make the costs of individual violent storms seem like very small change. The costs of the **greenhouse effect,** a worldwide warming trend caused by the addition of carbon dioxide and other heat-absorbing gases to the atmosphere, may exceed the resources of the collective world economies. The warming may cause the expansion of the heated oceans by at least 1 meter (39.37 in) and a significant melting of polar and high-mountain ice, adding enough water to the oceans to drown the highly populated coastal cities. What will it cost to replace a drowned New York, London, Amsterdam, Copenhagen, and Sydney? Consider Houston, Texas, for example. The city lies 80 kilometers (50 mi) inland, but it stands only 15 to 30 meters (50 to 100 ft) above the current sea level. Or, consider the plight of Delaware. The highest elevation in the state is 18 meters (60 ft). The loss of Delaware, a rich farmland, would be tragic, of course, but add to that the loss of the Netherlands, Bangladesh, and much of Denmark. Their entire populations would be displaced by an inundation of seawater.

A global warming trend could also modify the climates and, therefore, the productivity of agricultural regions. Whenever excessively hot weather continues unaccompanied by rains, soils dry out. America suffered a terrible drought in the 1930s, known

Figure 1.6   Meteor Crater, also known as Barringer Crater, in central Arizona, photographed from an altitude at 1,350 meters (4,500 ft).

as the Dust Bowl years, which wreaked havoc on the country's economy and the lives of many farmers (see box 8.3, "The Great American Dust Bowl"). Similarly, the drought of 1988 placed a severe burden on the farmers of the Midwest, and the lack of rain during the last two decades in Northeast Africa has inflicted hardship on millions of people in Ethiopia and Somalia. If the warming trend initiates a long-term drought, who will make up for the decrease in American food production, just when world population is increasing rapidly? The kitchen in our earth home may be useless without much to cook (see box 4.1, "Greenhouse Cooling?").

Weather scientists constantly monitor the composition, structure, and movement of the atmosphere. To acquire such data requires the use of familiar and unfamiliar instruments, from thermometers to remote sensing satellites. It also requires meteorologists to understand the interrelationships among the geography of our planet, incoming solar radiation, and atmospheric composition and processes. As a result of their continuing efforts, a specialized group of atmospheric scientists, called climatologists, have classified regional weather patterns. The last of the four chapters on weather discusses the holistic science of climatology. Since the formulation of this branch of earth sci-

ence, climatologists have pointed out the interrelationships among sun activity, latitude, altitude, temperature, precipitation, storms, soils, streams and groundwater levels, erosion, glaciation, vegetation, and agriculture.

Geology

Just as atmospheric scientists need instruments to "sense" the invisible air, scientists in the field of **geology** require tools for peering into those portions of Earth hidden by a covering of miles of unpenetrated hard rock. The practical importance of such "seeing" into Earth lies in scientific predictions of earthquakes and volcanic eruptions based on models of geologic interrelationships. Since World War II, earth scientists have worked with sophisticated apparatuses to determine the causes of geologic activity and features. Instruments used in the detection of nuclear bomb detonations have been applied to the study of Earth's interior, revealing a complex of layered spheres from a solid inner core to a relatively thin outer shell, called the crust.

The instrument that has provided the most knowledge of Earth's interior is the seismograph, which is sensitive to vibrations within the planet. The **seismograph,** discussed more fully

Box 1.1 Investigating the Environment

Radiative Forcing of Earth's Atmosphere

Scientists now know that a number of gases act like greenhouse glass, holding heat in an atmosphere the way glass in a greenhouse holds heat within a building. This process has been commonly called the greenhouse effect, but it is also known as *radiative forcing*. The latter term derives from the role of certain radiatively active gases. Although the greenhouse gases are transparent to incoming sunlight (shortwave radiation), they absorb outgoing longwave radiation (heat), forcing an additional heating of the atmosphere that would not occur in their absence. In fact, without the effect of these naturally emplaced gases, Earth's surface temperature would be 15.5° C (60° F) colder that its current average temperatures.

Water vapor is both the most abundant and the most efficient of the naturally occurring greenhouse gases. Evaporation from many water sources produces a constant supply of the odorless, colorless water vapor in our atmosphere. The atmosphere gets its water vapor from oceans, lakes, rivers, volcanic eruptions, geysers, soils, and plants. Other greenhouse gases also occur naturally, but human industrial and technological activities have increased these gases and added unnatural gases to our envelope of air.

Carbon dioxide (CO_2) is chief among the naturally occurring greenhouse gases that have been increased in Earth's atmosphere as a result of human activities. Since the preindustrial era (1750–1800), the amount of carbon dioxide in our atmosphere has risen 25%. During the past two centuries, the amount of carbon in our atmosphere has increased by 160 billion tons through the addition of carbon dioxide and other carbon-containing gases. When scientists exclude the effects of water vapor, they find that carbon dioxide now accounts for 66% of the radiative forcing by gases in our atmosphere (table 1.A). The sources of carbon dioxide include the burning of fossil fuels, cement production, and deforestation.

Other greenhouse gases that have increased since the rise of industrialization are methane (CH_4), nitrous oxide (N_2O), and chlorofluorocarbons (CFCs). The concentration of methane has almost doubled during the past two centuries, and nitrous oxide has increased by 8%. Methane, the major component of natural gas, is produced by anaerobic decomposition in biological processes, rice cultivation, intestinal fermentation in animals, and the decomposition of animal and municipal solid wastes. Methane is also released in coal mining.

Whenever deforestation strips the land of its plant cover, soils release additional nitrous oxide. Nitrate and ammonium fertilizers and the leaching of nitrogen fertilizers from soils into groundwater also increase this greenhouse gas.

CFCs, produced solely by human manufacture, are currently increasing in the atmosphere at the rate of 4% per year. We use these gases as refrigerants, aerosols, foam-blowing agents, and solvents. The primary CFCs in use are compounds that contain chlorine but no hydrogen atoms. Called "fully halogenated CFCs" for their lack of hydrogen, two of these gases are more important than the other CFCs in radiative forcing. CFC-11 (CCl_3F) and CFC-12 (CCl_2F_2) were both used extensively as aerosol propellants until the United States banned their use by 1980, but both gases will remain in the atmosphere for many years (table 1.A). Currently, companies are searching for CFC substitutes because of the 1987 Montreal Protocol, an international agreement that limits CFC use. In the United States, the 1990 Clean Air Act Amendments prohibit CFC production and use by the year 2000.

Nitrogen oxides, referred to as NO_x, are the two gases NO and NO_2 that are formed in lightning, by natural fires, in fossil-fuel combustion, and in the stratosphere from nitrous oxide. NO_x contributes to the greenhouse effect by increasing the formation of ozone, an effective greenhouse gas in the upper troposphere and lower stratosphere. In addition, a gas such as carbon monoxide (CO), though not directly a greenhouse gas, indirectly affects the amount of methane and ozone in the atmosphere. By reacting with the hydroxyl radical (OH), carbon monoxide limits natural reactions between OH and both CH_4 and O_3 that would remove those two gases from the atmosphere.

One of the problems our planet faces is the length of time required to remove the artificial greenhouse gases from the atmosphere. The residence time for some of these gases exceeds a century, indicating that there are long-term effects. Even if we were to stop emissions of carbon dioxides, for example, the gas we have already released to our atmosphere will affect radiative forcing for more than a century (table 1.A).

Apply Your Knowledge

1. Name the artificially produced greenhouse gases.
2. Which artificially produced greenhouse gases also occur naturally?
3. How does carbon monoxide indirectly affect radiative forcing?
4. How can the deforestation of Brazil's tropical rain forests contribute to radiative forcing?

Table 1.A

Global Contributions to Radiative Forcing by Greenhouse Gases Other Than Water Vapor and Atmospheric Residence Times

Greenhouse Gas	Percent of Radiative Forcing	Atmospheric Residence Time (Years)
Carbon dioxide	66	120
Methane	18	10.5
CFCs	11	55 (CFC-11); 116 (CFC-12)
Nitrous oxide	5	132

Source: U.S. Environmental Protection Agency, *States Workbook*. November 1992.

in chapter 9, was one of the principal instruments used by U.S. intelligence to "listen" for violations of the nuclear test ban treaty signed in the 1960s. The network of seismographic stations soon recorded an abundance of earth movements beyond those generated by atomic explosions. These natural motions, which included major earthquakes, established a pattern for the planet's interior based on how the seismic vibrations were distributed. The network of seismographs has now been refined for specific research into the structure of Earth's crust and deep interior. Now called **seismic tomography,** the study of both crust and interior through seismic wave analysis incorporates a number of seismic stations whose operators pool data to create a three-dimensional picture. Through this developing system of "seeing" Earth's interior, scientists are acquiring more detailed pictures of the structures and processes of the **lithosphere,** the planet's rocky outer layer, which includes the continents and the ocean floors.

Although geologists find seismic and other remotely sensed data valuable, to understand the lithosphere they rely heavily on the examination of tangible specimens. Fortunately, geologists have been able to explore and sample the continental rocks and a considerable area of the ocean floor. Their explorations have given us a great deal of information about the composition of the surfaces on which we live.

Thus, like the other earth science disciplines, geology has its own subclasses of information. Some geologists specialize in the study of minerals and rocks, others in the processes that shape the landscape, and still others in geologic hazards and the large forces at work within our planet (see box 9.2, "Earthquake Prediction: Where, When, and How Big?"). This text divides geology into the study of materials, processes, landforms, resources, and earth history.

Oceanography

Figure 1.1 provides evidence that from the perspective of space, no earth feature is more apparent than the oceans, which cover 71% of our planet's surface and contain more than 1 billion km³ (240,000,000 mi³) of water. The oceans are the source of water for the **hydrologic cycle,** which moves water as a liquid, solid, and gas; the oceans are efficient distributors of the sun's heat from low to high latitudes through surface currents; and they are the ultimate sinks for everything that runs off the continents. Keeping track of the movements of the oceans, their interactions with the atmosphere, their effects on the coastal environment, and their great varieties of plants and animals has required cooperation among oceanographers, meteorologists, and biologists. **Oceanography,** as you might have guessed, is a cooperative venture that aims to explain these interrelationships.

The surface waters of the oceans move in large currents driven by the general circulation of the atmosphere. In moving, they transport not only water but also heat from the equatorial regions. Tracing this flow of both water and heat has required observations at sea and from space, where satellites "image" the oceans in a variety of wavelengths. The satellite images, called "false color" *Landsat* and Thematic Mapper images, show oceanic

boundaries invisible to the naked eye. Through such imaging, earth scientists have been able to delineate various **eddies,** or spinning bodies of water, within the oceans and to plot the locations and movements of marine life (fig. 1.7).

"Seeing" under the water has also added to the understanding of earth systems. Recent adaptations of the sonar technology developed during World War II to oceanographic studies has revealed details as small as 100 meters (330 ft) on the deep ocean floor. New views of an extensive volcanic mountain chain that runs through the ocean now provide marine geologists and geophysicists with corroborating evidence that the seafloor newly forms and spreads along breaks or rifts. The theory of seafloor spreading, first promulgated in 1959, is now part of a larger, more encompassing theory called **plate tectonic theory,** the subject of chapter 10. *This theory is the most complete perspective, or model, that earth scientists have to unify earth studies.* The formulation of **plate tectonics** was only possible because of instruments that extend human senses to the invisible and the hidden.

THE PRACTICALITY OF EARTH STUDIES

As our only home, Earth is analogous to the houses in which we live. There is much to learn about a house, which the house itself can teach. The materials used in its construction are clues to its origin; the layout is a clue to the structure, and the daily processes, such as the entering and exiting of water through pipes, the weathering of paint and mortar, and the heaving of sidewalks after a winter and spring of freezing and thawing all provide

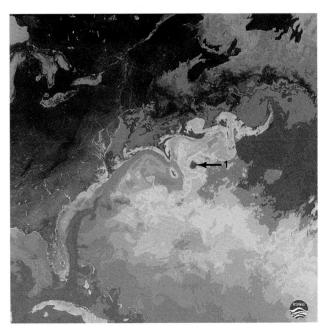

Figure 1.7   Sea surface temperatures imaged from an altitude of 960 kilometers (600 mi). The Gulf Stream flows along the eastern coast of the United States as a warm water current (red and orange). Temperatures diminish (yellow and green) as the current mixes with surrounding water.

clues about the house and its surroundings. The more a home-owner knows about a dwelling, the better prepared that owner is to anticipate problems and seek remedies: *ignorance is usually expensive and sometimes very dangerous*. Earth, like a house, reveals much about itself.

Costly Neighborhoods, Dangerous Circumstances

A good example of expensive ignorance is the location of San Francisco in an earthquake zone. The city's founders, unaware of the area's potential for violent tremors, established the community in what seemed to be an ideal setting. In the last century and in the first decade of the twentieth century, the city grew to more than 400,000 people. In 1906, the city was the Gem of the Pacific, but on April 18 at 5:12 A.M., a powerful earthquake shook the region and started fires that eventually burned through more then 10 square kilometers (4 sq. mi) of the city. More than 2,000 died, and most of the businesses were destroyed (fig. 1.8).

Of course, the survivors rebuilt the city, and as the decades passed, San Franciscans learned from scientists that other earthquakes of similar strength were possible. The information led to the architectural standards for "earthquake-proof" buildings. In 1989, the city felt it was prepared for an earthquake, and its residents and others around the United States may have joked about the "big one," a predicted earthquake as powerful as the quake of 1906. As the third game of the World Series between the Oakland Athletics and the San Francisco Giants was about to begin at Candlestick Park, the ground shook again. Sixty-two people lost their lives, most on the lower level of a double-deck highway in nearby Oakland (fig. 1.9). During the earthquake, which originated in the vicinity of Loma Prieta, just south of the Bay area, 3,757 people suffered injuries, 12,000 lost their homes,

Figure 1.9   Collapsed segment of the Nimitz Freeway (I-880) in Oakland, California.

367 businesses were destroyed, and property damage exceeded $7 billion according to government figures (see box 9.1, "Earthquake Engineering: Bracing for the Big One").

Knowledge about Earth Saves Lives

Its extensive damage made the October 17, 1989, earthquake in San Francisco one of the costliest natural catastrophes in American history. That remarkably few lives were lost, however, was partly the result of those architectural designs governed by geophysical and geological information that earth scientists had provided. In fact, earth scientists had a hand in saving other lives on the East Coast just six weeks prior to the earthquake. When Hurricane Hugo powered its way over South Carolina, it damaged or destroyed more than 35,000 homes, toppled trees worth an estimated $1 billion, and ruined about $500 million in crops (fig. 1.10). The 24 counties hit hardest by Hugo suffered $5 billion in damages. As in the West Coast tragedy, the work of earth scientists significantly reduced the number of casualties (18 died). Meteorologists and oceanographers had broadcast warnings of the storm's strength and direction, giving government officials sufficient time to arrange evacuations of those coastal areas most susceptible to the high winds and waters of the approaching hurricane (see box 5.2, "Hurricane Andrew: America's Most Destructive Storm").

The warnings of earth scientists aren't always heeded. When Mount Saint Helens showed signs of increased volcanic activity in 1980, the warnings of geologists were ignored by disbelieving and uncooperative people who did not understand the power such a volcano could generate. In particular, a longtime resident named Harry Truman refused to evacuate, claiming he would be there as long as the mountain was. Part of the mountain is gone now, obliterated in the May eruption. Harry Truman, unfortunately, is gone also. He did not seem to understand the nature of his home.

Figure 1.8   A portion of the devastation caused by the 1906 San Francisco earthquake. Some buildings collapsed, and others were gutted by fire.

Figure 1.10 Storm damage in coastal South Carolina from Hurricane Hugo in 1989.

Incomplete Knowledge

Even earth scientists cannot predict the intensity of earthquakes, volcanic eruptions, and some storms. On November 15, 1989, for example, meteorologists watched the development of a storm system as it approached the southeastern states. The National Weather Service issued a severe storm watch for several states, but no one knew that before this weather system moved over the Atlantic Ocean, its destructive winds would kill 29 people from Alabama to Canada. On that November day, the severe weather hit Huntsville, Alabama, at 4:37 P.M. as a tornado with winds estimated at 400 kilometers per hour (250 mph) destroyed 1,000 cars, 259 homes, 3 churches, 10 public buildings, and scores of electric power structures (fig. 1.11). In about ten minutes the storm fatally injured 17 people along its 26-kilometer (16-mi) path. Because weather systems move, they can devastate widely separated areas. While the people of Huntsville were assessing the damage on the following day, the same storm system generated unexpected high winds in Newburgh, New York, collapsing a cafeteria wall and killing seven children in an elementary school (see box 5.1, "Tornado: A Late Autumn Storm").

Our incomplete knowledge of Earth's processes is characterized by a statement a National Weather Service meteorologist made to reporters following the devastation of the Alabama storm. He described the Huntsville tornado as "a weatherman's nightmare. We couldn't see it on radar." In spite of many technological advances and scientific studies, we have much to learn before we fully understand our home planet. Earth scientists can enhance our knowledge about Earth.

Questions for Earth Scientists

Through the efforts of earth scientists, we now have an elaborate catalog of knowledge about our planet. We also know about conditions and processes that can change Earth's ability to support life. This introductory chapter has hinted at possible disasters, leaving us with many unanswered questions: Will our planet collide with a large comet or meteorite? Can we change the warming trend? Can we predict destructive storms with greater accuracy? Can we predict earthquakes? Earth scientists are working to answer these and other questions about our planet. Although they have only partial answers, they do know, through years of observation and research, the major causes of these phenomena.

Common sense tells us that knowledge is better than ignorance. On the level of our ordinary, daily activities, there is much

Figure 1.11   A small part of the destruction in Huntsville, Alabama, caused by a tornado on November 15, 1989.

that earth science can teach us that has practical meaning. What is the origin of our drinking water? of our gasoline? our gemstones and gold? aluminum and iron? What—and this is a very large question indeed—is the origin of Earth?

Questions for You

Even more important for you is a series of questions about your community. What is the cause of changes in your weather? How do the vast oceans affect your life? Why does your landscape take the forms you see? What is the composition of those forms, and what lies beneath their surfaces? What natural phenomena might threaten the stability and safety of your region?

Answers to these and other questions about your home come from the work of earth scientists: they explore, study, observe, and explain the processes, composition, features, history, and future of our house, the planet we call Earth. Study earth science for an obvious reason: A familiar home is not only a comfortable one; it may also be less costly and safer than an unfamiliar home.

Further Thoughts: A Holistic and Practical Science

Because earth science is an eclectic mix of ways to look at and understand our planet, it is a holistic science. Although individual earth scientists may strive for knowledge about a single feature or process outside of, on, or in the planet, the overall tendency of earth science observations and research is toward models of Earth that show the interconnectedness of its features, processes, and life.

Earth science also serves many practical purposes. By explaining how and why volcanoes erupt, the land quakes, and the atmosphere moves, for example, earth scientists lay the framework for safety. By exploring the planet with instruments that probe beyond the capabilities of human senses, they show us where we can mine the natural resources that support our technological societies. Finally, by revealing the past captured in the planet's features and by explaining the nature of the processes that have operated for billions of years, they show us the possibilities of our planet's future.

Significant Points

1. Earth is a unique planet because it supports life.
2. Earth scientists attempt to model the processes and features of our planet.
3. Scientists use instruments to detect whatever lies beyond human sensory perception. They can study, for example, the full extent of the electromagnetic spectrum, which lies beyond the wavelengths of visible light.
4. Four earth science disciplines—astronomy, meteorology, geology, and oceanography—are interrelated branches of knowledge.
5. Earth is subject to radiation from the sun and other sources, meteorite impacts, and other astronomical influences.
6. Our planet is a relatively small object by comparison with the great distances to other objects in our universe.
7. The study of the atmosphere has yielded clues about changes in worldwide weather patterns, given us warnings about violent storms, and led to predictions of future dangers.
8. Geologists have ascertained the nature of Earth's interior through the use of the seismograph, an instrument sensitive to ground vibrations.
9. Geologists rely on the analysis of specimens to ascertain the exact composition of the lithosphere.
10. Oceanographers study both the ocean basins and the waters they contain. Like the other earth scientists, they use instruments such as satellites to discern oceanic patterns and features.
11. The most significant earth science theory, called plate tectonics, was largely the product of numerous oceanographic studies.
12. Knowledge about Earth is very practical. Because we must rely on the planet to sustain our lives, we should understand the processes and features that affect us.

Essential Terms

model 4	ultraviolet (UV) radiation 7	greenhouse effect 8	hydrologic cycle 11
theory 5	ozone 7	geology 9	oceanography 11
electromagnetic spectrum 7	astronomy 8	seismograph 9	eddies 11
electromagnetic radiation 7	meteorology 8	seismic tomography 11	plate tectonics 11

Review Questions

1. How does scientific information gathering differ from sensory perception?
2. What is the ultimate goal of the earth sciences?
3. How does a scientific theory differ from speculation?
4. What led to the detection of the hole in the ozone layer of the stratosphere?
5. Why are multidisciplinary approaches to the study of Earth essential to earth scientists?
6. What evidence do we have that objects from beyond Earth's boundaries influence the planet?
7. How can atmospheric warming affect civilization?
8. What enables geologists to see into Earth's interior?
9. How can the oceans influence weather?
10. What recent natural disasters have produced fewer deaths than comparable past events because of earth scientists?

Challenges

1. How would you categorize the natural surroundings of your home? If you have observed your area's weather and the landscape for a number of years, you have at least a general idea about how hot, wet, flat, or hilly your neighborhood is. Would you make a comparison with some other area you have visited to establish your categorization?
2. What are the primary sources of your food? How would these sources be affected by changes in rainfall brought on by a global warming trend? Would your sources of meat be affected as much as your sources of grains, fruits, and vegetables?

3. What would happen to river mouths if global warming increases sea level? How would seaports be affected by such a change?

4. What characteristics or needs of life make living on other planets, even in artificially maintained Earth-like environments, difficult or impossible? Is there one characteristic or need you can identify as the most important?

5. What physical processes are at work to change your area?

6. How can the knowledge gained through earth studies help government officials protect citizens?

2

THE BIG PICTURE

Although we are earthbound, we cannot avoid the relationship between our planet and all that lies beyond its boundaries. Earth did not form in a void, and it both influences and is influenced by other bodies. For thousands of years humans have attempted to explain the composition, structure, and evolution of the universe and to understand the interrelationships among its many bodies. Today, our knowledge of bodies outside Earth is aided by sophisticated instruments, many of which have been launched into outer space.

AT THE END OF PART 2, YOU WILL BE ABLE TO

1. identify the methods astronomers use to measure the universe;
2. recognize the dual nature of light;
3. recognize and cite evidence for the significance of gravity and energy;
4. differentiate the components of matter;
5. explain the origins of the universe and Solar System;
6. compare the types and evolutionary stages of stars;
7. generalize from data the nature of the planets and other celestial bodies.

The Meditations of
Abu-al-ala Maárri (973-1057)
XVII

Two fates still hold us fast,
A future and a past;
Two vessels' vast embrace
Surrounds us—Time and Space.

2

THE UNIVERSE, MATTER, AND ENERGY

Earth as viewed from the surface of the Moon.

Chapter Outline

INTRODUCTION

Early on February 24, 1987, an infrequently observed event briefly changed a small portion of the sky over the Southern Hemisphere. A Canadian astronomer working in the Andes Mountains discovered an unexpectedly bright object in a nearby cluster of stars called the Large Magellanic Cloud. The bright spectacle was, in historical terms, unusual because very few such events have occurred relatively close to Earth. In fact, the last recorded object of similar brightness in Earth's vicinity occurred in 1604. To get some idea of just what is meant by "relatively close," you might consider the distance to the star cluster in which the bright object lay. The Large Magellanic Cloud is about 15,770,000,000,000,000 kilometers (9,780,000,000,000,000 mi) from our planet. As an astronomical event, the bright object was distant enough to appear small to the earthbound naked eye, but in its star cluster it was probably very significant. Regardless of its apparent size, the event was of enormous importance to scientists. It gave them a chance to study the catastrophic death of a star in an explosion called a **supernova.** The star that blew up is Sanduleak, pictured in figure 2.1 in the center of its shell of gases expanding into the surrounding space.

Because the supernova of 1987 occurred during our technological age, it became the first nearby disruption of a star to be recorded on sensitive instruments and observed through powerful telescopes. Previously studied supernovae have been too distant for accurate data collection on the scale presented by Sanduleak. To the scientists who study outer space, the event was important in three ways. First, its proximity allowed telescope views with a high degree of resolution, or clarity. Second, the observation was accompanied by the detection of very small particles called **neutrinos.** These virtually massless, high-speed particles pass through the planet all the time, rarely being detected because they do not interact with matter easily. Although they are hard to observe, neutrinos by the trillions probably passed through Earth at about the same time the astronomers first witnessed the light of the explosion. Third, the event was the subject of observation through a variety of instruments, such as radiowave antennae and infrared sensors.

To a novice in the study of outer space, the data released by the scientists who studied the 1987 supernova may serve to introduce the great distances encompassed by the universe. The supernova may also serve as a lesson about stars: they are *not* indestructible. This is significant because *our own sun is also a star,* and as a star, it will also change its character and undergo a kind of death throe. Although our sun's change will not mimic the supernova, it will entail an alteration radical enough to destroy our planet.

Big Numbers, Long Times

The destruction of Earth's sun is far off in the future, probably about 5 billion years hence, but a phase of the sun's development may jeopardize life in less than one-fifth that time. Big numbers associated with time and distance are common in astronomy, which is the study of the universe beyond Earth's

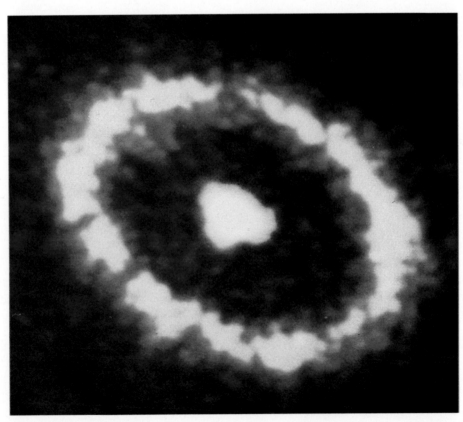

Figure 2.1 Sanduleak appeared to observers in the Southern Hemisphere as a supernova in the Large Magellanic Cloud.

confines. The quadrillions of kilometers to the 1987 supernova, for example, comprise a short distance by astronomers' standards. Objects that are tens of thousands of times farther than the Large Magellanic Cloud are known to exist. Even sizes are astoundingly large. Our sun, for example, is a little under 1,394,000 kilometers (864,000 mi) in diameter. The star whose explosion astronomers studied in 1987 was an estimated eight times larger than our sun.

Comprehending the great distances and sizes in the universe is an exercise in imagination for the residents of Earth, an oblate spheroid only 12,785 kilometers (7,927 mi) in diameter at its equator. Despite our advances in technology, most of us are bound to our planet—astronauts are very few in number. Our robot spacecraft are also mostly earthbound. Of the many space vehicles we have launched, most have merely orbited our planet. A few have traveled throughout the Sun's system of planets, and two, *Pioneer 10* (see chapter 1) and *Voyager I*, have left the Solar System.

Astronomical Questions

Sanduleak's abrupt end in a supernova is an indication that stars are not eternal. We live on a planet that is also destined to have an ending, which is tied to the fate of our sun. The idea of planets and stars having endings may seem like material for science fiction, but this occurs throughout the universe.

Before you learn of our sun's fate—and the fate of our planet—you should ask a series of questions about astronomy as a science. The great distances and sizes discussed require measurements, but most of the objects astronomers study are too far to visit for measurements or samples. How, then, do astronomers quantify the universe?

Once you have established how astronomers operate, you can begin to understand the great wealth of information they have gathered through centuries of observation. You can also learn about objects such as Sanduleak and our own sun and planets. Where and how, for example, did the universe, stars, and planets originate? Of what are they composed, and what is the origin of that material? How many are there? What is their behavior? Why are they located where we see them? And finally, What is their future?

MEASURING THE UNIVERSE

Because the distances and sizes in the universe are so great that measurements on the scale of Earth are inadequate, astronomers use the terms *light-year* and *parsec* to quantify the distances to the stars. A **light-year (lt-yr)** is simply the distance light travels in a single year, which, at approximately 300,000 kilometers (186,000 mi) per second, amounts to a little under 9.5 trillion kilometers (6 trillion mi). Astronomers also use a unit called a **parsec (pc),** which is 3.26 light-years. Such large numbers are staggering, especially when you consider that Sanduleak's supernova event was considered close, yet the Large Magellanic Cloud is more than 163,000 light-years (50,000 pc) away. In other words, Sanduleak exploded about 163,000 years ago, and

its light and neutrinos just recently passed by Earth. Measuring the distance to distant stars requires techniques involving triangulation and brightness. You will learn about both of these techniques in this chapter.

Astronomers also use the astronomical unit (AU) to measure distances within our Solar System. One AU is 150 million kilometers (93 million mi), or the distance between Earth and the Sun. Planets closer to the Sun than Earth are less than 1AU, whereas the farthest planet, Pluto, is 39 AU from the Sun (see appendix B).

In your study of astronomy, you should consider two other aspects of measurement: direction and motion. To understand these aspects, imagine yourself at the center of a sphere. Motion within that sphere can be perceived in several ways (fig. 2.2). Objects that move away from or toward you move along the radius of the sphere. Astronomers call such a positional change the *radial motion* of an object. When they determine an object's speed along the radial line, astronomers calculate the object's **radial velocity.** Of course, not every object moves along a radial line, and movements other than those directly toward or away from you appear to run along the arc of the sphere. The movements that change the position of an object across the hemispherical sky are known as *proper motions*. The rate at which a proper motion occurs is the **tangential velocity** of an object. These motions can be given in degrees, or fractions of degrees, of arc. Degrees are subdivided into minutes and further subdivided into seconds. The smallest proper motion, therefore, can be a fraction of a second.

Recognizing Motion

Proper motion is easily recognizable whenever there is a frame of reference or background against which the motion occurs. That frame of reference can be the horizon or some distant background of stars. Radial motion, however, occurs along the line of sight. If someone throws you a ball, you see it appear to increase in size as it approaches you. When you throw it back, you observe it decrease in size. Unfortunately, astronomers deal with objects that never get as close as a thrown ball and that lie at

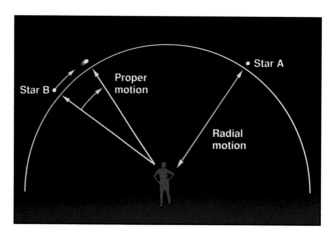

Figure 2.2 Star A exhibits radial motion, whereas Star B exhibits proper motion.

great distances, where a change in size is difficult to perceive. They need another method to determine radial motion and are fortunate to have one based on changes in the electromagnetic spectrum (see chapter 1).

The Doppler Effect

The wavelengths between .4 and .7μm (visible light) are not the only electromagnetic frequencies studied and used by astronomers. When they study the radiation of a distant object, astronomers measure changes in frequency, called the **Doppler effect.** To understand this effect, consider the changing pitch of a passing train, which not only sounds louder as it approaches but also sounds musically higher. By comparison, figure 2.3 shows a light source shining in all directions. As long as the source remains stationary, the wavelengths will be equal for an observer, but movement of a source toward and away from an observer changes the wavelengths. Whenever an approaching celestial

object emits light, that light shifts toward the shorter, blue end of the spectrum, just as the sound waves shorten and increase the pitch for an oncoming train. An object heading away from Earth has its light shifted toward the red end. The electromagnetic shift toward longer wavelengths derived from the radial motion of an object away from an observer is called the **red shift.**

Parallax

One method astronomers use to measure the distance to stars involves measurements of **parallax,** which is the *apparent positional change of an object that results from a change of perspective.* To visualize this apparent change, just hold up one finger steadily and look at it first with one eye, then with the other. Your finger appears to move along a plane even though it remains stationary.

To measure stellar parallax, astronomers need a greater distance between observation points than that between two eyes.

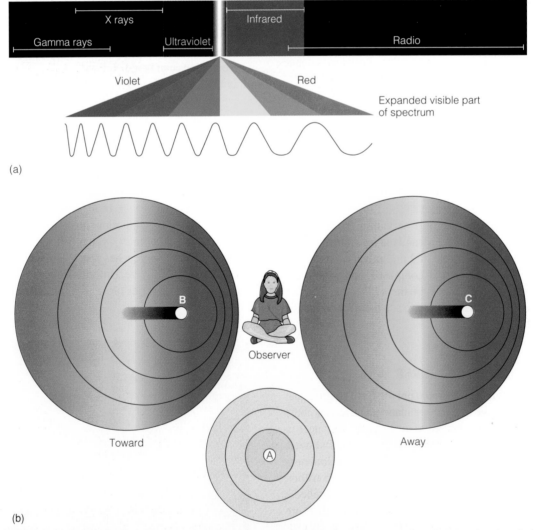

Figure 2.3 (a) Electromagnetic spectrum with the visible portion expanded showing longer wavelengths at red end and shorter wavelengths at violet end. (b) As a light source approaches an observer, the waves are compressed and the wavelengths shortened, emphasizing the violet end of the spectrum. As a light source moves away from an observer, the waves spread out, or lengthen, emphasizing the red end of the spectrum. This latter phenomenon is known as the "red shift" and is evidence that a light source is receding.

Box 2.1 Further Consideration

Red Shifted Galaxies: The Farther They Are, the Faster They Go

The relationship between distance from our galaxy and recession speed of distant galaxies can be revealed by examining their spectra. Nearby galaxies do not move away from the Milky Way as fast as galaxies farther out. Their movements can be seen in the shifts in the *H* and *K lines* resulting from calcium absorption (fig. 2.A).

The basis for understanding the relationship between the rate of recession and distance is known as *Hubble's law*. According to the law, the rate of recession equals distance multiplied by the quantity *H*, called *Hubble's constant*. Hubble's constant is a debated number, but many astronomers tentatively accept values around 15 kilometers per second per million light-years. Different values for Hubble's constant change the age and size of the universe. Figure 2.A shows distant galaxies and their rates of recession with respect to the Milky Way. Based on a constant of 15 kilometers/second/10^6 light-years, the distant Hydra is receding from our galaxy at 61,000 kilometers/second (37,820 mi/sec).

Apply Your Knowledge

1. What is the relationship between distance and recessional speed?
2. What is the method used to determine the direction of motion for a distant galaxy?
3. What is Hubble's constant?
4. Why is the value of Hubble's constant important to an interpretation of the age of the universe?

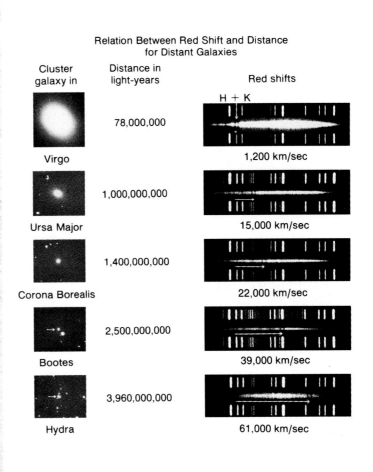

Relation Between Red Shift and Distance for Distant Galaxies

Cluster galaxy in	Distance in light-years	Red shifts
Virgo	78,000,000	1,200 km/sec
Ursa Major	1,000,000,000	15,000 km/sec
Corona Borealis	1,400,000,000	22,000 km/sec
Bootes	2,500,000,000	39,000 km/sec
Hydra	3,960,000,000	61,000 km/sec

Figure 2.A Spectra of galaxies. Photographs and spectra of galaxies in several rich clusters. The clusters range in distance from nearby (Virgo) to remote (Hydra). Prominent absorption features are indicated by arrows; these lines, known as the H and K lines, are due to calcium absorption. The lines in cluster spectra are very broad compared with lines in individual stellar spectra because of the motions of the stars in each galaxy that are contributing to the spectrum. The absorption lines are progressively red shifted as the distance to the cluster increases. This red shift is due to the Doppler effect and can be expressed in terms of the relative velocity of recession of the cluster. The recession velocity is found to increase linearly with distance to these clusters.

Stars are farther away than our fingers, and greater distances mean smaller changes in the parallax angle. To expand their perspective, therefore, astronomers use as their vantage points the opposite sides of Earth's orbit because it has a diameter of 2 AU (fig. 2.4). Unlike your looking at a finger, however, astronomers get a view of objects lying in different planes relative to Earth's orbit and shifting observation points. A distant star that lies in a plane perpendicular to our orbit would appear to move differently from one that lies on the same plane. At a perpendicular position, a star's movement would appear to be in a circle, while one in the same plane as our planet would appear to move back and forth in a straight line. Any star at an oblique angle to our orbit would describe an ellipse.

For the great distances represented by star positions, parallax measurements are made in seconds of degree of arc equated to parsecs. Even the parallax of close stars is equivalent to your recognizing the thickness of this book at 3.2 kilometers (about 2 mi) with unaided eyes. Because most stars lie at very

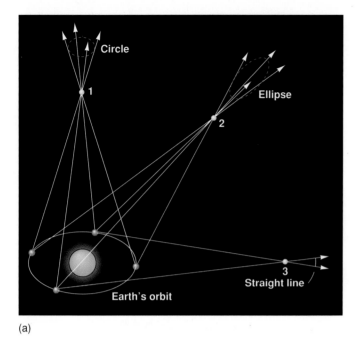

(a)

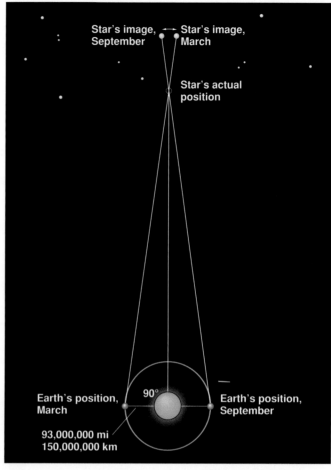

(b)

Figure 2.4 (a) Parallactic movements change shape with changing perspectives. (b) A stellar distance gauge. The astronomical term *parallax* refers to the way in which the position of a nearby star seems to shift against the motionless background of very remote stars as Earth orbits the Sun. By simple trigonometry, that shift can be used to calculate the star's distance. The star is photographed during the month of April and then six months later. The star's apparent movement across the celestial background is measured on the photographs as an angle, greatly exaggerated here. Halving that measurement yields one angle of a right triangle that has the star, the Sun, and Earth at its vertices. Because the Sun-to-Earth distance is known, the leg of the triangle representing the star-to-Earth distance is easily determined. With ground-based instruments, the technique is reliable only out to about 600 light-years.

great distances, the triangle becomes a very thin geometric feature that is difficult to measure. In fact, based on observations at 2 AU, parallax measurements within a 10% margin of error can be made for approximately 700 nearby stars. To measure distances to more distant stars, astronomers use a variety of methods, including stellar brightness.

Magnitude

The brightest object in our sky is the sun, but the scale of brightness astronomers use is based on the bright but distant stars Aldebaran and Altair. Norman R. Pogson devised this scale in 1856 to rank stellar **magnitude,** or brightness. On his scale, Aldebaran and Altair are magnitude 1 stars. Dimmer objects are represented by higher positive numbers. Since our sun *appears* to be brighter than either star, its magnitude must be expressed by a negative number, -26.5. This is the *apparent magnitude,* which results from the sun's proximity to Earth (table 2.1). In comparison with Sirius, the brightest star visible from the Northern Hemisphere, our sun appears ten billion times brighter.

In reality, however, the sun is not as intrinsically bright as any magnitude 1 star. Its *absolute magnitude,* or its brightness based on what it would look like if it were 10 parsecs away from us, is 5. The stars have intrinsic magnitudes; they are hot objects that produce their own bright electromagnetic radiation.

Variable Stars and the Doppler Effect

Some stars vary their magnitudes over short periods of 1 to 50 days. These **variable stars** have been classified according to their magnitude fluctuations, and one particular type, the *Cepheid Variables,* has been used in distance measurements. Their name derives from Delta Cephei, a giant star that contracts and expands at regular intervals. As Delta Cephei expands, it gives off more light, and as it contracts, less light. The process by which such an object produces its light will be discussed later, but its magnitude change is significant for measurements.

This shifting of frequencies can be used to interpret a Cepheid Variable's contraction and expansion—alternating radial motion respectively either toward Earth or away from it.

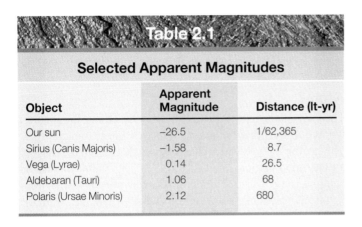

Table 2.1

Selected Apparent Magnitudes

Object	Apparent Magnitude	Distance (lt-yr)
Our sun	−26.5	1/62,365
Sirius (Canis Majoris)	−1.58	8.7
Vega (Lyrae)	0.14	26.5
Aldebaran (Tauri)	1.06	68
Polaris (Ursae Minoris)	2.12	680

The Cepheids may change size by as much as 10% within a matter of days. For enormous stars like the Cepheid Variables, that translates into millions of kilometers and a shift in frequency. If astronomers know the apparent (m) and absolute (M) magnitudes of a variable, they can calculate its distance (d) based on a well-established formula: $M = m + 5 - 5 \log d$.

Measuring Light and Other Radiation

Light decreases proportionally to the square of the distance from its source. This is a matter of experience for all of us. A flashlight held close to your face appears much brighter than one held a city block away. You can think of this as a change in the intensity of the light. A Cepheid Variable's visible, regularly changing magnitudes can be observed through a telescope or quantified through a process by which light striking an instrument called a **photomultiplier** yields a measurement of intensity by converting the light into a measureable electrical current. A device called a **spectrophotometer** acts like a prism to divide the light into colors (different wavelengths) before it enters the photomultiplier so that astronomers can look at the full spectrum of a star. The total (all the wavelengths) magnitude of radiation is called the *bolometric magnitude,* a quantity measured by these astronomical instruments.

The Nature of Light

Throughout this chapter, you have read about "brightness" and "light," two words that occur frequently in our everyday language. We have already considered brightness, but light itself is essential to astronomy, for this science began in ancient times as a study of the "lights" in the sky. But just exactly what is light? The electromagnetic spectrum shows light as a range of wavelengths, those falling between the shorter wave ultraviolet and longer wave infrared. Commonly, light is thought of as either rays or waves. If you think of light as waves, then you might make a comparison between light and waves on water. Water waves can interfere with one another if they come from different directions. In water waves, the high points (crests) and low points (troughs) can cancel or reduce each other's vertical dimension whenever they meet. Similarly, light waves can interfere with one another, and you can observe this interference as a set of parallel dark lines where the waves cancel each other.

To make your observation, just squint as you look at a light source through the very small space between two almost touching fingertips (fig. 2.5). Light observed in this fashion appears to be made of waves. There is a peculiar behavior of light, however, that makes it appear more like flows of discreet particles. These particles of light, called **photons,** are energy packages that can make a photomultiplier active. As the photons strike the tube-shaped instrument, they release negatively charged particles called **electrons,** which a detector can count. Astronomers, therefore, can objectively determine light magnitude. They can also observe other wavelengths in the electromagnetic spectrum with the aid of the spectrophotometer.

Light, therefore, has two identities. How we choose to observe light determines its identity. Whenever we want to observe it as waves, we see it act like waves. If we want to observe it act like particles, we see it act like particles. These are important considerations because much of astronomy is based on the interpretation of the light from sources that lie at distances too great for other kinds of study. For earth scientists, however, the physics of light composition is less important than its speed. *Nothing goes faster.* Its speed in a vacuum is a constant that can be used as a measuring stick for distance and as a universal clock.

Traveling at the speed of light, the messages from *Pioneer 10* take more than 7 hours to reach Earth. That means the information is about something that has already happened, just as the 1987 supernova Sanduleak occurred before *Homo sapiens* first walked on our planet. Light is our time machine, and it gives us, along with the rest of the electromagnetic spectrum, a way of reaching into the distant past and the origin of the universe.

Figure 2.5 Wave interference can be seen as small parallel dark lines between two almost touching fingers.

The Age and Size of the Universe

What do we see when we look through this "time machine"? First, we recognize that *the universe is, at the very least, indefinite*. No one has found an end, and similarly, no one has found a center. Second, the universe appears to be very old, with the most recent age estimates exceeding 12 billion years. Third, we see a visible universe that is apparently lumpy, with stars and groups of stars gathered in masses. Fourth, through the wavelengths of visible light we view a universe whose objects appear to be composed of matter. Because of this dominance and in preparation for the discussion of minerals in chapter 7, you should have a basic understanding of matter.

MATTER AND GRAVITY

The objects of our universe are composed of matter. In the simplest terms, **matter** is anything that occupies space and has mass. Matter also exerts an attractive force called gravity, as the mass of one substance pulls against the mass of another. Gravitational force increases with increasing mass and with decreasing distance. Thus, the pull of our planet's gravity is an attractive force that keeps you earthbound and prevents your drifting away into space. You recognize this force as your weight, and you would not weigh as much standing on the smaller mass of the moon. Your "moon weight," for example, would register only one-sixth of your "earth weight." Astronauts who have visited the moon have been able to move with relative ease despite wearing spacesuits and backpacks that would add 35 kilograms (77 lb) to their earth weight.

The book you are holding has mass. You know the book's mass as its weight, which is the product of gravitational attraction between the book and Earth. It also occupies space and has evident boundaries. The book is a solid, visible form of matter. The air around you is also matter. It is not, of course, visible like the book and does not appear to have definite boundaries, but you can feel it just by blowing on your hand. As a gas, the air we breathe also occupies space and has weight. In fact, quintillions (10^{18}) of kilograms of air encompass our planet. Liquid water is also matter. Its boundaries are perceived as the container in which it lies, regardless of what that container is. A bucket, a river channel, a lake, or an ocean basin can serve as the boundary for this form of matter.

The Composition and Structure of Matter

Our understanding of matter has changed over the centuries, but a common thread has often wound its way through the changing theories of matter. The connection among theories is the belief that different forms of matter have a common underlying makeup. Even the alchemists of the Middle Ages held this belief, as evidenced by their attempts to transmute other metals into gold. Their intuition told them that all substances have an inherent uniformity and that they might be manipulated to change forms.

In a sense this is true, because all matter is fundamentally composed of just three important types of particles: protons, electrons, and neutrons. Various combinations and arrangements of these three particles make up the solids, liquids, and gases of our universe.

Atoms

The three elementary particles join to make atoms, the smallest forms of matter that can maintain recognizable characteristics. The **proton** is a positively charged particle close to the mass of a **neutron,** a particle that carries no charge. Both protons and neutrons can be found in the corelike nucleus of atoms. Electrons lie outside the nucleus in "energy shells," where they orbit the nucleus. The energy shells are so called because energy is required to keep the electrons in their orbits. Within any given atom, there is a distinct segregation that keeps the protons and neutrons in the very dense central core and the electrons in their orbits. Because the electrons are moving in paths at velocities that enable them to make billions of orbits each second, there is a natural outward force, centrifugal force, that would normally cause them to fly away from the atom's nucleus. You can imitate this movement by whirling a key on a string then releasing your hold, letting both key and string fly. This outward force is counteracted in the atom, however, by the naturally attractive force between the opposing charges of the protons and electrons. All but the most distant electrons remain firmly associated with the atom.

The nucleus of the atom is held together by the **strong nuclear force,** which overcomes the natural repulsion of the crowded, like charges of the protons. This dense nucleus accounts for more than 99.9% of the atom's mass. The number of protons and neutrons is the **mass number,** a close approximation to the weight of the entire atom. Scientists also accept a categorization of atoms on the basis of the number of protons in the nucleus. Such a count of the protons yields the **atomic number** of an atom. Hydrogen, for example, with one proton in the nucleus has atomic number of 1.

All three particles are unimaginably light and small. A proton, for example, has a mass about one and a half septillionths of a gram. The dimensions of atoms are just as unimaginable as their weights. The diameters of about 40 million atoms would be required to stretch over a centimeter (more than 100 million per in). The diameter of an entire atom is about 100,000 times greater than its nucleus.

The three particles can be combined in different numbers, resulting in various atoms (fig. 2.6). Ninety-two combinations of these particles occur naturally, and they are called **elements.** In addition, scientists have been able to synthesize 11 other combinations, yielding a total 103 elements.

The differences among the atoms in figure 2.6 lie in the numbers of protons, neutrons, and electrons. Heavy atoms such as gold contain more protons and neutrons than lighter atoms. Regardless of their atomic number, however, atoms maintain an electrical equilibrium. The balanced numbers of protons and electrons within atoms makes them electrically neutral.

The elements can be grouped according to similar characteristics. The *periodic table of the elements* shown in figure 2.7 is an organization of atoms so that those in the vertical columns

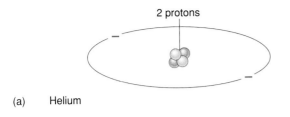

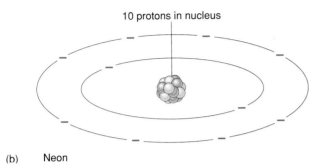

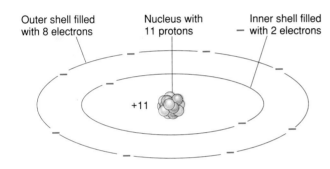

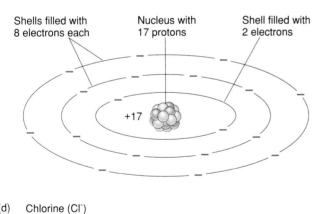

Protons   ◯ Neutrons   — Electrons

Figure 2.6 Simplified models of (a) helium atom, (b) neon atom, (c) sodium ion, and (d) chlorine ion.

have similar properties. In the far right column are the atoms helium, neon, argon, xenon, and radon. These elements have their outermost energy levels completely filled with electrons, and each one is nonreactive or inert. Because of their complete nature, these "noble gases" do not react with other elements. Having the outermost energy level full of electrons gives stabil-

ity to atoms, and in nature there is a mechanism for arranging this stability.

Combinations of Elements

All the elements other than the noble gases have incomplete outer energy levels, which can result in either the loss or gain of electrons in reactions with other atoms. These reactions form combinations of atoms called **molecules.** Understanding these combinations is important because they dominate the composition of our world.

Molecules are the smallest units of **chemical compounds,** which are the combinations of two or more atoms. The number of electrons in the outermost energy level is the control on molecular formation. A common substance such as salt is an example. In figure 2.8, common table salt is shown as a combination of sodium (Na) and chlorine (Cl). The sodium atom contains a single orbital electron in its outermost energy level. The great distance between the nucleus and that single electron enables the latter to overcome the attractive force of the protons and escape. With one less electron, the atom loses its electrical balance and becomes positively charged. This new form of sodium is an **ion,** which is an atom with an electrical charge . By contrast, the chlorine atom has seven atoms in its outermost energy level. To complete its outer shell, the chlorine atom simply picks up the single electron that abandoned the sodium atom or any free electron. With the addition of the new electron, the chlorine becomes a negatively charged ion because it contains more electrons than protons.

The two oppositely charged ions can now attract one another and form a molecule. This welding together of oppositely charged ions is called *ionic bonding* and is an important mechanism by which chemical compounds are created. Bonding can occur in other ways as well, such as by the sharing of electrons in outer shells by adjacent ions, an arrangement known as *covalent bonding* (*co,* meaning "with"; *valentia,* meaning "capacity"). This, too, has the effect of forming filled outermost electron shells and, consequently, of giving stability to the atoms. As you will see in chapter 11, this is the mechanism by which the water molecule forms.

Isotopes

Atoms can also alter their atomic mass units by the addition of neutrons to the nucleus, which is the same process by which all but one of the elements formed. Common hydrogen has a single proton in its nucleus, but two other forms of hydrogen are known. A second form of hydrogen, called **deuterium,** has a proton and a neutron in the nucleus (fig. 2.9). Such an atom differs not in atomic number but in atomic mass. The different atomic weight makes deuterium an **isotope** of hydrogen. Hydrogen has another isotope called **tritium,** an atom with two neutrons and one proton in the nucleus. Isotopes, therefore, are variants of an element based on mass number. Because the atomic number is based solely upon the count of protons and because that number does not change for isotopes of an element, the atomic number becomes the most fundamental characteristic for categorizing the elements.

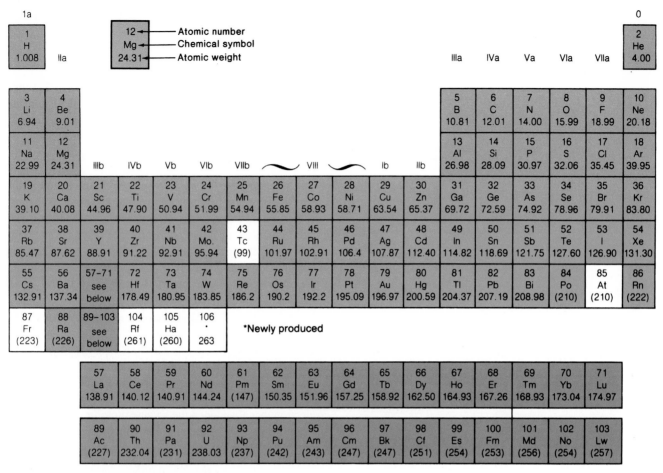

Figure 2.7 The periodic table of the elements.

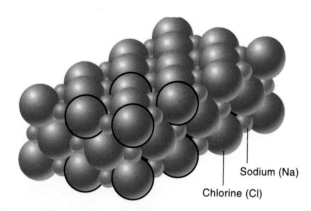

Sodium (Na)

Chlorine (Cl)

Figure 2.8 A molecular model for common table salt, halite.

Cold Dark Space

It is a common experience that we do not see light but rather light sources and whatever light illuminates. In space, light mostly travels through a vacuum, but it impinges on objects that allow it to pass on, reflect it, or absorb it. Between the widely scattered objects of the universe, however, light energy passes without illuminating anything; thus, space is dark, not light, and we observe either that which emits light or that which reflects it.

Outerspace is also cold. Our astronauts wear protective suits to provide them a livable environment and to insulate them from the cold whenever they emerge from their spacecraft for space walks. But what does "cold" mean?

Molecules, Temperature, and Energy

Of all the interrelationships we find on Earth and in the universe, the one between matter and energy is the most fundamental. Although it finds its expression in different forms, this relationship manifests itself in one commonly known way: **temperature.**

In a general way, everyone understands temperature. Weather forecasters tell us that the next day will be 21° C (70° F). We learn the scale they use by experience. We know that a day when the temperature reaches 35° C (95° F) is much hotter than one on which the temperature is only 0° C (32° F). The letters *F* or *C* are abbreviations of two scales. The **Fahrenheit** (F) scale uses 32° as the initial freezing temperature of water and 212° as the initial boiling point. The **Celsius** (C), or centigrade, scale uses 0° and 100°, respectively, for freezing and boiling. Both scales meet at −40°. Scales are used as measuring devices because they

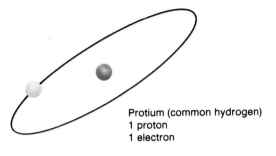

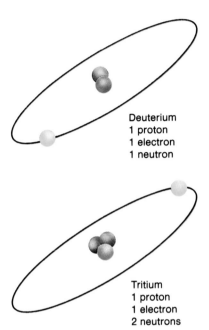

Protium (common hydrogen)
1 proton
1 electron

Deuterium
1 proton
1 electron
1 neutron

Tritium
1 proton
1 electron
2 neutrons

Figure 2.9 The three isotopes of hydrogen.

quantify the observable. Just what do the temperature scales measure?

Atoms are highly energetic entities. As you read earlier, electrons make high numbers of revolutions in their energy shells every second. Molecules possess similar motions. A water molecule, for example, written chemically as H_2O, breaks up and reforms hundreds of thousands of times every second. In addition to this internal molecular activity, molecules move and interact within their environment. This random motion is interpreted as **heat,** as molecules vibrate against one another. Faster motions among molecules generate greater heat.

Motion is a common form of energy release. In fact, **energy,** often defined as the ability to do work, usually means infusing some matter with motion. All forms of matter are associated with some motion. Earth orbits the sun; the wind blows, trees sway, people walk, and, at the unseen level of atomic vibration, electrons move in their shells. Because energy provides the mechanism that drives all earth processes discussed in this text, we will review the forms energy takes.

Potential energy is "stored" energy, or the energy that any body possesses because of its position or location. Water behind a hydroelectric dam, before it is sent through the turbines under the force of gravity to produce electricity, possesses potential energy. A boulder on a hillside also possesses this form of energy, as does a child poised on a diving board.

The energy of motion is called **kinetic energy.** Thus, when the water rushes through the turbines, it releases kinetic energy. Similarly, as the boulder rolls downhill or the child dives into the water, they release kinetic energy. On the molecular level, the vibrational motion of the individual ions releases kinetic energy. Thus, temperature is more specifically defined as the mean kinetic energy of molecules. In addition to the Fahrenheit and Celsius scales, the Kelvin (K) scale measures this kinetic energy from a hypothetical status of no molecular motion called **absolute zero.** The relationships among these three scales are shown in figure 2.10.

How hot can matter get? This question is similar to asking how fast molecules and ions can vibrate. Although outer space is a cold place, the cold is interrupted by intense concentrations of kinetic energy. One such concentration was the supernova Sanduleak, discussed as we began this chapter. Estimates of the energy inside the core of the exploding star seen in 1987 put its temperature during a 10-second outburst at 100 billion degrees Kelvin. Temperatures greater than those of Sanduleak were once the norm for our universe. Sufficient evidence now suggests that *the universe had a very hot origin.*

THE ORIGIN OF THE UNIVERSE

Astronomers have determined that the farthest observable sources of electromagnetic radiation are billions of light-years away. That also means, of course, that they are billions of years old. The most distant objects are also among the most energetic, and their red shifts indicate radial motions that make them appear to be moving away from us at a rate more than one-tenth that of light speed. *The universe appears to be expanding.*

Other evidence seems to support this expansion. In 1989, NASA launched *COBE (Cosmic Background Explorer),* a satellite that has confirmed a radiation that seems to be related to an expanding universe. The background radiation is a low-temperature phenomenon, colder than any commonly experienced on Earth. Unlike the radiation in your microwave or in your fireplace, this radiation could not be recorded on an ordinary household thermometer. It lies at a temperature of only 2.7 K. This radiation has no point source. Unlike a beam of light from a flashlight, which your eyes can follow back to its source, the background radiation seems to be originating almost equally from every direction. The most commonly accepted theory to explain both the apparent expansion of the universe and the background radiation is known as the **big bang theory,** the rapid expansion of superheated matter between 12 and 15 billion years ago.

The big bang theory is the temporal limit of science. The scientists who have worked on its description cannot be certain about the moment it occurred, but they can narrate what happened in the first microseconds of the expansion. The narration goes something like this: About 12 to 15 billion years ago, the universe was unformed. It existed as a superheated

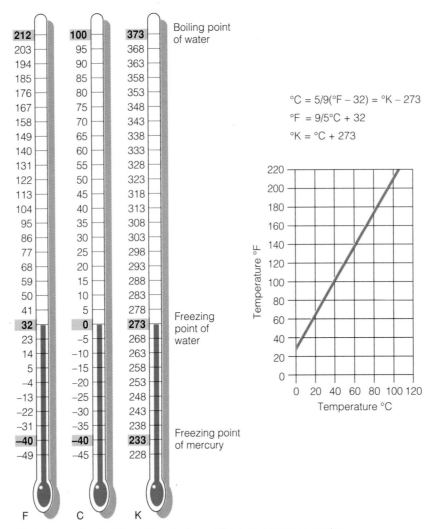

Figure 2.10 The relationship between the Celsius, Fahrenheit, and Kelvin temperature scales.

form of matter called **plasma,** occupying a significantly smaller volume than the present universe. A plasma is composed of the parts of atoms, but its high temperatures prohibit atom formation. The temperatures of the big bang plasma were probably greater than that of Sanduleak's supernova, possibly hundreds of billions of degrees. At those high temperatures, gravity, the strong nuclear force, and electromagnetism were probably not separate, and not even the elementary particles could exist.

The rapid expansion of the plasma was accompanied by a rapid temperature drop, allowing, in sequence, neutrinos, electrons, protons, and neutrons to form. During the big bang, an opposite kind of matter also formed. Called **antimatter,** it was composed of antineutrinos, positrons, and antiprotons. Whenever antimatter and matter collided, they annihilated each other. The matter of our universe, therefore, gives a mystery. Why did the universe not destroy itself just after the big bang? For some unexplained reason, matter dominated, and antimatter was generally annihilated. There might still be regions of antimatter today, but they would be virtually unrecognizable as such from Earth. Eventually, the subatomic particles of matter formed atoms and they, in turn, formed molecules (fig. 2.11). The expansion is generally perceived to

be a relatively uniform one because of the 2.7 K universal background radiation, sometimes called the "echo of the big bang." If the expansion were not approximately uniform, then what could account for the near uniformity of temperature of the background radiation?

A Lumpy Universe

Although the big bang is adequate enough to explain the origin of matter and the background radiation, the slight irregularities in the nearly uniform background radiation detected by *COBE* can explain the distribution of visible matter in the universe. The picture of the universe now emerging is one in which matter has been lumped together into the objects that we see around us, stars, and *aggregations of stars* called **galaxies.** Earth, for example, is composed of matter, and it belongs to a group of loosely similar bodies called the planets. The planets, in turn, belong to a larger system held together by the gravity of the sun, which is a large concentration of gases and plasma. The sun, though apparently widely separated from other suns, belongs to an aggregation of an estimated 100 billion suns called the **Milky Way Galaxy.** And the Milky Way is itself part of a group of galaxies. This lump of matter, covering millions of light years,

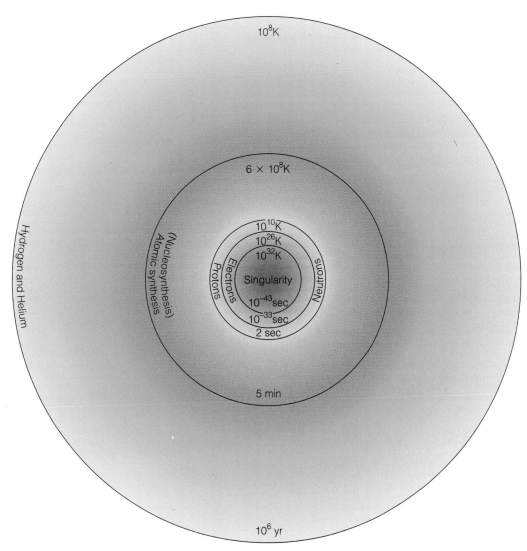

Figure 2.11 A model of the big bang. Units shown are time, not distance.

is just one of billions of similar lumps. When NASA astronomers announced in 1992 that the *COBE* satellite had detected minute variations in the temperature of the "echo" radiation of the big bang, they established the reason for the distribution of matter. These variations appear to explain the lumpiness; that is, *the big bang was not a perfectly uniform expansion.*

Galaxy Lumps

Individual galaxies like the Milky Way contain not only billions of stars but also dust and gases as interstellar ("between the stars") matter. Some of the smaller star aggregations have been called **globular clusters,** and the nearby Large Magellanic Cloud, which houses Sanduleak, falls into this category.

The explanations for galaxies that derive from the big bang model suggest that once matter began to mix out of the rapidly cooling plasma, gravity began to dominate its location. Smaller clusters of matter gradually pulled together to make larger clusters. This may not seem like a problem, but astronomers want to know why the matter of the universe is distributed in lumps. The big bang offers a solution to the problem of lumpiness. In this

cosmology, or model, of the universe, the lumpiness of matter results from the accumulation of increasingly larger galaxy clusters. This explanation of the universe relies on an initial density of electromagnetic radiation and plasma that would have been very high. At the great temperatures and densities of the big bang plasma, electromagnetic radiation could not have traveled through the opaque universe. With increasing expansion and declining temperatures, the universe became more transparent, and gravity became the dominating force that created the matter lumps. The current distribution of matter appears very much like random cracks in a pane of glass.

THE FORMATION AND STRUCTURE OF GALAXIES AND CLUSTERS

The galaxies most likely coalesced from lumps of hydrogen and helium, the two most abundant elements. Surveys of the *visible* universe reveal that hydrogen and helium comprise 99.9% of matter. Obviously, because they are the simplest, these two elements would have been the first to form after the big bang.

Under the increasing influence of gravity, the large, cloud-like lumps of hydrogen and helium began to rotate and contract, eventually reaching a few identifiable shapes. The spherical, smaller globular clusters represent one form, and the elliptical, larger galaxies represent another. Some of the possible galaxy shapes are shown in figure 2.12.

Galaxies can be large even by the standards of the universe, and our own galaxy's great dimensions prevent us from knowing its exact shape. Astronomers estimate the Milky Way has a diameter of 100,000 light-years and a core thickness of 10,000 light-years. Since we cannot get outside our own galaxy to look back—the distances are too great—we must look toward other galaxies to discern their general form and to surmise the shape of the Milky Way. The structure of galaxies can be generalized from the photo of the Andromeda Galaxy in figure 2.13. The structure includes a core, or bulge, where stars are most densely located. Even though they are relatively close, they are still billions to trillions of kilometers apart. Outside the core is the disk, which contains relatively young stars and gases. These young stars are known as *population I* stars. A halo encompasses the disk and contains clusters of older, *population II* stars with little or no interspersed gases. Finally, a large region probably at least as large as the visible galaxy lies outside all that is seen. This is the featureless region suspected to contain **cold dark matter,** so called because it is only indirectly detected.

Cold Dark Matter

The revolving disks of galaxies present a problem. They should rotate at varying speeds, with faster rates nearer the central core of each disk, but they do not. Instead, many disks show a uniform rotation speed throughout their visible mass. The only solution to this problem appears to lie in some missing mass, the assumed cold dark matter of the universe. Recent speculations about the cold dark matter suggest that it is the dominant form of matter, greater by far than the hydrogen and helium of the visible cosmos and making up as much as 95% of the universe. Its assumed low temperature may be near that of absolute zero, and it is not detected by any of our current astronomical instruments. Knowledge of the cold dark matter is, therefore, limited to what can be derived from the motions of the galaxies and its possible absorption and refraction of electromagnetic radiation.

THE FORMATION OF STARS

Each of the billions of galaxies contains millions to billions of stars. Our own galaxy, which is not unusual, has an estimated 100 billion stars. For certain, then, stars are common objects in the universe, and our own sun, whose structure is treated extensively in chapter 3, is such an object.

Our sun, which is similar to many stars, is composed chiefly of hydrogen and helium and is about 5 billion years old. Its large gravitational attraction holds in orbit nine planets and numerous smaller bodies that, together with the sun, make up the **Solar System.** The origin of this hydrogen ball and its attendant

Figure 2.12 Hubble classified galaxies by shape.

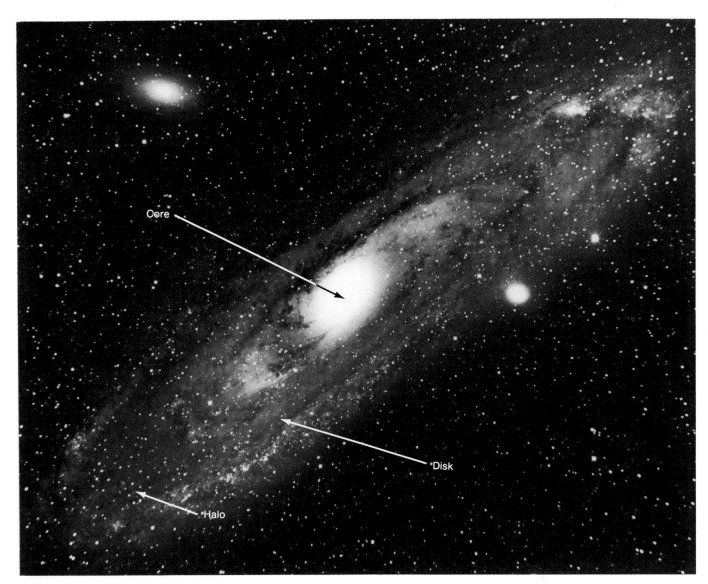

Figure 2.13   The Andromeda Galaxy.

bodies lies in a contraction of matter within one of the "arms" of the Milky Way's disk.

Nebulae and Star Formation

Stars arise from giant "clouds" of gases and dust called **nebulae,** which may be more than 1 parsec thick. Gravitational contraction, enhanced either by a shock wave from a nearby supernova's high speed, expanding sphere of **ejecta,** which is thrown material, or by magnetic fields, compacts the cloud. The cloud is initially cold, and as it begins to rotate its temperature rises. Because a nebula is vast, its rotational contraction to the size of our sun would spin parts of the cloud close to the speed of light if some force did not slow down its turning. Magnetic field lines of nearby gas clouds most likely direct some of the momentum (the "angular momentum") away from the cloud during its early stages of collapse, slowing the rotation. After about a million years, the larger nebula divides into several clouds.

Stages in Stellar Evolution

From its beginning to end, a star undergoes a series of changes dependent upon its initial mass. Generally, these changes fall into beginning, middle, and end stages in the life of a star.

The First Stage: Nebular Collapse   The collapse of a cloud with the mass equivalent to our Solar System eventually makes a very dense, opaque nebula that begins to radiate and absorb infrared energy. This first stage of stellar development produces a *protostar,* and it takes place beneath a thick, energy-absorbing blanket of dust and gas through which only radio waves easily pass. During its protostar stage, our sun probably covered thousands of astronomical units, but it continued to collapse.

Eventually, the protostar heats mechanically by kinetic energy through movements of gas and dust in the shrinking volume of the cloud, and, when its outer layers reach temperatures between 2,500 K and 3,500 K, begins to give off light that penetrates the surrounding gas and dust. Figure 2.14 shows the protostar and subsequent stages of stellar evolution.

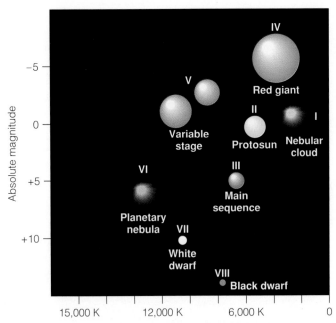

Figure 2.14 Stellar evolution based on the mass of Earth's Sun.

red giant, an enormous star with a relatively cool exterior. Some red giants have diameters approaching 10 AU, but our sun will probably swell to no more than 2 or 3 AU. Eventually, the outer part of a red giant, comprising at least 10% of the star's mass, will detach itself from the core, making a *planetary nebula* like the ring nebula pictured in figure 2.15.

During this last stage of development, the star's interior shrinks further, becoming a **white dwarf** only a few times larger than Earth. The small size of a white dwarf is not an indication of its mass, however. Except for the expulsion of a planetary nebula, the remnant core of a white dwarf may contain as much mass as our sun. The extraordinary gravitational compaction of matter into the stellar interior results in a star that has a density greater than anything in our Solar System. So much matter is packed into so little space in a white dwarf that a cubic centimeter of its matter would weigh at least a metric ton on Earth. Within this compact, hot white dwarf, ordinary gases like those found in our atmosphere cannot exist. At the very high temperatures found in the dwarf, electrons separate from atomic nuclei, producing a **degenerate gas.**

The Second Stage: Turning on the Solar Furnace   During the second stage of development a young star undergoes produces a fluctuating release of energy that makes it variable in appearance. As it strives to reach an equilibrium between gravitational collapse and expansion from its energy output, it turns into an intermediate, cool, variable stellar object known as **T Tauri** (named after a particular star). In a vast nebula from which many stars can form, T Tauri stars often appear in groups called T associations.

The mechanical generation of heat and light is not the main source of a star's energy. Deep inside the mass of gas—mostly hydrogen—the protons of hydrogen atoms are driven so closely together that they begin to join or fuse to form helium and release energy. This **fusion** of protons generates temperatures in excess of 15,000,000 K in the core of a *main sequence star* like our sun and supplies enough outward energy to halt the further gravitational collapse of the star.

The Third Stage: From Stability to Instability   As long as a main sequence star has an adequate supply of hydrogen for fusion, it remains relatively constant in size and energy output; thus, an equilibrium characterizes the third stage. The surface of our sun, for example, has not radically altered its radiation during recorded time, maintaining its approximate 6,000 K surface.

Remember that Sanduleak is evidence that stars are not eternal. Even our sun will change from its current, main-sequence stage to a larger, different-colored star. Whenever our sun has used about 60% of its hydrogen in fusion reactions, it will resume its initial collapse. As a result, its interior will become tens of millions of degrees hotter by compaction, possibly reaching 100,000,000 K, and it will turn helium into carbon during a stage known as the **helium flash.** The release of energy from helium fusion will do two things. First, it will halt the collapse. Second, it will make the outside layers of the sun expand into a

Figure 2.15 Ring nebula formed by the explosion of a variable star. The star's core, a white dwarf, is shown in the center.

The Hertzsprung-Russell Diagram

Astronomers categorize stars on the bases of their mass, size, color, and surface temperatures. Stars exhibit wide ranges of these properties. Our sun, for example, is an average star though it is large enough to encompass one million Earths. As large as that may seem, it is small by comparison with giant and supergiant stars, which may be larger than thousands of our suns. Other stars are more massive or brighter, and some are much hotter than our sun. A white dwarf, for example, may have a surface temperature of 12,000 K.

These properties can be visualized on the **Hertzsprung-Russell diagram,** usually shown as a graph but altered in figure 2.16 to emphasize stellar size and color . The diagram shows the relationship between star luminosities and surface temperatures. Figure 2.16 begins at the bottom with the ultimate gravitational collapse of a massive star, which cannot be halted by any energy output. Such a point is called a **black hole,** and it represents the end of a star about ten times more massive than our sun. When such a star begins its final gravitational contraction, it crushes itself into a gravitational point. Its gravity is evident, but nothing of the former star remains to be seen. Unlike objects that are visible because light can escape from them, the black hole has no escape velocity. In contrast, Earth's escape velocity is 40,000 kilometers (24,800 mi) per hour, the speed the astronauts aboard the *Saturn V* rocket had to attain to travel to the moon. With a gravitational attraction billions of times greater than we experience on Earth, a black hole is too powerful for light to escape; thus, it is black.

Farther up the scale are the main sequence stars. Their color and absolute magnitudes are close to that of our yellow sun, which is marked on the vertical line. With the red giants, however, both the color and the magnitude change, and the hot blue giants have magnitudes in negative numbers. Although the red giants have cooler surface temperatures than our sun, they are so much larger that their extensive surface area radiates more total energy.

In addition to the numerical designation given to each star on the basis of its luminosity and temperature, the key in the diagram also marks the spectral classes, specified by capital letters for a star's characteristic spectrum. Our sun, for example, falls into the spectral class G, giving it a surface temperature between 5,000 K and 6,000 K and a yellow color. The spectral type O stars can be a million times more luminous than our sun.

PLANETS AND THE NEBULAR HYPOTHESIS

Ancient astronomers were the first to notice that the tangential motions of some heavenly lights appear to be very rapid. Unlike the stars, which seem to be stationary, these "wanderers" change positions over the seasons. With the invention of the telescope, observers realized that these bodies were not the same as the stars. They do not lie at great stellar distances, and they follow nearby elliptical paths. We call these wanderers the planets.

Astronomers eventually catalogued eight planets, along with Earth, that orbit the sun. Two are closer than 1 AU, and the other six are farther away. The most distant planet, Pluto, has an orbit that keeps it an average distance 39 times farther from the sun than we are. The presence of the planets begs a question similar to the one about the universe. What is their origin?

As the sun was contracting from the nebular cloud, the flattened plane of the cloud began to dissociate itself into its own matter lumps (fig. 2.17). Small bodies moving through the dust and gas of the cloud began to collide and accumulate. The largest of these bodies probably had diameters no greater than a few tens of kilometers, and they have been appropriately called **planetesimals** ("little planets").

Even though the orbiting planetesimals traveled swiftly about the newly forming sun, their speed relative to one another was not great, and they were attracted by gravity into larger and larger masses. These larger bodies, called **protoplanets,** were bombarded by planetesimals as they swept through their orbital paths. Earth grew, therefore, from the *accretion of planetesimals.* This accretion began at the same time as the formation of the sun, about 4.6 billion years ago, and it most likely ended with a completed Earth by 3.8 billion years ago. The end of the planetesimal bombardment can be surmised from the 3.9-billion-year age for the oldest rocks found on our planet. Any older rock material was probably destroyed by the high-speed impacts of these gravitationally attracted bodies. Because the planetesimals accreted in the nebular cloud from which the sun also formed, this explanation for planetary formation is known as the **nebular hypothesis.**

In this text, some information on the planets is integrated with the discussion of Earth features and processes. The planets that have atmospheres, for example, differ from Earth in the composition, thickness, and structure of their envelopes of gas and clouds. Their atmospheres, however, obey the same physical laws to which Earth's atmosphere is subject. A comparison of atmospheres, therefore, is included in the discussion of atmospheric composition in chapter 4. Similarly, the volcanism with which we are familiar on Earth has its parallel in the history and current geologic activity on some of the other bodies in the Solar System. Thus, planetary geology is included in the discussion on volcanism in chapter 7. Appendix B summarizes the characteristics of the planets.

We may think that living on a planet makes us all authorities on the nature of planets. In reality, the planets have as many differences as they have similarities, and none, as far as anyone knows, can support life except Earth. As the planets accreted from the nebular cloud, those closer to the sun formed from an abundance of silicon and iron, becoming largely rocky bodies. The four closest to the sun are very dense spheres, and, because they are relatively small and Earth-like, are known as the **terrestrial planets** (from *terra,* Latin for "earth"). These interior planets, Mercury, Venus, Earth, and Mars, are small by comparison with all but one of the outer five. By contrast, four of the outer planets are very large. Called the **Jovian planets** (after the Roman god Jove, or Jupiter), these are large gaseous spheres with unknown interiors. The cloud covers on these planets are too thick for cameras and telescopes to penetrate.

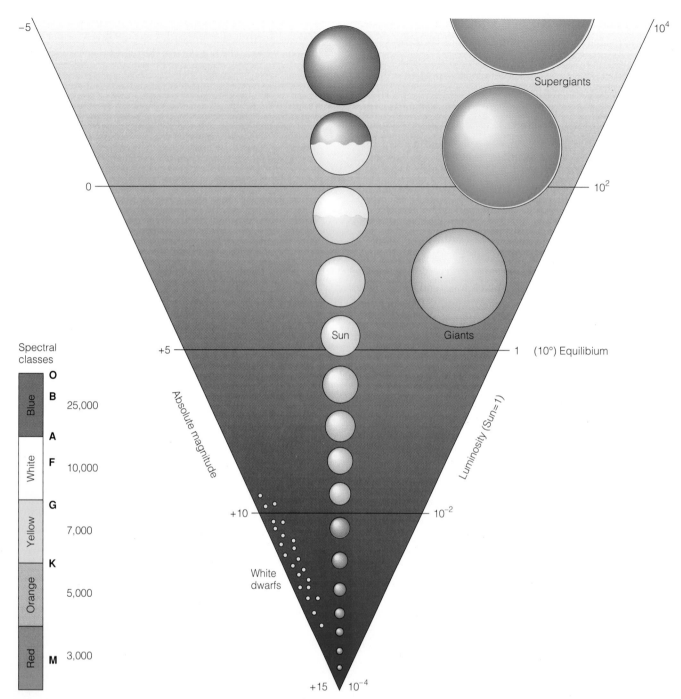

Figure 2.16 An adaptation of the Hertzsprung-Russell diagram that emphasizes size and color relationships. The size ordering does not indicate an evolutionary sequence.

Regardless of their distance from us and their clouds, we do know much about the planets because of observations made with the aid of telescopes and spacecraft. We have sent spacecraft to land on both Mars and Venus, and all but Pluto has been visited by instrument packages such as the *Voyager* craft. The following is a brief summary of what we have learned.

Mercury

Mercury, the smallest of the terrestrial planets and the closest to the Sun, was named for the fleet-footed Roman messenger to the gods. Because of its proximity to the Sun, it has the shortest and fastest orbit around the Sun (88 Earth days). Its period of rotation, however, is relatively long (58.65 Earth days), which leads to temperature extremes of 430° C on the day side to 180° C on the night side. The airless surface of Mercury is

Figure 2.17 The solar nebular hypothesis for the origin of the Solar System. *(1)* The solar nebula takes form. *(2)* It begins to rotate, flattens, and contracts. *(3)* The nebula is now distinctly discoidal and has a central concentration of matter that will become the sun. *(4)* Thermonuclear reactions begin in the primordial sun as material in the thinner area of the disk condenses and accretes to form protoplanets. *(5)* The late stage in the evolution of the Solar System in which planets are fully formed, and materials not incorporated in planets are largely swept away by solar wind and radiation.

heavily cratered, with one crater, the Caloris Basin, approximately 1,300 kilometers (806 mi) in diameter (fig. 2.18). Craters are more dominant on highland areas, which are thought to represent older terrain. Basinal areas appear to be filled with lava (although there are no other apparent volcanic features) and contain few craters, indicating a decreasing rate of meteorite bombardment and a more recent origin for this terrain. Experts suggest that the volcanic activity occurred early in the planet's history and probably ended some 3 billion years ago. Other surface features include landsliding and cliffs, called scarps, some as much as 3.2 kilometers (2 mi) high and hundreds of kilometers long.

Mercury possesses a magnetic field that is only 1% of the strength of Earth's, but its presence suggests that the planet has or had a liquid core, because such fields are attributed to convective currents flowing within a liquid core.

Venus

Venus, named for the Roman goddess of love and beauty because of its brightness in the night sky, is Earth's nearest planetary neighbor. Because of its almost identical size and density, it is often referred to as Earth's twin. Venus is perhaps best known for the 90-kilometer-(56 mi) thick layer of clouds composed primarily of CO_2 (96%) and N_2 (3.5%), with small amounts of sulfuric acid droplets. The greenhouse effect (see chapter 1) caused by the thick cloud cover as well as its nearness to the Sun and the atmosphere's composition imparts to

Figure 2.18   The highly cratered surface of Mercury.

Venus the highest planetary surface temperatures known in the Solar System (500° C). The thick cloud cover also gives rise to an enormous atmospheric surface pressure up to 90 times that of Earth's.

The *Magellan* spacecraft, placed into orbit around Venus in August 1990 after a 15-month journey from Earth, mapped as much as 90% of the Venusian terrain. Preliminary surveys from the *Magellan* and the earlier U.S. *Mariner* and Soviet *Venera* spacecrafts indicate a subdivision of surface terrain into lowlands and extensive gently rolling to rugged highlands. The massive Maxwell Montes (mountains) rise some 11,000 meters (36,000 ft) above the adjacent lowland terrain. The identification of volcanic terrain is still tentative, but the Magellan probe appears to have detected a lava flow from Maat Mons, the second tallest mountain on Venus (fig. 2.19). Additional data point to large impact craters and rift valleys. This varied surface is hot, dry, and almost certainly lifeless.

Mars

Named for the Roman god of war, Mars is the outermost of the terrestrial planets and is probably the best known and most written about of our sister planets. Although often referred to as the red planet, it varies in color from blue-green during the spring to brown in the fall. The seasonal differences result from an axis of rotation that is tilted to its plane of revolution. With a period of revolution approximately twice as long as that of Earth, Martian seasons are longer than Earth's. The Martian day, however, is approximately equal to ours, because its period of rotation is only slightly longer. Its distance from the Sun makes Martian temperatures relatively low, ranging from a high of 0° C at the equator during the daytime to −130° C at night.

The surface of Mars is known for a number of interesting features, some of which may be analogous to those on Earth. The *Mariner* orbiters and *Viking* landers sent back photos of a cratered, moonlike surface as well as data on Martian "soil" and rocks (fig. 2.20). When tested by *Viking,* the soils were found to be free of organic compounds and, thus, of any indication of life.

Figure 2.19 An enhanced *Magellan* image that shows an apparent lava flow associated with Maat Mons on Venus.

The Martian atmosphere is thin (less than 1% of that of Earth) and consists primarily of carbon dioxide. As thin as the atmosphere is, however, wind is an important surface activity and gives rise to spectacular dust storms and gigantic dune fields. Mars, in fact, possesses the largest sandy desert in the Solar System. The carbon dioxide atmosphere is also related to the famous polar caps, which expand and contract with the seasons.

Running water apparently played a major role in the early shaping of the Martian surface. Massive canyons, such as Valles Marineris, stretch for hundreds and even thousands of kilometers. Tributary dendritic (branchlike) canyons, meandering channels, and even braided channels are evident. Such features may indicate a time of milder climate during which rainfall occurred or large quantities of ground ice melted.

The interpreted history of Mars is one of early heavy meteorite bombardment (particularly of the uplands), followed by extensive volcanism. A large volcano, the shield volcano Olympus Mons, is 550 kilometers (341 mi) in diameter and 27 kilometers (16.75 mi) high and dwarfs any similar feature found on Earth (fig. 2.21). Olympus Mons stands on a topographic swell, which may be the upwarping of a "hot spot." Fractures and faults are found around the base of this uplift.

Jupiter

Jupiter, named for the son of Saturn and the king of all the gods in Roman lore, is the largest of the giant outer planets. Best estimates of the structure and composition of Jupiter attribute to it a core of iron-rich rock, covered by a thick layer of liquid hydrogen, and surrounded, in turn, by dense layers of clouds of hydrogen (mostly) and helium (fig. 2.22). It is not unlike the Sun in composition but would have to be some ten times larger to attain temperatures at which it would become self-radiating (i.e., a star). The thick atmosphere of Jupiter is subject to a complex circulation and storms. The most notable storm is the Great Red Spot, which is 40,300 kilometers (25,000 mi) long and 14,500 kilometers (9,000 mi) wide. Smaller storms traverse the planetary atmosphere at speeds of up to 360 kilometers (225 mi) per hour.

Figure 2.20 View of a rock-littered Martian landscape taken from the *Viking 2* lander.

Associated with Jupiter are 16 satellites, 4 of which exceed 3,000 kilometers (1,860 mi) in diameter. Analyses made during the passages of *Voyagers 1* and *2* indicate that all of the larger moons have ice-rock crusts, and that each also has its own peculiar features, such as Io, which has sulfur-burning volcanoes (fig. 2.23). Europa has a geologically young surface crisscrossed with an intricate system of intersecting lines that may represent faults or fractures in the icy surface. Ganymede has a cratered surface, "grooved" terrain, and a system of fractured mountains.

Saturn

Saturn, the ringed planet, was studied during the flybys of *Pioneer 11* and *Voyagers 1* and *2*. It was named for the Roman god of time because the planet had the longest period of change in the sky that ancient astronomers could recognize. The planet moves through a 29-year orbit around the Sun. Because of its great mass and distance from the Sun, Saturn actually generates more heat than it receives. As best as can be determined, Saturn possesses a thick, turbulent atmosphere that is 1,000 kilometers thick and consists primarily of hydrogen. Wind speeds within this atmosphere are estimated to be in excess of 1,775 kilometers (1,100 mi) per hour.

The most notable feature of Saturn has always been its rings (fig. 2.24). Three major rings are easily visible, but close-up photos reveal that these are each composed of 500 to 1,000 smaller rings, which, in turn, consist of billions of particles from the size of icy dust to the size of boulders. The rings are very thin in proportion to their width.

Seventeen moons orbit Saturn. The largest, Titan, is comparable in size to the inner planets, and its atmosphere will be discussed in chapter 4. Most of Saturn's moons are small, irregularly shaped, cratered, and consist of ice-rock mixtures.

Uranus

Uranus, named for the Greek god of the heavens, is one of the gas giants, about four times larger than Earth. Much of what we know of the planet comes from the *Voyager 11* flyby in 1986. Uranus apparently consists of a rocky core; a zone of liquid water, methane, and ammonia; and a thick cloud layer of hydrogen (87%) and helium (13%). The atmosphere, blue-green in color from methane, shows only a faint banding of its clouds (see fig. 4.1*a*). While estimated wind speeds of 360 kilometers (225 mi) per hour are known, Uranus is the only giant planet without oval storm systems. The planet possesses a magnetic field almost as strong as Earth's, and it is orbited by 15 moons, which show impact craters and fault scarps bounding rift basins. Uranus also possesses a set of rings, which was detected during the *Voyager* flyby.

Uranus differs from the other planets in that its axis of rotation is inclined 98° from the vertical, thus giving it a retrograde (or backward) rotation. As expected at this distance from the Sun, its period of revolution is long, some 84 years.

text continued on page 48

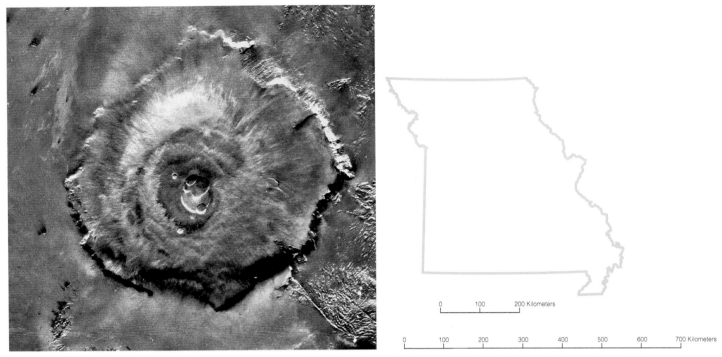

Figure 2.21 The Martian volcano Olympus Mons, one of the three largest volcanoes in the Solar System, is shown here with the outline of Missouri for scale. The volcano is about 550 km in diameter and 25 km higher than the surrounding Martian surface.

Figure 2.22 This photomontage of the Jovian System shows Jupiter and its four natural Galilean moons. Cloud bands parallel Jupiter's equator, each flowing in the opposite direction to the adjacent one. Io (*left*) is closest to Jupiter, followed by Europa (*center*), Ganymede (*lower left*), and Callisto (*lower right*).

Figure 2.23 Sulfur eruption from a volcano on Io, one of Jupiter's 16 moons. Photo was taken from *Voyager II* on March 4, 1979, at a distance of 500,000 kilometers (310,750 mi). Jupiter's gravitational pull on Io may be responsible for a tidal influence.

Box 2.2 Investigating the Environment

Earth's Moon

One celestial object we now know very well is the moon. Its diameter of 3,484 kilometers (2,160 mi) makes it smaller than Earth, and its mass is only 1/81 that of our planet. Because gravitational pull is related not only to mass but also to diameter, the moon exerts a pull 1/6 that of Earth's gravitational field. Thus, a 68-kilogram (150-lb) person would weigh 11.4 kilograms (25 lb) on the moon. This gravitational field is insufficient to hold gases, so the moon is an airless body.

As Earth's only natural satellite, the moon revolves around our planet in an elliptical orbit from **perigee,** or its closest position, to **apogee** or its farthest. If we consider the moon's orbit against the reference point of the sun, then one revolution, called the **synodic month** takes 29.5 solar days. Against the background of distant stars the moon's revolution takes 27.3 solar days, giving us the **sidereal month** (fig. 2.B). Because the moon rotates on its axis only once during a synodic month, observers on Earth see only one side.

One way we characterize the moon is by its so-called phases, or appearances. Throughout a synodic month, the moon appears to undergo a series of shape changes. These result from our view and from the revolution of the moon, which do not always allow us to see the complete sunlit side (fig. 2.C). During the new moon phase, the side opposite that which faces Earth is lighted, but the side facing Earth is in shadow. During the full moon phase, just the opposite occurs, as the side facing Earth is fully lighted. Between these darkened and lightened phases, the moon appears crescent and gibbous. A phase change from new moon toward full moon is called *waxing,* whereas a change from full moon to new moon is called *waning.* The moon appears in a new phase whenever it completes one-fourth of its orbit: thus, the phases change once every 7.38 solar days (fig. 2.D). During the quarter moon phases, an interesting visual phenomenon occurs as the light reflected from Earth partially lights the nighttime portion of the moon's disk.

Sometimes the moon's shadow falls on Earth to make a **solar eclipse.** On other occasions, Earth's shadow falls on the moon to make a **lunar eclipse.** Both lunar and solar eclipses block out some or all of the sun's

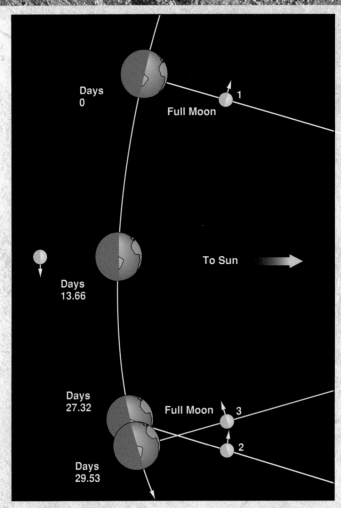

Figure 2.B   The difference between a sidereal month and a synodic month. The interval between two successive conjunctions of the Moon with a given star *(1-2)* is 27.32 mean solar days, or one sidereal month. The interval between two successive conjunctions of the Moon with the Sun, however, is 29.53 mean solar days, or one synodic month *(1-3)*. All angular relationships and relative sizes of Earth and Moon are to scale. Note that the Earth-Moon barycenter, not Earth's center, lies on the orbit.

light. On rare occasions, either Earth or the moon block out the full disk of the sun in a total eclipse (fig 2.E).

The moon is the only body beyond Earth that human beings have visited, and its proximity (average distance 381,555 km, or 236,564 mi) makes some of the moon's features visible to the unaided eye. What is this bright body that, though it gives off no light of its own, can be visible both day or night? Because the moon is close to Earth,

optical studies of the moon are as old as human eyes, and telescopes provided close-up views of its rugged surfaces long before Neil Armstrong and Edwin Aldrin landed their *Apollo 11* module on July 20, 1969. From the perspective of the unaided, earth-bound eye, the moon is a body of contrasting dark and light areas. The reason for the contrasts lies in the **albedo,** or reflectivity, of the moon's surfaces. For example, its plains, called **maria** because they looked

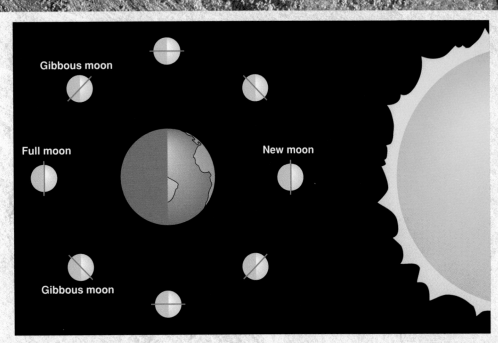

Figure 2.C Phases of the moon.

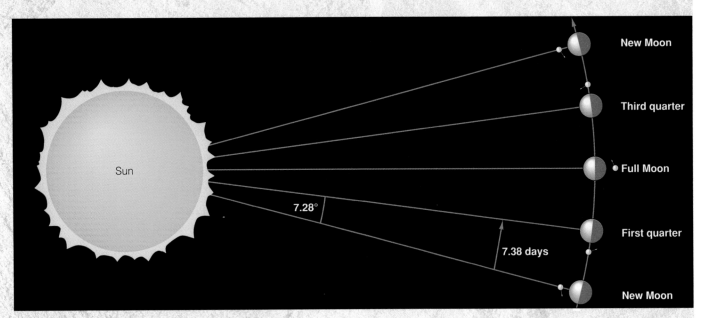

Figure 2.D One synodic month is the elapsed time between one new Moon and the next. At the new Moon *(bottom)*, Earth, Moon, and Sun are all aligned; 29.53 days later *(top)*, they are again aligned. The intervening phases of the Moon are separated from one another by one-quarter of this interval: 7.38 days, or approximately one week, during which time Earth moves through an angle of 7.28° along its orbit. Angles are not exaggerated.

Box 2.2 continued

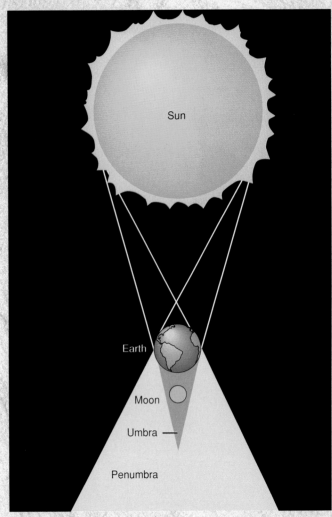

Figure 2.E Earth's shadow is broad enough to immerse the Moon if it is aligned with Earth and Sun during opposition.

Figure 2.F *Apollo 17* astronaut Harrison Schmitt collects a lunar sample.

like seas to Galileo, do not reflect light as efficiently as its highlands. From the perspective of two astronauts standing on the moon's surface, however, the body is planet-like, wet sand or ash colored, and not nearly as rugged as more distant perspectives from Earth indicate (fig 2.F).

When the *Apollo* astronauts returned with moon rocks, they added a new dimension to our knowledge. We now have samples that tell us not only something of the moon's composition but also something of the Solar System's history. Scientists dated the rocks the astronauts collected through a method known as **radiometric dating** (chapter 14). The rocks yielded various dates, with the oldest sample registering 4.4 billion years. Interestingly, this age agrees favorably with the estimated 4.6-billion-year age of the Solar System, but some moon rocks yield younger ages. Unlike Earth's rocks, those on the moon lie under airless and waterless conditions that prevent erosion on the scale present on Earth from occurring. Except for the changes caused by impacts from incoming micrometeorites, meteorites, and high-speed atomic particles, the moon's rocks have undergone comparatively little change over several billion years.

The impacts have been significant, however. Some of the rocks and lunar dust brought back had glassy beads on them, indicating that the rocks had been subjected to heat sufficient for melting and rapid cooling. Heat of this nature is associated with volcanic activity and impacts. Apparently, the two processes are related. Impacts of explosive force have cracked the moon's surface, dug craters, thrown ejecta hundreds of kilometers, and generated enough heat for **lavas** to form. The dark lavas, generally called *basalt,* contrast with the lighter-colored **anorthosite** of the highlands. Both basalt and anorthosite are **igneous rocks,** which means that they were once molten. Both types of rocks also occur on Earth. Because the basaltic lavas of the moon were not as **viscous** as those of Earth, moon lavas flowed widely after extruding through the **fissures** (or cracks) caused by impact.

Apply Your Knowledge

1. How many solar days are required for the moon to change from new moon to full moon?

2. During which phases do the moon, sun, and Earth lie in a straight line?

Box 2.3 Investigating the Environment

Will an Asteroid Hit You on the Head?

In December 1992, a dual asteroid called Toutatis missed Earth by a mere 3.54 million kilometers (2.2 million mi). Toutatis is composed of two, rough edged objects that are 2.6 and 4 kilometers (1.6 and 2.5 mi) in diameter (fig. 2.G). The objects belong to a group of thousands of bodies that pass through Earth's orbit and threaten to strike our planet. These objects, called near-Earth objects (NEO), can be as large or larger than Toutatis, and they may be either former asteroids, bumped from the asteroid belt, or the cores of comets.

One NEO that may collide with Earth is the comet Swift-Tuttle (fig. 2.H). Particles strewn throughout its orbit give us an annual August meteor shower. The comet itself was initially thought to be headed for a collision with Earth on August 14, 2126, but recent calculations predict a likely collision date in August 3044.

As you will read in chapter 14, the extinction of the dinosaurs 65 million years ago has been tentatively linked to the impact of a large NEO. If such an object hit the planet today, human existence would be threatened. On impact, an object the size of a football field would release the energy equivalent to a 10-megaton nuclear bomb. The energy released by the collision of an object the size of Toutatis or larger would be devastating.

Apply Your Knowledge

1. What is the name of the asteroid that almost struck Earth in 1992?

2. What comet may strike the planet?

3. Why is it difficult to detect many of the potential NEOs?

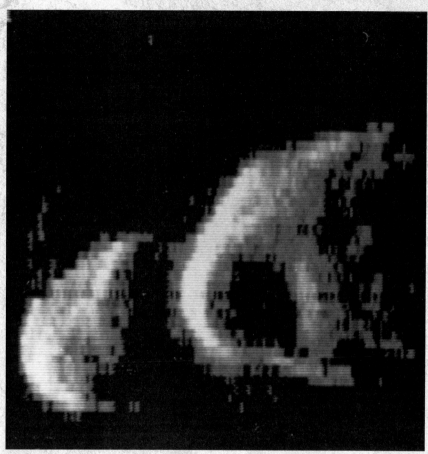

Figure 2.G   The NEO known as Toutatis is composed of two companion asteroids.

Box 2.3 continued

Figure 2.H   Comet Swift-Tuttle may strike Earth in August 3044. Here the thin tail of the comet can be seen.

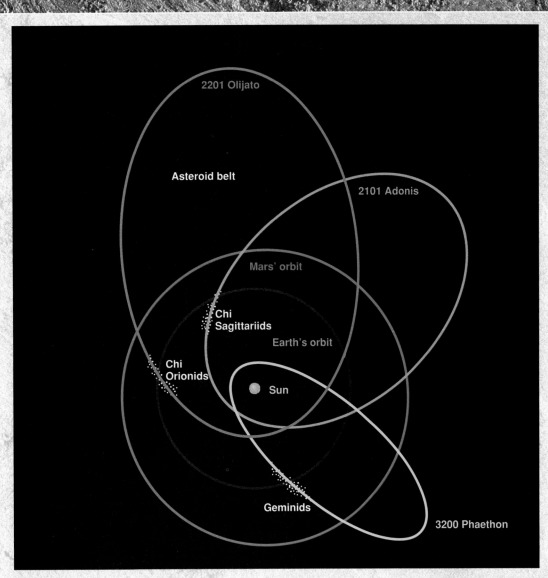

Figure 2.1 The highly elliptical orbits of objects near Earth take them on possible collision courses with the inner planets. The orbits shown here are noteworthy because of their associated meteor showers.

Figure 2.24 Saturn and its rings.

Neptune

Neptune, the fourth largest planet in the Solar System, is the densest of the giant planets. Long before it was actually observed, the presence of Neptune was predicted from perturbations it caused in the orbit of Uranus. Based on data from the 1989 flyby of *Voyager 11,* Neptune appears to be similar in composition and structure to Uranus, including an atmosphere rich in hydrogen and helium (fig. 4.1*b*). This atmosphere also contains a small amount of methane, which gives the planet a distinct bluish color. Consequently, Neptune is often referred to as the Blue Planet and is named for the god of the sea.

Whereas the atmosphere of Neptune is similar in composition and structure to that of Uranus, it differs in that it is characterized by clouds and numerous oval storm systems. Particularly prominent is a dark, turbulent zone near the equator known as the Great Dark Spot. That Neptune is approximately the same temperature as Uranus, but much farther from the Sun, suggests that it has an internal heat source.

Neptune has eight moons. Two of these, Triton and Nereid, have very eccentric orbits. The larger, Triton, is similar in size to the inner planets. It has an atmosphere of methane (probably) and an icy crust of nitrogen and methane. The surface is characterized by a series of dark streaks that are thought to have been formed by ice volcanoes driven by liquid nitrogen. In addition to its moons, Neptune also possesses several rings.

Pluto

Pluto, the outermost of the known planets, differs significantly from other planets in that its orbit lies some 17° outside of the plane of the ecliptic. The orbit is so highly eccentric that it actually brings Pluto inside the orbit of Neptune at times during its 248-year journey around the Sun. Pluto resembles the outer planets in composition and location but is most akin to the inner planets in size. The idea is widely held that Pluto may not have been a planet originally but rather was a moon of Neptune that escaped that planet's gravitational pull or was a large asteroid.

Pluto probably consists largely of ice, with a surface of frozen methane at a maximum temperature of −212° C (−350° F). Its atmosphere is thin and composed of methane and some heavier gases. Pluto has one moon, Charon, which is very close to the planet and has a diameter 40% as large as that of Pluto.

Speculation has occurred for a number of years about the existence of a tenth planet, planet X. Such speculation is the result of studies of the movements of Charon, which are too large and/or different to be accounted for by the gravitational attraction of Pluto. In addition, Pluto alone has too little mass to account for minor discrepancies in the orbits of the planets from Jupiter outward. Planet X, if it exists, must have a mass two to five times that of Earth and an orbit between 50 and 100 AU from the Sun, with a highly inclined plane of revolution similar to that of Pluto.

OTHER BODIES IN THE SOLAR SYSTEM

Three other types of bodies exist in the Solar System. These are comets, asteroids, and meteoroids. **Comets** are mixtures of ice and rock material that make brief visits to the inner Solar System on their highly eccentric orbits (fig. 2.25). They are characterized by a "tail" of gases that streams out from the solid "head" of the comet as it encounters high-speed particles emanating from the sun. The tail, therefore, always points away from the sun, regardless of the direction of the comet's motion.

Asteroids are solid, rocky bodies that occupy an orbit between Mars and Jupiter. They range in size from dust particles to more than 700 kilometers (434 mi) in diameter, and they number in the thousands. Because of their orbital position, scientists believe they are fragments that never accreted to form a true planet or they represent the remains of a disintegrated planet.

Meteoroids are rocky or metallic bodies traversing the Solar System by the millions and intersecting the orbits of Earth and the other planets, where they are influenced by gravity. As they come under the gravitational pull of Earth, for example, they fall through the atmosphere. Frictional heating makes them burn and glow, causing the streaks of light, or **meteors,** in the night skies. If a meteoroid is so large that it does not burn up completely, its remnant becomes a *meteorite* as it strikes the surface. Some meteorites contain carbon, and that has led to the speculation that life is not indigenous to Earth.

(a)

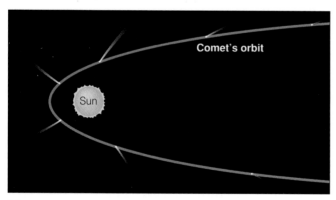

(b)

Figure 2.25 (*a*) Halley's comet. (*b*) The path of a comet in a highly eccentric orbit around the sun. Note that the comet's tail always points away from the sun.

Further Thoughts: Earth Is a Special Place

A study of the universe beyond Earth puts into perspective the special character of our planet. Although there appear to be other suns that have systems of associated matter, such as the star Vega, no one has yet found evidence of other planets with life. We do, however, have a good idea about our place in the universe (fig 2.26). The great distances between us and other solar systems preclude our traveling to the stars, leaving us with radiowave receivers and other sensors as our only means of detecting any signs of intelligence beyond our solitary home.

As a small planet that revolves around an average star, Earth's special character is largely determined by its relationship to the sun. In the next chapter, you will read about the sun as the primary influence on the movement of our atmosphere. In the form of electromagnetic radiation, energy produced in the sun's interior strikes our planet's surfaces, warming them and the air that overlies them. Increased warming means increased molecular motion, expansion of the atmospheric gases and surface waters, and decreased density. The lighter, warmed air rises, often carrying with it great quantities of evaporated water, transferring the sun's energy into atmospheric movements that drive the weather systems.

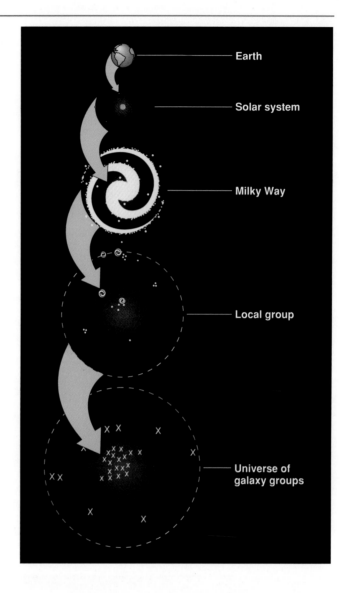

Figure 2.26   Earth's place in the universe.

Significant Points

1. In addition to other changes, the supernova Sanduleak demonstrates that stars change their shapes. Since our sun is also a star, it, too, will change, ultimately undergoing a kind of death in about 5 billion years.

2. By comparison with the rest of the universe, Earth is a very small object. In their study of objects beyond our planet, astronomers use large measuring units, such as the light-year, parsec, and astronomical unit.

3. Radial and proper motions help astronomers understand the relationships among the objects beyond Earth. Paralax measurements yield the distances to about 700 stars.

4. Brightness, called magnitude, provides clues to the distance of very distant stars.

5. Light acts like both waves and particles. The unmatchable velocity of light is 300,000 kilometers (186,000 mi) per second.

6. Matter is anything that has mass and occupies space. The fundamental building blocks of matter are protons, neutrons, and electrons.

7. Ninety-two types of naturally occurring atoms, called elements, are arranged by atomic number in the periodic table. Humans have synthesized other elements.

8. Elements can combine to make chemical compounds.

9. The kinetic energy of molecules is expressed as temperature.

10. The universe originated about 12 to 15 billion years ago when a superheated plasma expanded in an event known as the big bang.

11. A lumpy universe is filled with matter aggregations called galaxies and globular clusters. These are collections of stars, gases, and dust, plus suspected cold dark matter. Our galaxy, the Milky Way, appears to have older,

population II stars in its core and a halo of clusters. The spiral arms of the galaxy's disk house younger, population I stars and abundant gases.

12. Within galaxies and clusters, large nebulae collapse and divide to become stars and Solar Systems.

13. The evolution of a star begins in a nebular cloud, and it runs through beginning, middle, and end stages. A star can take various forms, including the initial protostar and variable T Tauri forms. The main sequence star is a form like our sun. After a star such as our sun uses up its fusion fuel, it changes in sequence into a red giant, planetary nebula, and a white dwarf. Larger stars can become black holes.

14. The nebular hypothesis accounts for the formation of the planets through the accretion of planetesimals. Nine planets are known to orbit our sun, though some suspect a tenth planet may also belong to our Solar System. Numerous small bodies, called comets, asteroids, and meteoroids, also populate our Solar System.

15. Two types of planets inhabit the Solar System: terrestrial planets, which are small, dense bodies close to the sun, and Jovian planets, which are large, gaseous bodies farther removed.

Essential Terms

supernova 20	parallax 22	neutron 26	nebulae 33
light year (lt-yr) 21	magnitude 24	molecule 27	fusion 34
parsec (pc) 21	variable star 24	isotope 27	nebular hypothesis 35
astronomical unit (AU) 21	photon 25	temperature 28	comets 50
radial velocity 21	electron 25	big bang theory 29	asteroids 50
tangential velocity 21	matter 26	galaxy 30	meteoroids 50
Doppler effect 22	atom 26	cold dark matter 32	
red shift 22	proton 26	Solar System 32	

Review Questions

1. What is parallax? How do astronomers use it?
2. What is the speed of light, and how is it used to establish a measuring scale?
3. What is the difference between radial and proper motion?
4. How does a shift in wavelength reveal the direction of movement for a distant star or galaxy?
5. Why do some stars appear to vary in magnitude?
6. What are the fundamental components of matter?
7. What are chemical compounds? What is covalent bonding?
8. How does atomic number differ from atomic mass number?
9. What discovery led to the big bang theory?
10. What is the echo of the big bang? How did the *COBE* satellite confirm the big bang?
11. What is a galaxy? What forms do galaxies take? How do the star populations vary within galaxies?
12. What are the dominant forms of matter in the visible universe? What kind of matter may actually dominate the composition of the universe?
13. How do stars evolve? What is their primary source of energy in the main sequence stage?
14. What four planets could be classified as terrestrial? Why do astronomers call Jupiter a "failed sun."
15. How could the surface temperature on Venus serve as a warning to us?

Challenges

1. How far in kilometers (or miles) is an object that lies 2 parsecs from Earth?
2. What might happen to life on our planet if a large comet or asteroid were to impact on Earth's surface?
3. What obstacles would hamper us from communicating with life outside our Solar System?
4. How are we solar-powered beings?
5. What difficulties would we encounter if we wanted to establish a colony on the other planets? What hardships would we encounter on the moon?

3

THE INVISIBLE ENVELOPE

*E*arth has a dynamic atmosphere that transposes solar energy into the changes we call weather. Our daily activities and fashions reflect the influence of those changes. At times the influence is minor, requiring us to add or subtract an article of clothing for comfort. But on occasions the influence is substantial, as it is, for example, during violent storms.

AT THE END OF PART 3, YOU WILL BE ABLE TO

1. describe the source of energy that drives atmospheres;
2. identify and cite evidence for the seasons;
3. identify the variable compositions of atmospheres;
4. distinguish the structures of atmospheres;
5. integrate the quantifiable elements of weather;
6. classify weather patterns;
7. discuss the influence of weather on life.

from Ode to the West Wind
by Percy Bysshe Shelley (1792–1822)

Drive my dead thoughts over the universe
Like withered leaves to quicken a new birth!
And, by the incantation of this verse,
Scatter, as from an unextinguished hearth
Ashes and sparks, my words among mankind!
Be through my lips to unawakened earth
The trumpet of a prophecy! O, Wind,
If Winter comes, can Spring be far behind?

3

PRELUDE TO WEATHER

Double rainbow developed during mountain showers on the island of Maui, Hawaii.

Chapter Outline

INTRODUCTION

The effects of weather are direct and significant. We dress for it, plan picnics around it, suffer from its extremes, and construct our buildings for protection from it. We alternately engage in indoor or outdoor activities as the weather allows, and we attempt to predict atmospheric conditions for days, weeks, and even seasons. In spite of our efforts, many predictions, regardless of the methods we use to obtain them, often yield inadequate warnings of rapid atmospheric change. On June 14, 1990, such an unpredictable change deluged the eastern Ohio town of Shadyside, filling a local stream called Wegee Creek to flood stage and devastating the small community. The individual storm that caused that flooding had its origin in a series of storms that moved across the Midwest and gathered strength as they approached eastern Ohio. Although the forecasters knew that storms were on their way through the area, they did not know that one particular cell of the storm system would produce the localized downpour that filled Wegee Creek with 4 to 7 meters (15 to 25 ft) of water and killed dozens of Shadyside's inhabitants.

Storms are inseparable from water in some form. While still suffering from a long-term drought in February 1992, California's Los Angeles and Ventura counties suffered from flash flooding when heavy rains fell along the Malibu coast and the Santa Monica Mountains (fig. 3.1). In some areas, such as Woodland Hills, the floodwaters raced along at 56 kilometers (35 mi) per hour. By the time the waters had receded, they caused damages estimated at $23 million and killed nine people.

Liquid water, however, is not the only devastating product of storms. Communities throughout Canada, the northern United States, and many mountainous regions are sometimes buried by heavy snowfalls. In the American Midwest, large hailstones have damaged property and crops.

The worst storms, however, are associated with coastal and island populations, where seawater rises to inundate the land. In 1970, a powerful storm not only flooded the Ganges River but also drove water from the Bay of Bengal over low-lying Bangladesh (then called East Pakistan), killing more than 300,000 people. Giant storms also occur in the Pacific Ocean, where they are called typhoons, and in the Atlantic Ocean, where they are known as hurricanes. To their victims these storms are terrible events. For earth scientists these occasional storms, however severe and tragic, punctuate the processes that continuously operate on our planet.

The Study of Weather

Daily changes among gases, called **weather,** are inescapable because they result from a continuous planetwide movement of the **atmosphere,** which is the sphere of gases that envelopes our planet. No place on Earth's surface is exempt from such change. Even the daily heat of deserts, such as those in New Mexico, Arizona, and California, is followed by cool, and sometimes cold, nights. The deserts also get infrequent but often torrential rains that scour the land. Keeping track of the ubiquitous changes in weather requires a worldwide network of observers and observation platforms housing sophisticated instruments, but the invisible atmosphere still surprises us with unpredicted storms and droughts.

Producing a model of atmospheric composition and processes to provide accurate forecasting is one task of **meteorologists,** or weather scientists. Their efforts to understand the invisible envelope of air surrounding the planet have saved lives and informed most of us about the causes and aspects of daily weather conditions.

If you live in the United States or Canada, then you are well advised to learn the natural mechanisms that drive a storm system from the northern Pacific Ocean across the 48 contiguous states and Canadian provinces and out into the Atlantic. The weather in Seattle on Monday may well be a clue to weekend conditions in New York, and a hurricane over Brownsville, Texas, may be the cause of rainy weather in Brownsville, Pennsylvania, within a week. The snowstorm that whitens the streets of Buffalo, New York, may also cover the roads in Montreal.

Televised Weather Lessons

One of the reasons local television meteorologists show more than the area's weather is that they are aware of the interrelationship between the geography of a country and the movements and changes of the air over the country. A well-informed viewer can learn much about the local forecast by looking at the big picture of an entire country's weather. In the usual format, the forecaster shows a view of a large landmass or an entire country from space and then a series of regional and local radar images, which define rainfall and snowfall patterns. Typically, we also receive information of little interest to many, such as barometric pressure and tendency, upper air conditions and patterns, and wind direction. Although this information may seem irrelevant, it is actually the basis of the forecast. By studying this and the next two chapters, you will become familiar with the terms meteorologists use. You will also understand why forecasters give information about the weather conditions at great distances from the local area and at considerable altitude above the local surface.

The Weather Story

Weather observers and forecasters are coauthors of a daily or hourly story. Observers collect details on atmospheric conditions and funnel their information into a network of the National Weather Service, which in the United States is centralized in Suitland, Maryland. The National Weather Service then tracks the movements of air on a series of maps, which act as a story line for the nation's weather. This story line can provide clues to the science of weather, but it does not give the entire saga of Earth's atmosphere and atmospheric processes. Understanding the details in the short story of daily or weekly weather is covered in chapters 4 and 5. With these details in hand, you can understand the global patterns of weather discussed in chapter 6, but even the global tale lacks some of the elements of the weather story.

Figure 3.1   After six years of drought, Los Angeles and Ventura counties in California were subjected in 1992 to their worst flooding since the 1930s. These winter rains caused more than $20 million in damages and killed nine people. Although the 1992 storms produced abundant rainfall, they did not relieve the drought conditions.

To understand the full epic of weather, you need first to take into account Earth's origins in the nebular cloud (chapter 2), the composition of the atmosphere, and the relationship between Earth and the object most directly responsible for atmospheric changes: the sun. *It is the energy of the sun that is the principal influence on the weather.*

THE SUN

The most influential body in the Solar System is the sun. Its gravity holds the system together, and its energy illuminates the planets. To comprehend the full story of Earth's atmosphere, you should understand the source and nature of the energy that drives the various gases in the processes we call weather.

Fusion and Electromagnetic Radiation

The reactions generating the energy that eventually drives Earth's atmosphere occur in a process known as proton-proton fusion (fig. 3.2). Although other fusion reactions are possible,

the proton-proton reaction is the dominant process of the sun's interior. When two protons are driven very closely together under the intense heat of the sun's core, the strong nuclear force binds them. As figure 3.2 shows, one of the protons decays into a neutron, a positron, and a neutrino, and the departure of the latter two transfers kinetic energy away from the reaction site. Now, the remaining proton and the newly formed neutron are free to combine with another proton, making an isotope of helium. Under the same intense heat and pressure that initially drove the protons together, helium isotopes then combine, releasing two protons with about half the energy obtained in the fusion reaction.

One of the problems most of us have in visualizing this reaction lies in our considering protons and other subatomic particles as separate entities with well-defined boundaries, something like baseballs. The protons that combine do so when they overcome the repulsion they mutually offer because of their positive charges. The analogy most often used to describe this fusion is one given by the German-born American scientist Hans Albrecht Bethe, who envisioned a "tunneling" by one proton into the vicinity of another. With enough protons attempting this

Figure 3.1 *continued*

joining under the great pressures of the sun's interior, some accomplish it. The production of kinetic energy in the sun's core is just the first step in driving Earth's atmosphere through electromagnetic energy. The energy changes form as it travels from the sun's interior to our planet.

Structure of the Sun and Transposition of Energy

The energy produced by the sun must travel through the sun's layered structure and across the approximate 150-million-kilometer (93-million-mi) vacuum of space to reach our planet. Traversing those obstacles of sun and space changes the form in which the energy is transferred.

1. Two hydrogen nuclei (protons) are driven close together in the sun's core. One proton decays to a neutron and a nucleus of deuterium is formed.

2. The deuterium nucleus and a single proton fuse to form helium ($_2He^3$).

3. Two helium ($_2He^3$) atoms combine to form an atom of common helium ($_2He^4$). Two protons are lost in this process and a significant amount of energy is released.

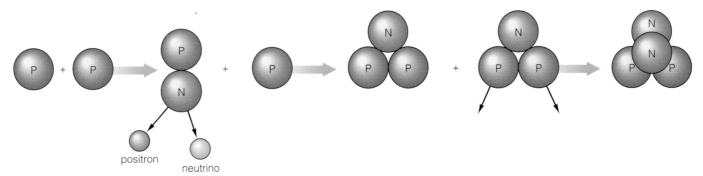

positron   neutrino

Figure 3.2 The proton-proton fusion reaction is a three-step process that occurs under great pressure and high temperature.

Fusion generates energy in the form of shortwave electromagnetic radiation, and it begins a sequence of atomic-scale events that produces a spectrum of energy, including gamma rays, X rays, radio waves, ultraviolet and infrared radiation, and visible light. It is this radiation that eventually makes its way throughout the Solar System and that, today, strikes the planetary surfaces and atmospheres. In addition to electromagnetic radiation, the sun also produces *corpuscular* (from the Latin for "little body") *radiation,* or charged atomic particles, which make up the **solar wind.** The solar wind eventually interacts with Earth's upper atmosphere and its magnetic field, which directs the corpuscular radiation toward our planet's magnetic North Pole.

The Core and Radiative Zone

The sun's energy is generated deep inside its almost million mile diameter sphere. The central portion of the sun, about one-half the solar diameter, is occupied by a *core,* or solar furnace, in which each second the fusion reaction converts 400 to 600 million tons of hydrogen into helium and 4 million tons of matter into kinetic energy. The reaction occurs inside a plasma that exceeds 15,000,000 K. The core is also under tremendous pressure, probably more than 200 billion times that which we experience under our planet's atmosphere (fig. 3.3). At the diffuse boundary of the core, temperatures drop a few millions of degrees and no fusion occurs. The radiation from the sun's interior then continues outward through a zone that has the density of water. This zone, called the **radiative zone,** extends from the core to about seven-tenths of the way to the sun's surface.

Within the zones of the sun neutrons make their way to the surface, but positrons collide with electrons, producing shortwave electromagnetic radiation in the form of gamma rays. The gamma rays move to the radiation zone, where they are converted to the less energetic X rays and ultraviolet rays. As this energy passes to the surface, or **photosphere,** a transformation of energy occurs to longer wavelengths, producing visible light.

The Convective Zone

Within the upper 30% of the sun's radius is a highly opaque zone, and the radiation initiates a convective activity, where masses of the sun's matter rise and fall as they carry and lose the heat generated by fusion. In this **convective zone,** continuous plumes of gas rise toward the sun's surface to release energy. The time taken by the energy of fusion to pass through the various solar zones may exceed 10 million years.

The Sun's Surface Zone and Atmosphere

Above the sun's surface lies its atmosphere, the only visible part of the sun. The surface zone and atmosphere have three identifiable regions: the *photosphere, chromosphere,* and *corona.*

The Photosphere and Sunspots   The 6,000 K photosphere is the brilliant surface we see. It is this surface that houses **sunspots,** dark depressions in the photosphere with temperatures below 4,000 K (fig. 3.4). A typical sunspot is dark only because it does not emit as much light as the surrounding photosphere. Actually, if we were to see one by itself at night, we would observe an object ten times brighter than a full moon. Sunspots are also associated with strong magnetic fields, and scientists suspect that the magnetism is somehow responsible for the cooler nature of the spots. Additionally, their relationship to the sun's magnetic field is tied to a cycle between maximum and minimum periods of sunspot activity. Periods of greatest sunspot numbers occur on an average of every 11.1 years. Because researchers have now identified reversals in the polarity of the sun's magnetic fields, scientists also speak of a

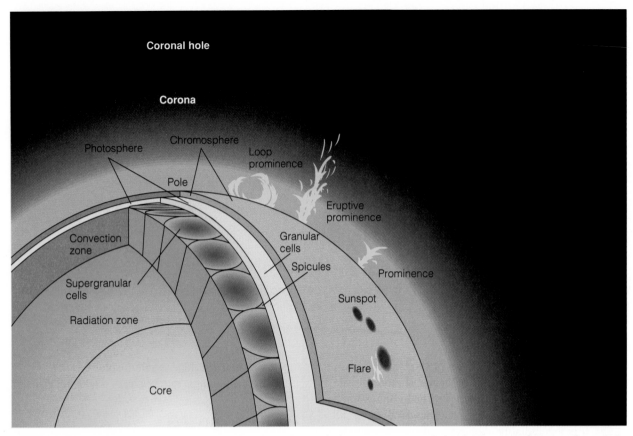

Figure 3.3 Structure of the sun. Energy generated in the core through the proton-proton chain reaction must first pass through the radiation zone, which is the thickest of the sun's spheres. Within the radiation zone the electromagnetic wavelengths increase toward visible light. Turbulent cells of hot gas then carry energy through the convection zone toward the photosphere, which releases energy to the Solar System.

22-year cycle linked to those shifts in polarity. Because sunspots can be tracked in their journey across the sun's surface, scientists now know that our star rotates. The rotation is a differential movement, with low latitudes moving faster than the higher ones. From 10°N of the solar equator to 10°S, the sun rotates once every 27 days.

The Chromosphere and Spicules The **chromosphere** is a thin reddish rim from which **spicules, solar flares,** or **prominences** appear to erupt (fig. 3.5). The eruptions vary in size. Spicules are the smallest and most common of these outward movements of solar gas. Prominences are the largest.

The Corona The outermost part of the sun is a very-low-density, uneven shell called the solar corona. The corona is observed when the sun is blocked by the much closer and smaller (but intervening) moon during an eclipse. The corona has a density only a hundred-billionth that of Earth's atmosphere, but it is very hot, reaching temperatures above 2,000,000 K. Both the great temperature of the corona and its shape appear to be the product of magnetism. Observations made by astronauts aboard *Skylab* revealed that the corona is also punctured by "holes" through which stream charged particles called the solar wind.

Corpuscular Radiation: The Solar Wind

The production of charged particles in the fusion reactions of the sun leads to a "wind" of plasma, composed of highly energetic protons and electrons. The wind moves out from the sun at high speed, encountering the planets as it moves through the Solar System. On Earth, the wind collides with molecules and atoms in the atmosphere, generating a light called the **aurora borealis,** or the northern lights. In high latitudes, the aurora borealis is a colorful display of wavelike movements. People in Canada and Alaska are fortunate to see these colors frequently. Farther to the south the aurora loses its colors, and at the latitude of New York City it is rarely seen. Every so often, however, a particularly brilliant aurora occurs. In 1991, observers saw a color display in the state of New York. Ordinarily, when the aurora is visible in the lower 48 states, it is white.

According to studies made aboard *Skylab*, the solar wind is associated with the *coronal holes.* Coronal holes are 500,000 K cooler than the rest of the corona, and the cooler temperatures seem to be related to the generation of the solar wind.

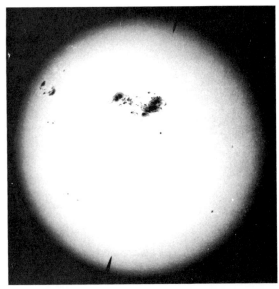

Figure 3.4 Sunspots appear dark because they are cooler than the background of the photosphere against which they are seen.

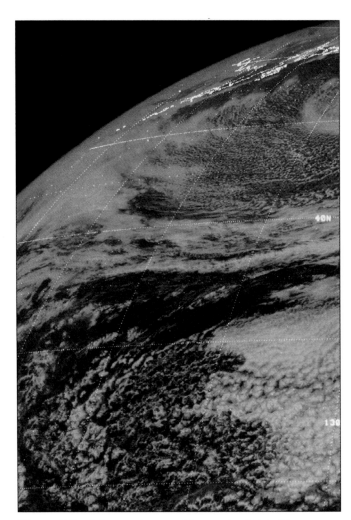

Figure 3.5 Solar prominences follow paths dictated by magnetic fields. Often, these paths form arcs of brilliant plasma.

EARTH-SUN RELATIONSHIPS

The sun's electromagnetic radiation takes about 8 minutes to reach Earth. When it arrives, it meets a round planet tilted to the plane of its orbit (fig. 3.6). Both the roundness and the tilt play important roles in the transfer of energy from the sun to our planet. To understand fully how that transfer works, you should learn about the relationship between the shape and orientation of Earth and the incoming electromagnetic radiation.

Earth is not a stationary planet, and its constant motion is an important control on incoming solar radiation. Because we are bound by gravity to our planet, we do not feel its motion, but our planet is traveling very fast in its journey around the sun. It also turns like a spinning top.

Rotation

The days result from Earth's **rotation,** which is a turning on its axis (fig. 3.6). This turning occurs at variable speeds, depending on your position on the surface of the planet. Quito, Ecuador, located on the equator, makes a bigger circle during a single turn than does Chicago, Illinois. Thus, Quito must move faster than Chicago to make one rotation. In one day Quito must travel the full circumference of our planet, or more than 38,700 kilometers (24,000 mi).

The importance of the Earth's turning is that the sun does not light just one "side" of the sphere: the rotation distributes the energy *every 24 hours.* The "side" turned away from the sun loses energy, effectively preventing an out-of-control buildup of heat from occurring. The rotation gives us day and night and the basis for our hours (see table 6.2).

Revolution

Earth also revolves around the sun in an elliptical *orbit.* The time for this orbit is 1 year, or 365.25 days. This **revolution** takes our planet as close as 147.25 million kilometers (91.5 million mi) from the sun and as far as 152 million kilometers (94.5 million mi) away from the solar furnace. At its closest approach to the sun, Earth is said to be at **perihelion.** When it is farthest from the sun, Earth is at **aphelion** (fig. 3.6). As Earth now moves in its orbit, aphelion occurs during the Northern Hemisphere's summer; perihelion, during its winter.

Box 3.1 Further Consideration

Time Zones

Whenever Earth completes a single, complete turn about its axis, one Earth day has passed. Since ancient times, humans have attempted to divide this turning into convenient, trackable units. Today, we divide one rotation into 24 time zones, giving us a 24-hour day (fig. 3.A).

Because a circle—and, thus, a sphere—is divided into 360°, each of the time zones corresponds to 15° of rotation ($360 \div 24 = 15$). With an arbitrary starting point on Earth's surface, each 15° segment makes a 7.5° spread on either side of the center of a time zone. Since Earth turns from west to east, the 15° time zones equate to 15° of longitude, and the time of each zone is 1 hour later in an eastward direction.

If the 0° longitude (a meridian that runs through Greenwich, England) is a starting point, and it is 5 P.M. along that meridian, then at 7.6°W of the Greenwich meridian, the time is 4 P.M. Along the length of the longitudinal line that runs 7.6°E of the Greenwich

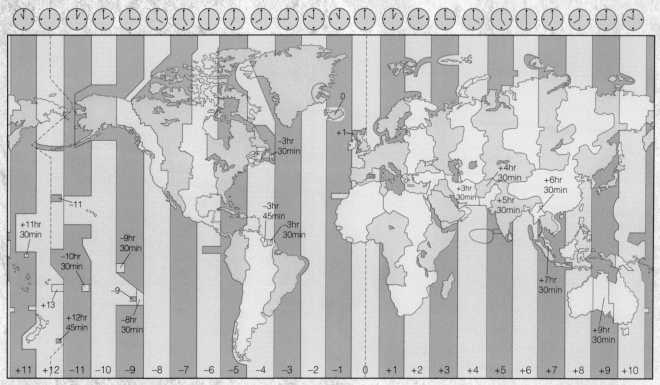

Figure 3.A   Standard time zones of the world.

Tilt and the Seasons

Earth tilts 23.5° to the plane of its orbit. The direction of tilt does not vary as the position of the planet changes during its revolution, making the North Pole alternatively lean toward and away from the sun (fig. 3.6).

Earth's greater proximity to the sun in winter rather than in summer may seem surprising. You would expect temperatures to be warmer when Earth is closer to the source of its atmospheric heat. The reason for the apparent discrepancy lies in Earth's 23.5° tilt. This tilt, or inclination, plus its spherical shape determine the intensity of the solar radiation at the surface. The explanation may seem complex, but it is, in fact, relatively simple. Just imagine shining a flashlight held perpendicular to a surface as you see in figure 3.7a. The beam of light, which makes a right angle with the surface, concentrates its energy in a small

circle. If you incline the flashlight, the circle of light becomes first an oval and then a parabola, as figure 3.7b and c show. The same effect occurs over the curved surfaces of a tilted Earth. Figure 3.7d shows the light from the sun as parallel arrows. Those arrows strike the tangents of the curved surface at different angles. As the angle approaches 90°, the energy is more concentrated per unit area.

During the year, the angles of the sun's rays change over the surface of the Earth (fig. 3.6). Tangents to Earth would reveal a shifting angle and either increasing or diminishing energy per unit of area. During the Northern Hemisphere's summer, the North Pole "leans" toward the sun while the South Pole "leans" away. The result is a greater concentration of the sun's energy in the Northern Hemisphere than in the Southern Hemisphere. The circumstance reverses during the Northern Hemisphere's winter. The two hemispheres always lie in opposing seasons: fall-spring meridian, the time would be 6 P.M. The center of the time zone that includes the 7.6°W meridian is along the 15°W meridian. If Philadelphia lies at 75°W of Greenwich, it lies in the middle of the fifth time zone west of Greenwich. If the 0° meridian's time is 5 P.M., Philadelphia's time is 12:00 P.M. Philadelphia will require five more hours to reach the position of Greenwich relative to one rotation.

The time zones marked on maps do not, however, reflect a pure 15° division. Population densities and political boundaries, such as the borders of territories, states, and countries, are used as time zone variants to allow citizens a common time reference. Thus, the time zones are not always straight north-south lines (fig. 3.B).

The solar day—the passage of the sun during the daylight hours—is divided into two units. Before the sun is halfway in its journey across the sky of the central time meridian of a zone, the time is A.M., or *ante* (before) *meridian* (A.M.). After the sun reaches the midpoint of its journey across the central meridian's sky, the time is *post* (after) *meridian* (P.M.). The western portion of a time zone does not have the same sun-time as the eastern portion because the sun rises in the east. The standard time for the zone, however, ignores the actual relationship of the zone's east and west boundaries to the sun.

Apply Your Knowledge

1. What time is it in Los Angeles when it is 12:00 P.M. in New York City?
2. What time is it in Miami, Florida, when it is 12:00 P.M. in Perth, Australia?

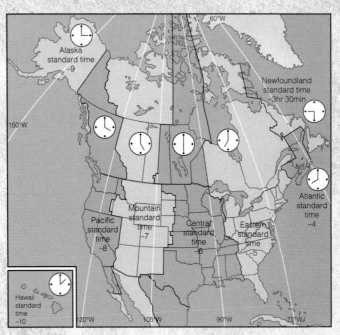

Figure 3.B Standard time zones of Anglo-America.

and summer-winter. The cause of the seasons, therefore, is not the distance of the Earth from the sun but rather *the tilt of the planet to the plane of its orbit,* an imaginary elliptic sheet that would run from the center of the Earth through the center of the sun.

A further examination of figure 3.6 reveals that the tropics, lying between 23.5°N latitude and 23.5°S latitude, would receive more energy during the year than would higher latitudes. The concentration of energy in low latitudes results in higher heat values for these latitudes than for the high latitudes, where the sun angle is greater.

Latitude, therefore, or distance from the equator, is the most significant control on the amount of heat available for the atmosphere. Between the Tropic of Cancer (23.5°N) and the Tropic of Capricorn (23.5°S) the *noon* sun angle never varies more than 47° from directly overhead. At higher latitudes, 23.6°

to 90°, the angle varies to a greater extent. During the two hemispheres' respective winters, the latitudes greater than 66.5° are not lighted by the sun for 2 to 6 months. The absence of incoming solar radiation, or *insolation,* results in a net loss of heat from the planet's dark (or nighttime) surfaces.

THE BIG PICTURE: A HYPOTHETICAL STORM CROSSES THE CONTINENT

Although Earth receives only two-billionths of the sun's energy output, enough insolation arrives at the planet's surface to power the atmosphere into a series of changes that we call weather. The many observers and observing instrument platforms that gather information on these changes help forecasters tell an important story about atmospheric conditions. Each

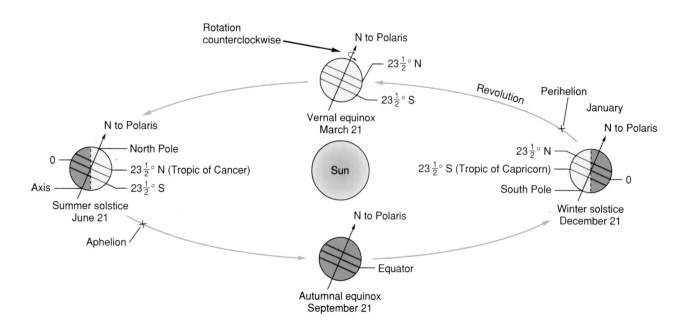

Figure 3.6   Earth-sun relationships. Solstices and equinoxes actually vary from the twentieth to the twenty-second of the respective months.

day's weather is only a subplot of a larger tale, and your local weather may originate thousands of kilometers away. Although we have not yet covered the details of meteorology, you may find it instructive to follow a hypothetical storm system as it crosses the North American continent. You are already familiar with much of this story because forecasters have given you bits and pieces of similar tales and shown you similar maps during the daily television news. Once you have followed the hypothetical storm, you will read about meteorology in the context of typical atmospheric changes.

An Autumn Storm System Is Born Along the Northwest Coast

Somewhere over the North Pacific Ocean during November a great mass of water vapor and air rotates counterclockwise around a relatively warm center of rising air. The rotation continues to pick up speed as the entire mass migrates southeastward like a giant top spinning across a floor. The growing storm, as viewed from space, takes on the appearance of a large comma. It approaches British Columbia, Washington, and Oregon with steadily increasing wind speeds, some close to 120 kilometers (75 mi) per hour near the center of the system. Such a system would appear as a swirling band of clouds similar to that shown in figure 3.8.

With the approach of the storm, meteorologists in the Northwest predict rain. These rains come as forecasted when the storm initially enters the country by rising over the coastal mountain ranges of western Oregon and Washington. A heavy rainfall accompanies the movement over the steep increase in elevations,

with the Olympia Mountains, the highest elevations along Washington's coast, receiving the heaviest downpours. The rains also drench the valleys from Puget Sound southward through the Willamette Valley, and both Portland and Seattle forecasters track the same storm system.

East of the valleys lie the high Cascades, a line of north-south trending volcanoes that includes Mount Saint Helens and Mount Hood. With elevations in excess of 3,000 meters (10,000 ft), the Cascades present a formidable obstacle to the southeastward-moving air. In the vicinity of the volcanoes, the air rises rapidly, drops in temperature, and loses more of its supply of water vapor as rain. As the air continues to rise, it precipitates snow instead of rain on the higher elevations, those above 2,100 meters (7,000 ft), where the temperatures on this cold November day fall below freezing. High winds whip the snow into a low-visibility blizzard that sweeps over the high plateau country of eastern Washington and Oregon and nearby sections of southern Idaho, southwestern Montana, and western Wyoming. Countrywide, American meteorologists report, "It is a bleak November day in the Northwest." At the same time Canadian meteorologists note that this mass of airborne moisture will soon move southeastward and exit British Columbia.

Eventually, the storm picks up more cold air, which moves southward from Canada and joins the swirling system like flotsam drawn into a whirlpool. The Canadian air is drier than the Pacific air and, coupled with the storm's movement away from the ocean as it crosses the rough surfaces of the western states, weakens the system (fig. 3.9). The storm wends its way toward New Mexico, where newscasters announce the coming of a weak, winterlike storm. By the time the storm reaches the state,

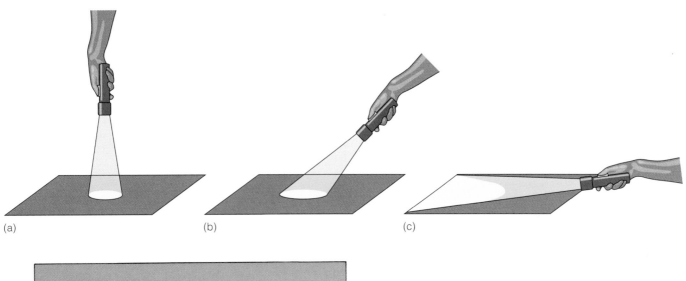

(a)　　　　　　　(b)　　　　　　　(c)

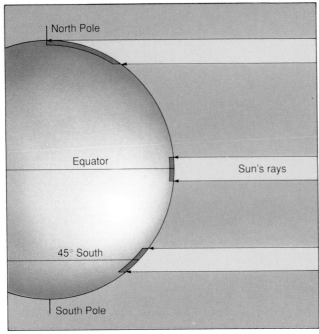

North Pole

Equator

Sun's rays

45° South

South Pole

(d)

Figure 3.7 Diagrams *(a)* through *(c)* show an increasing areal coverage by the same amount of energy, lessening the intensity per unit area. Diagram *(d)* shows this same effect on the surface of Earth for the equinox.

it is almost devoid of the precipitation and strong, circular winds that plagued the Northwest.

The System Enters the American Southwest

Now in the region where Native Americans of the Southwest tell legends of the firebird Phoenix, the storm begins to revitalize like the mythical bird. The infusion of life into the dying storm comes in the form of a group of high-altitude, high-velocity winds collectively called the **jet stream.** The jet stream provides the storm with new circulation, moisture, and a driving energy that guides it into an eastward track.

The initial support of the jet stream drives the rejuvenated storm to the east and results in increased snowfall over the higher elevations of northern New Mexico and southern Colorado. The storm then moves farther eastward with greater strength and

speed. Snow begins to spill onto the hill country of western Texas and onto the panhandles of both Texas and Oklahoma. Now the storm center lies over the Great Plains, where it generates a merger of two radically different kinds of air masses: the cold, dry air of the northern American Midwest and the warm, moist air of the Gulf of Mexico. Eastern and midwestern forecasters now begin to issue warnings that violent storms are probable. This hypothetical weather system now appears to be very similar to the 1989 and 1990 systems that caused death and destruction in Huntsville (chapter 1) and Shadyside.

The merger of the two types of air—warm, moist and cold, dry—creates a series of thunderstorms as the cold air wedges itself beneath the warm air. Some of these thunderstorms develop into severe storms with high winds, large hail, and in some areas, tornadoes. Emergency warnings flash across television screens of the lower Midwest, which is south of the storm

Figure 3.8   A storm system that enters the continent from the Pacific Ocean appears as a large comma-shaped cloud band. The movement of the system is west to east, or left to right in this space photo.

center, where the intensity of the system is greatest. The National Weather Service of the United States issues two different sets of warnings: south of the storm center it announces a severe thunderstorm and tornado watch; while north of the center, it issues a winter storm watch. Forecasters along the storm track base their predictions on their area's position north or south of the jet stream, the boundary between winter- and summerlike storms. Residents of Kansas and southern Nebraska brace themselves for snow, sleet, and freezing rain.

The storm center continues to move through Oklahoma and into central Arkansas. Just north of the center residents encounter narrow bands of freezing rain and then, to the north of the center about 160 to 240 kilometers (about100 to 150 mi), a zone of sleet. In Kansas and southern Nebraska, snowfall ranges between 15 and 30 centimeters (6 and 12 in). The strengthened storm now drives the snow into near-blizzard conditions. Farther to the north the snow diminishes, and forecasters predict cold but relatively dry conditions from southeastern Saskatchewan to northern Wisconsin.

The System Influences the South and the East

Hourly forecasts warn of winter storms headed toward the Northeast as the jet stream changes the storm center's direction (fig. 3.10). Additional moisture from the Gulf of Mexico reenergizes the storm system, which now heads along the western slopes of the Appalachians toward Ohio. On the southern and eastern sides of the storm center, thunderstorms and showers move toward the Northeast, but on the northern side of the center and the jet stream, winter storms begin to drop heavy snows from southern Illinois to southwestern Pennsylvania. All across the country forecasters point out the relationship between the elevations and nature of the continent's surfaces, or its **topography,** and the moving storm. Now, they note the storm's encounter with the highlands of the Appalachians, where the air must once again rise to pass over the higher elevations.

As the storm center reaches Kentucky, it has distanced itself from the source of moisture, and it begins to weaken again. Almost simultaneously, however, another storm system forms along the coast of North Carolina near Cape Hatteras. Supported by the northeastward-trending jet stream, the new storm increases in strength almost in inverse proportion to the old storm's

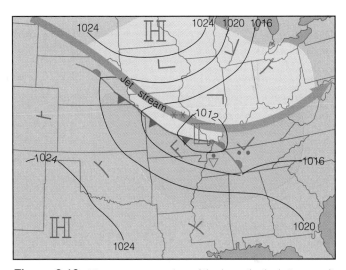

Figure 3.10 The map expression of the hypothetical storm as it passes over the American Midwest (See Symbol Key with figure 3.9.)

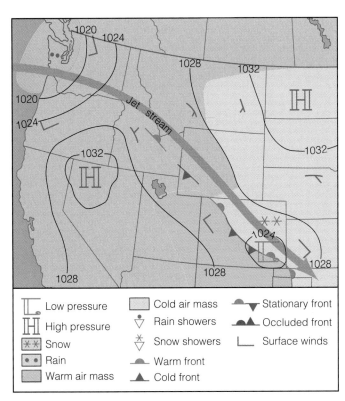

Figure 3.9 The map expression of the northwest coast of the United States as a storm system enters the continent. All the terms used in the map and in the description of the hypothetical storm are explained in chapters 4 and 5.

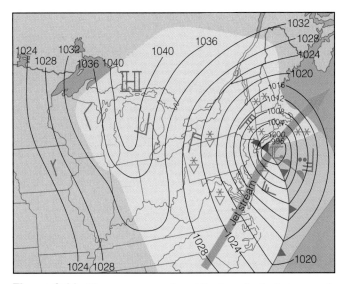

Figure 3.11 The map expression of the hypothetical storm as it exits the continent and moves over the Atlantic Ocean. (See Symbol Key with figure 3.9.)

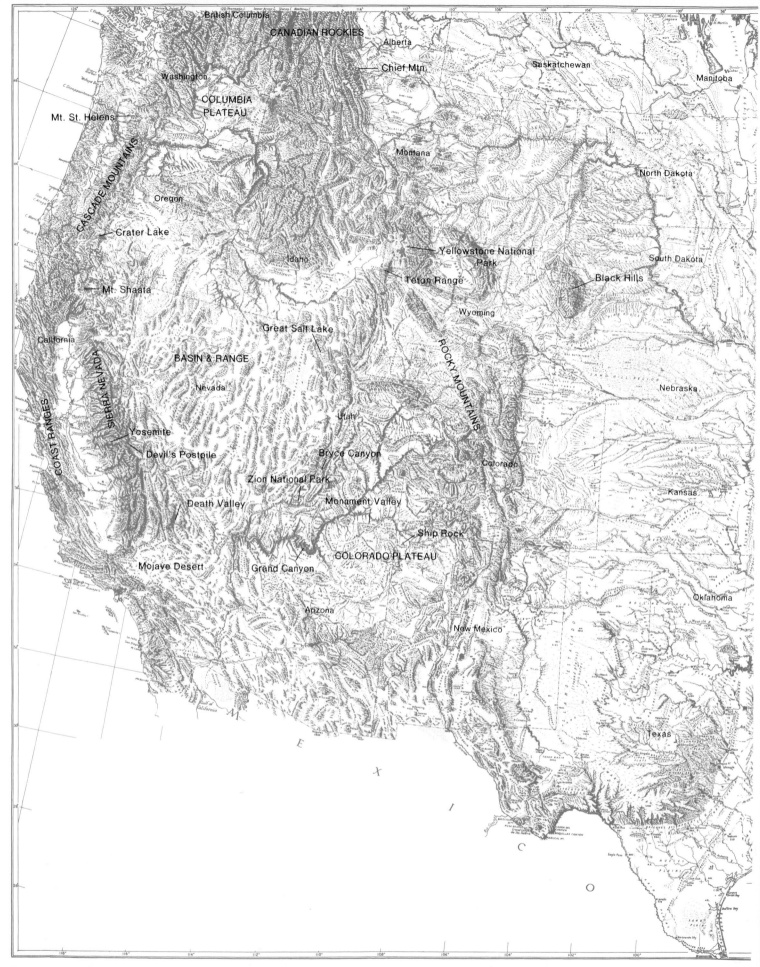

Figure 3.12 Raisz physiographic map of the 48 contiguous states and southernmost Canada. Copyright by Erwin Raisz 1957, reprinted with permission by Raisz Landform Maps, Melrose, MA 02176.

CANADIAN SHIELD

Laurentian upland of low hills and many lakes

Ontario

Quebec

Newfoundland

New Brunswick

Maine

Minnesota

Wisconsin

Michigan

ADIRONDACK MOUNTAINS

Vermont

New Hampshire

Massachusetts

New York

Rhode Island

Connecticut

Iowa

Pennsylvania

New Jersey

Ohio

Illinois

Indiana

Maryland

Delaware

West Virginia

Missouri

Virginia

Kentucky

APPALACHIAN MOUNTAINS

North Carolina

Tennessee

Arkansas

OUACHITA MOUNTAINS

South Carolina

Mississippi

Alabama

Georgia

Louisiana

Florida

Mississippi Delta

Miami

LANDFORMS OF THE UNITED STATES

by ERWIN RAISZ

Sixth revised edition, 1957

Scale   50   100   150   200 Miles

100   100 Kilometers

decrease in intensity. Forecasters along the eastern seaboard note a transfer of energy from the old to the new storm system, and the heaviest snowfall shifts to the east (fig. 3.11). North Carolina, Virginia, and Maryland are successively buried under heavy snows. As much as 45 centimeters (18 in) fall on Washington, D.C., and Baltimore, Philadelphia, New York, and Boston all feel the effects of blizzard and near-blizzard conditions.

Having worked its way across the continent, the system brushes by Nova Scotia and Newfoundland on its way over the Atlantic Ocean. In its passage, this system would have affected the daily activities of millions of people.

Meteorology and Climatology

The hypothetical storm just described contains all the characteristics of a composite system, one that could occur but seldom does. Because such a storm system is possible, its path and characteristics reveal something about weather studies: they require knowledge of the kind of surface over which the air passes in addition to information about the atmospheric conditions from the surface to high altitudes. The weather is a product of short-term atmospheric processes in specific geographic areas, and forecasters usually limit their predictions to 12- to 48-hour periods in typical "today-tonight-tomorrow" forecasts.

The rapid changes our fluid and invisible atmosphere can make require constant vigilance by weather observers. The weather is essentially the current conditions of the atmosphere at a given locality, and only by understanding how those conditions arise can a forecaster predict the impending weather.

When earth scientists look at the history of weather for a specific place, they derive its **climate,** or the aggregate weather for a geographic locale. Climatologists study regional and global patterns of weather to discern their causes and effects over long periods, such as months, seasons, years, or even centuries and millennia.

Aggregate weather or climate should not be confused with average weather. To limit the study of climate to average temperatures and precipitation for a place is to miss the extremes that climatologists take into account. For example, to the casual observer, Des Moines, Iowa, and San Francisco, California, may have apparently similar average annual temperatures—in the 10° to 16° C (50° to 60° F) range—but they do not have the same type of climate. In fact, even slight differences in average temperatures indicate significant variations between the two locales. The San Francisco climate is influenced by the Pacific Ocean. Its relatively warm winters and cool summers give it a yearly average temperature of 13.6° C (56.5° F). Des Moines, landlocked in the central part of the country, undergoes greater extremes of temperature than San Francisco and has hot summers and cold winters.

Interrelationship: Geography and Weather

Knowledge of the kinds of surfaces over which air moves is essential to an understanding of both weather and climate. The geography of the North American continent provides a variety of landforms, which influences conditions in the lower atmosphere. Good weather forecasters, in fact, are reliable because they know the influence Earth's surfaces have on the moving air.

The landform map of North America shown in figure 3.12 reveals its eastern and western mountain systems, its broad central plains and lowlands, and its coastal plains. Each of these landforms governs, in part, the type of weather it experiences and passes on to its neighboring landform in a general west-to-east pattern across much of the continent. Mountain ranges, for example, tend to be locales where moving and rising air masses lose moisture as either rain or snow, making the air drier on eastern slopes than on western slopes. Bodies of water, like the Great Lakes, add moisture to the passing air. Thus, the Buffalo, New York, region is known for its heavy snowfalls, with snowfall amounts diminishing to the southeast of lakes Erie and Ontario.

Further Thoughts: A Complex Atmosphere

With the preliminary view of weather given in this chapter, you are now ready to explore the details of Earth's atmospheric components and processes. In the next three chapters, you will read about the special composition of this planet's air and how it differs from the atmospheres of the other planets. You will also learn about the significant relationships and interactions among the atmosphere, landforms, and water bodies.

Chapter 4 provides a perspective on the structure of the atmosphere, and the following chapter explains why the lower atmosphere produces weather. In chapter 6, you will read about climate.

It should now be obvious that the study of weather involves more than just research into air movements. Meteorologists are concerned about composition and atmospheric energy, as well as air movements. Their apparent inability to predict weather precisely on occasion is a reflection of the many interrelationships they must consider before forecasting weather.

Significant Points

1. Weather is the condition of the atmosphere at a given locality or region.
2. Meteorologists, or weather scientists, collect data on the daily movements among the gases of the atmosphere and centralize their information to establish a model of recognizable atmospheric phenomena. In the United States, the National Weather Service uses this information to make predictions.
3. The sun is the ultimate source of energy for atmospheric movements. Its energy arrives at our planet in two forms: electromagnetic radiation and corpuscular radiation (solar wind).
4. The sun derives its power from the proton-proton fusion reaction in its core, where hydrogen forms helium under temperatures in excess of 15,000,000 K. The fusion releases kinetic energy when subatomic particles leave the reaction site.
5. The sun has a layered structure. Outside the core is a radiative zone where shortwave gamma rays are produced. This energy migrates to the next zone, called the convective zone, where plumes of gases continuously rise to release energy at the sun's surface.
6. The sun's atmosphere has three regions. The photosphere is the part of the sun we see. It contains cooler regions called sunspots. These sunspots are related to magnetic fields. The chromosphere is the source of spicules and solar flares. The corona, visible only during an eclipse, is an uneven plasma of 2,000,000 K with cooler "holes" through which the solar wind "blows."
7. When its orbit takes our planet closest to the sun, Earth is at perihelion. Our elliptical orbit's farthest point from the sun is called aphelion. Aphelion occurs during the Northern Hemisphere's summer.
8. The tilt of Earth to the plane of its orbit is the cause of seasons.
9. The rotation of Earth about its axis distributes the incoming solar energy over the planet's surfaces, gives day and night, and serves as the basis for our time measurements.
10. The jet stream is a changing group of high-altitude, high-speed winds. It influences the movements of surface weather systems.
11. Weather is influenced by topography and surface composition. In part, air derives its weather characteristics from the surfaces over which it passes. Water bodies can add moisture to a weather system. Mountains and plains affect weather systems differently.
12. Aggregate weather for a locale or region over a period measured in decades is called climate.

Essential Terms

weather 56	radiative zone 59	chromosphere 60	rotation 61
atmosphere 56	photosphere 59	spicule 60	revolution 61
meteorologist 56	convective zone 59	solar flare 60	perihelion 61
solar wind 59	sunspot 59	prominence 60	aphelion 61

Review Questions

1. How does weather alter the human environment?
2. In what general direction do weather systems move across North America?
3. What is the primary source of energy for Earth's atmosphere?
4. What is the source of the sun's energy?
5. What is corpuscular radiation?
6. How is the sun layered? (Draw and label a diagram, showing zonation in the sun.)
7. What is the aurora borealis?
8. What are the main influences on seasonal temperature variations?
9. How does the jet stream affect surface weather?
10. What were the major changes that occurred as the hypothetical storm crossed the continent?

Challenges

1. Why are weather predictions important to farmers?
2. How does weather affect the flow of goods and services in your area?
3. What area's or city's weather acts as a precursor of your area's or city's weather?
4. What landform boundaries define the region for which your local forecasters predict weather?
5. How would your life change if no weather predictions were available?
6. Why are the North and South Poles shrouded in darkness during the winter months?
7. Why does the Northern Hemisphere have summer when the Southern Hemisphere has winter?
8. How would the intensity of radiation from the sun change between aphelion and perihelion?
9. What might cause an increase in intensity in the aurora borealis?
10. How would a change in Earth's tilt affect seasons?

4

ORIGIN, STRUCTURE, AND CHARACTERISTICS OF THE ATMOSPHERE

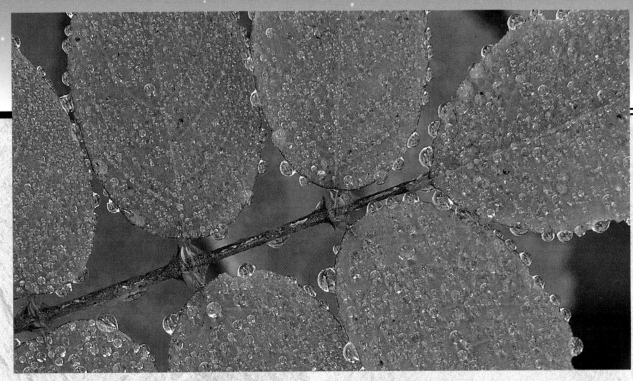

Dew forms from condensed water vapor, one of the components of our atmosphere.

Chapter Outline

INTRODUCTION

Earth's atmosphere may appear to be unique in its composition and structure, but in some ways it is similar to the other atmospheres in the Solar System. Although the amount and proportion of gases may differ from planet to planet, *atmospheric gases everywhere obey the same fundamental physical and chemical laws*. The thick, blue atmospheres of Uranus and Neptune contain methane (fig. 4.1), but that gas is no different from the CH_4 found on Earth, where it is subject to heating, cooling, and the planet's gravitational pull. The different temperatures and abundances of methane found on all three planets reflect only a difference in degree and not in kind. Similarly, carbon dioxide, which is the primary constituent in the atmosphere of Venus but comprises less than 1% of Earth's gases, acts like a greenhouse gas regardless of its place in the Solar System. The proportion of this gas in the respective atmospheres of the two planets makes a significant difference in extent of the greenhouse effect (chapter 1). If Earth incorporated Venus-like quantities of CO_2 in its atmosphere, our planet's surface would be too hot to support life.

PLANETARY ATMOSPHERES

Before you proceed in your study of Earth's weather, you should have a general idea about the nature of atmospheres. Are the gases of the other planets similar to those of Earth? Are those atmospheres thicker or thinner than Earth's? Do other planets have weather? Is that weather similar in any way to Earth's? With a background in planetary atmospheres, you can then delve into the workings of our own gaseous covering. To answer these questions, we must rely on studies of the other planets, which have provided data on the probable origin, composition, and motions of their atmospheres.

Atmospheric Compositions

Earth's envelope of air, which allows jet airplanes like the Concorde to fly at altitudes approaching 16 kilometers (10 mi), is actually a relatively thin sphere of gases. The Jovian planets, by contrast, have very thick atmospheres. Jupiter's multicolored clouds conceal a planet dominated by an atmosphere of mostly hydrogen and helium possibly hundreds of times thicker than Earth's atmosphere. Thick, rapidly moving clouds lying throughout Jupiter's upper atmosphere prevent us from knowing just how thick its gaseous layer really is. We do know that the great mass of this largest planet in the Solar System allows it to hold the lighter gases more effectively than the smaller Earth and the other three inner planets. The gravitational pull of Jupiter's mass creates an increasingly more dense atmosphere toward its interior, and liquid molecular hydrogen (H_2) may lie under tremendously high pressures only 965 kilometers (600 mi) below the outer gases. In contrast, the terrestrial planets are quite different simply because they are smaller, mostly solid bodies closer to the sun. The lightest gas, hydrogen, is easily lost to space in the uppermost regions of Earth's atmosphere, and water vapor, a con-

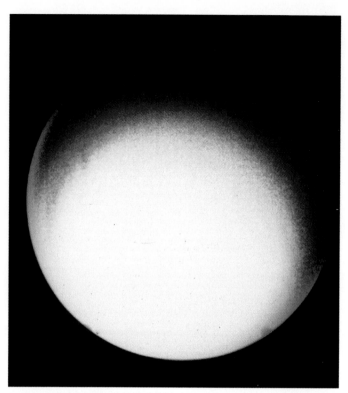

(a)

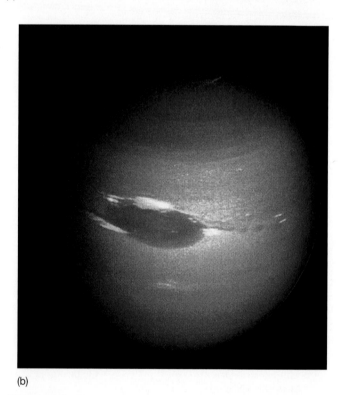

(b)

Figure 4.1   The "blue" planets: *(a)* Uranus and *(b)* Neptune.

stituent of our planet, appears to have escaped the confines of Mars billions of years ago.

Three of the four relatively small, inner planets may have had initial atmospheres of comparable size and composition, while Mercury, which is virtually airless today, probably lost its envelope of atmosphere when it was formed. Gaps in our knowl-

Box 4.1 Further Consideration

Greenhouse Cooling?

In chapter 1 you learned that the atmosphere is subject to radiative forcing because of the accumulation of greenhouse gases. Ironically, the artificial production of greenhouse gases, particularly CO_2, may cool both the mesosphere and the thermosphere. If CO_2 doubles, according to George L. Siscoe of the U.S. Air Force Geophysics Laboratory and Timothy L. Killeen of the University of Michigan, the mesosphere (80–100 km, or 50–62 mi) will cool nearly 10° C (18° F), and the thermosphere (above 100 km, or 62 mi) will cool almost 50° C (90° F).

If the gases of the heterosphere continue to cool, their density will increase (fig. 4.A). Within as little time as a century, the heterosphere may shrink by compressing itself against the upper reaches of the homosphere. The ramifications of this compressed heterosphere are not fully understood, but scientists are attempting to explain the possible effects through sophisticated computer models. R. G. Roble of the High Altitude Observatory of the National Center for Atmospheric Research in Boulder, Colorado, suggests that these models predict a considerable change in the vertical structure of the atmosphere by the end of the next century. Roble notes that the temperature near 50 kilometers (31 mi) has already decreased by 3° C to 4° C (5.4-7.2° F) during the past 10 years. As the temperature of the upper atmosphere continues to drop, chemical and physical processes throughout the entire atmosphere may undergo changes. These changes may affect climate in ways that are adverse to human life.

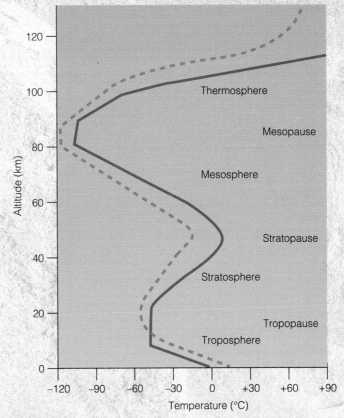

Figure 4.A The present vertical temperature structure of the atmosphere (solid curve) and the predicted temperature structure at the end of the twenty-first century, when the concentrations of various trace gases are expected to double.

Apply Your Knowledge

1. Using figure 4.A, determine where the temperature structure of the atmosphere will change the most.

2. If the atmosphere contracts when it cools, what happens to it when it is heated?

3. What would be the effect of a shrinking atmosphere on Earth-orbiting satellites?

edge about the other planets have led to numerous speculations about their atmospheres, and, until the 1970s, Mercury was thought to have a thin veil of sodium and helium. Interestingly, this innermost planet looks much like our cratered moon, which we now know to leave a trail of sodium atoms as it orbits Earth.

Venus and Mars have atmospheres that now differ radically in structure and composition from Earth's. Both, however, may hold clues to Earth's initial atmospheric composition. The dominant gas of Earth's current atmosphere is nitrogen, which makes up 78% of the air we breathe (table 4.1). The dominance of this gas is surprising when you consider the composition of atmospheres on Venus and Mars. These two planets have atmo-

spheres with compositions dominated by carbon dioxide, and many scientists believe that the early atmosphere of Earth was also rich in CO_2.

The atmosphere of Venus is considerably denser than Earth's, as evidenced by the pressure it exerts on the Venusian surfaces. On Earth you experience a sea-level average pressure of 1.04 kilograms/centimeter^2 (14.7 lb/in^2) as the overlying atmosphere compacts itself under gravity. If you were to stand on Venus, you would literally be crushed by a pressure 90 times greater than that to which you are accustomed, or nearly 3/4 ton per square inch. The heavy atmosphere of Venus is 96% carbon dioxide. Mars also has a carbon dioxide-dominated atmosphere,

Table 4.1

Characteristics and Makeup of the Planetary Atmospheres

Planet or Moon	Color	Thickness	Main Constituents	Storms	Winds
Venus	White with dark lines	Thick	CO_2 (96%), N2 (3.5%), H_2SO_4	?	Probable
Earth	Blue, rainbow colors	Thin	N_2 (78%), O_2(21%), Ar, CO_2	Yes	Yes
Mars	Red	Thin	CO_2	Dust storms	High speed
Jupiter	Yellow, brown, red	Thick	H_2, He	Great Red Spot	High speed
Saturn	Yellow, tan, blue, green	Thick	H_2	Yes	High speed
Titan	Reddish	Thick	N_2, Ar, Ch_4	No	Light
Uranus	Blue-green	Thick	H_2 (87%), He (13%), CH_4	?	?
Neptune	Blue	Thick	H_2, He, CH_4	Great Dark Spot	Yes
Triton	?	Thin	CH_4	?	?
Pluto	?	Thin	CH_4	?	?

but, unlike the air of Venus, it is not very dense, about 100 times less dense than Earth's.

Typically, a planetary atmosphere is densest near the planet's surface. The *proportion of individual gases found in the lower part of an atmosphere is relatively uniform horizontally because of mixing.* Near the solid or liquid surfaces of the planets, all atmospheres lie under the weight of overlying gases. Compressed by the great weight above them, the gaseous molecules of the lower atmosphere interact, exchanging whatever energy they absorb from their environment. The mechanisms of energy exchange will be discussed later, but it is important to realize that the energy exchange among molecules imparts a mixing that is characteristic of the homogeneous, or compositionally uniform, lower atmosphere.

Table 4.1 presents the main components of the planetary atmospheres. Whereas instruments on automated spacecraft have sampled atmospheres on Mars and Venus, scientists have determined the gases of the other planets by analyses of their spectra and by deduction (chapter 2). Astronomers, for example, can deduce that neither Uranus nor Neptune houses abundant free oxygen in its atmosphere. Because methane reacts with oxygen to produce water and carbon dioxide, any free oxygen would be bound by the following reaction:

$$CH_4 + 2\,O_2 \rightarrow 2\,H_2O + CO_2$$

Atmospheric Origins

How did the planets obtain the components of their atmospheres? The composition of each planet's atmosphere is the product of several influences, including its point of origin in the nebular cloud from which the Solar System formed. The distribution of elements in the nebular cloud was an important control on the planetary components. Iron, for example, is a prominent component in the makeup of the terrestrial planets, but it does not exist as a gas in their atmospheres. Hydrogen and helium are important constituents of the Jovian bodies.

Another influence on the evolution of atmospheres was planetary size, because mass determines the strength of the gravitational field that holds an atmosphere. Gravity coupled with surface temperatures may give even a relatively small body the ability to hold an atmosphere of moderately heavy gases. With its cold temperature of $-179°$ C ($-290°$ F), Saturn's moon Titan, the second largest moon in the Solar System, has a gravity strong enough to hold nitrogen, argon, and methane. Chemical reactions among the elements that produce compound gases like CH_4 further sorted the gases that were gravitationally bound to each planet. For some of the planets, geological processes, such as volcanism, added gases from the interior, and on Earth, biological processes served to redistribute, cycle, and alter some elements and compounds.

Just how the proportions of gases on Earth developed is partly veiled in a hidden past, but the proliferation of life over the planet's surface has had a profound effect on the atmosphere's composition. Through **photosynthesis,** plants convert carbon dioxide, water, and sunlight into carbohydrates. Through **respiration,** animals assimilate oxygen. Whereas the former process gives off oxygen, the latter releases carbon dioxide, and both processes have altered radically the nature of Earth's initial atmosphere. The preponderance of carbon dioxide on Venus and Mars might be taken, therefore, as indirect evidence that no photosynthesis takes place on either planet.

Through a variety of experimental and exploratory methods, earth scientists have repeatedly attempted to discern the composition of Earth's atmosphere prior to the influence of life. Air from the relatively recent past, about 12,000 years ago, has been found trapped inside glacial ice. Much older air, captured in a bubble inside a sample of amber (hardened sap) from the age of the dinosaurs, has given some clues to the atmosphere's long-term composition. Surprisingly, the sample held air that was 9% richer in oxygen than the air we breathe.

The flyby of Titan by *Voyager I* in November 1980 resulted in sufficient data collection for speculation that Saturn's largest moon has an atmosphere somewhat like that in which life began on a young Earth. Titan's atmosphere contains acetylene (C_2H_2) and hydrogen cyanide, two gases that may have played a key role in the generation of organic molecules. The red moon also lacks molecular oxygen (O_2), making it a reducing, rather than an oxidizing, environment. Such an *anoxic* (oxygenless) atmosphere seems to have been necessary for life's compounds to form. After the rise of life on our planet, the balance of early component gases began to tip as plants absorbed and converted carbon dioxide into food and released molecular oxygen.

Atmospheric Motions

A significant aspect of Earth's atmosphere is motion. The air of the lower atmosphere constantly moves and mixes. This motion is the product of energy transfer as the sun's energy is converted into heat differentially along Earth's surfaces and is transferred to the atmosphere. Heat energy drives an atmosphere, but only when that heat is unevenly distributed. Thermodynamic studies tell us that heat flows from hotter to colder areas or from hotter to colder substances. A common experience, such as touching the handle of a hot pan, proves this. The energy flow from pot to hand is called **conduction,** which is a transfer of heat energy from one substance to another in contact with it. In gases, the transfer of energy adopts a different method. Heated gases rise. Whenever gases incorporate more energy, they do so through more rapid molecular motion. This motion results in an expansion of a heated gas; the gas expands, occupying a larger volume and, as a consequence, becomes lighter. Less dense than the surrounding gases, the heated gas rises and takes its energy to a higher altitude, losing energy to the environment during its climb. This vertical heat transfer by a moving fluid is called **convection** (fig. 4.2).

Just about all atmospheres undergo uneven heating, which manifests itself in air movements including some 548 kilometer (340 mi) per hour winds on Saturn. One atmosphere that appears to be rather even in its temperature, however, is that of Titan, where a relatively uniform, very cold temperature predominates at the moon's surface. As a result of the evenness of temperature, Titan probably has little atmospheric movement. On Earth, these atmospheric movements are well documented, and they contribute to our weather.

Although uneven atmospheric heating is the key to air motion, not all planets receive their heat energy from a single source.

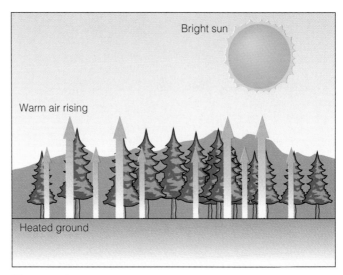

Figure 4.2 Convectional transfer of heat.

Jupiter, for example, gives off more energy than it receives from the Sun, and Saturn has a suspected heat source in a helium rain that may have fallen more than a billion years ago. Solar radiation that strikes the upper atmosphere of Jupiter impinges upon a very thin layer of molecules, where little energy transfer can occur. Remember that closely packed or compressed gas molecules are more efficient at distributing energy than are widely separated molecules.

The motions of atmospheric gases depend on the energy influx derived from five possible sources:

1. solar radiation,
2. planetary radiation (infrared radiation),
3. fluid convection (the rising of a heated gas),
4. friction (by compression of descending gases), and
5. condensation (changing from a gaseous to a liquid state).

Each of these sources of atmospheric energy is discussed in this and the next chapter. The movements generated by each source operating alone or in conjunction with the others can be quite spectacular. As terrible as hurricanes and typhoons are for residents of our planet, each of Earth's big storms, which serves to move atmospheric gases and redistribute heat, is tiny by comparison with stormlike movements in the upper atmospheres of the Jovian planets.

Jupiter and its neighbor Saturn have stormlike activity as swirling clouds called *eddies* move rapidly against the background of other clouds. The scale of these eddies is so large that the most famous Jovian storm, the long-lived Great Red Spot, rivals the size of our planet (fig. 4.3). A sequence of still photos taken once every 2 days on *Voyager I*'s approach to the planet reveals an animated vortex of reddish gases so dense that neighboring bands of clouds are deflected around the oval, counterclockwise rotation of the storm (fig. 4.4). Similar, though smaller, swirling cloud systems have been observed elsewhere on Jupiter, on Saturn, and on Neptune.

Figure 4.3 The Great Red Spot, more than 25,000 kilometers (15,500 mi) long, is the largest of Jupiter's storm systems.

STRUCTURE OF EARTH'S ATMOSPHERE

The pull of gravity concentrates the atmosphere over Earth's surface, making the lower part of the atmosphere more dense than the upper part. In fact, about one-half of the atmospheric mass lies in the lowest 6.5 kilometers (4 mi). Weather scientists recognize this gravity-generated distribution as one basis for a classification scheme that divides Earth's air into two spheres whose mutual boundary lies at about 80 kilometers (50 mi) above the ground. The lower of these two is the dense **homosphere,** so named because its composition is relatively uniform. The uniformity results from the vertical and horizontal mixing of crowded gases, which, even on a molecular level, frequently collide. The outer, less-dense layer is the relatively unmixed **heterosphere,** a zone where widely spaced, and generally lighter, molecules rarely collide and where mixing is infrequent. The lighter gases of the heterosphere, then, extend from 80 kilometers (50 mi) to the top of the atmosphere and do not exhibit the movements characteristic of the lower, heavier gases of the homosphere.

Figure 4.5 shows the extent of both segments of the atmosphere. It presents another classification of atmospheric zones not based on gravity separation and mixing but rather on temperature. Another basis for classifying the spheres are temperature and **air pressure,** which is dependent on the amount of air above any level, since pressure is a function of weight. Death Valley, the lowest point on the North American continent, experiences greater air pressures than the summit of Mount McKinley simply because it has more overlying air. At the approximate altitude of 5.5 kilometers (18,000 ft), the pressure is about half that at Earth's sea level surface.

The temperatures throughout the spheres vary widely and follow different trends. In two of the spheres temperatures generally rise with increasing altitude, whereas in the other two spheres temperatures fall with increasing altitude.

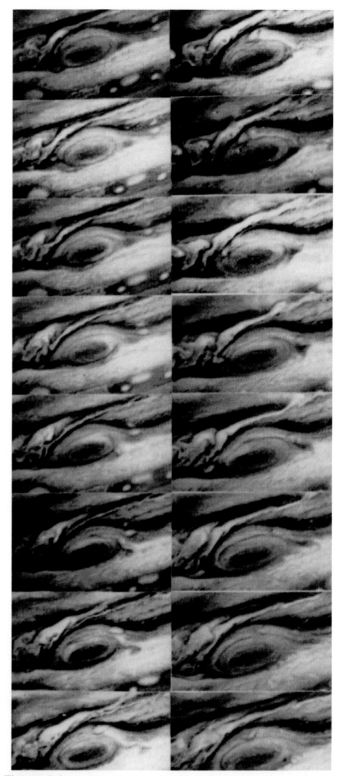

Figure 4.4 Images taken by *Voyager I* at approximately 40-hour intervals show the counterclockwise rotation of the Great Red Spot of Jupiter. Small clouds approach the spot from the east and circle it in six to ten days.

Meteorologists call the drop in temperature with increasing altitude a **lapse,** and they refer to a rise in temperature with increasing altitude as an **inversion.** The fluctuating upper boundary layer of each sphere is characterized by a steady temperature,

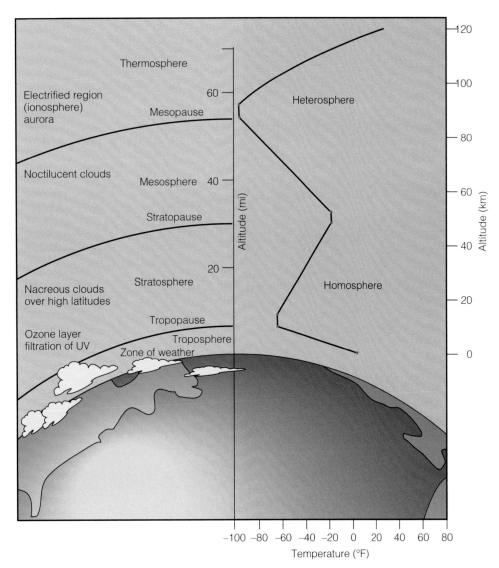

Figure 4.5 The layered structure of the atmosphere.

where no change occurs with increasing altitude, giving rise to the term **pause.** Thus, the **stratosphere,** shown in figure 4.5, ends at the **stratopause;** the **troposphere** ends at the **tropopause.**

The dense, lowest 16 to 24 kilometers (10 to 15 mi) of atmosphere is the zone that most concerns us because it is the **weather sphere.** Of course, the troposphere is the region of the atmosphere that is of greatest importance to humans and the focus of this text. It is the region directly responsible for maintaining life on Earth's surface, and it affects life forms with its daily and seasonal changes. The stratosphere, however, has recently drawn the close attention of scientists because its boundaries encompass the ozone layer, which appears to be under attack by industrially produced gases, such as the CFCs (chapter 1).

ZONE OF WEATHER

As you can see in figure 4.5, the troposphere is part of the homosphere. In this dense, lower layer of the atmosphere lie overcrowded molecules that vibrate and collide at great velocities in random motions. The energy for this movement derives chiefly from the sun. As you learned in the previous chapter, the spread of this radiation is unequal over the planet's surface for two reasons: Earth is tilted to the plane of its orbit, and its surface is curved.

The differential heating that results from these two controls on the concentration of solar energy is reflected in the uneven heating of the molecules in the troposphere. Electromagnetic radiation from the sun strikes Earth's surfaces, where it is reradiated into the surrounding envelope of gases, heating them by transferring solar energy. Wherever air molecules receive more heat energy, they vibrate faster. Faster molecular motion is synonymous with higher temperatures. The concentration of solar radiation on a hot summer day and subsequent reradiation produce

Box 4.2 Investigating the Environment

Ozone-Depleting Compounds and Stratospheric Ice

Earth's stratosphere houses the important ultraviolet-absorbing gas ozone. The relatively recent (1980s) discovery of a "hole"—actually, a lessening—in the ozone layer over Antarctica has increased speculation that some mechanism speeds up the process of depleting the ozone layer.

Ozone-depleting compounds (ODCs) are trace gases that release chlorine and bromine into the stratosphere. ODCs effectively deplete stratospheric ozone because they are capable of surviving the journey through the troposphere until they encounter the stratosphere, where they release their chlorine and bromine atoms (fig. 4.B).

The CFCs that are partially responsible for the depletion of the ozone are not the only ODCs manufactured by humans (table 4.A). Other gases, including the halons (compounds with one or more bromine atoms) and compounds containing hydrogen also act as ODCs.

According to Owen P. Toon of NASA's Ames Research Center and Richard P. Turco of UCLA, the CFCs that reach the stratosphere are broken apart by ultraviolet light, releasing chlorine that exists in a free state or that reacts with ozone to form chlorine monoxide (ClO) (fig. 4.C). The chlorine eventually reacts with other gases, and the new compounds have no effect on the ozone. In a reaction with atmospheric methane, chlorine forms gaseous hydrochloric acid (HCl), and in a reaction with nitrogen dioxide (NO_2), it forms chlorine nitrate ($ClONO_2$). Both of these compounds become reservoirs for chlorine atoms. Some process, however, appears to release the chlorine from the reservoirs, enabling it to destroy more ozone.

The process appears to take place in large (10–100 km, or 6–60 mi, long) lens-shaped, iridescent clouds. These "nacreous" (mother-of-pearl) clouds remain stationary even though high winds can move through them. Typically associated with very cold temperatures (−83° C), nacreous clouds often form as high winds that reach into the stratosphere form waves over mountainous areas. An atypical formation of the clouds, however, appears to occur at higher temperatures (−78° C), particularly over Antarctica. The clouds, composed of ice crystals of varying size, have been called polar stratospheric clouds (PSCs). Within these clouds, the reservoirs of chlorine are prevented from forming as nitric acid (HNO_3) is precipitated.

The chief contribution of the clouds to ozone depletion lies in the removal of nitrogen, an element that reacts with chlorine to make a molecule that does not react with ozone. The process of freeing chlorine to react with ozone begins in the Antarctic spring, as sunlight dissociates molecular chlorine into atomic chlorine. The chlorine atom destroys ozone by taking on one of the oxygen atoms, leaving an oxygen molecule behind. Chlorine monoxide (ClO) then reacts with atomic chlorine to make Cl_2O_2, which sunlight breaks apart, freeing the chlorine to destroy additional ozone. Bromine in the stratosphere also removes an oxygen atom from the ozone molecule to form bromine monoxide (BrO) and molecular oxygen. When the bromine monoxide encounters chlorine monoxide, a reaction frees the bromine and chlorine atoms and produces molecular oxygen. The bromine and chlorine are free to react with additional ozone molecules.

Apply Your Knowledge

1. What ODCs are CFCs?
2. What are the chlorine-reservoir molecules?
3. How do chlorine and bromine destroy ozone?

Table 4.A

Artificially Produced ODCs

Fully-Halogenated CFCs (containing no hydrogen)

CFC-11 (CCl_3F)
CFC-12 (CCl_2F_2)
CFC-113 (CCl_2CClF_2)
CFC-114 ($CClF_2CClF_2$)
CFC-115 ($CClF_2CF_3$)

Halons (containing bromine)

Halon 1211 ($CBrClF_2$)
Halon 1301 ($CBrF_3$)

Other ODCs

HCFC-22 ($CHClF_2$)
Methyl Chloroform (CH_3CCl_3)
Carbon Tetrachloride (CCl_4)

---

molecular collisions at a greater rate than the less-concentrated solar rays produce on a cold winter day.

Pressure

Another influence on the proximity and motions of air molecules is the sheer mass of the atmosphere. The tremendous weight exerted by the quadrillions of kilograms of overlying air make the molecular collisions in the tropospheric gases frequent and violent. Each second, molecular collisions impinge on gases and planetary surfaces, and these motions can be recorded as air pressure. Density, pressure, and temperature are, therefore, interrelated in Earth's troposphere. Close to the surface, pressures are relatively high, molecules are crowded, and reradiation transposes solar energy. Farther up in the troposphere, pressures are lower, molecules are less crowded, and the source of reradiation is farther away.

The interrelationships among the pressure, heat source, and temperature are even more evident when you consider atmospheric movements. Heated molecules move rapidly, collide frequently, occupy a greater volume, and become lighter. As the heated air rises, pressures are not as great as they were below. Under lower pressures, air collisions diminish, and temperatures fall. Falling temperatures mean more compact, more slowly

Figure 4.B Stratospheric clouds can form over the polar regions if the air cools sufficiently during the winter. Nitric acid trihydrate clouds are visible as thin, dark orange layers. The water-ice clouds appear whitish. The clouds, such as these over Stavanger, Norway, help to initiate the chemical reactions that destroy ozone.

vibrating molecules with increasing weight per unit volume. The product is a heavy, cold air mass that descends toward the surface, displacing the air below. In this way, the gases of the troposphere constantly intermix through a vertical exchange. Winds also mix the troposphere, and, like vertical movements, they result from pressure differences. Wind and pressure are discussed later in this chapter.

The bottom of the troposphere comes in contact with the planet's surface, almost three-quarters of which is water. It is the interrelationship between the air and water that effectively moves most of the heat energy that the atmosphere indirectly derives from the sun.

Water and Heat Transfer

It's easy to see the importance of water vapor in the atmosphere. Hot, humid days make us more uncomfortable than hot, dry days. On a larger scale, the introduction of water vapor into the air can turn a pleasant day into a stormy one. In the hypothetical storm system described in chapter 3, for example, water vapor from the region near the Gulf of Mexico reenergized the storm.

Water vapor absorbs, holds, and transfers heat energy in measureable units known as calories. A calorie is 1/100 of the amount of heat energy needed under Earth's standard sea-level atmospheric pressure to raise the temperature of 1 gram of water

Box 4.2 continued

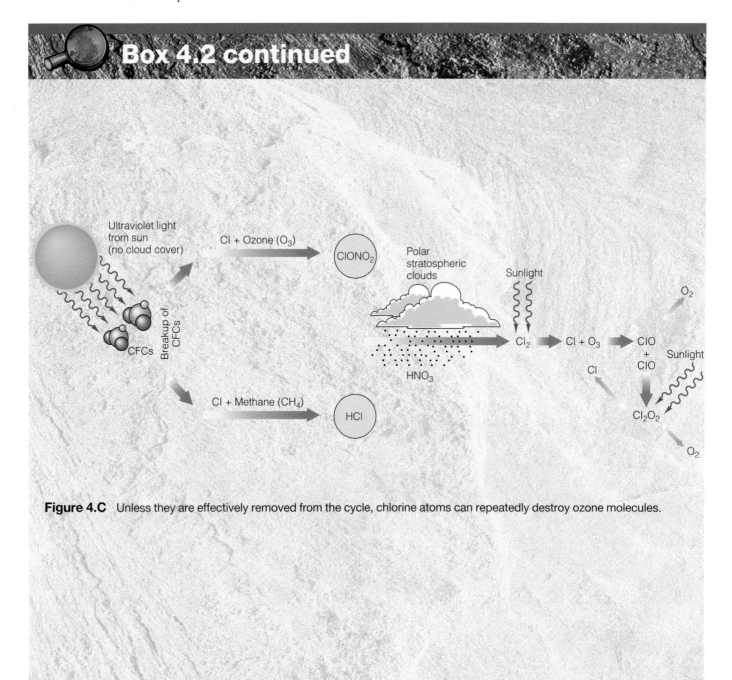

Figure 4.C Unless they are effectively removed from the cycle, chlorine atoms can repeatedly destroy ozone molecules.

from 0° C to 100° C. The mechanisms through which these quantifiable heat absorption and transfer processes occur are

1. **evaporation,** the heat-absorption process that turns liquid water into water vapor;
2. **condensation,** the heat-release process that turns water vapor into liquid water; and
3. **sublimation,** the process by which ice turns into water vapor or vice versa.

In evaporation, liquid water must absorb 640 calories per gram to turn into gaseous water. That heat energy is stored as the **latent heat of vaporization** by the more rapidly moving gas molecules. Those molecules remain a gas as long as they

hold that energy, but a loss of that energy returns the water vapor to the more slowly vibrating, liquid state. To acquire the energy necessary for vaporization, water molecules rob energy from neighboring molecules. The surrounding environment cools as a result. This process accounts for your being chilled after stepping out of a shower, bath, or swimming pool. Evaporation along the surfaces of your body robs the skin of heat energy. Thus, evaporation is a cooling process.

By contrast, condensation is a heating process. As the gaseous water turns once again into liquid water, it releases its latent heat into the surrounding environment. When evaporated water migrates in air currents and then condenses, it transfers heat from one place to another.

A third process also serves to transfer heat. It is possible for water vapor to turn directly into ice, or for ice to turn into water vapor. That process, which usually occurs at high altitudes, at high elevations, and on cold winter days, is sublimation. The most common sublimation process is the formation of ice crystals at very high altitudes.

With almost three-fourths of the planet's surface covered by oceans, water vapor is readily available to the troposphere and plays an important role in the **heat budget,** or heat transfer, over the planet. Its presence in the air is, however, variable, amounting to as much as 4% of the volume of a given air mass. Obviously, air over water has a better opportunity to acquire moisture than air over land, and, therefore, masses of air can be thought of as being either moist or dry.

AIR MASSES

If you think of the troposphere as a three-dimensional model, you can appreciate not only its horizontal and vertical components but also its depth. The troposphere contains an everchanging series of **air masses,** each with its own short-lived characteristics. Air masses of different sizes rise to different heights, move at different rates, maintain their own patterns of temperature and pressure, and contain different amounts of moisture, depending on their origin over either land or water. Except for their clouds, they are invisible; nevertheless, meteorologists can describe and classify them.

Our hypothetical storm system in chapter 3, for example, evolved in the northern Pacific, where it emerged from a cool but moist region. It carried the moisture it picked up from the surface waters of the Pacific into southern British Columbia and the northwestern United States, where rains fell over Washington and Oregon. The system's point of origin is a **source region** for the development of cool, moist air masses called **maritime polar (mP) air** (fig. 4.6).

Other source regions provide the troposphere with air masses with different characteristics. Air masses that develop over land are generally drier than those that originate over water. Therefore, during summers a region such as the American Southwest is a source for **continental tropical (cT) air.** Temperatures are relatively high over states such as Arizona and New Mexico. Both are landlocked and unable to provide an extensive source of water, so the air becomes dry (continental) and warm (tropical).

Once the air masses become homogeneous units, they leave their source regions at varying rates. They alter their characteristics as they encounter other air masses, and they eventually mix with the **ambient air** (surrounding air) and with the new air mass. Thus, the hypothetical storm system, which developed from maritime polar air, lost most of its moisture by the time it crossed a great expanse of land and moved into New Mexico.

An air mass is either cooler or warmer or drier or wetter than the ambient air and any neighboring air masses. This is an important aspect of the weather. *Relative temperatures and pressures are the characteristics that determine how air masses interact.* Since cold air is denser than warm air, for example, a colder air mass will wedge itself under a lighter warm air mass. Although weather forecasters may announce the absolute or quantitative values for temperatures and pressures in a given air mass, they make their predictions largely on the relative differences among air masses.

FRONTS

The leading edge of a moving air mass is called a **front.** Fronts are the troposphere's lines of battle in the shifting wars among air masses. The leading edge of a relatively cool, dry air mass will force its way beneath warmer air and drive the warmer air upward. Additionally, cold air masses generally move faster than the less-dense warm air masses, and often the front of the former overtakes the front of the latter and lifts it.

Because the size and density of air masses differ, these units of the troposphere are in constant conflict throughout their short lives. The conflicts among air masses produce one of three results:

1. a denser air mass plows beneath a less-dense mass;
2. a lighter air mass overrides a denser one; and
3. two air masses temporarily ward off each other's advance until a mixture of the two occurs.

These results, in turn, produce four kinds of fronts, which meteorolgists classify on a relative basis (fig. 4.7):

1. a cold front, the leading edge of a cold air mass;
2. a warm front, the leading edge of a warm air mass;
3. a stationary front, the boundary between two relatively motionless air masses; and
4. an occluded front, a front that has been lifted, leaving a gap, or occlusion, between it and the ground.

QUANTIFIABLE CHARACTERISTICS OF AIR MASSES

Because scientific endeavors require objective analyses, meteorologists look for ways to measure characteristics of the atmosphere. The daily movements of air masses in the troposphere keep weather observers busy identifying the changing characteristics of the air. To define any one air mass as maritime polar or continental tropical, scientists constantly measure a specific set of **weather elements,** or quantifiable characteristics (table 4.2). These elements are the measurable properties of the atmosphere. A number of reliable instruments serve as the tools of weather observers. In chapter 5, you will read about these elements as they relate to weather mapping and forecasting.

Surface Temperature and Its Measurement

What are the hottest and coldest surface temperatures ever recorded on our planet? On September 13, 1922, Azizayah (or Azizia), Libya, experienced the highest atmospheric temperature officially recorded at Earth's surface. On that day the temperature reached 57.7° C (136.4° F). Just 9 years earlier

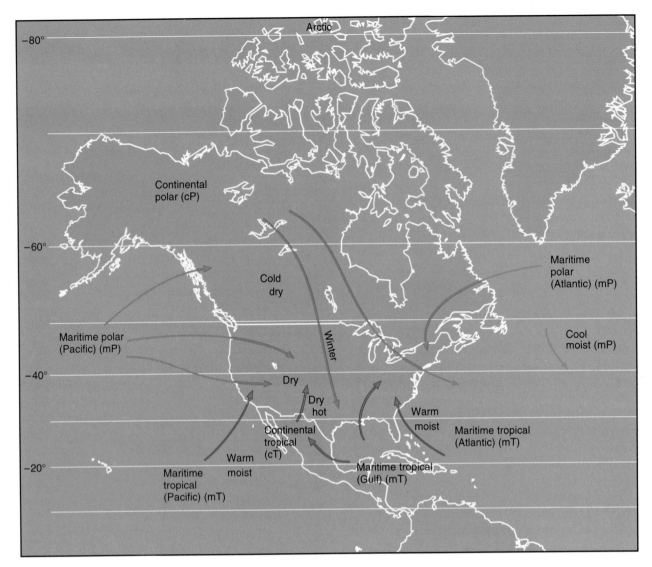

Figure 4.6   Source regions for North American air masses.

Death Valley, California, experienced the hottest day ever recorded in North America, 56.6° C (134° F). At the other end of the troposphere's heat spectrum are the cold temperatures of Antarctica, where in August 1960, scientists felt the sting of −89.44° C (−129° F). The February 3, 1947, temperature of −62.77° C (−81° F) at Snag, Yukon, in Alaska, makes that date the coldest day on record in North America.

For meteorologists, record keeping is the key to understanding the weather. Daily observations, therefore, are part of the regimen of meteorology, giving us temperature ranges. Ordinarily, the range of temperatures over 24 hours is relatively moderate. On some days, the high in a particular area may not vary more than a few degrees from the daily low. In other areas, humans must adjust to a wide range of temperatures during a single day. The greatest change in temperature ever recorded for one day occurred in Oleminsk, Siberia, where people arose to a very brisk −60° C (−76° F), but by the late afternoon they sweltered under 45° C (113° F).

To measure temperature, scientists use a very common instrument, the **thermometer.** Thermometers are filled with either mercury or alcohol and measure temperatures by expansion and contraction of their liquid in a sealed tube. Heated, the liquid expands and rises in the tube, with calibrated markings representing degrees. Cooled, the liquid contracts, falling into a reservoir well or bulb at one end. A *maximum thermometer,* used by weather observers to record the highest temperature on a given day, has a constriction in the glass tube to prevent the mercury from returning to the reservoir when the temperature falls and the liquid contracts. A *minimum thermometer,* used to record the daytime low, has a thin plastic pin inside the alcohol-filled tube that prevents the liquid from rising if the temperature subsequently rises. The minimum thermometer uses alcohol because its freezing point of −130° C (−202° F) is lower than mercury's freezing point of −39° C (−38.2° F).

A **thermograph** uses an ink-filled needle connected by a rod to two dissimilar metals curved into a spring. As the metals contract or expand, the rod rises or falls, allowing the ink to draw a line on graph paper attached to a rotating drum. Thermographs keep continuous records of temperature fluctuations electrically

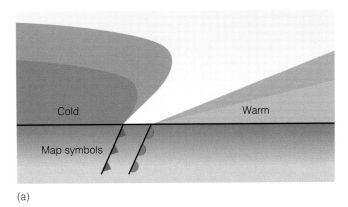

(a)

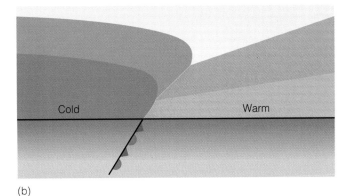

(b)

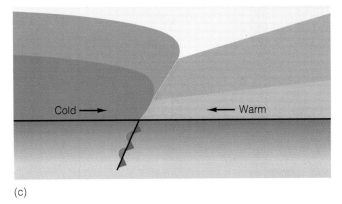

(c)

Figure 4.7 The four types of fronts and their map symbols: *(a)* cold and warm fronts, *(b)* occluded front, and *(c)* stationary front.

or mechanically, allowing observers to check the inked graph paper at their convenience.

Surface Pressure and Its Measurement

Atmospheric air pressure is exerted by the random movements of the compressed fluids we call gases. The great weight of air that lies on Earth's surfaces provides more, however, than a simple crushing action. The force that compressed gases exert is as much a sideways action as a downward one. Rapidly moving gas molecules strike any surrounding molecules they encounter, including other air molecules. For this reason, a balloon expands with the increasing pressure of added, and therefore, compressed air. If the pressure inside the balloon were not equal in all directions, the balloon would collapse at the site of diminished

Table 4.2		
Quantifiable Weather Elements and the Instruments to Measure Them		
Weather Element	**Selected Instruments**	**Unit of Measurement**
Temperature	Thermometer Thermograph	Degrees Celsius or degrees Fahrenheit
Pressure	Aneroid barometer Mercurial barometer Barograph	Inches of mercury or millibars
Humidity	Hygrometer Sling psychrometer Psychrometer Hydrograph	Percent or grains or grams
Wind	Weather vane Anemometer	Compass direction, knots, mph, meters per second
Precipitation	Rain gauge Tipping bucket	Millimeters or centimeters or inches

pressure. An analogy that may make this clear is a punctured bucket of water (fig. 4.8). The water squirts out the sides of the bucket because the pressure from the weight of the water forces it against the insides of the bucket.

As you learned earlier in the chapter, the pressure at Earth's surface is approximately 1.04 kilograms/centimeters^2 (14.7 lb/in^2). At this pressure, every 0.09 meter^2 (1 ft^2) of surface holds 969.2 kilograms (2,116.8 lb), regardless of its orientation: horizontal, inclined, or vertical. That force of about a ton would seemingly crush each of us, except that it is an evenly distributed pressure over and around all surfaces. The air inside your mouth, therefore, maintains a pressure that balances the air outside your cheeks.

Measuring air pressure in kilograms per square centimeter (or pounds per square inch) gives us a good sense of the amount of force the atmosphere exerts on surfaces. Meteorologists, however, prefer to use another method to quantify pressure. Because the pressure of our atmosphere can push a confined fluid up a pipe, meteorologists use a glass tube that looks very much like a large thermometer. It is called a **mercurial barometer** and is calibrated in both inches and millibars. The standard sea-level atmospheric pressure on our planet equals 29.92 inches of mercury in the tube, and this, in turn, equals 1,013.2 millibars (thousandths of a bar). The mercurial barometer shown in figure 4.9 is easy to read. Whenever air pressure decreases, the mercury falls in the tube. With an increase in air pressure, the mercury rises.

Although 1,013.2 millibars (29.92 in) is the standard pressure at Earth's sea-level surfaces, it represents an average only. Higher and lower pressures have been recorded. The lowest pressure on record was measured at sea just northwest of Guam in 1958, when the mercury fell to 877 millibars (25.9 in). By contrast, observers in Barnaul, Russia, registered a pressure of 1,079

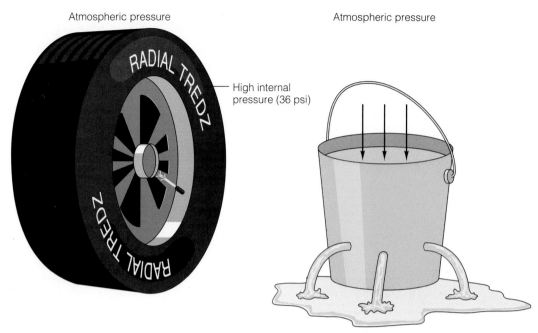

Figure 4.8 Fluids (air and water) flow from areas of high pressure to areas of low pressure.

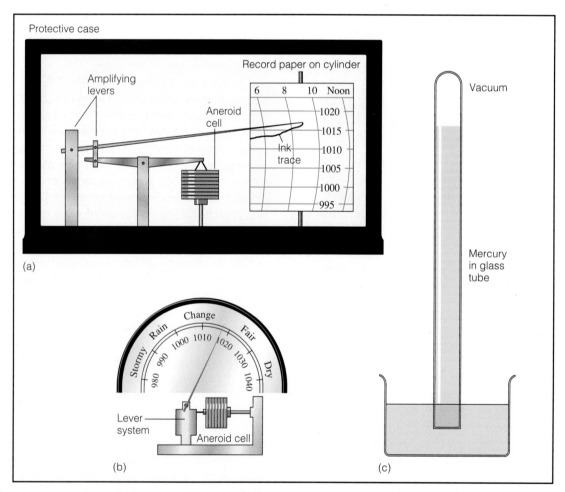

Figure 4.9 Important meteorological instruments: *(a)* barograph, *(b)* aneroid barometer, and *(c)* mercurial barometer.

millibars (31.89 in) on January 23, 1900; it was the highest surface pressure ever recorded.

Because the mercurial barometer is long, heavy, and cumbersome, many commonly used barometers measure pressure by another method. The **aneroid barometer** employs a corrugated, airless metal can that contracts and expands with changes in pressure. Whenever the can alters its shape, a needle connected to the center moves along a gauge calibrated in both inches and millibars. Often the face of the gauge is marked "fair" or "stormy," as is the aneroid barometer in figure 4.9. The barograph uses an aneroid barometer to keep a running record of pressure changes.

Humidity and Its Measurement

Meteorologists use a variety of instruments to measure the amount of water vapor (gaseous water) in Earth's atmosphere. They also categorize their measurements to emphasize different relationships between the atmosphere and the water vapor it holds. Most weather forecasters quantify the moisture in an air mass by referring to the **relative humidity.** The term gives us a percent ratio between the actual amount of water vapor present and the vapor carrying capacity of the air mass.

For most of us, the relative humidity is only of significance on a hot, very humid day. We feel discomfort on such a day because little evaporation of water vapor occurs on our skin. In winter, a high relative humidity may have the opposite effect. On cold days, we need to conserve our bodily heat, and a lack of evaporation on our skin keeps heat loss down. The body recognizes by its comfort or discomfort the importance of temperature to relative humidity. On a more objective scale, the temperature of an air mass controls its relative humidity: warmer air has a greater carrying capacity than cooler air. Thus, more water vapor must be present in a warm air mass for it to have the same relative humidity as in a similar quantity of cool air.

Because the United States National Weather Service recognized our need to know the apparent weather, or how we will feel, it implemented in 1984 a heat index scale that relates humidity to human perception.

Meteorologists also measure the **absolute humidity** of an air mass by determining the density of water vapor. Just as census takers can tabulate the number of people living in a given square mile, so meteorologists can determine the amount of water vapor in a parcel of air. Typically, absolute humidity is given as grams of water per cubic meter of air. Both relative and absolute humidity ratios change with alterations in the air parcel. If the air parcel's temperature rises, then the air expands, occupies a larger volume, and reduces the absolute humidity. Similarly, with a rise in temperature the carrying capacity of the air parcel increases, and the relative humidity declines. Because both measurements can change with the addition of heat and the subsequent expansion, another measurement gives a direct ratio dependent only on the amount of water put into or taken out of an air parcel. That quantity is the **specific humidity,** which reflects the ratio between the weight of the water vapor and the weight of the air parcel. It is a weight-to-weight, or a mass-to-mass, ratio expressed in either grams of water vapor per kilograms of air (g/kg) or in grains (0.065 g).

The instruments used to measure humidity depend on the absorption or loss of water and on evaporation. **Hygrometers** use water absorption or accumulation to register humidity. Each of us is equipped with a personal hygrometer in the form of hair. As our hair absorbs water, it expands and lengthens. Attached to a graduated scale, the hair, once calibrated to a standard, yields a reading of humidity. More accurate than the **hair hygrometer** and requiring less attention to calibration is the **electrical absorption hygrometer,** which uses a thin electrolytic film containing a water-absorbing salt. As the salt absorbs water, electrical resistance across the film changes, and a reading of humidity is made. A **psychrometer,** an instrument composed of two thermometers, one with a cloth sock around the mercury bulb, uses evaporation to measure humidity. Once the cloth is soaked with water, evaporation cools the bulb, and the mercury falls in the tube to reflect the loss of heat. The wet bulb temperature can then be compared to the temperature of the dry bulb thermometer. Greater differences between the wet bulb and dry bulb temperatures reflect greater evaporation and drier air. To speed up the evaporation process, some meteorologists employ a **sling psychrometer.** Spinning the two thermometers forces more air past the wet bulb, shortening the time required for evaporation (fig. 4.10).

Precipitation and Its Measurement

Measurements of snow and rain are usually given in inches, millimeters, and centimeters. A *rain gauge* is simply a calibrated tube opened on one end. Placed vertically, it catches falling rain. Some rain gauges work by mechanical or electrical recording devices connected to a *tipping bucket.* The bucket contains two funnel-shaped buckets that pivot and spill when they fill. Each overturn of the buckets sets off an electrical signal that records both the amount and the time of the precipitation. Still other gauges use floats connected to recording instruments. Although the amount of rainfall is significant, its intensity is also important. A 2.5-centimeter (1-in) rainfall covers an acre with more than 75,700 liters (20,000 gal) of water. If those gallons fall over 24 hours, their erosional impact on the soil is less than that of a similar quantity falling in 1 hour.

Measurements of snowfall require a measuring stick, but weather observers usually translate inches of snow into their liquid water equivalent. Because of the space between snow crystals, an inch of snow does not equate to an inch of water. Melting the snow caught in a funnel yields the liquid water equivalent.

Precipitation amounts vary greatly around the world. There are some sites in the Atacama Desert in Chile where precipitation has never been recorded. In contrast, Henderson Lake, Canada, holds the North American total annual rainfall record of 665 centimeters (262 in). The highest total annual rainfall on Earth occurred at Cherrapungi, India, where observers measured 2,646 centimeters (1,041.8 in). Reunion Island in the Indian Ocean experienced the heaviest daily total rain on March 15, 1962, when 187 centimeters (73.6 in) fell. The wettest place in

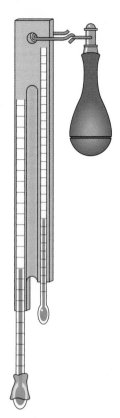

Figure 4.10   A sling psychrometer.

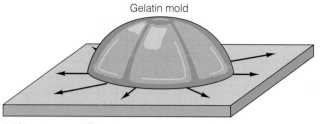

(a) No pressure difference

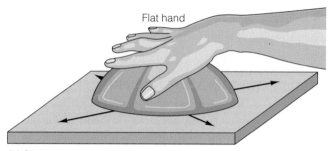

(b) Slight pressure

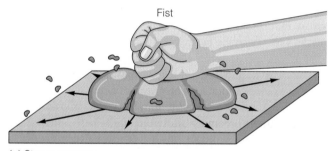

(c) Strong pressure

Figure 4.11   Differences in pressure result in varying movements of gelatin. *(a)* Gelatin bulges outward under its own weight. *(b)* Slight pressure causes oozing flow. *(c)* Strong pressure causes rapid movement.

the United States is Mount Waialeale, Hawaii, where an average 1,199 centimeters (472 in) fall each year.

Snowfalls vary as much as rains, and those who live in equatorial regions see snow only at higher elevations. The greatest yearly snowfall occurred at Paradise Station, Washington, during the 1955–56 winter, when 2,540.8 centimeters (1,000.3 in) fell. A 1-day snowfall of 193 centimeters (76 in) on April 14, 1921, gives Silver Lake, Colorado, the world record.

Wind and Its Measurement

Wind results from pressure differences because air—and any other fluid—flows from areas of higher pressure to areas of lower pressure. As the difference between pressures increases, wind speeds increase. You might visualize the movement of air by a thought experiment. If you were to press on the top of a gelatin dessert, it would ooze away from the pressure of your hand. A slow application of light pressure results in a slow oozing. If you punched the top of the gelatin mold, the dessert would fly outward at high speed. The difference between the slow outward ooozing and the rapid flying of the dessert lies in the relative difference between the force exert-

ed by your hand and the surrounding force of the air pressure (fig. 4.11).

Meteorologists measure both wind speed and wind direction. Wind speed can be measured by an **anemometer,** an instrument that uses three cups attached to a rotating axle. The cups capture the wind, spin, and turn the axle. The rotations of the axle can then be transferred to a counter or a dial, giving the wind speed in knots, miles per hour, or meters per second. As you learned earlier, wind speeds in excess of 484 kilometers (300 mi) per hour have been observed on Saturn. Earth's highest recorded winds reached 363 kilometers (225 mi) per hour on Mount Washington in April, 1939. Tornado winds probably meet or exceed this figure, but the danger presented by these violent storms makes accurate measurements difficult and estimates must suffice.

Further Thoughts: A Complex and Dynamic Atmosphere

All atmospheres undergo some dynamic changes as energy is transferred from a planet's surface or as it arrives from the sun. If we judge by wind speed alone, we find that

some planets have atmospheres far more dynamic than Earth's. The Giant Red Spot on Jupiter and the high-speed winds of Saturn and Mars, for example, would be devastating to life on our planet. Nevertheless, incidents of violent weather on Earth demonstrate that its atmosphere is in a state of constant flux.

Armed with knowledge of the atmospheric components and the physical laws that govern them, meteorologists have established practical atmospheric models that serve as the bases for everyday and long-term predictions. Keeping track of the atmosphere's dynamic complexity requires constant vigilance because humans cannot escape the effects of turbulence, heat, and precipitation. The next chapter establishes the meteorologic use of this knowledge and the model that represents Earth's weather.

Significant Points

1. Most of the planets have atmospheres, but each planet's envelope of gases evolved under different circumstances. On Earth, photosynthesis changed the composition of the early atmosphere by removing carbon dioxide and adding free oxygen.

2. Dominated by nitrogen, Earth's atmosphere contains, in addition to oxygen, significant amounts of carbon dioxide, water vapor, argon, and other gases. This collection of gases is both stratified and dynamic.

3. The dynamism of the lower atmosphere, called the troposphere, produces weather, the short-term atmospheric conditions at any locale.

4. Conditions recorded over long periods are the data for climatology, which is the study of aggregate weather.

5. Above the troposphere lies the ozone-rich stratosphere, which filters incoming ultraviolet light. Two more atmospheric zones, identifiable by temperature and pressure characteristics, lie above the stratosphere: the mesosphere and the thermosphere.

6. The rarefied air of the upper reaches of the atmosphere also serves as a shield against shortwave radiation.

7. Each of the "spheres" ends in a "pause," a transition zone marked by a steady state temperature.

8. The atmospheric component most influential in the formation of weather is water vapor. Its introduction into a mass of air can turn a nice day into a stormy one, and it is largely responsible for the transfer of heat.

9. Fluctuating heat, measured in degrees Fahrenheit or Celsius as temperature, is one of the dynamic weather elements, which also include pressure, wind, humidity, and clouds.

10. The troposphere is the site of moving air masses that acquire identifiable characteristics. Air masses of varying sizes rise to different heights, move at different rates, maintain their own temperature and pressure patterns, and contain different amounts of moisture. Each of these masses originates in a source region from which it obtains its initial characteristic temperatures and moisture. Some are maritime, or moist, masses because they develop over water; others are continental because they evolve over land. Air masses can also be polar or tropical, depending on their latitude of formation.

11. As these masses move, their leading edges, called fronts, collide, causing various weather conditions.

12. Air masses undergo changes wrought by various influences, including the movement of high-altitude winds called the jet stream, and they produce a continual series of changing weather conditions.

Essential Terms

photosynthesis 76	lapse 78	heat budget 83	ambient air 83
respiration 76	inversion 78	air mass 83	front 83
conduction 77	pause 79	source region 83	relative humidity 87
convection 77	evaporation 82	maritime polar (mP)	absolute humidity 87
homosphere 78	condensation 82	air 83	specific humidity 87
heterosphere 78	sublimation 82	continental tropical (cT)	
air pressure 78	latent heat of vaporization 82	air 83	

Review Questions

1. What is the dominant gas of Earth's current atmosphere? Were the constituent gases of the atmosphere ever different from those we breathe today?

2. What is the standard sea-level pressure?

3. Why is the lower atmosphere called the homosphere?

4. How has photosynthesis affected the composition of the atmosphere?

5. What characterizes the upper boundary of each atmospheric zone?

6. What characteristics help define the boundaries of air masses?
7. What is a source region?
8. What is the significance of relative air temperatures?
9. How does water control Earth's heat budget?
10. What are the four kinds of fronts?
11. How do fronts interact?
12. What are the quantifiable weather elements?

Challenges

1. What might happen to the distribution of agricultural regions like America's wheat or corn belts if Earth's atmospheric temperatures rise?
2. What photochemical reactions take place in our atmosphere?
3. What explorations are currently taking place or planned for our Solar System? What difficulties would space vehicles or space travelers have in acquiring information about a hard surface on the Jovian planets?
4. Would a piece of paper ignite in the high temperatures of the thermosphere?
5. Why would the tropopause lie at higher altitudes over the equator than over the North Pole?

5
UNDERSTANDING THE TROPOSPHERE

Tornado at Binger, Oklahoma on May 22, 1981.

Chapter Outline

INTRODUCTION

You now know that solar energy is the ultimate driving force behind air movements of the troposphere. The ways the sun's energy is converted to atmospheric motion all depend on the *transparency of the atmosphere to electromagnetic radiation.* After crossing the near-vacuum of space, the sun's predominantly shortwave radiation encounters a number of atmospheric obstacles on its way to our planet's surface. For some wavelengths our atmosphere is almost opaque; thus, life on the surface is protected from the harmful, very short wavelengths of X rays and gamma rays. The envelope of gases blanketing Earth also selectively filters out other frequencies, including most short ultraviolet rays. In varying amounts, other frequencies, such as visible light and the longer wavelengths of the radio waves, pass through the gas molecules, particles of dust, water vapor, and water droplets of our atmosphere to reach the surface, energizing both life and weather systems.

SOURCES OF ATMOSPHERIC HEAT ENERGY

Our planet is large enough to intercept only one two-billionths of the total energy output of the sun, and of that fraction, less than half reaches the surface to be redistributed as heat by infrared radiation and by Earth's two moving surface fluids: air and water. As figure 5.1 shows, 34% of the incoming solar radiation is reflected to space from atmospheric dust, gases, and clouds and directly from Earth's surfaces. The dust, gases, and clouds of the atmosphere also absorb 19% of the incoming solar energy. Another 28% of the radiation is diffused, scattered, and reflected toward Earth's surface, where, with the 19% that directly penetrates the atmosphere, it heats the land and waters. The 47% (19% + 28%) of the radiation that reaches the surface through reflection and absorption undergoes a transformation from shortwave to longwave, or heat, radiation and eventually warms the atmosphere. Remember that Earth radiates in longer wavelengths because it is a cooler body than the sun. This earth radiation is doubly significant. First, Earth, rather than the sun, directly heats our air; and second, weather is, ultimately, the mechanism by which the atmosphere moves this heat derived from the surface. Atmospheric warming is primarily a result of familiar processes: radiation, conduction, convection, and wind.

Radiation and Atmospheric Opacity

Our atmosphere is almost opaque to thermal infrared energy coming from the surface, but about 20% of the terrestrial radiation escapes to space. Both water vapor and carbon dioxide serve to absorb infrared radiation, and other gases, such as CFCs and methane, also intercept infrared energy from the surface. Among the gases that retard the loss of heat from Earth's surface, water vapor plays the dominant role, but its influence is mostly limited to the lower atmosphere, where contact with the oceans increases specific humidity.

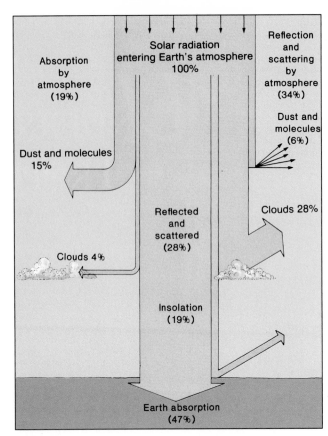

Figure 5.1   Distribution of incoming solar energy.

The Role of Conduction in Moving Heat

Although gas is a poor conductor of heat, the atmosphere does come into contact with Earth's surfaces, where heat is transferred between molecules in contact. Essentially a daytime and warm season process, conduction between surface and air does not play a major role in the redistribution of heat energy throughout the atmosphere. As a mechanism for transferring heat, convection plays a role greater than conduction in the heating of the troposphere's upper layers and in moving heat over large areas of Earth's surface.

The Roles of Convection and Advection in Moving Heat

The vertical redistribution of heat energy through convection currents is a process more important than conduction because it is related to other air movements. When heated air rises, carrying its heat energy aloft, it displaces cooler air, which sinks. Convection also allows replacement air to flow horizontally along Earth's surface in a movement called **advection,** which we commonly call **wind.** A complex set of interrelated movements between convection and advection currents is responsible for the day-to-day changes in weather and for the redistribution of heat around the planet. Vertical and horizontal airflow combine to move the atmosphere locally and regionally in wheel-like cells. These cells, sometimes referred to as **con-**

vection cells, are unevenly distributed across Earth's surface, and the imbalances between rising and sinking air masses generate winds on both large and small scales. The largest cells power the general circulation of the atmosphere, creating advection currents that sweep across the planet's surface, redistributing heat.

THE GENERAL CIRCULATION
OF THE ATMOSPHERE

The uneven heating of Earth's surface creates pressure imbalances that result in winds. On the largest scale, these winds circulate the gases of the troposphere.

Pressure Gradient

Figure 5.2 reveals how a pressure change can produce advection of different intensities. Mercurial barometers A and B show the pressure readings for two sites on the surface. The mercury at site A stands at 1,024 millibars (30.2 in), while that at site B fills the barometer to 1,012 millibars (29.8 in). Because the distance between the two sites is relatively great, the slope between the tops of the two mercury columns is slight. That slope represents a **pressure gradient.** The difference between the barometric readings at sites C and D is greater than those at A and B; and, consequently, the slope, or gradient, is also greater. Because air flows from areas of higher pressure to areas of lower pressure, advection would occur in both instances. However, because C and D register a larger difference and a steeper pressure gradient, the wind between these two sites would be greater.

In an extension of the image of a pressure gradient, meteorologists think of high pressure as *ridges* and of low pressure as *troughs.* Thus, advection is an air movement comparable to water flowing off a hill and into a valley.

Over Earth's surface, very large ridges of high pressure feed air into neighboring pressure valleys. Chief among these ridges are the high-pressure systems of the subtropics, which form as air descends from altitudes near the tropopause, as figure 5.3 illustrates. These ridges lie between two low-pressure areas, one in the equatorial region, called the **equatorial low,** and another in the subpolar zone. The **subtropical highs** feed replacement air along the surface, generating very large, persistent wind systems. Smaller pressure gradients lie between the **subpolar lows** and the **polar highs,** but enough of a ridge exists to establish another wind system at high latitudes in both hemispheres.

The Coriolis Force

A rotating Earth influences the direction of moving fluids across its surface. The influence stems from the various speeds different parts of the planet travel in a daily journey around the axis of rotation. By analogy, the rate of spinning is like the car tire rotation shown in figure 5.4. The outside of the tire has a greater distance to travel than does the edge of the hubcap; yet, both make

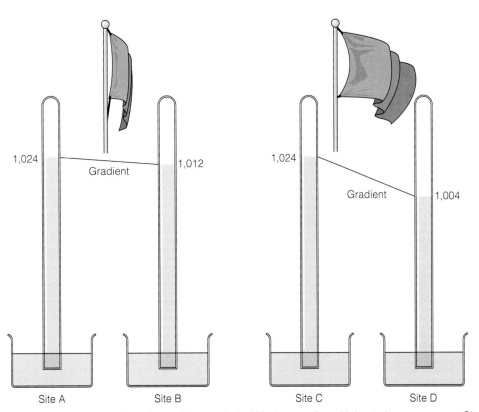

Figure 5.2 Steeper pressure gradients produce faster-flowing winds. Winds move from higher to lower pressure. Site A: Chicago. Site B: Buffalo. Site C: Mobile. Site D: New Orleans.

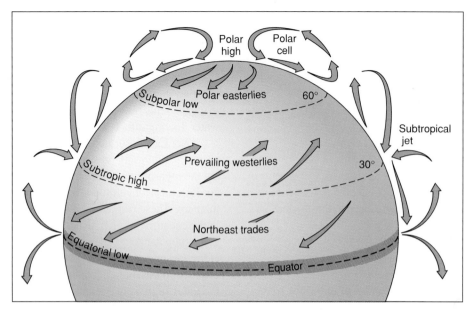

Figure 5.3 Pressure and atmospheric circulation in the Northern Hemisphere.

Figure 5.4 The outside of a car tire must travel a greater distance than the inside near the hubcap to make one rotation.

their respective turns in the same time. Similarly, during every 24 hours a point on the equator must travel a greater distance than a point at 40°N. The rotational rate for the former is more than 1,609 kilometers (1,000 mi) per hour, while that of the latter is closer to 1,126 kilometers (700 mi) per hour. Any object "at rest" on the equator is thus traveling at a higher rate toward the east than a similar object "at rest" at higher latitude. A latitudinal motion between locations with different rotational speeds in-

duces a deflection, or bending, called the **Coriolis force,** either to the right or to the left. In the Northern Hemisphere, there is a right-hand deflection, whereas the opposite occurs in the Southern Hemisphere. The strength of the Coriolis force is great enough that a moving car, if not for the friction with the road and the driver's compensatory adjustments of the steering wheel, would also veer from a straight path.

Cyclonic and Anticyclonic Circulation

The large convection cells that produce surface high and low pressures and their resulting winds are subject to the Coriolis force. In the Northern Hemisphere, the ascending air of the equatorial and subpolar lows rotates in a **cylonic** (counterclockwise) fashion, and the descending air of the subtropical highs rotates in an **anticyclonic** (clockwise) fashion as seen from space. You can use your right hand to demonstrate the direction of this advection in the Northern Hemisphere. Figure 5.5 shows a hand descending toward the ground and turning clockwise. The fingers, representing the flow of air, point outward and encounter the ground as an obstacle to further descent. Similarly, a mass of cold, descending air spreads outward whenever it encounters Earth's surface. By contrast, the palm-up right hand of the figure represents the rising, deflected air of a low-pressure system in the Northern Hemisphere. As the hand (air) ascends, it turns to the right (counterclockwise), and the fingers (winds) flow inward, to fill the low-pressure area with replacement air.

The result of deflection is a system of winds trending out of either the east or the west. Along the equator, the *northeast* and *southeast trade winds* (or *trades*) blow from the east. At higher latitudes, the other sides of the subtropical highs produce a westerly wind (one that blows from the west) called the *prevailing*

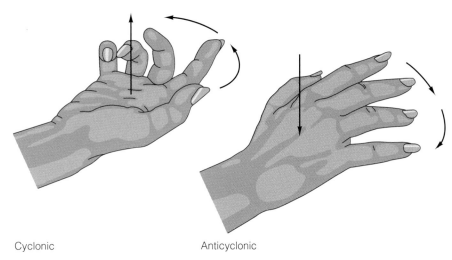

Cyclonic   Anticyclonic

Figure 5.5   Analogs of Northern Hemisphere cyclonic and anticyclonic rotation.

westerlies, which, in the Northern Hemisphere, passes over the bulk of the world's population. In the high latitudes, the weakest of the surface systems, the *polar easterlies,* spin out of the polar highs (fig. 5.3).

Other Influences on Winds

When both the Coriolis force and the pressure gradient are equal, a state of *strophic balance* occurs, producing an *ideal* horizontal air movement called a *geostrophic wind.* Such a wind flows perpendicularly to the pressure gradient (fig. 5.6). With balanced forces around high- and low-pressure systems, the resultant airflow is called a *gradient wind,* which flows along either a curved clockwise or counterclockwise path (fig. 5.7). Another force acting on air closest to Earth's surface is *friction.* The landscapes and surface waters impede the free flow of air, producing a drag that influences the direction and velocity of winds (fig. 5.8). Under the influence of *gravity,* denser air falls downslope (see katabatic winds below).

REGIONAL AND LOCAL WINDS

Both the continents and the oceans affect the direction of Earth's surface winds. Mountains, lowlands, and ice caps on land can act more or less as dry, frictional barriers that impede the free flow of air along the surface. With less frictional resistance, oceans present broad largely uninterrupted surfaces over which the air flows.

One of the most predictable of winds influenced by the distribution of land and water and their resultant pressure systems is the **sea breeze.** Sea breezes develop whenever air pressure is low over the neighboring land and high over the water (fig. 5.9). Primarily an afternoon and early evening wind, the sea breeze is the movement of air from the cooler, descending air over the water toward the warmer, ascending air over the land. Because the heating of the land's surface is a daylight phenomenon, nighttime produces an opposite, seaward-flowing wind called a **land breeze.** This wind develops as the land rapidly loses its heat, while the adjacent water maintains its accumulated heat. Whereas the relatively cooler land produces a local high pressure, the warmer water heats the overlying air, causing it to rise and produce a relative low pressure.

Many areas of the world experience local winds that have an effect on the local landscape. These winds generally form in mountainous or hilly regions where uneven heating and cooling during certain seasons creates a warming condition. On the **lee side** (side opposite that hit by the prevailing wind) of the Rocky Mountains in the United States, a downslope wind compresses and warms itself as it approaches elevations of greater pressure. The result is the **chinook winds** of late winter and early spring. Because the warmth of the chinook winds increases evaporation and melts snow, Native Americans coined its name for "the snow eater." A similar wind occurs in the Alps, where it is called the **foehn.**

Generally, in regions where local elevations vary greatly, air drainage downslope sometimes creates weak winds. These are responsible for low night temperatures in the adjacent valleys. Whenever these downslope winds cover large areas for relatively long periods, they are called **katabatic winds.** In Alaska and northern Europe, where glaciers lie high in the mountains and higher plateaus, these winds can be stiff breezes, blowing steadily for days and weeks at a time. Similarly the Adriatic Sea is the recipient of katabatic winds called **bora.** The same process produces the *mistrals* of the French Mediterranean. Note that with all these local and regional winds, special conditions take precedence over the general circulation of the atmosphere.

MOISTURE AND PRECIPITATION

In chapter 3, you learned that the kinds of surfaces over which air moves largely determine its moisture content. Air that moves over water is more likely to be moist than winds that move over land. *It is important to understand that advection*

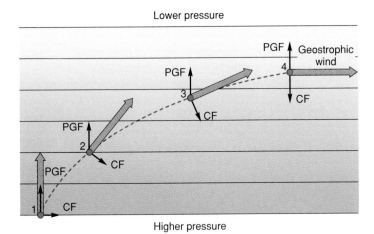

PGF   Pressure gradient force
 CF   Coriolis force

Figure 5.6 Geostrophic wind.

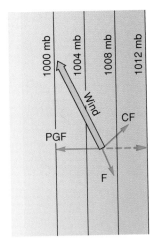

Winds at 30° N cross
the isobars toward the
low pressure

PGF   Pressure gradient force
 CF   Coriolis force
  F   Friction

Figure 5.8 Surface wind. PGF = pressure gradient force, CF = Coriolis force, F = friction.

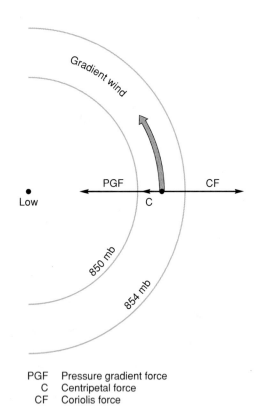

PGF   Pressure gradient force
  C   Centripetal force
 CF   Coriolis force

Figure 5.7 Gradient wind.

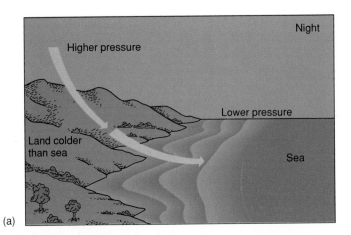

(a)

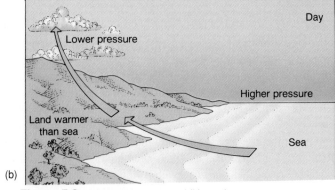

(b)

Figure 5.9 (a) Land breeze and (b) sea breeze.

is just as responsible for moving moisture into the atmosphere as it is for redistributing heat. Moisture moves, however, on a two-way street. Moisture that becomes incorporated into the atmosphere eventually finds its way back to Earth's surfaces as some form of precipitation.

Sources of Moisture

Almost three-quarters of Earth's surface is covered with water. The air that comes into contact with the great expanses of oceans, large lakes, and extensive river systems picks up mois-

ture along the air-water interfaces. Even plants, which pull water from soils, and the soils themselves, supply water to the atmosphere. The process by which plants give off water is called **transpiration.** On a hot summer day, for example, the leaves of a large oak tree might transpire more than 300 liters (80 gal) into the atmosphere.

Evaporation

In chapter 4, you learned that evaporation is a cooling process. The energy required to evaporate water is stored as the **latent heat of vaporization,** which can be returned to the environment through condensation. Significantly, the amount of moisture already present in an air mass determines, with the air temperature, the rate at which evaporation—and transpiration—takes place. Motionless air, however, regardless of the presence of these two quantifiable elements, is not as efficient an evaporator as is moving air. This is why advection is so important in the transfer of moisture from Earth's surfaces to the atmosphere. The principle behind this is as simple as what makes a hair dryer work. Moving air interchanges more molecules and, through more numerous molecular collisions, transfers more heat energy than does stationary air.

Although evaporation data are sketchy for the world's many surfaces, scientists have arrived at a few general statements about evaporation. Oceans are, obviously, more important than lands in adding moisture to the atmosphere. The zone of maximum evaporation from ocean surfaces occurs between latitudes 10° to 20°N and S of the equator. Over the continents, the maximum evaporation zone lies between the equator and 10° both north and south, especially because of the transpiration from the dense foliage of the tropical rain forests.

The Hydrologic Cycle

The **hydrologic cycle** is a recurrent process that moves moisture into and out of the atmosphere and through the life forms, landforms, rocks, soils, and standing water bodies of Earth's surface. Figure 5.10 shows the interrelationships required for the hydrologic cycle to be complete.

The bulk of the moisture that precipitates onto the continents derives from the oceans. Although local **evapotranspiration** (from evaporation + transpiration) does supply moisture, it does not compare in quantity with that produced by evaporation from the seas.

Moisture in Air Masses

Air masses can be thought of as great invisible rivers that transport vast amounts of water vapor between sea and land or vice versa. Because of the general circulation of the atmosphere, humid, tropical, maritime air masses travel poleward, where they

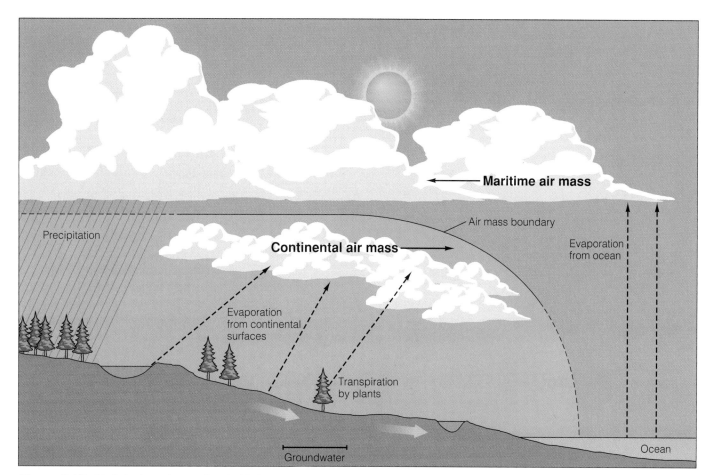

Figure 5.10 Hydrologic cycle correlated with the air-mass cycle.

become cooler, lose most of their moisture through precipitation in one form or another, and ultimately transform themselves into polar continental air masses. In the reverse direction and process, polar continental air masses move toward the equator, warming as they travel to lower latitudes, picking up moisture from soils, lakes, rivers, and vegetation. A continental air mass that migrates from Canada over the breadth and length of the Mississippi River's drainage area would evaporate more water than the great river would discharge in a comparable time. Eventually, continental polar (cP) air acquires enough moisture and heat to become maritime tropical (mT) air, precipitating moisture over the oceans.

The Dew Point

Warm air can hold more moisture than can an equal amount of cold air. To demonstrate this, visualize two sponges with different capacities. Cold air is like a somewhat compressed sponge; it has a reduced capacity for moisture. Warm air is an uncompressed sponge. Both sponges can hold moisture, but the uncompressed sponge has a greater capacity than the compressed sponge (fig. 5.11).

Note that both sponges can have their capacities diminished by further compression. It is a similar process that initiates condensation and the return of moisture to the surface. Whenever an air mass is cooled, it acts like a squeezed sponge. The air, its volume decreased by lowered temperatures, begins to lose its moisture, first through condensation and then through precipitation.

The temperature at which an air mass becomes saturated like a squeezed sponge at the limit of its capacity is called the **dew point.** The relative humidity of an air mass at its dew point is 100%. If an air mass is cooled to its dew point above the initial freezing temperature (32° F, or 0° C), *dew* forms. You can make dew simply by removing a cold can from the refrigerator

and placing it in a warm, humid room. As the air of the room comes in contact with the sides of the can, it cools and loses moisture, which condenses as liquid drops on the sides of the can. Dew points reached below freezing produce water in the solid and thin ice layers called **frost.**

Air masses with high relative humidity require only slight cooling to reach their dew points, while those with low relative humidity require significant cooling to reach their dew points. Condensation within and subsequent precipitation from an air mass is, therefore, dependent on the mechanisms that lower air-mass temperatures.

ATMOSPHERIC COOLING

Several mechanisms cool tropospheric air. Air that lies close to the surface can lose heat energy by conduction wherever the land or water is colder than the overlying air. Ordinarily, such cooling results in small-scale condensation. Another possible mechanism for cooling air is atmospheric mixing. As two dissimilar air masses merge, they lose their original temperatures and relative humidities, and the warmer of the two cools. This, too, is a relatively inconsequential mechanism. The dominant mechanism of air-mass cooling and condensation is **lifting.** As air masses either rise or are raised, they lose heat to the surrounding environment in a process called **adiabatic cooling.**

Rising air encounters lower pressures, and it expands. This leads to cooling. Air released from a car tire provides an analogy. The air pressure inside the tire exceeds that outside the tire. When the air is released through the valve, it expands, and it feels cool even if the tire is heated by travel.

As the unsaturated air rises through the atmosphere, it cools at the relatively rapid rate of 10° C per 1,000 meters (or 5.5° F per 1,000 ft in English units) until it reaches its dew point. Such a rate is called the **dry adiabatic rate,** and it is a simple manifestation of cooling by expansion. Upon reaching its dew point, lifted air loses both heat and moisture through condensation. Since condensation is a heating process, the cooling rate decreases with further lifting. After it reaches its dew point, lifted air cools at the **saturated,** or **wet, adiabatic rate,** which varies with temperature, but is about 6° C per 1,000 meters (or 3° F per 1,000 ft in English units).

A parcel of air that rises is not in equilibrium with the ambient air because it is warmer and less dense. Such rising air is said to be **unstable air.** An unstable air mass continues to rise and move away from its original position until it matches the temperature of the ambient air. Whenever the ambient and air parcel temperatures match, a **stable equilibrium** exists with little or no vertical movement. If a parcel of air is colder (denser) than the surrounding air, it is **stable** air and has a tendency to return to its original position. Throughout the troposphere, as you learned in chapter 4, the ambient air undergoes a lapse, and instability in the atmosphere arises only when an air parcel is warmer at any altitude than the surrounding air (fig. 5.12).

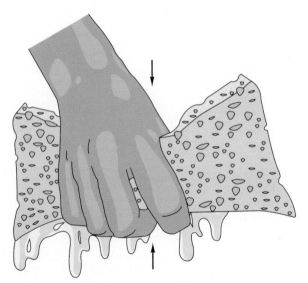

Figure 5.11 Squeezed sponge analog to cooled air. As the sponge is squeezed, it looses its capacity to hold moisture. Air that is cooled similarly undergoes a reduction in capacity.

CONDENSATION NUCLEI

Condensation takes place only on surfaces, such as the cold can removed from the refrigerator. In the atmosphere, abundant microscopic particles, generally called dust and composed of industrial pollutants, (particularly sulfurous compounds derived from the combustion of fossil fuels), volcanic ash, windblown soil and organic particles, pollen, and especially sea salts, provide condensation surfaces. Some of these surfaces attract water vapor. These tiny particles, called **hygroscopic,** or **condensation, nuclei,** must be present for condensation to occur.

FOG

Condensation in the air close to Earth's surface occurs as dew, frost, or **fog.** Fog is basically a cloud at ground level. Different kinds of fog have varied origins. In North America a **radiation fog** occurs mostly during the fall and winter, when the air is cooled because of loss of heat through radiation. The ideal circumstance for a radiation fog is a clear night with no wind; clear skies allow radiation to escape without encountering a cover of clouds. Ordinarily, a radiation fog forms in inland areas, in river valleys, and over low-lying areas with moist ground. Dispersal of the fog occurs with an increase in wind intensity or with the warming of the ground by the sun.

An **advection fog** can develop over either land or sea. Usually associated with winter and spring, this type of fog forms whenever warm air cools as it passes over cold land.

Dissipation occurs with a change of airflow and the heating of the land. Advection fogs are common over the seas during spring and early summer. At sea, and particularly along coastal areas, warm air cools as it moves over the cool sea surface. Whenever the wind shifts or the air warms, the advectional fog dissipates.

Frontal fogs are associated with air-mass fronts, where warm air masses come into contact with colder air. There is no season during which such fogs are frequent. The fog disappears as soon as the front passes. **Upslope fogs** occur at higher elevations with the adiabatic cooling of rising air.

CLOUDS

Clouds are composed of water droplets and, at great heights, ice crystals. Clouds are reflections of the physical processes taking place in the atmosphere and thus are indicators of weather conditions. They also serve as building blocks of weather systems.

Lifting Mechanisms

Three kinds of lifting produce clouds (fig. 5.13). The first type is **convectional lifting.** Thermally induced, convection means unstable air must rise through greater altitudes, where it cools, loses its ability to hold its moisture as it reaches the dew point, and forms clouds. Another type is **orographic lifting,** which occurs as air rises to pass over such topographic highs as plateaus and mountains. Orographic lifting is a mechanically induced process that moves air to higher altitudes, where cooling takes place. A third type of lifting is dynamically induced, requiring the leading edge of cold, denser air to plow under and lift warmer, lighter air. This type is **frontal lifting.**

Because air stability controls thermally induced lifting, the vertical development of clouds is dependent on the degree of instability in a given parcel of air. Stable air tends to produce **stratiform,** or layered, **clouds,** but unstable air produces **cumuliform,** or accumulated and vertically developed, **clouds** (table 5.1). Above the stratiform clouds, a layer of warmer air caps the vertical extent of cloud formation (fig. 5.14).

Meteorologists categorize clouds according to their shapes. Those with layered structures are called stratus, whereas those with ball-like accumulations are called cumulus. Both terms can serve either as prefixes or suffixes (fig. 5.15). The terminology for cloud types derives from a scheme devised in 1803 by British naturalist Luke Howard. In Howard's classification, a puffy cloud is *cumulus* (heaped); a sheetlike cloud is *stratus* (layered); a wispy cloud is *cirrus* (curly); and a rain-bearing cloud is *nimbus* (dark and rainy). Four cloud divisions based on height include high clouds (*cirro*); middle clouds (*alto*), low clouds, and, finally, vertically developed clouds. A cumulonimbus cloud is a vertically developed, billowy, rain-producing cloud, whereas a nimbostratus is a layered, rain-bearing low cloud.

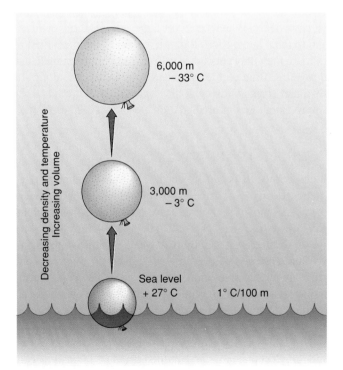

Figure 5.12 Dry adiabatic expansion.

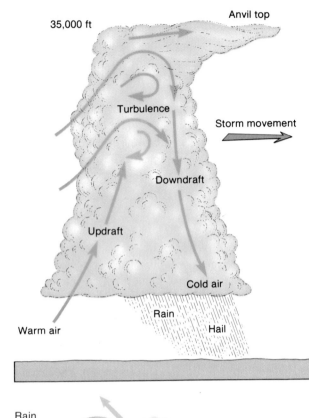

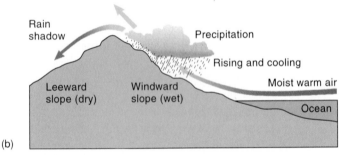

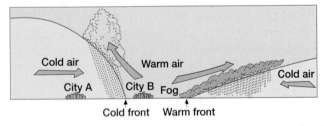

Figure 5.13 Three types of lifting mechanisms: (a) convectional lifting, (b) orographic lifting, and (c) frontal lifting.

Cloud Mechanisms Responsible for Rain and Snow

Rain and snow originate in clouds of different temperature characteristics. In *warm cloud* formation, condensed water forms droplets that collide and coalesce as they fall through the cloud. Because of a slight buoyancy, the smaller droplets fall as fast as the larger droplets, ensuring collisions and droplet growth. Other influences on the formation of raindrops are *updrafts* (upward-moving air) and *electrical charges*. This last influence manifests itself after a flash of lightning, with which large droplets appear to be associated. Even without lightning, clouds establish elec-

Altitude (in feet)	Type	Description
High, 20,000+ (cirro)	Cirrus (Ci)	Wispy clouds of ice crystals
	Cirrostratus (Cs)	Thin, whitish sheet; milky sky; sometimes a halo forms
	Cirrocumulus (Cc)	Mackeral sky
Middle, 6,500-20,000 (alto)	Altostratus (As)	Uniform, sheet cloud
	Altocumulus (Ac)	Flattened globular masses
Low, surface-6,500	Stratocumulus (Sc)	Large globular mass with flattened section
	Stratus (St)	Low uniform layer of clouds
	Nimbostratus (Ns)	Rain-bearing, layered clouds
Vertically developed, 1,000 to the top of the troposphere	Cumulus (Cu)	Thick, dense cloud with vertical development
	Cumulonimbus (Cb)	Heavy mass of cloud with great vertical development

Table 5.1 — **Cloud Types**

trical fields, and these fields appear to have a strong influence on droplet formation.

Condensation in cold clouds results in the formation of ice particles. As additional water condenses preferentially on the ice particles, they become the hexagonally shaped stars familiar to us as snowflakes. When they grow large enough, the snowflakes begin to fall. On their way through the cloud and the air below the cloud, they may encounter warmer temperatures, melt, and fall as rain. If the temperature lies below freezing along the snowflakes' paths of descent, they maintain their form and fall as snow. Both types of precipitation are shown in figure 5.16.

MIDDLE-LATITUDE CYCLONES

Middle-latitude (between 35° and 65°) weather is characterized by a succession of pervasive, alternating low- and high-pressure systems that move from west to east along the path of the prevailing westerlies. The movement of these pressure systems is responsible for the great variability of weather in the United States and Canada because they occur along zones of convergence and conflict between polar and tropical air masses. Thus, understanding these systems of counterclockwise and clockwise air movements, which direct air masses into and out of a region, is the key to interpreting most of North America's weather.

The daily mapping of those movements requires forecasters to identify the boundaries and internal structure of pressure centers. On weather maps, the National Weather Service uses millibar measurements to mark the air pressures that define cyclones

(a)

(b)

Figure 5.14 (a) Cumuliform clouds with flat bottoms and dome-shaped tops indicate vertical air movement. (b) Stratiform clouds.

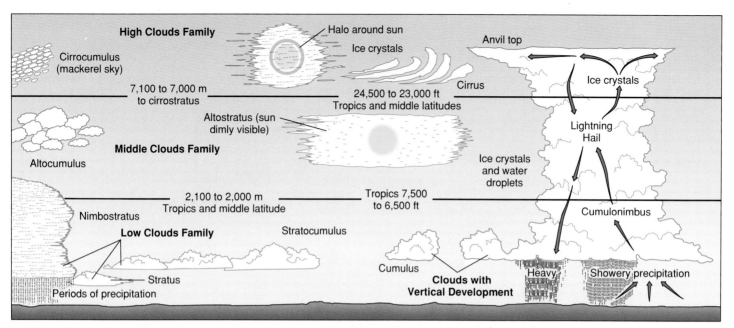

Figure 5.15 Cloud types.

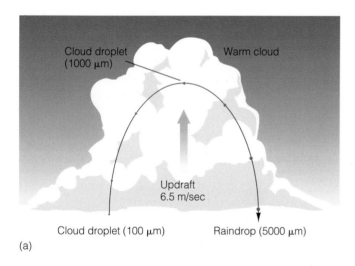

(a)

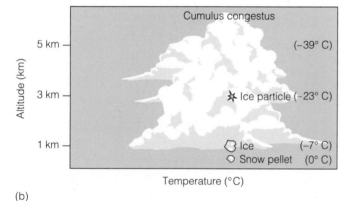

(b)

Figure 5.16 (*a*) Warm cloud precipitation. (*b*) Cold cloud precipitation.

and anticyclones. An important mapping symbol is the **isobar** (*iso,* meaning equal, and *bar,* meaning pressure), which is a line that connects sites with identical pressure readings. *Isobars never touch* because they represent different pressure readings, and they mark those readings in arbitrary intervals, such as the 3 millibar units shown in figure 5.17.

Size, shape, speed, and direction vary for each cyclone, but some average characteristics are well known. Typically, the long axis of a low is about 100 times greater than its vertical thickness. Some cyclones are known to reach altitudes of about 10 to 11 kilometers (about 6 or 7 mi). Over North America, a winter cyclone might have dimensions of 1,048 by 1,935 kilometers (650 by 1,200 mi). Usually, these systems move relatively fast, with summertime velocites reaching a little over 30 kilometers (20 mi) per hour, and the wintertime cyclones racing over the surface at speeds approaching 50 kilometers (30 mi) per hour. As they trend generally eastward, cyclones follow any one of nine different paths across southern Canada and the United States. The tracks largely determine how much moisture is associated with each cyclone. One that moves along the Alberta track passes over broad expanses of land until it reaches the Great Lakes, where it can pick up moisture (fig. 5.18). Because air in motion about the cyclone comes from different directions, air masses meet along the various kinds of frontal boundaries, which were introduced in chapter 3. As they encounter each other, these air masses interact along their mutual and changing boundaries, producing by frontal lifting most of the inclement weather experienced in North America.

By the turn of the twentieth century some patterns of inclement weather were apparent. For a long time scientists had known that precipitation was related to decreases in barometric pressure. Based on this information and on their surface observations, some Scandinavian scientists developed a model of middle-latitude cyclones. The stages in this model trace a cyclone from formation to dissipation.

In the initial stage of the model, two dissimilar air masses meet along a straight-line boundary (fig. 5.19). As the cyclone develops, this boundary bows and twists into two, then into three kinds of fronts. The energy for this process comes from converging air and from the heat released during condensation. In the final stage a loss of this energy leads to the dissipation of the cyclone.

If cyclones, in general, go through this process, why do some cyclones, in particular, generate more severe storms than others? The answer to this question lies in the relationship between the upper and lower portions of the weather sphere. At the surface, this relationship expresses itself in the development of fronts. Meteorologists represent fronts associated with cyclones and anticyclones by lines with attached symbols. The symbols lie on the leading side of the front. Although the air is invisible, its frontal shapes reveal themselves in certain types of clouds and patterns as seen from space. The constantly shifting air around the cyclone molds the accompanying cloud system into the shape of a large comma as seen from above. Precipitation can accompany the clouds, depending on the type and the instability of the air lifted along the fronts.

PRESSURE PATTERNS ALOFT AND VERTICAL AIRFLOW

In chapter 3, you learned that a high-speed wind called the jet stream flows in a general west-to-east trend across North America. The position of this air stream and the atmospheric pressures associated with it influence the development of cyclones (cyclogenesis) by directing air vertically toward or away from the center of a surface pressure system.

By sending aloft an aneroid barometer, weather observers can map the pressure patterns of the upper air and trace the winds driven by the pressure gradient. Atmospheric pressure can change at any given altitude just as it changes along the surface,

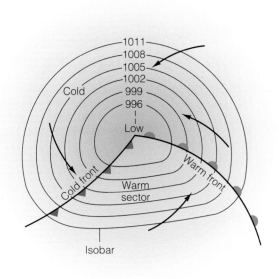

Figure 5.17   An isobar map indicating a low pressure center. Interval is three millibars.

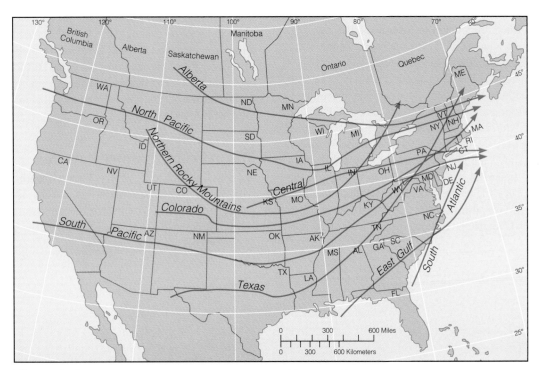

Figure 5.18   Typical cyclonic paths across the United States and southern Canada.

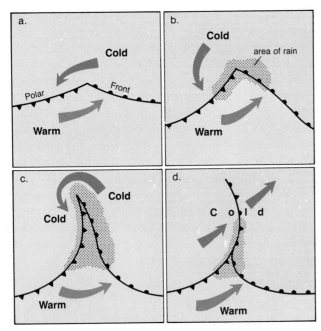

Figure 5.19 Formation of waves and fronts.

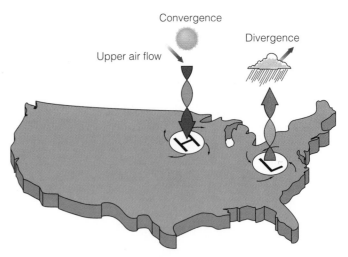

Figure 5.20 Convergence and divergence can intensify a cyclone or an anticyclone. Whenever divergence aloft is greater than convergence at the surface, a low intensifies. Its cartographic expression yields closer spacing between isobars.

and the upper-level air may flow where the barometer reads only 500 millibars. Because half of the atmosphere's mass lies below an altitude of approximately 6 kilometers (under 4 mi), the pressures associated with the jet stream are about half that of surface values. At this high-altitude, lower-pressure environment, winds move faster on average than surface winds, partly because they encounter no hard rock or water to rob them of energy through friction. The *upper-level winds* trend along a path parallel to the isobars, converging and diverging (flowing together or apart) in shifting locations.

Although the winds associated with a cyclone converge on the low at the surface, they begin to diverge as they rise (fig. 5.20). Whenever the winds aloft are directed away from Earth's surface by the pressure gradient, the surface pressure decreases as upward flow increases. The ultimate effect of divergence aloft is to intensify the low at the surface. Such an intensification depends on a neighboring anticyclone. The replacement of air that exits the bottom of the anticyclone must come from above; thus, the strengthening of both systems is an interdependent process. Figure 5.20 also shows air converging aloft to replace that lost by advection at the surface of the high. For systems in the Northern Hemisphere, the greatest increase in cyclone intensity occurs when the surface storm has the proper upper air support, which is an upper-level trough of low pressure west of the sur-

face low. Without this upper-level support, fully developed storms weaken.

FULLY DEVELOPED CYCLONES

Over many years of observations, meteorologists have framed the nature of a typical cyclone. Figure 5.21 shows the pattern most often seen in a fully developed cyclone of the middle latitudes. The cold front generates a narrow band of showers. This band can develop as a zone of intense precipitation and storm activity called a **squall line.** Behind the cold front, cP air moves counterclockwise, catching the warm front to the east and lifting it to make an occluded front. Maritime tropical air moves north or northeastward ahead of the cold front. As this warm, moist air encounters cooler air, it rises to make a warm front. The air north of the warm front is composed of modified air that cooled as it moved northward.

Warm Front Weather Patterns

A warm front is a long, gently sloping boundary between overlying warm, mT air and underlying cooler air (fig. 5.22). The gentle incline of this boundary has an average slope ratio of 1:150. Well ahead of the surface contact the warm air rises to

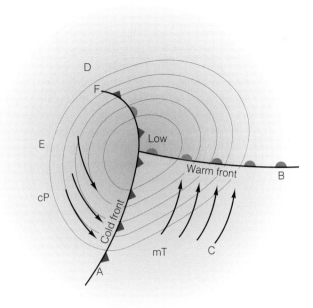

Figure 5.21 A fully developed cyclone in a map view: (*A*) marks the cold front; (*B*) is the warm front; (*C*) is mT air; (*D*) is the original center of the low, (*E*) is cP air, and (*F*) is the occluded front.

form cirrus clouds. Altostratus clouds or low stratocumulus clouds lie farther back, followed by lower, nimbostratus clouds, which bring prolonged light precipitation under stable air conditions. Some warm fronts are accompanied by relatively unstable conditions, producing storms intermixed with light precipitation. Figure 5.22 shows both stable and unstable conditions along a warm front.

Cold Front Weather Patterns

Cold fronts are relatively steep slopes (1:50) where rapidly advancing cold air forces warm air upward. The gradient of a cold front can be four times steeper than that of a warm front. This abrupt boundary is the site of cumulonimbus formation and violent weather, often in the form of squalls, thunderstorms, or even tornadoes. Although the precipitation is heavy, it is usually of shorter duration than that of the warm front. Behind the cold front, stratus, stratocumulus, and cumulus clouds are frequently followed by clear skies and fair weather (fig. 5.23).

THUNDERSTORMS

Severe thunderstorms form in unstable air where the wind speed increases rapidly with increasing altitude, often in the vicinity of the jet stream. As a thermodynamic machine, a thunderstorm converts the potential energy of the latent heat of vaporization into the kinetic energy released through condensation. Moist air, which is unstable, rises rapidly as violent vertical currents within the storm. Thunderstorms are capable of releasing torrential downpours of rain or heavy snows. Gusty winds, lightning, and, of course, thunder are all typical of these storms.

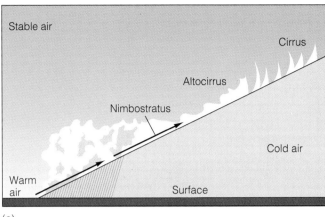

(a)

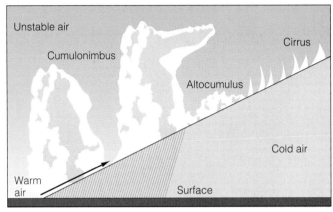

(b)

Figure 5.22 A warm front is a long, gentle slope with a leading edge of high cirrus clouds and a lower, trailing edge associated with periods of prolonged precipitation, which is usually light, as in (*a*), or with heavy precipitation where the air is unstable, as in (*b*).

Thunderstorms evolve through three stages (fig. 5.24). In the first stage, converging surface winds center on a low, then rise to carry moisture through condensation levels. In the second stage, rain, snow or hail begin to fall within the well-developed vertical cloud while higher up, in the cloud "towers," ice begins to form. This mature stage of thunderstorm development brings a downdraft and lightning. In the third stage, the downdraft destroys the parent cyclonic cell. Along the surface, convergent winds shift to become divergent winds, cutting off the updraft and effectively blocking the energy source for the system. Without further updrafts, the storm dies and precipitation stops.

Thunderstorms may reach the tropopause in their mature stage. The great vertical development is generated by lifting. Ordinarily, more thunderstorms occur in low latitudes than in high latitudes, making convection the most important lifting mechanism. Figure 5.25 shows the distribution of thunderstorms by frequency of occurrence for various latitudes.

Lightning and Thunder

For some as yet unexplained reason, electrical charges in a thunderstorm separate during the second stage of development, making the upper cloud layers positively charged and the lower layers negatively charged. Precipitation may, however, play a key role in this polarization of charges. Whenever the electrical potential gradient becomes sufficiently imbalanced, a lightning flash occurs. The temperature of the lightning may reach 10,000° C (18,032° F), causing rapid heating and expansion of the neighboring air. This expansion results in thunder, the noise of lightning. Because the speed of light is so much greater than the speed of sound, thunder is heard after the lightning flash. A rule of thumb that tells the observer how far away a lightning strike has occurred relies on this temporal difference. Every 3 seconds between the appearance of the lightning and the sound of thunder represents 1 kilometer between the observer and the lightning (or 5 seconds for every mile of distance between the observer and the lightning).

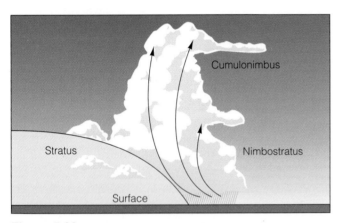

Figure 5.23 The structure of a cold front.

Tornadoes

The most violent storms on this planet are tornadoes, which are funnel-shaped clouds accompanied by winds in excess of 322 kilometers (200 mi) per hour. This storm activity is centered on an intense low pressure cell, which is related to a squall line of severe thunderstorms associated with a middle-latitude cyclone. Generally, a rapidly rotating updraft within a very unstable air mass creates a vortex of air that descends to the surface. Figure 5.26 shows this vortex within a cumulonimbus cloud. Once the funnel reaches the ground, a true tornado forms and moves rapidly. As they travel between 40 and 60 kilometers (25 and 35 mi) per hour, tornadoes sometimes skip or jump over the surface, rather than moving smoothly along in contact with the ground. The general trend of movement for these violent storms is southwest to northeast.

Because tornadoes are dangerous, our knowledge of them is still incomplete. Observations of tornadoes are often limited to chance sightings, though some weather scientists have made an art of chasing tornadoes. Through their work and that of other scientists, we now know that the energy for tornadoes derives from a collision between two distinctly different air masses.

In North America, the colliding air masses are the cold, dry, and stable Canadian air designated as continental polar and the warm, moist, and unstable Gulf of Mexico air called maritime tropical. Because these air masses are capable of colliding throughout the year, tornadoes can occur in any season from southern Canada to the Gulf of Mexico. Within North America, the region that produces most tornadoes is the interior plains. Over these plains many air-mass collisions

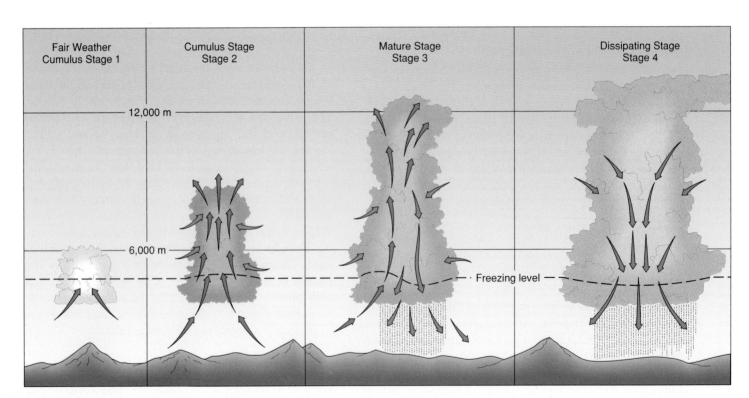

Figure 5.24 Stages in the development of a thunderstorm.

occur. In particular, the United States experiences 75% of its tornadoes between March and July.

The violence of tornadoes is sometimes tragic. During a 16-hour period on April 3, 1974, more than 100 tornadoes crossed the American Southeast, leaving a path of destruction, injuring more than 600 people, and taking 350 lives. The damage left behind by a passing tornado results from high winds, flying debris, and the great pressure differences between the center of the vor-

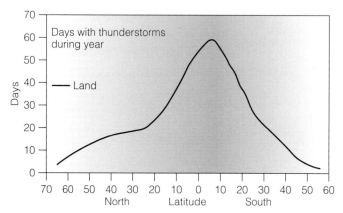

Figure 5.25 Latitudinal frequency of thunderstorms.

tex and the interior of buildings. Literally, buildings fly apart in response to the pressure gradient.

OCCLUDED FRONTS

Cold fronts move faster than warm fronts. In the proximity of the cyclonic center, they often catch and undermine warmer air, separating a warm front from the surface by a layer of denser, colder air. The result is an occlusion, or a gap between the warm front and Earth's surface. This type of air boundary is called a *cold occlusion* (fig. 5.27a). It is also possible for the air behind the cold front to be less dense than the air preceding it. In figure 5.27b, the warm front boundary stays in contact with the surface, while the cold front overrides it. This produces a *warm occlusion.*

Both types of occluded fronts are accompanied by precipitation. The amount and type of precipitation depend on the air temperature and the rate of lifing. If a cold occlusion passed over your location, you would experience a drop in temperature. A warm occlusion's passing would have the opposite effect. Often the precipitation associated with an occluded front, regardless of type, is prolonged and heavy.

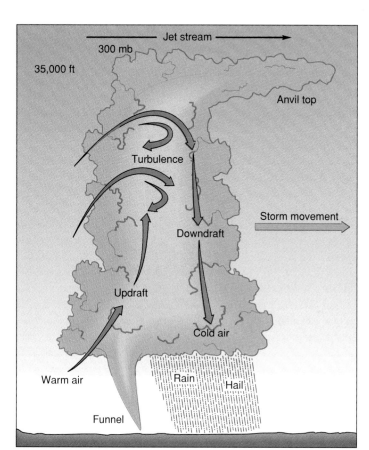

Figure 5.26 The vortex within a cumulonimbus cloud that results in a tornado.

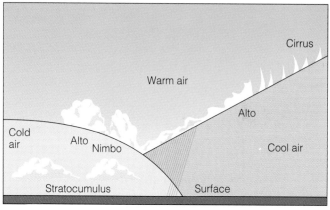

(a)

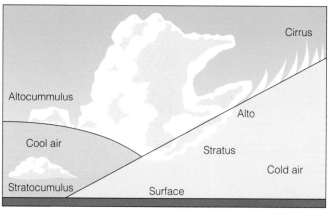

(b)

Figure 5.27 In a cold occlusion (a), the warm front is occluded, while in a warm occlusion (b), the cold front is occluded.

Box 5.1 Investigating the Environment

Tornado: A Late-Autumn Storm

Tornadoes are known as the most destructive storm systems to move across the face of Earth. Although such systems can occur at any time of the year, most people do not associate tornadoes with late autumn, a time of year when many air masses show stability over the United States.

During a 3-day period from Saturday, November 21, to Monday, November 23, 1992, an unusual late-autumn thunderstorm system ripped a destructive path across nine southeastern states. According to the National Weather Service, 80 tornadoes were sighted, with at least 45 twisters touching down in a 24-hour period ending at 7:00 A.M. Eastern Standard Time on Monday.

The unseasonable weather was caused by a collision between moist air from the Gulf of Mexico and drier, cooler continental air. The maritime air first moved over Texas and Louisiana in association with a low-pressure system on Saturday, then it moved on to Mississippi, Alabama, and Georgia. By early Monday, the storm system had passed over North Carolina, before passing out to sea.

In the wake of the storm, tornadoes were responsible for 25 deaths, including 15 in Mississippi, 5 in Georgia, and 1 in each of the other states touched by the system. The destructive force injured over 300 persons and caused power outages, snapped trees, and destroyed property.

The tornadoes terrorized the population of the region. One eyewitness to the event said, "The sky turned black, and there was a thunderous roar lasting for 30 minutes. The tornado then appeared, hugging the ground for miles, sweeping away everything in its path." A government official touring the storm's aftermath stated, "The area looks like a war zone."

Apply your Knowledge

1. Where did the low-pressure system originate? Why?

2. Why is the appearance of mammatus clouds at the base of a severe thunderstorm a good initial indicator of the potential for tornado formation?

3. The weather maps in figures 5.A, 5.B, and 5.C are summaries of observations. Why is the southeastern part of the United States susceptible to tornado activity at any time of the year?

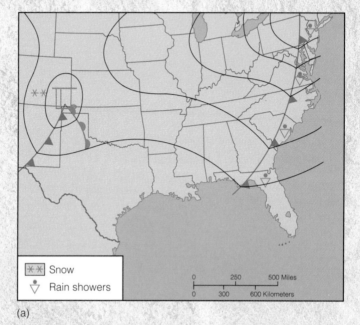

Figure 5.A Saturday, November 21, 1992. (Source: NOAA Daily Weather Maps.)

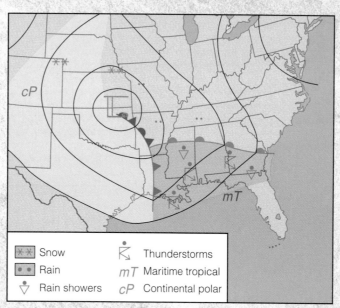

Figure 5.B   Sunday, November 22, 1992. (Source: NOAA Daily Weather Maps.)

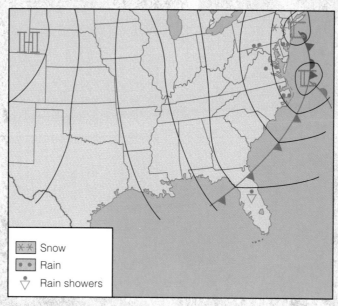

Figure 5.C   Monday, November 23, 1992. (Source: NOAA Daily Weather Maps.)

Box 5.2 Investigating the Environment

Hurricane Andrew: America's Most Destructive Storm

No natural disaster in American history has been more devastating than the August 1992 storm designated Hurricane Andrew. The total costs of Andrew's destruction have now exceeded $20 billion, mostly to Dade County, Florida. The storm severely damaged more than 60,000 residences in a path 48 to 80 kilometers (30 to 50 mi) wide. Some communities lost more than 90% of their buildings. In all, the storm displaced 250,000 people, ruined businesses and public facilities, and killed more than 20 people. Florida was just a brief passage for Andrew as the storm continued to move to the west and northwest, eventually striking Louisiana west of New Orleans (fig. 5.D).

Evacuations of both southern Florida, including the Florida Keys, and southern Louisiana saved lives, but the unexpected strength of Andrew caught many by surprise. After witnessing the April 23 devastation of southern Florida, the people of the Mississippi Delta knew that evacuation was the only short-term protection they could rely on for safety. In the aftermath of the storm, weather scientists discovered that Andrew's winds reached strengths powerful enough to move steel beams weighing tons. At 265 kilometers

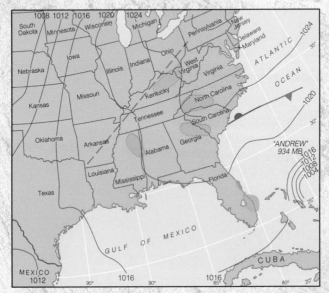

(a) Sunday, August 23, 1992

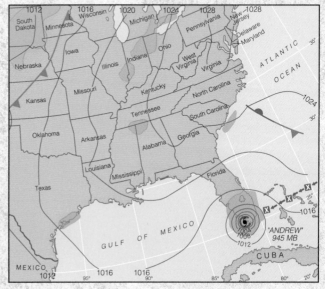

(b) Monday, August 24, 1992

(c) Tuesday, August 25, 1992

(d) Wednesday, August 26, 1992

Figure 5.D The sequence of weather maps shows the movement of Hurricane Andrew from just east of the Bahamas to landfall in southern Louisiana. All are as of 7:00 A.M. EST of the day indicated: (a) Sunday, August 23, (b) Monday, August 24, (c) Tuesday, August 25, and (d) Wednesday, August 26, 1992. (Source: NOAA Daily Weather Maps.)

(164 mi) per hour the recorded winds were high, but only the damage such as that shown in figure 5.E let meteorologists know that wind speeds may have exceeded 323 kilometers (200 mi) per hour. In the coastal waters of the Gulf of Mexico, where oil drilling platforms stood in the storm's path, oil companies suffered damages estimated at $1.5 billion. Hurricane Andrew had an accompanying storm surge that probably reached 18 meters (60 ft) around some of the platforms. About 200 of these costly rigs were put out of commission with major damage.

Andrew's path was watched closely by the National Hurricane Center, which ranks storms according to their power. Andrew appears to have reached the highest ranking (5), making it a rare hurricane. When Andrew struck Florida, it stayed over land for only a short period (fig 5.D). Ordinarily, hurricanes lose strength as they pass over land. Because Florida is not very wide, Andrew passed from the warm Atlantic waters into those of the Gulf of Mexico and gathered renewed strength as it approached Louisiana.

You can see in figure 5.D that Hurricane Andrew lay east of the Bahamas on Sunday, August 23, 1992. The storm hit Homestead and surrounding communities in southern Florida on Monday, then quickly moved into the Gulf of Mexico. Pressure readings for the hurricane were not exceptional, and in fact, they seemed relatively high, as the marked isobars reveal. Monday's pressure for the center of the storm registered 945 millibars. By Wednesday, Andrew's lowest reading was 975 millibars.

Bands of clouds associated with the hurricane produced tornadoes in Louisiana that cut narrow swaths of destruction across the landscape west of New Orleans. The devastation in Florida was so widespread that hurricane winds, not tornadoes, must have done the damage.

Apply Your Knowledge

1. Where did Hurricane Andrew originate? Why?

2. The weather maps are summaries of observations. What is the role of the National Hurricane Center, and how effective is its warning system?

Figure 5.E   Typical scene of the devastation caused by Hurricane Andrew, August 1992.

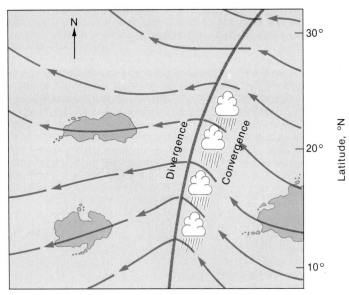

Figure 5.28 The isobaric expression of a tropical depression looks like a wave.

TROPICAL CYCLONES

The most familiar classification of tropical cyclones is based on wind speed. A *tropical disturbance* is characterized by poor circulation and mild breezes. Its isobaric expression is open, (fig. 5.28). A *tropical depression* produces higher wind speeds, usually about 56 kilometers (35 mi) per hour, and it turns under one or two closed isobars, showing increased intensity. A *tropical* **storm** is a more violent cyclonic movement, with wind speeds between those of a depression and 119 kilometers (74 mi) per hour. On a weather map, a tropical storm exhibits closed isobars.

Earth's biggest atmospheric storms are **hurricanes** (also called *typhoons*). You have already read how one particular storm devastated the Carolinas in chapter I. These atmospheric events are so large that they receive identifying names, indicating the importance the National Weather Service affixes to them. Wind speeds associated with hurricanes are usually well over 120 kilometers (75 mi) per hour, with some hurricanes accompanied for brief periods by 290 kilometers (180 mi) per hour winds.

Observations of many hurricanes have enabled meteorologists to label the main characteristics of these storms.

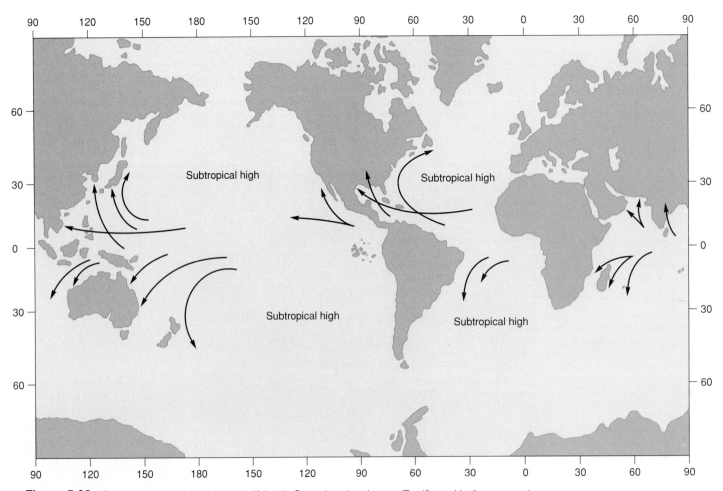

Figure 5.29 Common tracks of hurricanes. (Atlantic Ocean) and typhoons (Pacific and Indian oceans).

Usually, they occur in late summer and early fall in the Northern Hemisphere because of the heat accumulated in warm water areas during the high sun period. The excessive latent heat of vaporization eventually becomes the enormous kinetic energy of these storms. Unlike the smaller cyclonic systems to their north, hurricanes ordinarily have no anticyclonic companion, and hurricanes also yield lower barometric readings. A low between 914 millibars (26.9 in) and 931 millibars (27.5 in) is not uncommon.

The tracks of hurricanes pass over and by some highly populated areas in the Western Hemisphere. The Windward Islands of the Antilles are extremely vulnerable, as are the large islands of the Greater Antilles. Texas is frequently hit,

and a 6.2-kilometer (10-mi) seawall at Galveston is testimony to that community's resolve not to be inundated by the storm surge of seawater that rises in response to the lower barometric pressure associated with hurricanes. The common tracks of hurricanes and typhoons are shown in figure 5.29. Table 5.2 compares the characteristics of cyclones, tornadoes, and hurricanes.

Anticyclones and Fair Weather

Cells of high pressure with roughly concentric isobars and, in the Northern Hemisphere, clockwise circulation are often accompanied by fair weather. Anticylones are generally larger than middle-latitude cyclones; they also move across the surface more slowly and have smaller pressure gradients. They are common to the middle latitudes both north and south of the equator, particularly during the warm season.

Beside light winds or even calm conditions, summer anticyclones produce clear skies and fair weather cumulus clouds. With little cloud cover, these systems allow radiational cooling during the night. Winter anticyclones are usually continental in origin. Characteristically stable, dry air is sometimes stagnant, allowing pollutants to accumulate. Like the summer anticyclones, they produce clear night skies, but the temperatures are usually much colder.

Two summer anticyclones are of climatic significance for the United States. One located in the western Atlantic and called the Bermuda High funnels air into the country from the

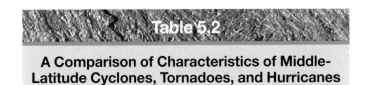

Table 5.2

A Comparison of Characteristics of Middle-Latitude Cyclones, Tornadoes, and Hurricanes

	Cyclone	Tornado	Hurricane
SHAPE	Egg	Funnel	Disc
PRESSURE GRADIENT	Moderate	Very steep	Steep
DIMENSION	Large area (1,000+km)	Small area (10+m)	Intermediate area (300+km)
ORIGIN	cP-mT conflict	cP-mT conflict	Latent heat
SEASON	Winter	Spring	Late summer, fall

Table 5.3

Atmospheric Changes Resulting from Frontal Passage

Weather Element	Before Passing	While Passing	After Passing
Typical Weather Associated with a Cold Front			
Winds	South-southwest	Gusty, shifting	West-northwest
Temperature	Warm	Sudden drop	Colder
Pressure	Falling steadily	Sharp rise	Rising steadily
Clouds	Increasing Ci, Cs, then either Tcu or Cb	Tcu or Cb	Often Cu
Precipitation	Short period of showers	Heavy showers of rain or snow sometimes with hail, thunder, and lightning	Decreasing intensity of showers, then clearing
Visibility	Fair to poor in haze	Poor, followed by improving	Good except in showers
Dew Point	High; remains steady	Sharp drop	Lowering
Typical Weather Associated with a Warm Front			
Winds	South-southeast	Variable	South-southwest
Temperature	Cool-cold	Steady rise	Warmer
Pressure	Usually falling	Leveling off	Slight rise, followed by fall
Clouds	In this order: Ci, Cs, As, Ns, St and fog; occasionally Cb in summer	Stratus-type	Clearing with scattered Sc; occasionally Cb in summer
Precipitation	Light to moderate rain, snow, sleet, or drizzle	Drizzle	Usually none; sometimes light rain or showers
Visibility	Poor	Poor, but improving	Fair in haze
Dew point	Steady	Slow rise	Rise, then steady

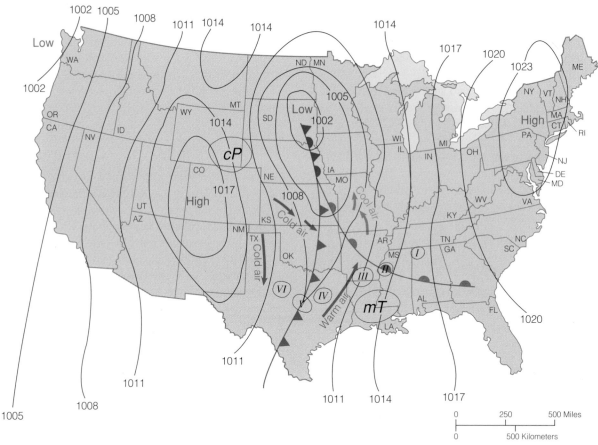

Figure 5.30 Movement of a typical cyclone through the central United States. (Source: NOAA Daily Weather Maps.)

Gulf of Mexico and the region of warm water around eastern Florida. The other summer anticyclone is located along the coast of California, where it blocks the eastward movement of cyclones and diminishes precipitation over southern California. Both of these high-pressure systems ordinarily fade with the changing sun angle. In California, the result is winter rains in the vicinity of Los Angeles. The influence of these and other anticyclones on climate is discussed more fully in chapter 6.

A CYCLONE

A cyclone that moves into the north central United States is typified in figure 5.30. Table 5.3 shows what weather can be expected along several lines passed by the system. Unlike the hypothetical storm you read about in chapter 3, this is a real storm system. You should be able to pick out the regions of cP and mT air and the fronts associated with the cyclonic movement. Above this surface system lies a divergent wind system that intensifies the cyclone.

Further Thoughts: Meteorology and Safety

As this chapter illustrates, Earth's dynamic atmosphere constantly distributes energy over the planet's surface. The expression of this distribution is the daily weather. At times this expression is an exhibition of great power, as it is in thunderstorms, tornadoes, and hurricanes. Because we now understand most of the mechanisms that drive the atmosphere and their quantifiable components, we can recognize tropospheric patterns on the bases of many observations.

Meteorologists can now predict violent or otherwise dangerous weather with greater accuracy and more lead time than was once possible. The practical consequence of knowledge about fronts, storms, and icy and foggy conditions

is a safer world. As they did during hurricanes Hugo and Andrew, weather predictions save lives. In conjuction with technicians and government agencies, scientists have invented elaborate warning systems that make our lives safer. Some airports now have weather detecting systems that warn pilots of powerful downdrafts that produce the wind shear that has been known to cause crashes. Along coasts, meteorologists issue small craft warnings whenever storms produce dangerous wave conditions. In winter, meteorologists issue warnings about hazardous driving conditions caused by snows and icy rains. When we heed these warnings, we reduce the level of risk to property and life.

Significant Points

1. Solar energy is the ultimate driving force behind the atmosphere, but radiation from Earth's surface is the immediate source of atmospheric heat.
2. Atmospheric warming is primarily attributed to radiation, conduction, convection, and advection. Of these processes, the last two are the most significant.
3. The planet's surface is unevenly heated. Warmer surfaces heat the overlying air, causing it to rise in convection currents.
4. Air lifted by convection, by orographic barriers, and by fronts can cool adiabatically to reach the dew point. The dew point is the temperature at which a parcel of air becomes saturated. Condensation takes place on microscopic surfaces in the atmosphere called condensation, or hygroscopic, nuclei.
5. The atmosphere circulates through large convection cells, which are governed in their movement by centrifugal and centripetal forces, by friction, by the pressure gradient, and by the Coriolis force. Moving fluids trend to the right in the Northern Hemisphere and to the left in the Southern Hemisphere.
6. The atmosphere circulates through the trade winds, the prevailing westerlies, and the polar easterlies.
7. Steeper pressure gradients mean winds with greater velocities.
8. In the Northern Hemisphere, cyclonic action is a counterclockwise movement; anticyclonic action is a clockwise movement.
9. Middle-latitude cyclones are associated with stormy weather; middle-latitude anticyclones are associated with fair weather.
10. Local winds, such as the chinook of the lee side of the Rockies, prevail over the general circulation. Katabatic winds are downslope winds.
11. Sea breezes prevail during the day, and land breezes prevail during the night.
12. Evapotranspiration from the continents provides air masses with some moisture, but the chief source of moisture for the atmosphere is the oceans.
13. Clouds can be classified by their altitude and their shape. Layered clouds are called stratiform, while nonlayered clouds are called cumuliform.
14. Three types of fronts can produce precipitation. A cold front is associated with squall lines, including thunderstorms, and tornadoes when the air is very unstable. A warm front is often the moving site of prolonged precipitation. An occluded front can combine the effects of both kinds of fronts.
15. Tornadoes are violent vortices with funnel shapes and high winds.
16. Hurricanes are enormous storms that develop in late summer and fall over tropical waters and then move, in the path of the trade winds, across the ocean, sometimes encountering land.
17. Hurricanes derive their energy from the latent heat of vaporization, which becomes kinetic energy during condensation.
18. Lightning occurs when opposing electrical charges accumulate on opposite ends of a cloud. Most lightning is cloud to cloud.

Essential Terms

advection   92
pressure gradient   93
cyclonic   94
anticyclonic   94
strophic balance   95
geostrophic wind   95
sea breeze   95
land breeze   95

katabatic wind   95
transpiration   96
latent heat of vaporization   97
hydrologic cycle   97
evapotranspiration   97
dew point   98
dry adiabatic rate   98

saturated, or wet, adiabatic rate   98
hygroscopic, or condensation, nuclei   99
advection fog   99
frontal fog   99
upslope fog   99
convectional lifting   99

orographic lifting   99
frontal lifting   99
stratiform clouds   99
cumuliform clouds   99
isobar   102
squall line   104
tropical storm   112
hurricane   112

Review Questions

1. Illustrate the distribution of solar radiation. Indicate the average percentage for radiation:
   a. absorbed directly by the atmosphere;
   b. reflected or scattered;
   c. absorbed by Earth's surface;
   d. absorbed by the atmosphere after being reflected or reradiated from Earth.
2. Explain how heat is transferred by conduction, convection, and radiation.
3. How does advection differ from convection?
4. What is pressure gradient, and how does it influence air motion?
5. Identify the locations of Earth's pressure belts.
6. Explain the cause of the Coriolis force.

7. Illustrate the general circulation of planetary winds as they would appear on a rotating Earth.
8. Describe the various types of atmospheric motion and give an example of each.
9. Sketch the pertinent aspects of the hydrologic cycle.
10. What role do hygroscopic nuclei play in the precipitation process?

11. Describe the physical properties of clouds.
12. Distinguish between warm cloud and cold cloud precipitation.
13. Illustrate the life cycle of a typical wave cyclone.
14. Identify the weather pattern associated with a warm front.

Challenges

1. Follow your local weather forecasts for 1 week, and compare them with your actual weather observations.
2. What landforms influence your local winds?

3. Compare barometric readings with local weather patterns.

Bee plant in Black Rock Desert in northwestern Nevada.

Chapter Outline

INTRODUCTION

Earth's troposphere envelopes and supports life. Two of its gases, oxygen and carbon dioxide, are essential for life's primary processes: respiration and photosynthesis. Even sea animals and plants benefit from the ocean-air exchange of these gases. Because the interface of air, land, and ocean is also a dynamic zone, we experience frequent events that characterize the ephemeral nature of atmospheric conditions. The zone of weather also subjects life to both small and large stresses, such as hurricanes and droughts. With its changes in temperature, moisture, and pressure, the lower atmosphere undergoes the hourly changes called weather. These continuous changes significantly influence Earth's surface.

The atmosphere is a major influence on the natural landscapes of Earth's surface and on the life that inhabits those landscapes. If you have traveled beyond your local area, you may have noticed just how Earth's atmosphere affects landscapes and vegetation. One common observation is that plant species are not universal. Their growth is limited by specific needs for water, temperature, and sunlight, three constituents of the weather. No one expects a palm tree, which is not a plant indigenous to the East Coast of North America, to grow and flourish in New York's Central Park or in the environs of Montreal. Palms require relatively warm weather year round. Thus, when the residents of Florida planted their first imported palms, they had a good idea that the weather was conducive to these plants. Similarly, no one expects a California redwood tree to grow in Las Vegas. The dry, hot conditions in Nevada's desert are unfavorable to the growth of redwoods, which require cool, moist weather.

Differences in plant species are not the only changes you might notice from one landscape to the next. The colors of soils also change. If you travel from southern Florida to southern central Canada, you could see soil color change from a very light tan (or even a dirty white) in Florida to a red in the South, then to a brown and even black in Indiana, only to lighten somewhat at the border of the two countries. These color changes are indicative of varying soil types, and all soils are largely dependent on the regional, predominant atmospheric conditions.

Animal behavioral patterns also reflect the influences of weather elements. Polar bears, for example, live in cold areas unlike the environments of their more southerly cousins, the black bears. Polar bears roam the pack ice of the Arctic and do not look for honey in beehives as black bears do. Bees, because they rely on flowering plants, do not live in the same area as polar bears. The weather of a region apparently influences the evolution of variations within species.

Unlike other animals, humans can adapt to virtually every weather condition on Earth through technology. Our ability to control our environment through clothing, housing, and air conditioning makes us an exception to the controls that limit other animals' distribution. Yet, as the examples of storm damage given in the preceding chapters indicate, humans are also often at the mercy of atmospheric conditions. The human experience suggests that the elements of weather influence not only the natural landscape but also the character of human life.

CLIMATOLOGY

The composite weather, or the generalization of weather patterns, is **climate.** As an *average* of weather conditions, climate is, specifically, the historically collective states of the troposphere for a given area over a particular interval. The area could be as small as a farm, where the *microclimate* could be studied. It could also encompass a large metropolitan area, yielding the *local climate.* Or, it could cover several states or provinces, in which case the composite of weather would be the *macroclimate.* Over long periods, weather observers gather detailed information on the temperature, humidity, precipitation, prevailing wind conditions, pressure, and storms. Observers collate their data into daily, monthly, and annual averages. The averages and the extremes of individual weather conditions with their associated causes portray a climate. This portrayal is the work of climatologists, and their study of climatic conditions is climatology.

Climatologists have adopted several ways to study long-term weather. Physical climatologists study physical laws as they relate to such weather elements as the average energy input and the subsequent conversions of energy to temperature, precipitation, and atmospheric motion. *Regional climatologists* apply physical principles to geographical units to produce an understanding of the variations or patterns of climate over an area. A few specialized branches of climatology deal with time, scale, and place. For example, *paleoclimatology* is the study of ancient climatic conditions by methods such as examining the thickness of tree rings for signs of wetter or drier years and analyzing the oxygen isotope ratios in sea shells as temperature indicators and clues to glacial epochs.

CLIMATIC ELEMENTS

The purpose of this chapter is to present a survey of climatic elements and their effects. Climatologists such as Wladimir Köppen have given us ways of making sense of the varying weather patterns around the world. Köppen, in 1918, and his successors examined the quantifiable weather elements, particularly *temperature* and *moisture,* to determine a means of classifying and explaining the atmospheric conditions that prevail in different areas. As a result of their efforts, we now know why we can expect changes in vegetation as we travel through various regions. We know, also, how to classify rain forests, deserts, savannas, and tundras. Because of their work, climatologists can explain the associations among soils, vegetation, and weather patterns.

CLIMATIC CONTROLS

Through meticulous observation, climatologists have defined a set of influences that govern soil, vegetation, and weather patterns. The cause-and-effect relationships between the weather elements and various climatic controls, *which determine the range of temperature and the amount and type of precipitation in a given area,* now appear obvious to anyone familiar with climatology. Table 6.1 summarizes seven climatic controls and their simplified effects on climate.

The world's climates appear to be very diverse. Cool weather predominates the high latitudes; warm weather, the low latitudes. While most of the eastern half of North America is relatively humid, much of the western half has relatively dry weather. Some sections of the continent, like the Montreal region, have severe winters; others, like the islands near Charleston, South Carolina, have mild winters. It may seem obvious that wherever high elevations occur, the weather is usually very cool or even cold, regardless of the season. This regional variability results from one or more of the climatic controls. Understanding climate, therefore, requires knowledge of the seven controls listed in table 6.1 and explained in the following sections.

Latitude and Sun Angle

The chief influence on climate is **latitude,** or the position in degrees north or south of the equator (0°). The sun's energy arrives at the planet's curved surface, where, because of Earth's tilt to the plane of its orbit, it is mostly concentrated in the low latitudes. The incoming solar radiation (insolation) that falls on the surfaces of our world amounts to 6 months of daylight each year. As you learned in chapter 3, the intensity of that daylight is dependent on the **angle of incidence** for the insolation at any latitude (see fig. 3.6). This angle is the number of degrees from overhead where the sun appears at any time of day. As the sun appears higher in the sky, its electromagnetic radiation is more concentrated on the surface of any locale. Because the tropics experience a higher sun on average than do the higher latitudes, more solar energy is available to warm low-latitude surfaces. These surfaces warm the air above them. As figure 6.1 shows, the insolation for three widely separated latitudes in the Northern

Table 6.1

Climate Controls and Their Effects

Control	Effect
Latitude and sun angle	Influences the amount of energy available for troposphere
Distribution of land and water	Influences the rapidity of heat gain or loss from Earth's surfaces and the availability of moisture
Altitude	Influences temperature and adiabatic changes
Orographic barriers, Semipermanent highs and lows	Lessen moisture content Influence wind direction and air-mass development and movement
Ocean currents	Influence pressure systems plus the amount of moisture available to an area
Storms	Transfer energy and moisture between and across latitudes and longitudes

Hemisphere can vary in both intensity and yearly distribution. This annual difference for 40°N and the equator means a concentration of insolation during June and early July for the more northern latitude, while the equator experiences two peak insolation periods, the equinoxes.

In table 6.2, you can see just what the difference in angle of incidence means for the daylight available at each of several locations during the cold and warm seasons. The tilt of Earth to the plane of its orbit prolongs summer daylight for both poles (90°N and S), but during this continued daytime, the poles still experience cold weather. In high latitudes, the angle of the sun is low even during summer, and insolation is not as intense as it is at lower latitudes, which undergo a daily period of night. Whereas the period of daylight is relatively constant in the tropics, the fluctuations between the lengths of day and night can be great for middle and high latitudes.

Latitude controls both the daily and the seasonal influx of solar energy at Earth's surface. Since the dynamism of the troposphere is primarily the result of insolation transferring solar energy to the surface and atmospheric gases, latitude is the most important climatic control.

Distribution of Land and Water

Water is an efficient holder of heat, but it requires a great amount of energy to change its temperature. In our practical, everyday experiences, we encounter this phenomenon whenever we try to boil water. The old saying, "A watched pot never boils," refers to the apparently slow process of boiling water, which requires extended contact between the pot and a heating element or flame. Land, by contrast, heats up and cools off relatively fast. The difference between the two substances becomes evident when you consider the flow of heat through them.

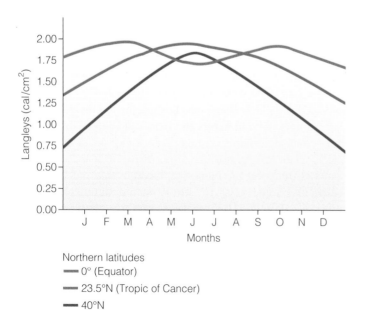

Northern latitudes
— 0° (Equator)
— 23.5°N (Tropic of Cancer)
— 40°N

Figure 6.1 Annual distribution of insolation.

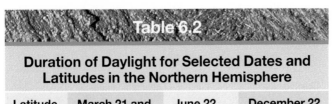

Table 6.2

Duration of Daylight for Selected Dates and Latitudes in the Northern Hemisphere

Latitude	March 21 and September 22	June 22	December 22
0°	12 hr	12 hr	12 hr
10°	12 hr	12 hr 35 min	11 hr 25 min
20°	12 hr	13 hr 12 min	10 hr 48 min
30°	12 hr	13 hr 56 min	10 hr 04 min
40°	12 hr	14 hr 52 min	9 hr 08 min
50°	12 hr	16 hr 18 min	7 hr 42 min
60°	12 hr	18 hr 27 min	5 hr 33 min
70°	12 hr	2 mon	0
80°	12 hr	4 mon	0
90°	12 hr	6 mon	0

When the sun shines on the ocean, much of its energy is used in the evaporation of the water, a process that leaves surrounding water cooler. The transfer of energy that turns a liquid water molecule into a gaseous one is reversed whenever the gas returns to a liquid state through condensation. In the hydrologic cycle, that reversal occurs in the atmosphere every time water droplets form. The removal of heat energy through evaporation keeps the sea relatively cool compared to drier land areas, where much less water is available for evaporation. Because water molecules are tightly bound to one another through electrical attraction, they require a large input of energy to break their bonds (chapter 11). Water, therefore, can store a greater amount of energy than can the land. In addition, water is free to move convectively, allowing it to heat to greater depths than land. Land may be heated to depths counted in centimeters while water, which allows more penetration by electromagnetic radiation, may be heated to depths counted in meters. The effect of this is that inland locations undergo greater temperature ranges than do coastal areas.

Altitude

A May blizzard would not be unusual on the top of New Hampshire's Mount Washington, which stands over 1,800 meters (more than 6,000 ft) above sea level. At 6,194 meters (20,320 ft) above sea level, the peak of Mount McKinley has a constant covering of snow and ice (fig. 6.2). Quito, Ecuador, lies on the equator, but its average annual temperature is 12.8° C (55° F), reflecting its more than 2,743-meter (9,000-ft) altitude. As you learned from the graph of the atmosphere's structure in chapter 4, temperatures generally decrease through the troposphere (see fig. 4.5). Experiencing temperature changes as you hike up a tall mountain is very much like the temperature changes that occur from the equator toward the poles.

Orographic Barriers

The troposphere is influenced by the topography. Whenever a highland interrupts the surface, air must rise to bypass the barrier. The subsequent windward rise and leeward fall cools, then heats, the air. Those mountains that lie along coasts, such as the Coast Range of western North America, can modify maritime air masses before they reach the continental interior. Collectively called *orographic barriers,* highlands strongly influence the leeward climate in their immediate vicinity as katabatic, evaporating, and drier winds descend to make desert conditions.

One of the more telling instances of orographic influence occurs in the western United States. There, more than 125 centimeters (60 in) of precipitation fall annually on the windward sides of the Sierra Nevada. The shielded leeward lowlands to the east of the mountains have a desertlike climate (fig. 6.3).

Semipermanent Highs and Lows

Seasonal high- and low-pressure systems influence the flow of air across large regions. Called *semipermanent highs* and *lows,* these systems migrate, strengthen, and fade seasonally because of the changing sun angle. Under the dictates of the Coriolis force, winds swirl in clockwise and counterclockwise patterns around the pressure centers, like the Bermuda High, a warm weather control off eastern North America (fig. 6.4a). Along the coast of southern California, a summertime high-pressure system usually blocks maritime air from Los Angeles, then moves southward during winter to allow rain to enter the country (fig. 6.4b).

The positions of the semipermaenent high- and low-pressure systems govern the surface winds. Thus, the prevailing westerlies and the polar easterlies migrate with the pressure changes. These two belts of opposing winds meet in the middle latitudes, where a kind of atmospheric conflict rages at their boundaries. Their changing positions cause a fluctuation in air temperatures over the continent. When the polar winds are stronger, a mass of cold air pushes toward the equator. The northward movement of warmer air signals the victory of the prevailing westerlies in their flow toward the poles. The alternating victories of one air system over another largely determine the weather for the central and eastern sections of the United States and southern Canada. These changes have been averaged and plugged into the classification of climates so that, for example, we can expect the Bermuda High to move maritime tropical air into the eastern half of the United States during summer. Residents of the southeastern seaboard and of the interiors of Alabama, Tennessee, Georgia, and the Carolinas know they must dress for warm, humid summer weather. When the Bermuda High weakens and shifts eastward toward the Azores, the flow pattern shifts. Figure 6.5 shows two maps, one with typical cP air migrating southward and the other with mT air moving northward under the influence of the Bermuda High.

Another influence of the semipermanent highs and lows is the monsoonal event that occurs yearly in the Indian Ocean

Figure 6.2 Clouds veil Mt. McKinley, the highest peak in North America.

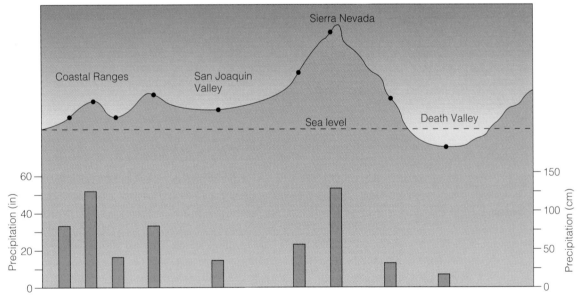

Figure 6.3 The influence of topography on precipitation.

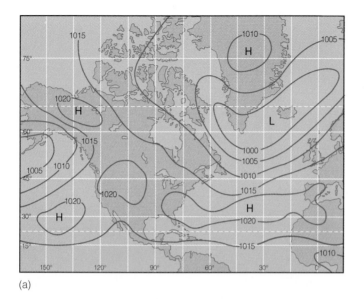

(a)

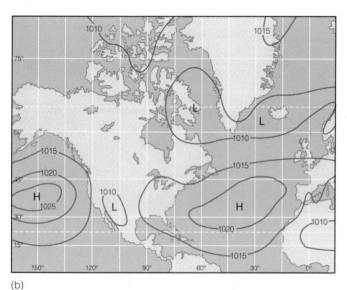

(b)

Figure 6.4 (a) The winter position of semipermanent high-pressure systems that influence North America. (b) The summer position of semipermanent high-pressure systems that influence North America.

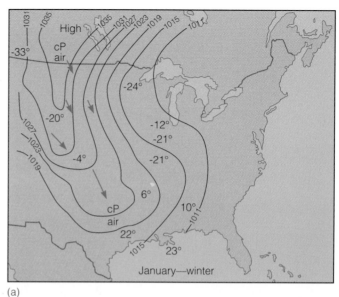

(a)

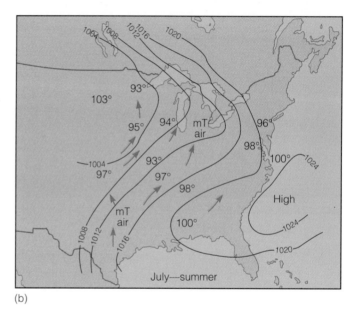

(b)

Figure 6.5 (a) Typical cP air mass moving southward. (b) Typical mT air mass moving northward under the influence of the Bermuda High.

and on the Indian subcontinent. A semipermanent high develops over Asia during the cold winter of the continental interior. During the subsequent warming of summer, the high fades and a low-pressure system develops west of India. The low funnels air off the Arabian Sea to the east, where it encounters the mainland of India and the orographic barrier of the Deccan Plateau. Within the region other areas, particularly Bangladesh off the Bay of Bengal, suffer the consequences of a season of abundant rainfall. In Bangladesh, the waters often inundate much of the lowland. In recent years, extensive monsoon floods have temporarily displaced millions of that country's citizens, isolating them on a few dry places, such as on elevated railroad tracks. Monsoon rains are actually important in bringing many lands the moisture necessary for agriculture.

South Floridians would be short of fresh water if it were not for the seasonal rains.

Semipermanent highs can be associated with either polar or tropical air masses and, in turn, with either cold or warm weather. Vigorously moving cold highs originating in Canada are likely to bring lower-than-normal temperatures, especially in winter, to the United States. Tropical highs, such as the Bermuda High of summer, are associated in the United States with above-normal temperatures and increased humidity. Diminished or even absent cloud development and fair weather are prevalent characteristics of both cold and warm highs.

By contrast, low-pressure systems are likely to bring higher-than-average temperatures in winter and lower temperatures in summer. As a general rule, the convergent air movements typical of a low-pressure center produce condensation, cloud cover, and precipitation.

Near 30°N and S of the equator there occurs a series of anticyclonic systems usually called the **subtropical highs.** Because of the unequal heating of land and water, a considerable modification in the distribution of these anticyclones can be observed over a year. These deviations are most apparent in January and July, when one hemisphere experiences summer and the other winter. A close examination reveals that the highs are well developed and exist separately over the oceans during summer in each hemisphere. In addition to the seasonal changes in intensity, the pressure belts shift their latitudes. A subtropical anticyclone of the Northern Hemisphere might shift about 10° between summer and winter. This seasonal shift corresponds to the direction of the migration for the vertical rays of the sun. The maps in figure 6.6 show Earth's distribution of surface pressure systems for both January and July. Note the shift of the land-based high between the Northern and Southern Hemispheres. The cold air that develops over the continents during winter sinks to make semipermanent highs that fade when the summer returns and a prolonged and higher sun heats the land.

The dominant subtropical highs that influence North America are the Bermuda High and the California High. On either longitudinal side of these highs, the air acquires differing characteristics. As a rule, the eastern side of the high is characterized by stable, dry air, whereas the western side is usually the site of unstable, wet air. Along the two coasts of the continent this relationship is evident as California experiences a seasonal dry period and the southeastern United States undergoes a wet period.

Ocean Currents

The surfaces of the oceans are waters in flux. Large clockwise and counterclockwise *gyres,* or whirlpool-like movements, circle the oceans. Because water holds heat well, those currents that flow away from the equatorial regions transfer heat to higher latitudes. The return flows, from high latitude to low latitude, serve as cold surfaces from which the air obtains its characteristics.

Ocean currents generally follow the prevailing winds, and they are influenced by the Coriolis force, turning right in the Northern Hemisphere and left in the Southern. This movement sets up a dominant pattern of cold currents along the western coastlines of the major landmasses and warm currents along the eastern coasts. Like rivers of water within the oceans, these

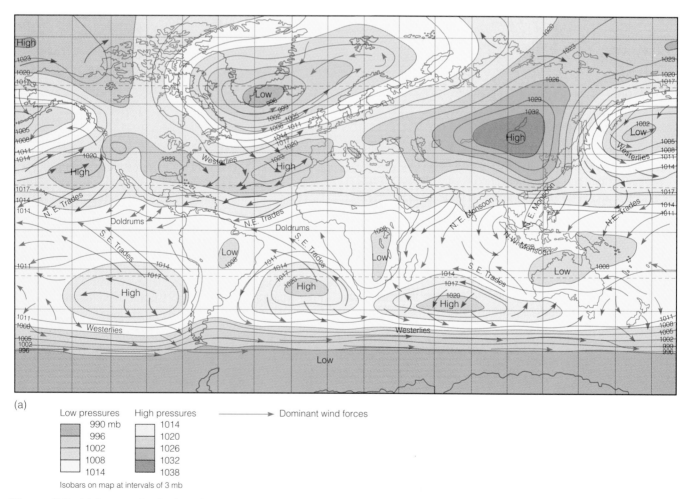

(a)

Low pressures
	990 mb
	996
	1002
	1008
	1014

High pressures
	1014
	1020
	1026
	1032
	1038

→ Dominant wind forces

Isobars on map at intervals of 3 mb

Figure 6.6 (a) January distribution of pressure.

continued

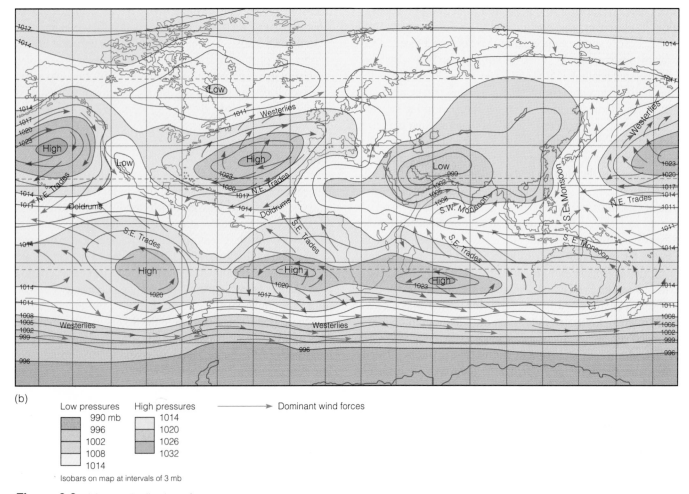

(b)

Low pressures	High pressures	⟶ Dominant wind forces
990 mb	1014	
996	1020	
1002	1026	
1008	1032	
1014		

Isobars on map at intervals of 3 mb

Figure 6.6 (b) July distribution of pressure.

currents can be distinguished by flow rate, width, depth, and chemistry (chapter 11). Because they are surfaces over which air must pass, they can be distinguished by the characteristics they imbue to the troposphere. Generally, *cold ocean currents make coastal weather cool and dry. Warm currents generate warm, moist coastal climates.* For example, part of the North Pacific Drift, a high-latitude current that wends toward Alaska and Canada, becomes the Alaska Current, a warm current that moderates the climate of North America's northwest coast (fig. 6.7). The water that does not spiral northward as the Alaska Current turns to the south, eventually becoming the cold California Current and influencing the weather of Oregon, California, and Baja California.

You learned in chapter 3 that air masses obtain their characteristics from the surfaces over which they develop and that cold surfaces are associated with high pressure and little precipitation. In the Southern Hemisphere, a significant flow of cold surface water passes the western coast of Peru and Chile, producing one of the driest of coastal desert regions, including the high Atacama Desert, portions of which may not receive any rainfall for several decades at a time. The Peru, or Humboldt, Current often extends from 4°S to 30°S (fig. 6.8).

Storms

Storms occur in relatively set patterns and are regular occurrences in many areas. From late summer to late fall, the Caribbean and the Gulf of Mexico are often on the paths of hurricanes. In Japan and the Philippines, yearly typhoons lash the coasts with torrential downpours. Storms also occur regularly around Cape Horn. In each area, storms provide moisture and transfer heat. In fact, the peoples of the Pacific Orient largely depend on typhoons to bring large quantities of fresh water to the islands. In most equatorial land regions, the climate is characterized by almost daily thunderstorms, usually occurring in the afternoon and evening. These storms support a lush growth called the *rain forest,* which is discussed later in this chapter.

GLOBAL PATTERNS OF TEMPERATURE

The distribution of Earth's mean annual temperatures for January and July appears in figure 6.9. These temperatures are informative because they represent the extremes of opposite seasons. In the following discussion, you will learn some of the more important tendencies of this temperature distribution.

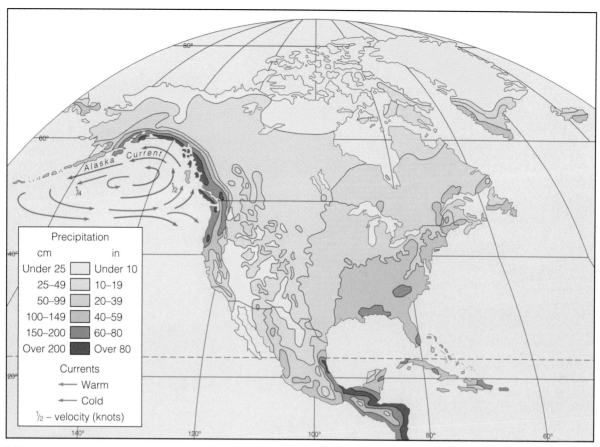

Figure 6.7 The Alaska Current, part of the North Pacific Drift, is a high-latitude current that flows toward Alaska and western Canada.

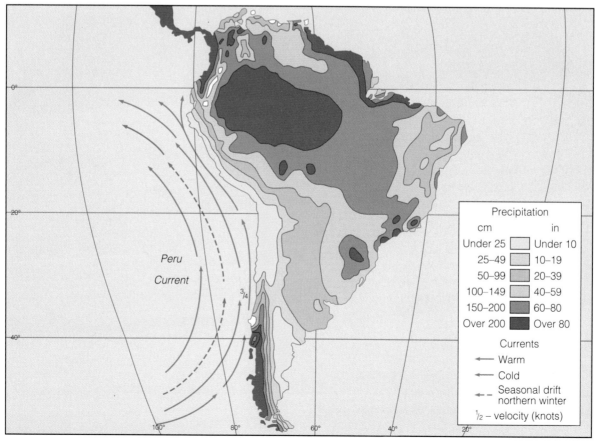

Figure 6.8 The Peru Current.

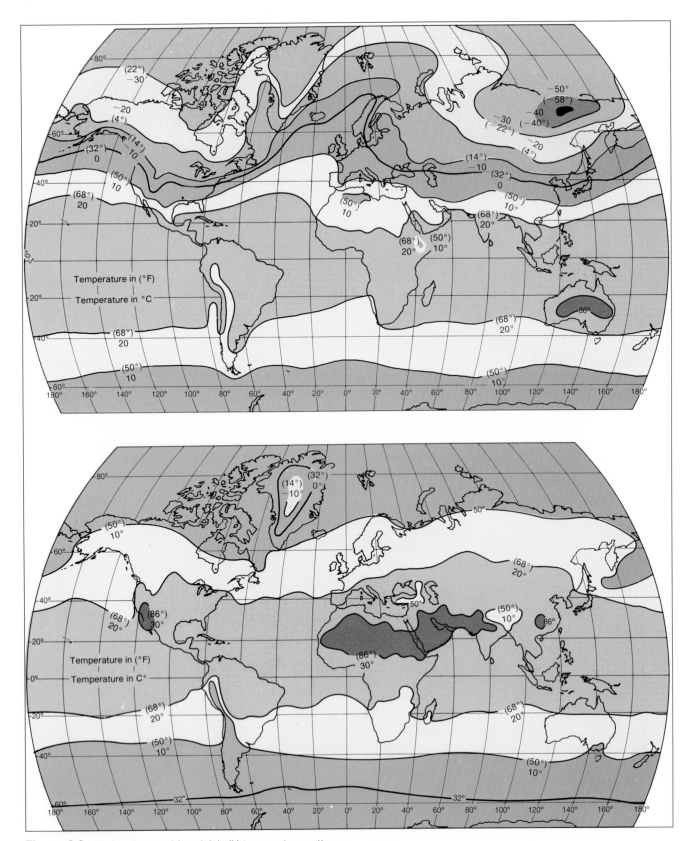

Figure 6.9 Global January (*a*) and July (*b*) temperature patterns.

When you examine the temperatures in figure 6.9, you can see a marked latitudinal shift of temperature between the two months. This reflects the migration of the noon overhead sun. Meteorologists map temperatures by plotting and then connecting equal temperatures with lines called **isotherms** (*iso,* meaning equal and *therm,* meaning heat). With changing values in insolation, the isotherms migrate. That migration is greater, as you may expect, over the land than it is over water. Land, of course, heats up and cools off to greater extremes than does water, and the lowest temperatures occur over the large land areas of Asia and North America in January. The change in isotherms is manifest in another pattern. In the Northern Hemisphere, the July isotherms bend poleward over the continents, whereas they bend equatorward in January. The January isotherms for the oceans, which do not cool as much as the land, bend poleward. Just why do the isotherms adopt this bend?

The cold weather of North America during winter extends far into the south central portions of the continent. Even Texas experiences cold temperatures as the rapidly cooling land surfaces lose heat in the absence of a high sun and in the presence of cP air masses. The cold tempertures of southward moving air masses coupled with radiational cooling during a season of longer nights and shorter days make the southern interior of the continent as cool as its more northern coastal areas. By contrast, Montana, another interior region far from the temperature-moderating influence of large water bodies, heats rapidly in summer, with the overlying air reaching temperatures more common to the lower latitudes. If you follow the isotherms for both seasons from the coasts to the interior, you will, in effect, trace a cold air bulge southward during winter and a warm air bulge northward during summer. You will also trace the extremes of surface cooling and heating in the heart of the continent.

In the Southern Hemisphere, the isotherms follow the parallels of latitude more closely than they do in the Northern Hemisphere. The principal reason for this lies in the land-water distribution. The Southern Hemisphere is chiefly an ocean hemisphere, and the uniformity of surfaces is mirrored by the latitudinal distribution of temperatures.

As noted earlier, temperature ranges are also affected by the distribution of land and water. In general, the annual *range of temperature,* which is the difference between the mean temperatures of the warmest and coldest months, is greater at higher latitudes (fig. 6.10). The range is also greater over land than over water, as you might expect. The smallest ranges occur at both low and high latitudes. The largest ranges are mostly confined to the Northern Hemisphere's continents, whose January-July temperature fluctuations reflect the rapid and extensive heating and cooling of land surfaces.

GLOBAL PATTERNS OF PRECIPITATION

Precipitation is the other important climatic element. Its significance is evident in the kinds of studies necessary for climatic groupings. In their attempt to classify weather patterns, for instance, climatologists collect various data about moisture, such as the average total amount of precipitation at a locale, the regularity of its occurrences, and the seasonal variability of rain and snowfall. Their interest in the data is more than scientific because all life is dependent on water. The surface ecologies of Earth are

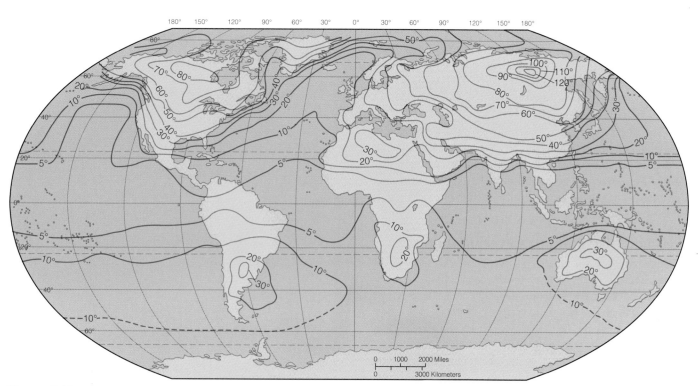

Figure 6.10 Annual range of surface temperatures.

directly tied to the amount of precipitation and the moisture that is subsequently available to support life. Answering just why certain areas have more abundant precipitation than others has meant identifying patterns and their causes.

Figure 6.11 shows a simplified distribution of precipitation in annual amounts for every 20° of latitude. The precipitation occurs as zonal or latitudinal patterns strongly bound to the general circulation of the troposphere. Practically all precipitation is the result of cooled ascending air masses. Therefore, precipitation is most abundant wherever air rises and least abundant wherever it sinks.

Keep in mind that global precipitation patterns depend on ascending and descending air. Ascending air carries moisture aloft, where lower air pressure causes the air to cool. The lower air temperatures cause the moisture to condense and become available for cloud formation and precipitation. Descending air warms as it reaches lower altitudes, and the heating inhibits cloud formation and limits precipitation by giving the air a greater capacity to hold its moisture. The four zones depicted in figure 6.11 can be divided into two relatively wet (I and II) and two relatively dry (III and IV) patterns.

Zone I represents the wettest areas of the world, the equatorial to tropical regions that cover 20°N to 20°S latitude. Air rising in broad belts along the equator contains large amounts of water vapor because of high temperatures, warm surface waters, and high rates of evaporation. The rising warm air has a great capacity for moisture, but the ascent to higher altitudes decreases its holding power by lowering its temperature, and condensation inevitably follows the rise.

The other wet pattern falls in zone II. Poleward beyond 30°N and S lies the prevailing westerlies and accompanying frequent precipitation. With numerous cyclonic systems, air rises throughout the westerlies because of frontal lifting.

Zone III encompasses one of two dry patterns. At the polar limit of the equatorial region is the zone of subtropical high pressure. The predominantly descending air throughout this zone dictates an anticylonic pressure distribution. The result is a rather dry climate region with little cloudiness, favoring the development of deserts and semidesert climates.

The second dry pattern falls within zone IV, a region in very high latitude with abundant descending air. The sinking cold air at the poles contains little water vapor, and the mechanisms by which the low humidity might be condensed and eventually precipitated are generally lacking.

Seasonal Distribution of Precipitation

The precipitation of summer falls into warmer air than that of winter. Because warm air has a greater capacity to hold moisture than cold air, the moisture that falls in winter is not as easily evaporated as that which falls in summer. Cold winter air does not have the heat energy available for the efficient evaporation of water molecules.Therefore the seasonal distribution of precipitation is a second important consideration climatologists make when they study the effect of moisture. Within most of the four zones of figure 6.11 there is a sea-

sonal regularity, an important aspect of precipitation. In some of the zones, precipitation occurs evenly with a seasonal balance (fig. 6.12).

In general, humid conditions prevail all year under the equatorial low, in the westerly winds, and in the subpolar lows. By contrast, dry conditions are prevalent in the polar regions and the subtropical highs throughout the year. Seasonal precipitation occurs outside these zones, particularly in the tropical region north and south of the equator because of the shifting trade winds at the so-called *Intertropical Convergence Zone* (ITCZ).

Modifications to the Latitudinal Precipitation Zones

The actual distribution of climates is the product of temperature and precipitation modified by the seven climate controls. Figure 6.13 shows a *longitudinal modification of the zonal patterns* largely based on the following:

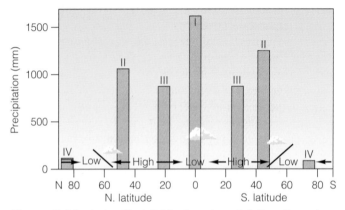

Figure 6.11 A simplified distribution of precipitation. Zones I and II are relatively wetter zones, whereas zones III and IV are relatively drier zones.

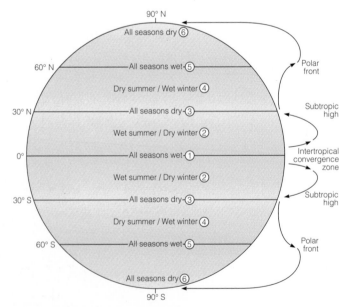

Figure 6.12 Seasonal distribution of precipitation (generalized).

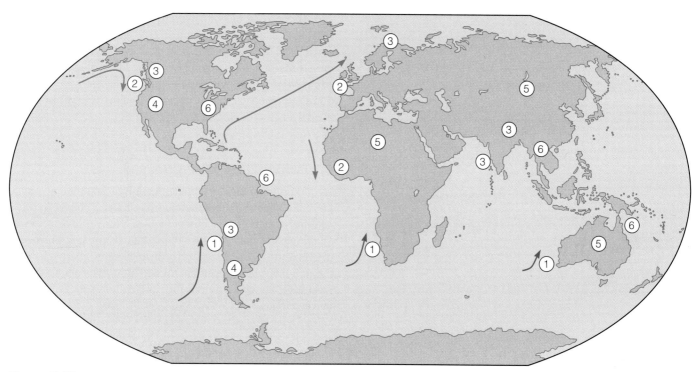

Figure 6.13 Modifications to zonal precipitation patterns. Numbers correlate to those in text.

1. dry coastal areas adjacent to cold ocean currents;
2. humid coastal areas adjacent to warm currents;
3. humid windward sides of highlands;
4. dry lee sides of highlands;
5. dry continental interiors;
6. coastal areas not chiefly influenced by either cold or warm currents.

THE CLASSIFICATION OF CLIMATES

The word climatology derives from the Greek *klima* ("region"), a related word of *klinein* ("to slope"). The significance of this origin may already be obvious to you. The composite weather patterns occur *regionally,* and they are primarily controlled by sun angle, a product of the curvature of a planet that just happens to be tilted, or sloped, to the plane of its orbit.

The ancient Greeks were, in fact, influential in establishing a common way of classifying weather patterns. In figure 6.14, the world is marked according to the Greeks' designations based on temperature and the astronomical phenomena of sun angle and planet orientation. Because they believed their own climate to be an ideal one, they called it "temperate." Beyond the hot deserts across the Mediterranean they deduced that even hotter regions made human habitation impossible. These hot areas they called "torrid." From the experiences of travelers and their own wintertime experiences with occasional cooler winds blowing out of the north, they believed in a northern "frigid" zone.

The Greeks had done well in recognizing that climate varies and that its variety has an apparent cause—sun angle. Their cat-

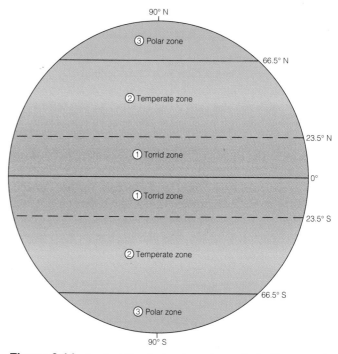

Figure 6.14 Ancient Greek climate system adapted to a round Earth.

egories failed, however, to account for longitudinal variations and the different controls identified by scientific observers. Because the ancient Greek climate system does not account for variations within the generalized zones of figure 6.14, it is a relatively useless model of climate. By contrast, modern classifications of climates have evolved from scientific observation and data collection and have resulted in quantitative approaches to weather—and climate—studies.

Table 6.3

Letters Used by Köppen and Trewartha and Their Designations

Capital Letters	Designation
A	Tropical climates
B	Dry climates
C	Low midlatitude climates
D	High midlatitude climates
E	Polar climates
S	Semiarid
W	Arid
T	Tundra
F	Ice cap
H	Highland
Lowercase Letters	
a	Hot summers
b	Warm summers
c	Cool summers
d	Cold summers, cold winters
f	Precipitation in every month
w	Dry winter
s	Dry summer
m	Monsoonal precipitation

The purpose behind Köppen's classification was a practical one: he attempted to devise an objective classification scheme that would be applicable to climatic variability. For Köppen and other climatologists, there are differences in the way one region or another is "torrid." Some hot areas are dry; others, wet. The significance of these differences becomes apparent when climatologists look at vegetation, since plants reflect the character of the local, and regional, weather. Another climatologist, C. W. Thornthwaite, advantageously used vegetation types and distribution to define climates further and eliminate deficiencies in the Köppen system. Thornthwaite perceived a need for relating precipitation and evaporation to plant growth, a need not met by Köppen. Yet, what Köppen did is worth studying because no special knowledge is required to understand his classification.

To simplify the great variety of weather patterns, Köppen based his classification on *the monthly and annual means of temperature and the total monthly and yearly amounts of precipitation for a locale.* Because he worked with such easily understood data, he made climate relatively easy to comprehend. In his attempt to make order from the apparent chaos called weather, Köppen employed five major climate types, all represented by a capital letter. Generally, these letter symbols mimic the global pattern perceived by the Greeks in that they exhibit a latitudinal trend (fig. 6.15). With the refinements by Köppen and G. T. Trewartha, these lettered climates now stand for more than simple latitudinal designations (table 6.3).

Köppen and Trewartha also qualified these major designations to account for both temperature and precipitation, using lowercase letters to complement the capital letters. Armed with these additional letters, you can now read the Köppen labels. A Cfb climate would be a middle-latitude climate with precipitation all year and with a cool summer. Box 6.1 gives a more complete listing of the Köppen terminology.

CLIMATE DESCRIPTIONS

By plotting monthly average temperatures and total precipitation on a *climagraph,* climatologists model a climate type. In addition to their graphs, climatologists write prose descriptions that frame the characteristics of individual climates. The rest of this chapter presents specific information about climates through maps of its representative regions; lists of characteristics for the climatic subdivisions; and climagraphs of representative stations.

TROPICAL (A) CLIMATES

Throughout the year, hot and humid climates prevail in a belt that extends from about 15° to 25°N and S of the equator (fig. 6.16). These type A climates can be distinguished from all other types by their year-round high temperatures and abundant rainfall.

The yearly mean temperature in the tropical moist climate is generally between 21° C and 27° C (70° F and 80° F). The annual range of temperature is small and considerably less than the daily variation. Slight seasonal variations in the angle of the sun's rays and the length of day are the primary controls on temperature, resulting in a yearly temperature range that is less than the daily range.

The amount of precipitation in the tropical climates is quite significant because it rarely falls below 88.9 centimeters (35 in) annually. Instability in the tropical air masses usually peaks during the late afternoon and evening, resulting in the numerous convectional thunderstorms that supply the A climates with abundant precipitation. In some A climates, hurricanes also produce considerable precipitation.

Because rainfall varies in amount and distribution with the time of year, with the locale, and from year to year, precipitation patterns are the basis for subdividing the tropical climates. All of the A climate subdivisions receive sufficient moisture to be classified as humid climates. Some of these are represented by excessively wet weather patterns.

The relationship between the total precipitation for the year and the amount that falls during the driest month determines the different varieties of A climates. When there is no drought period, the climate is tropical rain forest (Af). If an area displays a definite drought during one season and rainfall during another, the climate is tropical savanna (Aw). In some locations, there is a brief drought and significant rainfall throughout the remainder of the year, producing a tropical monsoon climate (Am).

Box 6.1 Further Consideration

Classification of Climates

Köppen recognized five major climatic groups, which correspond to five principal vegetation groupings. He designated a capital letter for each of the five. Further subdivisions of the major climates are represented by other mostly lower-case letters. The climate formula modified by Trewartha is descriptive, giving the characteristics of the average monthly and annual temperatures and the total monthly and yearly precipitation. It includes also the seasonal trend of these two elements. The five groups are as follows:

A: Tropical rainy climates with no cool season; the average temperature of the coolest month is above 18° C (64.4° F).
Af: Tropical rain forest climate; **f** = rainfall of the driest month is at least 6 centimeters (2.4 in).
Aw: Tropical savanna climate; **w** = distinct dry season in winter or low sun period. There is at least 1 month with less than 6 centimeters (2.4 in).
Am: Tropical monsoon climate; **m** = monsoon-type climates are intermediate between Af and Aw, resembling Af in the amount of rainfall and Aw in the seasonal distribution. Precipitation of the driest month is less than 6 centimeters (2.4 in), but it is more than the amount computed from the following formula:

$$a = 3.94 - r/25$$

where *a* is the rainfall of the driest month and *r* is the annual rainfall.
B: Dry climates; the common characteristics of the B climates is their aridity. The potential evaporation is greater than the rainfall. Köppen's formulas for identifying dry climates involve not only the annual average temperature and the total annual rainfall but also the season of maximum precipitation.
BW: Arid climate or desert
BS: Semiarid climate or steppe

Table 1.A

	Boundary Between BS and Humid Climates	Boundary Between BW and BS
Rainfall evenly distributed throughout year	$r = 0.44t - 8.5$	$r = \dfrac{0.44t - 8.5}{2}$
Rainfall maximum in summer (70% or more of yearly total falls in warmest 6-month period)	$r = 0.44t - 3$	$r = \dfrac{0.44t - 3}{2}$
Rainfall maximum in winter (70% or more of yearly total falls in coolest 6-month period)	$r = 0.44t - 14$	$r = \dfrac{0.44 - 14}{2}$

Other letters used with the B designation are:
h = average annual temperature over 18° C (64.4° F).
k = average annual temperature under 18° C (64.4° F).

In Köppen's formulas for identifying a steppe (BS) and a desert (BW) margin, *r* is the annual rainfall in inches and *t* is the average annual temperature in degrees Fahrenheit.
C: Mesothermal climates; the average temperature of the coldest month is below 18° C (64.4° F).
Cf: No distinct dry season; the driest month of summer receives more than 3 centimeters (1.2 in).
Cw: Dry winter; at least ten times as much precipitation in the wettest month of summer as in the driest month of winter.
Cs: Dry summer; at least three times as much precipitation in the wettest month of winter as in the driest month of summer, and the driest month of summer receives less than 3 centimeters (1.2 in).

A third set of letters used with *both* C and D climates is the following:
a = hot summer; average temperature of the warmest month over 22° C (71.6° F).
b = cool summer; average temperature of the warmest month under 22° C (71.6° F).
c = cool short summer; less than 4 months over 10° C (50° F).
d = cool, short summer with coldest winter month averaging below -38°C (-36.4°F)

D: Microthermal; the average temperature of the coldest month is below 0° C (32° F), and the average temperature of the warmest month is above 10° C (50° F).
Df: cold climate with humid winter
Dw: cold climate with dry winter
E: Polar climates; the average temperature of the warmest month is below 10° C (50° F)
ET: Tundra climate; average temperature of the warmest month is below 10° C (50° F) but above 0° C (32° F)
EF: Perpetual frost; average temperature of all months is below 0° C (32° F)

Apply Your Knowledge

1. What would a Csa climate represent?
2. What would a BSk climate represent?
3. Can a desert be cold?
4. What differentiates the C and D climates?

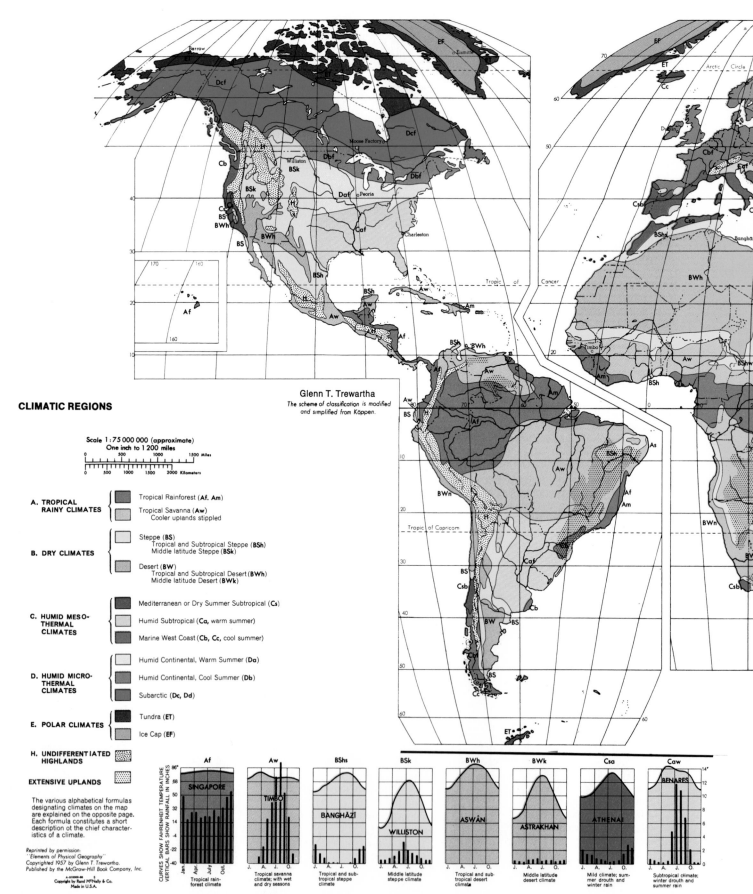

Figure 6.15 The Köppen classification as modified by Glenn T. Trewartha. (From Glenn T. Trewartha, *Elements of Physical Geography.* Copyright ©1957, Glenn T. Trewartha. Reprinted by permission of McGraw-Hill Book Company, Inc.)

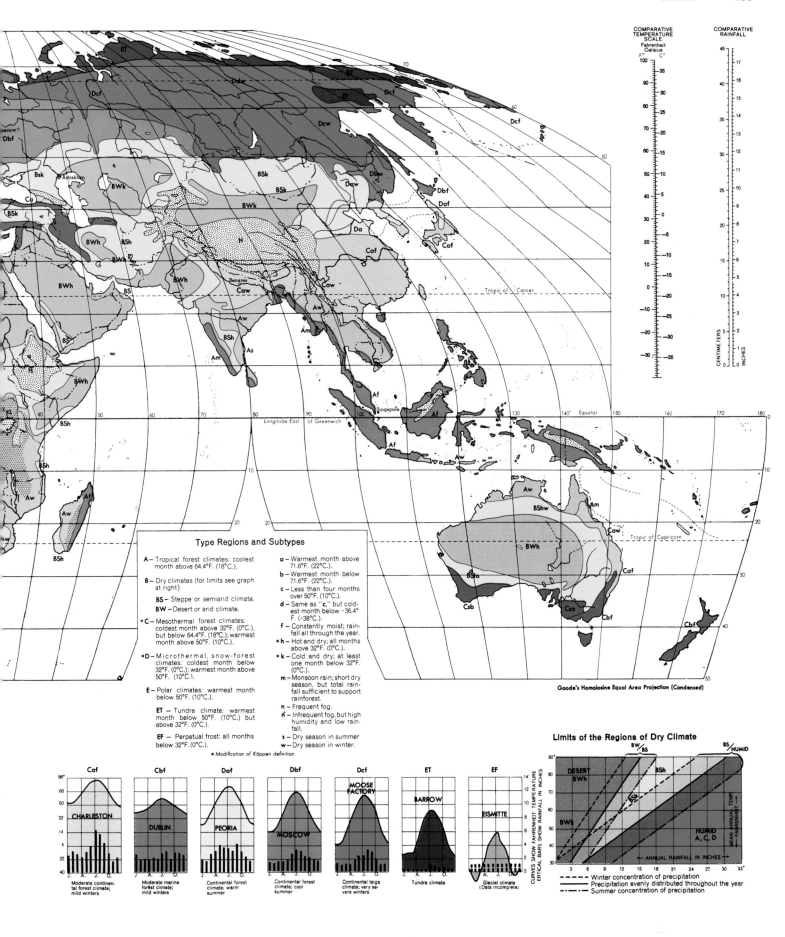

COMPARATIVE
TEMPERATURE
SCALE
Fahrenheit
Celsius

COMPARATIVE
RAINFALL

Goode's Homolosine Equal Area Projection (Condensed)

Type Regions and Subtypes

A – Tropical forest climates: coolest month above 64.4°F. (18°C.).

B – Dry climates (for limits see graph at right)

   BS – Steppe or semiarid climate.

   BW – Desert or arid climate.

*C – Mesothermal forest climates: coldest month above 32°F. (0°C.), but below 64.4°F. (18°C.); warmest month above 50°F. (10°C.).

*D – Microthermal, snow-forest climates: coldest month below 32°F. (0°C.); warmest month above 50°F. (10°C.).

E – Polar climates: warmest month below 50°F. (10°C.).

   ET – Tundra climate: warmest month below 50°F. (10°C.) but above 32°F. (0°C.).

   EF – Perpetual frost: all months below 32°F. (0°C.).

* Modification of Köppen definition

a – Warmest month above 71.6°F. (22°C.).

b – Warmest month below 71.6°F. (22°C.).

c – Less than four months over 50°F. (10°C.).

d – Same as "c," but coldest month below -36.4° F. (-38°C.).

f – Constantly moist; rainfall all through the year.

*h – Hot and dry; all months above 32°F. (0°C.).

*k – Cold and dry; at least one month below 32°F. (0°C.).

m – Monsoon rain; short dry season, but total rainfall sufficient to support rainforest.

n – Frequent fog.

n' – Infrequent fog, but high humidity and low rainfall.

s – Dry season in summer.

w – Dry season in winter.

Cbf
DUBLIN
Moderate marine forest climate; mild winters

Caf
CHARLESTON
Moderate continental forest climate; mild winters

Daf
PEORIA
Continental forest climate; warm summer

Dbf
MOSCOW
Continental forest climate; cool summer

Dcf
MOOSE FACTORY
Continental taiga climate; very severe winters

ET
BARROW
Tundra climate

EF
EISMITTE
Glacial climate (Data incomplete)

CURVES SHOW FAHRENHEIT TEMPERATURE
VERTICAL BARS SHOW RAINFALL IN INCHES

Limits of the Regions of Dry Climate

DESERT
BWh

BWk

BSk

BSh

HUMID
A, C, D

MEAN ANNUAL TEMPERATURE
FAHRENHEIT

ANNUAL RAINFALL IN INCHES

- – – Winter concentration of precipitation
——— Precipitation evenly distributed throughout the year
– · – · – Summer concentration of precipitation

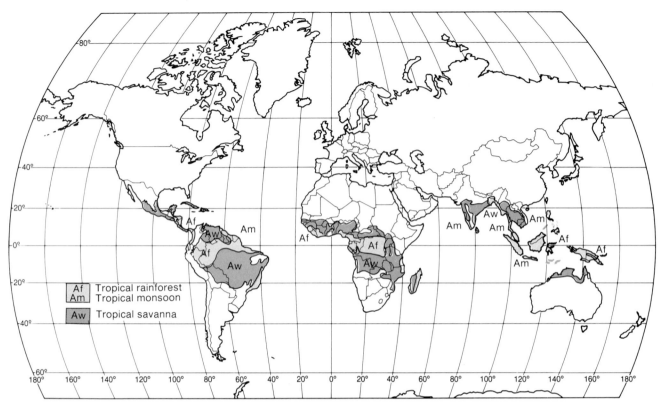

Figure 6.16 Type A climates.

Figure 6.17 A tropical rain forest.

Tropical Rain Forest (Af)

Tropical rain forests lie in the region of the equatorial low and the windward slopes of the trade winds (fig. 6.16). Table 6.4 summarizes the characteristics of the tropical rain forest climatic type. The trees in tropical rain forests may reach 45 meters (150 ft), forming a high canopy that contains the additional foliage of *epiphytes,* plants that grow on other plants (fig. 6.17).

Iauarete, Brazil, is encompassed by a tropical rain forest. Its equatorial location yields a heavy, evenly distributed pattern of rainfall (fig. 6.18). Typical of Af climates, the station exhibits a very small annual temperature range.

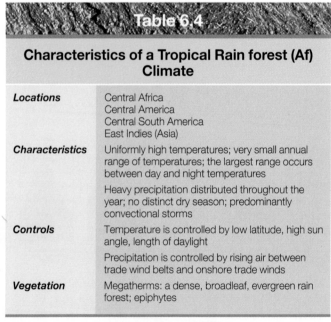

Table 6.4	
Characteristics of a Tropical Rain forest (Af) Climate	
Locations	Central Africa Central America Central South America East Indies (Asia)
Characteristics	Uniformly high temperatures; very small annual range of temperatures; the largest range occurs between day and night temperatures
	Heavy precipitation distributed throughout the year; no distinct dry season; predominantly convectional storms
Controls	Temperature is controlled by low latitude, high sun angle, length of daylight
	Precipitation is controlled by rising air between trade wind belts and onshore trade winds
Vegetation	Megatherms: a dense, broadleaf, evergreen rain forest; epiphytes

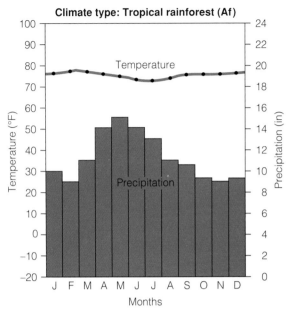

Climate type: Tropical rainforest (Af)

Station name: Iauarete, Brazil
Total precipitation: 138.03 in (350.6 cm)
Annual temperature: 77° F (25° C)
Range: 2.5° F (1.4° C)

Figure 6.18 Climagraph of Iauarete, Brazil.

Tropical Savanna (Aw)

Located poleward of tropical rain forests, the tropical savannas are transition zones between humid and dry conditions (fig. 6.16). Table 6.5 presents the major characteristics of an Aw climate, and figure 6.19 shows the type of grasslands called *savanna* commonly found under these climatic conditions.

A tropical savanna can be found at Cuiabá, Brazil, located 15°S of the equator (fig. 6.20). The station displays a seasonal rainfall pattern and three temperature periods: the hot, wet season during the rains (December through March); the cool, dry season at the time of low sun (April through August); and the hotter dry season just preceding the rains (September through November).

Figure 6.19 A tropical savanna.

Table 6.5

Characteristics of a Tropical Savanna (Aw) Climate

Locations	Northern Australia South and southeast Asia Central America Coastal South America
Characteristics	Temperature shows seasonal variations with highest temperatures coinciding with time of highest sun; small annual temperature range
	Precipitation shows a seasonal distribution with moderate annual amounts; precipitation amounts peak during high sun and diminish with the low sun drought
Controls	Temperature is controlled by tropical location, sun angle, and length of day
	Precipitation is controlled by the migration of the ITCZ; wet season is associated with mT air and convectional showers; dry season is dominated by a subtropical high
Vegetation	Savanna grasslands: growth of tall grasses with widely spaced trees

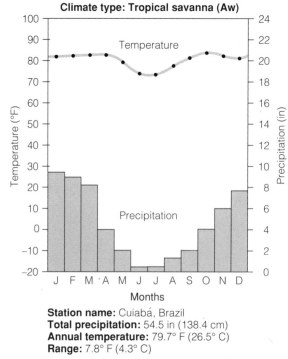

Climate type: Tropical savanna (Aw)

Station name: Cuiabá, Brazil
Total precipitation: 54.5 in (138.4 cm)
Annual temperature: 79.7° F (26.5° C)
Range: 7.8° F (4.3° C)

Figure 6.20   Climagraph of Cuiabá, Brazil.

Figure 6.21   Environment associated with a tropical monsoon climate.

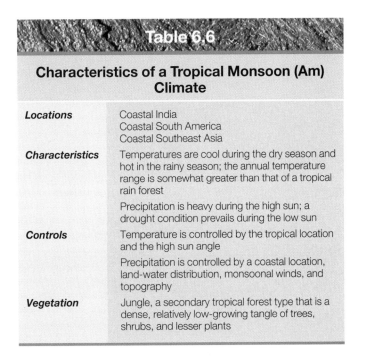

Table 6.6

Characteristics of a Tropical Monsoon (Am) Climate

Locations	Coastal India Coastal South America Coastal Southeast Asia
Characteristics	Temperatures are cool during the dry season and hot in the rainy season; the annual temperature range is somewhat greater than that of a tropical rain forest
	Precipitation is heavy during the high sun; a drought condition prevails during the low sun
Controls	Temperature is controlled by the tropical location and the high sun angle
	Precipitation is controlled by a coastal location, land-water distribution, monsoonal winds, and topography
Vegetation	Jungle, a secondary tropical forest type that is a dense, relatively low-growing tangle of trees, shrubs, and lesser plants

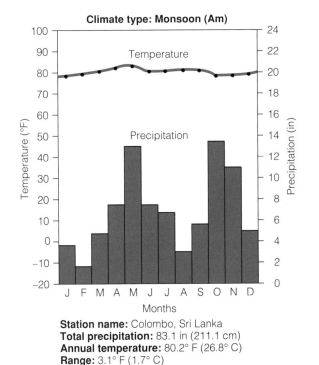

Station name: Colombo, Sri Lanka
Total precipitation: 83.1 in (211.1 cm)
Annual temperature: 80.2° F (26.8° C)
Range: 3.1° F (1.7° C)

Figure 6.22 Climagraph of Colombo, Sri Lanka.

Tropical Monsoon (Am)

The tropical monsoon climate is a modification of the tropical rain forest. Located in the equatorial precipitation belt, this variation occurs mainly near coastal regions where monsoon winds flow onshore (fig. 6.16). These seasonal winds govern the yearly distribution of precipitation. Table 6.6 gives the major characteristics of the Am climate, and figure 6.21 shows an environment associated with a tropical monsoon climate.

Colombo, Sri Lanka, lies in a tropical monsoon climate. The station is located in an area where the monsoon wind system is particularly well developed. The total annual rainfall is heavier than the average for the other tropical climate subtypes (fig. 6.22). Colombo's heaviest rainfall occurs during the warm season. The dry season is neither as dry as nor as long as that of the tropical savanna.

DRY (B) CLIMATES

Aridity is the common characteristic of B climate types. Climatologists define these dry climates as patterns in which a meager precipitation falls in a region with a potential for excessive evaporation. Simply put, evaporation exceeds precipitation.

B climates are widely distributed (fig. 6.23). They are especially concentrated, however, in the vicinity of the subtropical high-pressure belts centered on 30°N and S latitudes. Other locations include regions adjacent to cold currents, the lee sides of mountain ranges, and continental interiors. Dry climates stretch over a wide latitudinal range and occupy a variety of continental positions. Compared to other climates in the same latitude, B climates usually show a wide annual temperature range. Some B climates are cold, whereas others are hot. In either instance, the heat is adequate to provide sufficient energy for extensive evaporation.

Based on precipitation, B climates fall into two patterns. In some areas, B climates are almost devoid of precipitation, and what does fall is insufficient to resupply that lost to evaporation. Such an arid climate is called a desert. Wherever precipitation falls in amounts abundant enough to support a sparse vegetative cover, a variety of B climate known as a *steppe* occurs.

Steppe (BS) or Semiarid Climate

A transitional, semiarid climate that borders the true desert and the humid climates is a steppe (fig. 6.23). Table 6.7 outlines the characteristics of the BS climate. Semiarid climates do support plant growth, but the vegetation is sparse by comparison with that in humid climates (fig. 6.24).

At 39°N, Denver, Colorado, sits within a middle-latitude steppe. The summers are hot (22.2° C, or 72° F), whereas the winter is cold (0° C, or 32° F). The result is a large temperature range (22.2° C, or 40° F). The total annual precipitation is 36.3 centimeters (14.3 in). The yearly precipitation is usually lower than that of a tropical savanna with the rainfall maxima falling in spring and summer (fig. 6.25).

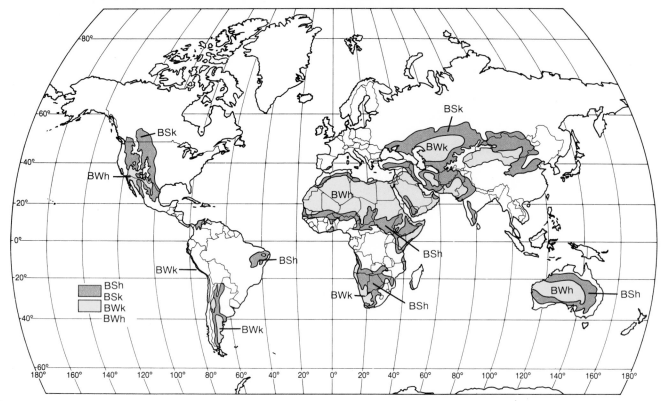

Figure 6.23 Distribution of B climates.

Figure 6.24 Steppe climate—native grasses of the Great Plains.

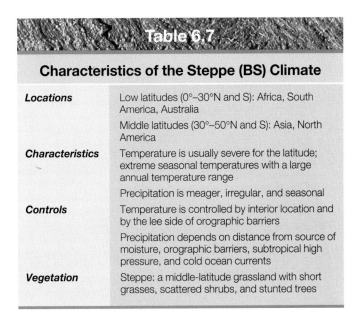

Characteristics of the Steppe (BS) Climate

Locations	Low latitudes (0°–30°N and S): Africa, South America, Australia
	Middle latitudes (30°–50°N and S): Asia, North America
Characteristics	Temperature is usually severe for the latitude; extreme seasonal temperatures with a large annual temperature range
	Precipitation is meager, irregular, and seasonal
Controls	Temperature is controlled by interior location and by the lee side of orographic barriers
	Precipitation depends on distance from source of moisture, orographic barriers, subtropical high pressure, and cold ocean currents
Vegetation	Steppe: a middle-latitude grassland with short grasses, scattered shrubs, and stunted trees

Desert (BW) Climate

The desert, or arid, climate is distributed widely in both the tropical and middle latitudes. Nearly all continents have deserts (fig. 6.23). The characteristics of deserts appear in table 6.8. Arid climates are home to *xerophytic* plants. These plants store water efficiently by inhibiting evapotranspiration (fig. 6.26).

Yuma, Arizona, is situated in a low middle-latitude desert. Its warmest month is July, and its coolest is January. The precipitation is meager (fig. 6.27).

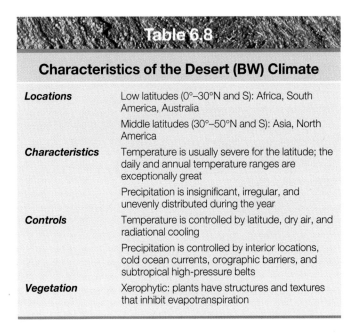

Characteristics of the Desert (BW) Climate

Locations	Low latitudes (0°–30°N and S): Africa, South America, Australia
	Middle latitudes (30°–50°N and S): Asia, North America
Characteristics	Temperature is usually severe for the latitude; the daily and annual temperature ranges are exceptionally great
	Precipitation is insignificant, irregular, and unevenly distributed during the year
Controls	Temperature is controlled by latitude, dry air, and radiational cooling
	Precipitation is controlled by interior locations, cold ocean currents, orographic barriers, and subtropical high-pressure belts
Vegetation	Xerophytic: plants have structures and textures that inhibit evapotranspiration

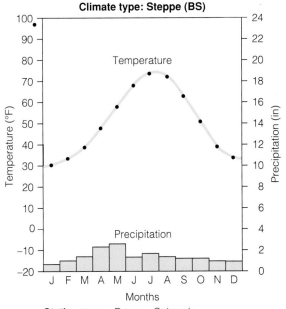

Station name: Denver, Colorado
Total precipitation: 13.8 in (35.1 cm)
Annual temperature: 50.7° F (10.4° C)
Range: 41.7° F (23.2° C)

Figure 6.25 Climagraph of Denver, Colorado.

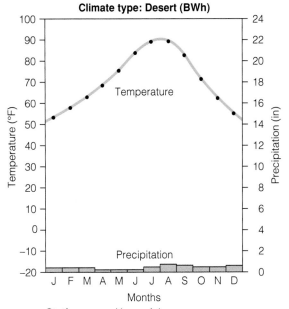

Station name: Yuma, Arizona
Total precipitation: 3.27 in (8.3 cm)
Annual temperature: 71.8° F (22.1° C)
Range: 37.6° F (20.9° C)

Figure 6.26 Climagraph of Yuma, Arizona.

Figure 6.27 Xerophytes—plants designed to survive within zones of low precipitation and high evaporation—can be classified as annuals or perennials. Annuals have a short life cycle, growing, reproducing seed, and dying in one season. Perennials survive over longer periods of time by conserving water until the rains come. Water is stored in the structure of these Joshua trees on the Mojave Desert of California.

HUMID MESOTHERMAL (C) CLIMATES

The humid *mesothermal* (C) climates occupy the subtropical latitudes and some higher, middle-latitudinal positions under the influence of the prevailing westerlies, particularly on the western sides of continents (fig. 6.28). Because of the prevailing winds, marine influence is stronger on the western sides of continents than it is on the eastern sides. Poleward of the tropical climates, the mesothermal climates lack the constant heat of the tropics and the arid conditions of the B climates. Seasonal changes are the most evident characteristic of these climates.

Winter temperatures of the mesothermal climate group are relatively mild. The average monthly winter temperature is above 0° C (32° F). Latitude is an important temperature control. Marine influences are weaker on eastern coasts and stonger on western coasts.

Precipitation patterns vary among mesothermal climates. The humid mesothermal always has sufficient annual precipitation to distinguish it from the dry climates. Precipitation joins temperature as quantifiable characteristics that determine the subtypes of these climates. These subtypes include the humid subtropical (Cfa), the marine west coast (Cfb), and the dry summer subtropical (Cs).

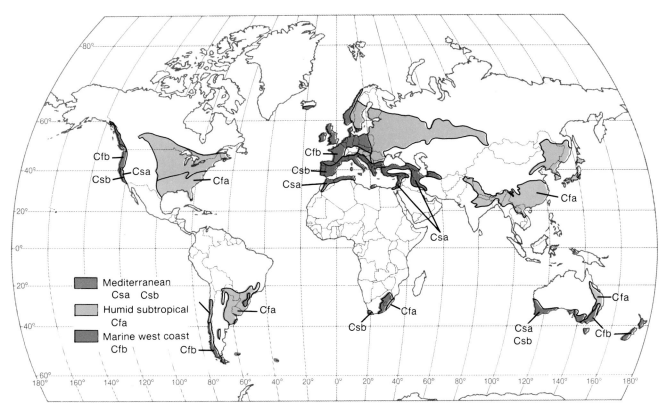

Figure 6.28 In Mediterranean (Csa, Csb) climates, summers are warm and dry, followed by cool, mild, rainy winters. Humid subtropical (Cfa) climates have hot, rainy summers and cool, wet winters. Marine west coast (Cfb) climates are in the storm track of the middle-latitude cyclone.

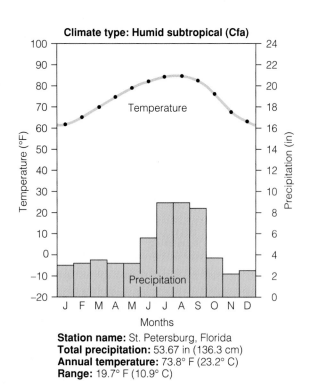

Station name: St. Petersburg, Florida
Total precipitation: 53.67 in (136.3 cm)
Annual temperature: 73.8° F (23.2° C)
Range: 19.7° F (10.9° C)

Figure 6.29 Climagraph of St. Petersburg, Florida.

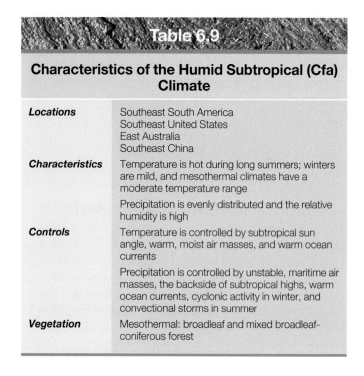

Characteristics of the Humid Subtropical (Cfa) Climate

Locations	Southeast South America Southeast United States East Australia Southeast China
Characteristics	Temperature is hot during long summers; winters are mild, and mesothermal climates have a moderate temperature range
	Precipitation is evenly distributed and the relative humidity is high
Controls	Temperature is controlled by subtropical sun angle, warm, moist air masses, and warm ocean currents
	Precipitation is controlled by unstable, maritime air masses, the backside of subtropical highs, warm ocean currents, cyclonic activity in winter, and convectional storms in summer
Vegetation	Mesothermal: broadleaf and mixed broadleaf-coniferous forest

Humid Subtropical (Cfa) Climate

The humid subtropical climate is located at the equatorward margins of the middle latitudes between 25° and 35°N and S (fig. 6.28). Typically, this subtype of mesothermal climate is situated on the eastern side of landmasses. Table 6.9 gives the characteristics of the Cfa climate. A broadleaf forest is typical of this mesothermal climate.

St. Petersburg, Florida, has mild winters and long, hot, and humid summers (fig. 6.29). The precipitation pattern displays a maximum in summer (June through August).

Dry Summer Subtropical (Cs) Climate

The dry summer subtropical climate is modeled on a significant distribution of its features on lands surrounding the Mediterranean Sea. For this reason, this climate type is also known as a Mediterranean climate. The Cs climates occur in the subtropical latitudes along the western sides of continents adjacent to cold ocean currents (fig. 6.28). The characteristics of this climate type appear in table 6.10. The typical vegetation of a Cs climate is called Mediterranean scrubland (fig. 6.30).

Santa Monica, California, is a community with a cool, mild, rainy winter. Its summer is warm and dry (fig. 6.31). Nearby ocean waters moderate temperatures and maintain a small annual temperature range.

Figure 6.30 Mediterranean scrubland.

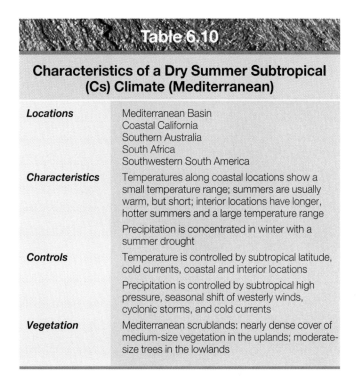

Table 6.10

Characteristics of a Dry Summer Subtropical (Cs) Climate (Mediterranean)

Locations	Mediterranean Basin Coastal California Southern Australia South Africa Southwestern South America
Characteristics	Temperatures along coastal locations show a small temperature range; summers are usually warm, but short; interior locations have longer, hotter summers and a large temperature range
	Precipitation is concentrated in winter with a summer drought
Controls	Temperature is controlled by subtropical latitude, cold currents, coastal and interior locations
	Precipitation is controlled by subtropical high pressure, seasonal shift of westerly winds, cyclonic storms, and cold currents
Vegetation	Mediterranean scrublands: nearly dense cover of medium-size vegetation in the uplands; moderate-size trees in the lowlands

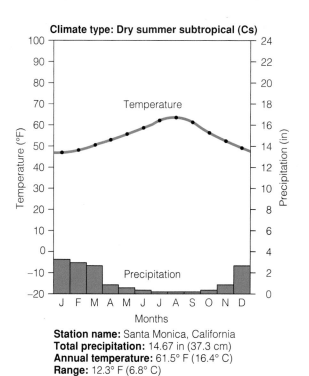

Climate type: Dry summer subtropical (Cs)

Station name: Santa Monica, California
Total precipitation: 14.67 in (37.3 cm)
Annual temperature: 61.5° F (16.4° C)
Range: 12.3° F (6.8° C)

Figure 6.31 Climagraph of Santa Monica, California.

Marine West Coast (Cfb) Climate

Marine west coast climates are characteristically mild. They occupy positions on the western or windward sides of middle-latitude continents, poleward of 40°N and S, where the onshore westerly winds bring predominantly marine conditions to the land (fig. 6.28). Table 6.11 lists the characteristics of the Cfb climate. Stands of Douglas fir are typical of Cfb vegetation (fig. 6.32).

With its elevation of 335 meters (1,100 ft), Valsetz, Oregon, is located in the coastal ranges of Oregon. The city receives a heavy winter precipitation that reflects orographic uplift and cyclonic activity. This station has a contrasting summer precipitation minimum (fig. 6.33). The shift of the subtropical high northward affects rainfall. The marine influence causes abnormally mild winters and cool summers.

Figure 6.32 Stands of Douglas fir typical of marine west coast (Cfb) climates.

Table 6.11

Characteristics of a Marine West Coast (Cfb) Climate

Locations	Northwestern Europe Pacific Northwest of United States British Columbia Australia South Africa
Characteristics	Temperatures are cool during summer and mild during winter; the temperature range is moderate
	Precipitation shows an even distribution, but there are frontal storms that can be heavy with orographic uplifting; cloudiness is common
Controls	Temperature is controlled by latitude and the marine influence
	Precipitation is controlled by marine influence of onshore westerly winds that blow over cold currents
Vegetation	Mesothermal: middle-latitude forest; the North American Cfb is a principal source of high-grade softwoods, such as Douglas fir, California redwoods, and hemlock

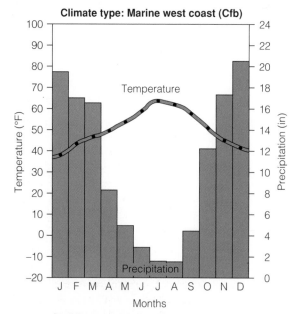

Station name: Valsetz, Oregon
Total precipitation: 14.67 in (37.3 cm)
Annual temperature: 61.5° F (16.4° C)
Range: 12.3° F (6.8° C)

Figure 6.33 Climagraph of Valsetz, Oregon.

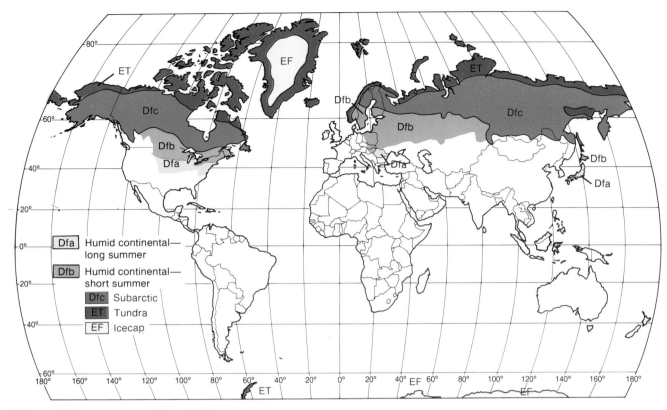

Figure 6.34 Distribution of D and E climates.

Humid Microthermal (D) Climates

Humid, *microthermal* climates are located in continental interiors and along eastern coasts poleward of the C climates in the Northern Hemisphere (fig. 6.34). Colder, longer winters and greater seasonal temperature ranges characterize these severe winter climates. Because the D climates result from extensive interior areas and from coasts without strong maritime air influences between 40° and 65°N, they do not occur in the Southern Hemisphere.

Close similarities among the subtypes of D climates produce classifications with subtle variations. Generally, precipitation varies by increasing from west to east and from north to south. The subtypes include those identified by long summers (Dfa), short summers (Dfb), and two subarctic environments (Dfc and Dfd).

Humid, Continental Long Summer (Dfa) Climate

Eastern Europe, central United States, and central China are extensive, middle-latitude continental areas with long, hot summers and severe winters. Precipitation in these areas is either evenly distributed throughout the year or concentrated in warmer months. The temperatures of Dfa regions result from length of day, sun angle, and the absence of maritime air masses. Precipitation occurs during cyclonic activity. The Dfa climate supports a mixed forest (fig. 6.35).

Humid, Continental Short Summer (Dfb) Climate

The humid, continental short summer climate of the high middle latitudes occurs on both North America and Europe. Although its seasonal temperature range is great, the Dfb climate has a warm, not a hot, summer. The winter cooling of the continent is a strong temperature control. Precipitation patterns mimic those of the Dfa, but the forests are dominated by conifers (fig. 6.36).

Figure 6.35 A mixed forest in autumn.

Figure 6.36 Conifer forest typical of humid, continental, short summer (Dfb) climates.

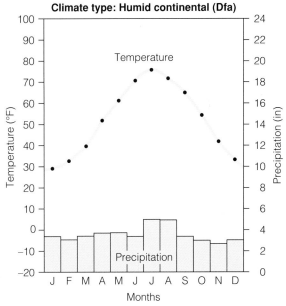

Figure 6.37 Climagraph of Allentown, Pennsylvania.

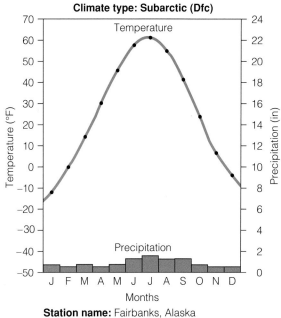

Figure 6.39 Climagraph of Fairbanks, Alaska.

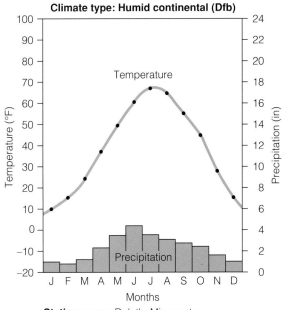

Figure 6.38 Climagraph of Duluth, Minnesota.

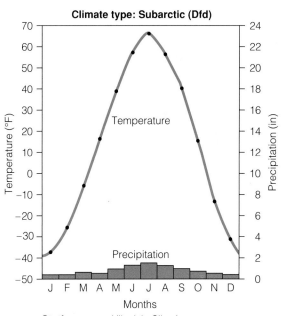

Figure 6.40 Climagraph of Vilyuisak, Siberia.

Subarctic (Dfc and Dfd) Climates

Two variations of high middle-latitude climates occur in northern Canada, Alaska, northern Asia, and northern Europe. Both are typified by larger temperature ranges and short, cool summers, and both are influenced by high-pressure systems and cold, continental air masses. The vegetation of these D climates is called the *taiga,* a moist coniferous forest.

Allentown, Pennsylvania, has a typical Dfa climate (fig. 6.37). Duluth, Minnesota, which lies farther north and west of Allentown, has a Dfb climate (fig. 6.38). Its interior location prevents maritime air from moderating its temperatures. Fairbanks, Alaska, has the short and cool summers of a Dfc climate (fig. 6.39). Located beneath a great winter high-pressure system is Vilyuisak, Siberia, the site of a Dfd climate (fig. 6.40).

POLAR (E) CLIMATES

Some areas on Earth's surface have continuously cold and dry weather (fig.6.34). Bitter winter temperatures and the lowest mean annual temperatures mark the E climates as the most severe on our planet. Even during summer, all sites within E climates experience low temperatures controlled by low sun angles and the energy absorption required to melt ice and snow. E climates have a low specific humidity that produces a meager precipitation. The Köppen system divides these climates on the basis of temperature. Where the warmest summer temperatures average 0° C (32° F), the climate supports the arctic tundra. Colder areas are covered by an ice cap.

The Tundra (ET) Climate

The tundra lies in a region with no true summer (table 6.12). Temperatures briefly remain above freezing during the warm season, permitting a ground cover of mosses, lichens, and grasses (fig. 6.41). During most of the year these plants lie under a covering of snow.

Cape Pembroke in the Falkland Islands has an E climate influenced by maritime conditions. The seasonal temperatures differ from those under continental influences. Cool summers and

Table 6.12

Characteristics of the Tundra (ET) Climate

Location	Arctic fringe
Characteristics	Temperatures are cool during high sun months; under maritime influences temperature ranges are small; under continental influences temperature ranges are large
	Precipitation is meager with a warm season concentration
Controls	Temperature is controlled by latitude, low sun angle, length of daylight, and continental or maritime location
	Precipitation is controlled by high pressure, low evaporation, and cool to cold air with little capacity for moisture
Vegetation	Tundra

Figure 6.41 Tundra vegetation.

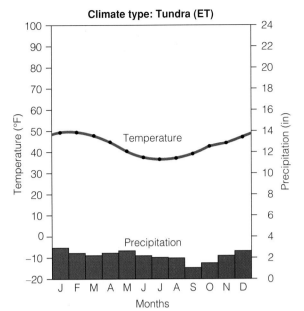

Climate type: Tundra (ET)

Station name: Cape Pembroke, Falkland Islands
Total precipitation: 26.3 in (66.8 cm)
Annual temperature: 42.7° F (5.9° C)
Range: 12.6° F (7.0° C)

Figure 6.42 Climagraph of Cape Pembroke, Falkland Islands.

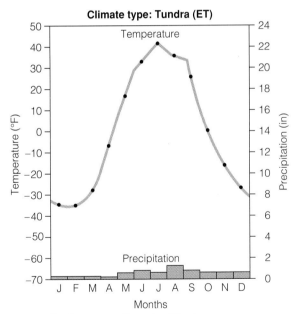

Climate type: Tundra (ET)

Station name: Sagastyr, Siberia
Total precipitation: 3.3 in (8.4 cm)
Annual temperature: 1° F (–17.2° C)
Range: 77° F (42.8° C)

Figure 6.43 Climagraph of Sagastyr, Siberia.

cold winters produce a small temperature range. Precipitation is enhanced by the maritime air (fig. 6.42). Sagastyr, Siberia, is a tundra under continental influence. Annual temperature ranges are large because of bitterly cold winters. Precipitation is meager (fig. 6.43).

Ice Cap (EF) Climate

Ice cap environments have small temperature ranges and very cold temperatures (table 6.13, fig. 6.44). A meager precipitation falls as snow and turns to ice. Continental ice sheets, such as those on Greenland and Antarctica, reach thicknesses in excess of 3.2 kilometers (2 mi).

Eismitte, Greenland, is located at 3,030 meters (9,941 ft) on top of a thick ice sheet (fig. 6.45). Its interior location and high altitude prevent temperatures from rising above freezing. Eismitte experiences a large annual temperature range (36° C or 65° F) and meager annual precipitation (11 cm, or 4.3 in).

Table 6.13

Characteristics of the Ice Cap (EF) Climate

Locations	Central Greenland Antarctica
Characteristics	Temperatures are the lowest on Earth; coldest temperature ever recorded on Earth's surface was –59.7° C (–127° F) near South Pole
	Precipitation is meager and nearly always falls as snow during the warmer months; cyclonic activity generates most of the snowfall
Controls	Temperature is controlled by the high albedo of snow and ice, low sun angle, dry air, clear skies (radiational cooling), semipermanent high pressure, long dark period
	Precipitation is controlled by high pressure, low evaporation, and cold air moisture capacity

Figure 6.44 Polar ice cap.

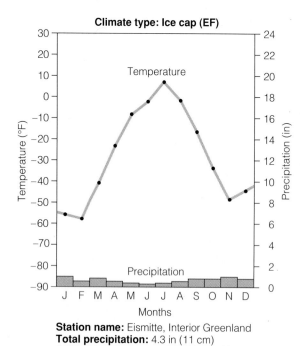

Climate type: Ice cap (EF)

Temperature

Precipitation

Months

Station name: Eismitte, Interior Greenland
Total precipitation: 4.3 in (11 cm)
Annual temperature: −22° F (−30° C)
Range: 65° F (36.1° C)

Figure 6.45 Climagraph of Eismitte, Greenland.

HIGHLAND (H) CLIMATES

Highland climates, represented by an H on climate maps, have highly varying temperature and moisture characteristics. As you learned in the section on altitude earlier in this chapter, the vertical component of climate is in many ways a mirror of the latitudinal climate zones. At higher elevations, more polelike conditions are experienced. Highland climates are influenced by latitude, land-water distribution, orographic lifting and adiabatic changes, and, of course, altitude.

Box 6.2 Investigating the Environment

El Niño and the Southern Oscillation

El Niño is a warming of the surface waters of the southeastern Pacific Ocean. A 2° C to 3° C (3.6° F–5.4° F) warming off the coasts of Peru and Ecuador usually occurs at Christmastime; thus, it is "The Child." As an event, El Niño occurs irregularly every 2 to 7 years, but when it is accompanied by a warming of more than 6° C (10.8° F), it significantly influences weather throughout the world, overriding the climatic stability that underlies the Köppen system.

The Southern Oscillation is a fluctuation in the pressure systems of the western Pacific, where southern Asia is dominated by alternate high and low pressure. Because the Southern Oscillation cycles over approximately 30-month periods, it has been quantified in an index by Sir Gilbert Walker, who observed the changes in pressure as early as 1924. Walker developed an index for the Southern Oscillation that is calculated by subtracting atmospheric pressure in the western Pacific from the pressure of the eastern Pacific. Whenever the index value is positive, the difference between east and west is higher than normal (fig. 6.A). Both El Niño and the Southern Oscillation (usually referred to as ENSO) affect climate and human activities. Successful farming in India has been linked to the Southern Oscillation, and fishing success can be interrupted in the Pacific with the occurrence of El Niño.

The warm water associated with El Niño appears to impact on global weather patterns. El Niño has been associated with drier weather in southeastern Africa, India, the western Pacific, and northern South America. During an El Niño event, the central Pacific and southeastern United States experience wetter than normal conditions. The 1982–83 El Niño was responsible for floods in California and an intensification of the African drought. An El Niño during the winter of 1991–92 appeared to be responsible for drenching Los Angeles, particularly during three storms during a 10-day period in February, when more than 20 centimeters (8 in) of rain generated floods and mudslides that killed several people. In some mountainous areas of southern California, more than 38 centimeters (15 in) of rain fell.

Apply Your Knowledge

1. What is the El Niño?
2. What is the Southern Oscillation?
3. During what time of year does the El Niño occur?
4. How frequently does an El Niño occur?

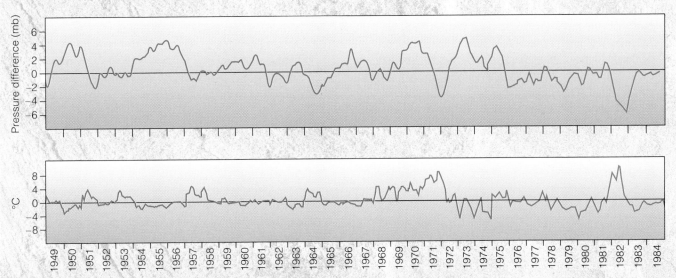

Figure 6.A Historical records of the Southern Oscillation index *(top)* and of the sea-surface temperature off the Peruvian coast *(bottom)* show the two variables are linked, although the association is not perfect. The Southern Oscillation index is calculated by subtracting the atmospheric surface pressure at Darwin, Australia, from the pressure at Easter Island. When the index is positive, the east-west pressure gradient is greater than normal and the trade winds are stronger; conversely, a negative index indicates weak trade winds. The graph shown here is based on monthly averages of the surface pressure and has been statistically "smoothed." The graph of the sea-surface temperature at Puerto Chicama, Peru (7.7°S), shows the departure from normal monthly average temperatures. According to an analysis of these two and other variables by William H. Quinn of Oregon State University, strong Niños occurred in 1957–58, 1972–73, and 1982–83. Moderate Niños occurred in 1953, 1965, and 1976–77. Except for the late-starting 1982 event, the anomalous temperatures of the sea surface off Peru have tended to begin around March, after the Southern Oscillation index has fallen from positive values and the trade winds have slackened. Late in the year the sea surface cools; then the warming resumes, climaxing early the following year. Some Niños, such as that of 1972–73, were preceded by an intensification of the trade winds. A wind buildup in 1974, however, was not followed by a Niño, and the 1982 event began without a buildup. (From "El Niño" by Colin S. Ramage. Copyright © June 1986 by Scientific American, Inc. All rights reserved.)

Further Thoughts: Climatology Is Holistic

Climatology is a model of interconnected systems. It binds sun, air, ocean, and land in an encompassing explanation of ongoing processes, and it simplifies the complexity of weather variability over Earth's surfaces. Because many weather observers have kept accurate records of local weather conditions for decades, climatologists have been able to classify weather patterns on the basis of temperature and precipitation.

Each of the four earth science disciplines offers a unification of principles, processes, and features. In astronomy, the ultimate unification lies in identifying the nature of distant and nearby celestial objects and in explaining the initial formation of the universe in the big bang. In meteorology, the unification lies in climatological classifications. But what about the solid Earth? Does it lend itself to a similar classification scheme? Can the mountains and valleys and their rocks also fit into a unifying model? At the end of part 4, you will encounter a geological analog of climatology in your study of plate tectonics.

Significant Points

1. The atmosphere is a major influence on natural landscapes and life.
2. The composite of weather, or the generalization of weather patterns, is climate, which is based on data collected over a minimum of 30 years.
3. Temperature and moisture are the primary weather elements that climatologists use to classify tropospheric patterns.
4. Cool weather predominates in high latitudes and warm weather in the low latitudes because distance from the equator is the major control on climate. The angle of incidence for insolation changes with changing latitude, so the sun shines with less intensity on an area at higher latitudes than at lower latitudes.
5. Semipermanent high- and low-pressure systems influence the longitudinal distribution of climate patterns.
6. Warm ocean currents transfer heat from low latitudes to higher latitudes. Where cold currents lie off the western sides of continents, dry conditions prevail.
7. Temperature ranges and extremes are affected by the distribution of land and water surfaces.
8. The A, C, and D climates are relatively wet, whereas the B and E climates are relatively dry.
9. Storms transfer energy and moisture latitudinally and longitudinally.
10. Altitude offsets the effects of latitude.
11. Orographic barriers rob the air of its moisture on their windward sides.
12. The distribution of land and water surfaces determines whether or not air masses are moist or dry.

Essential Terms

climate   118
latitude   119
angle of incidence   119
subtropical highs   123

isotherms   127
mesothermal   141
microthermal   145

Review Questions

1. How does the atmosphere influence life?
2. What is climate?
3. What are the two weather elements upon which climate is determined?
4. How is sun angle related to the position and strengths of semipermanent high-pressure systems?
5. How does the Alaska Current influence climate?
6. What is the difference between temperature ranges over land and water?
7. What are the climatic controls?
8. What do the capital letters A, B, C, D, and E stand for in the Köppen-Trewartha climatic system?
9. What type of climate supports the growth of epiphytes?
10. Why are Mediterranean-like conditions found along western North America?

Challenges

1. Identify three factors that could change future global precipitation patterns.
2. What are some of the reasons that make it difficult to predict whether Earth's climate will become cooler or warmer?
3. What human activities are contributing to the potential changes in temperature and precipitation patterns?
4. Identify the major controls of climate that influence your geographic area?
5. Explain how fluctuations in solar output and the Earth-sun relationship may account for climatic changes.

4

GEOLOGICAL COMPOSITION AND PROCESSES

*I*n the eighteenth century James Hutton, called by many the father of geology, recognized that our planetary surface is dynamic. Since the time of his observations and insights, we have learned much about the composition of and processes that shape the solid and not-so-solid Earth. Aided by extensive laboratory and field work and by remote sensing, geologists have discovered much about the solid components of our planet, their distribution, and their interactions.

AT THE END OF PART 4 YOU WILL BE ABLE TO

1. identify the variable compositions of minerals and rocks;
2. infer the distribution of elements in and on our planet;
3. distinguish the surface processes that shape the landscape;
4. identify and illustrate the internal processes that shape the planet's ocean basins and continents.
5. recognize how plate tectonic theory accounts for numerous geological phenomena.

The Brocade of the Mountain

from Tokaido by Ichiryusai Hiroshige (1797–1858)

As little by little the autumn mists
Cleared away in the skies,
I saw the mountain's bright brocade
Woven up before mine eyes.

7

EARTH MATERIALS

Crater Lake, Oregon, a caldera formed by the explosive eruption and collapse of Mount Mazama approximately 6,600 years ago.

Chapter Outline

INTRODUCTION

The solid Earth upon which we walk and from which we obtain our building materials and mineral resources is composed of various kinds of rock. The rock runs deep beneath our feet, in some places as much as 80 or 100 kilometers (50 or 60 mi) before a somewhat different type of material is encountered. The general nature and broad origins of rock are well known, and the hundreds of varieties resolve themselves into a few clusters of similar types. Rocks, in turn, are composed of combinations of atoms and molecules that form substances called minerals. Like rocks, the thousands of identified varieties of minerals can be separated into a small number of groups based on chemical and physical similarities.

Almost everyone has heard of minerals in association with foods; for example, cereal boxes that proclaim the product to be "full of essential vitamins and minerals." Such "minerals" include zinc oxide, manganese sulfate, and potassium chloride, and they represent combinations of elements that allow our bodies to maintain their forms and functions. They are essential to the well-being of organisms, but they are not, in themselves, organic. While it is increasingly commonplace to supplement foods with "minerals" during processing, nature has been doing this for millions of years, and organisms through countless generations have evolved to a state of equilibrium with and dependence on what nature provides. To the earth scientist, however, the term "mineral" has a much broader meaning.

MINERALS

Minerals are nature's building blocks. Rocks are made of minerals, and rocks form Earth's outer shell. Earth scientists—specifically, mineralogists—have named and described more than 3,000 different minerals, and they are always on the lookout for new varieties. Some minerals, such as halite and quartz, are common and familiar; others, like gem minerals, are rare and either of great value or unfamiliar (or both). Regardless of their abundance or scarcity, they all share certain features. To understand these common features, we must examine the definition of a mineral. *A mineral is a naturally occurring inorganic element or compound, with a definite internal arrangement of ions* (i.e., a solid) *and a chemical composition that is fixed or that varies within narrow limits.*

Minerals are "naturally occurring" because they form as a result of natural processes rather than as a product of human laboratories and factories. Thus, we would exclude from consideration as minerals such materials as steel, cement, glass, and, if we carried this line of reasoning to its logical conclusion, synthetic rubies, even though such rubies would be identical to those that form naturally.

In legal documents, such as deeds, and in newspaper and magazine articles, coal, oil, and natural gas are often referred to as mineral fuels. Such fuels represent the remains of once-living organisms (e.g., microscopic marine organisms or trees and ferns in ancient swamplands). Consequently, they are *or-*

ganic (the result of life processes) in origin rather than *inorganic*. Mineral fuels are often considered part of a country's or a state's mineral resources, but under the strict definition, they are not minerals.

Because a mineral can be either a single element or a molecular compound, mineralogists express a mineral's composition precisely by using standard chemical symbols. For example, a single molecule of the mineral halite, or common rock salt, is written as NaCl, meaning that it consists of a single ion of sodium (Na^{1+}) and a single ion of chlorine (Cl^{1-}). Figure 7.1*a* represents an ionic model of NaCl. The ions, shown in their relative sizes, are held together by the attractive force of their opposite electrical charges. The expanded view shown in figure 7.1*b* depicts the ions in their relative positions and also isolates the basic structural unit that is repeated to form the mineral.

Composition and Classification

While there are 92 naturally occurring elements, only 8 form the great bulk of the minerals at or near Earth's surface (table 7.1). Perhaps surprisingly, the most abundant of the 8 is oxygen, which, when uncombined with other elements, occurs as a gas. In combination with other elements, however, oxygen helps to form a variety of minerals, some of which are valuable natural resources (discussed in chapter 13). An example of a valuable oxygen mineral is magnetite, an important source of iron. A more familiar oxygen mineral is ice.

The chemical nature of minerals also provides earth scientists with a convenient method to sort them into smaller groups, a common procedure known as classification. Some minerals are formed of ions of a single element. These are the uncombined, or *native, elements* (fig. 7.2*a*).

Some minerals are formed by combining various elements with oxygen, the *oxides;* with sulfur, the *sulfides;* or with one of the family of halogen elements (chlorine, iodine, fluorine, bromine), the *halides.* Other minerals are formed by combining certain elements with the carbonate complex ion, CO_3^{2-}; the sulfate complex ion, SO_4^{2-}; or the phosphate complex ion, PO_4^{3-} (fig. 7.2*b*).

The Silicate Minerals

Easily the most diverse and abundant of all the groupings, however, is the family of minerals known as the *silicates*. These minerals combine various elements with a basic building block of one ion of silicon and four ions of oxygen, known as the silicon-oxygen tetrahedron (fig. 7.3*a*). Minerals of increasingly complex composition and structure can be formed when two or more of these tetrahedra share oxygen ions (fig. 7.3*b, c,* and *d*).

Included within the silicate group are such common minerals as quartz and feldspar (fig. 7.4). In fact, so common are these minerals that the silicates are collectively known as the rock formers. More than 90% of all minerals found in nature belong to the silicate group, and these minerals, in turn, account for more than 90% of the rocks in Earth's outer shell.

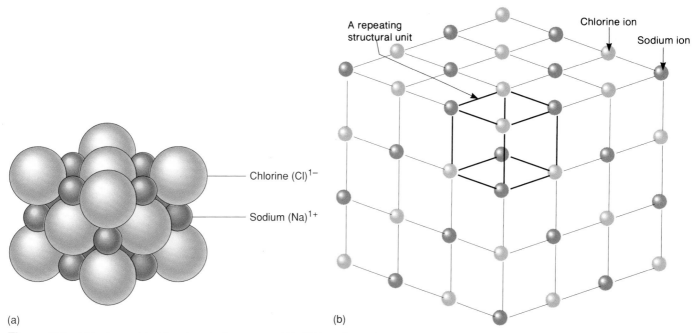

Figure 7.1 (*a*) Ionic model of the chemical compound NaCl (the mineral halite). The ions, shown in their relative sizes, are held together by the attractive force of their electrical charges. (*b*) Lattice model of NaCl expanded to show the positions of the Na and Cl ions but not their relative sizes.

Figure 7.2 Mineral specimens: (*a*) graphite (carbon), an example of a native, or uncombined element; and (*b*) an intergrowth of the light-colored carbonate mineral calcite ($CaCO_3$) and the dark-colored halide mineral fluorite (CaF_2).

Mineral Formation

Minerals form in nature by a process known as **crystallization.** That is, individual ions present in natural solutions join together to form molecules of solid compounds that grow larger by the addition of more ions. Some crystals form in the moderate temperatures at Earth's surface. Water molecules, for example, join together in the atmosphere to form the ice crystals we call snowflakes. Evaporating seawater causes dissolved ions of sodium (Na^{1+}) and chlorine (Cl^{1-}) to come close enough to attract one another and form solid particles of halite. Most of the common minerals, however, form from neither air nor water solu-

tions but rather from rock solutions, that is, liquid masses formed from mineral matter at high temperatures deep within Earth. This liquid material is known as **magma.**

As magma cools, the individual ions that have been moving about in the magma lose energy and slow down. As the ions slow, they come together and attach themselves to one another in a regularly repeated arrangement. This regular pattern of ions is known as a *crystal lattice* and it continues to grow as long as the appropriate ions and sufficient space are available (fig. 7.1*b*). Eventually, the growing mineral is large enough to see first with the microscope and then with the naked eye. This orderly internal arrangement of ions is the principal characteristic of solids;

Table 7.1

The Most Common Elements in Earth's Crust

Element	Average Weight Percent
Oxygen	46.6
Silicon	27.7
Aluminum	8.1
Iron	5.0
Calcium	3.6
Sodium	2.8
Potassium	2.6
Magnesium	2.1
Other elements	1.5

Source: Data from B. Mason and C. B. Moore, *Principles of Geochemistry*, 4th ed. Copyright ©1982 John Wiley & Sons, Inc.

thus, minerals are solids. The internal ionic arrangement also gives rise to a regular external geometric form (e.g., a cube or a pyramid) that is bounded by smooth, plane surfaces. Such a feature is known as a **crystal,** and every mineral has its own characteristic crystal form that may serve as an important identifying characteristic. Normally, however, good crystal form is only developed where the crystal grew in a void or within a liquid mass so that its growth would not be impeded by other solid crystals (fig. 7.5).

Ionic Substitution

The particular internal arrangement of ions for any mineral is governed by three factors: the availability of certain ions, the size of those ions (ionic radius), and the atomic charge on each ion. Any ion must be able to fit into the space for it in the growing crystal lattice. In addition, when crystal growth is complete, the resulting mineral must be electrically neutral. That is, just as many positive charges as negative charges must have gone into its formation.

The silicate mineral plagioclase feldspar can be used to illustrate this concept. Plagioclase exists in a sodium-rich variety ($NaAlSi_3O_8$) and a calcium-rich variety ($CaAl_2Si_2O_8$). This variation occurs because the calcium ion and the sodium ion are very similar in size, and either can fit in the appropriate place in the crystal lattice. That is, they can physically substitute for one another, a process known as **ionic substitution.** Calcium and sodium do not, however, carry equal electrical charges. The sodium ion has a charge of +1, while the calcium ion carries a charge of +2. To substitute calcium for sodium indiscriminantly would lead to a mineral that is not electrically balanced. Consequently, if calcium is to substitute for sodium, a second, compensating substitution must occur that will return the mineral to electrical neutrality. The substitution of aluminum for silicon does that. The two ions are of similar size, but aluminum has a charge of +3 and silicon has a charge of + 4. Thus, a paired substitution of calcium and aluminum (total charge of +5) for sodium and silicon (total charge of +5) leaves the resulting mineral electrically neutral.

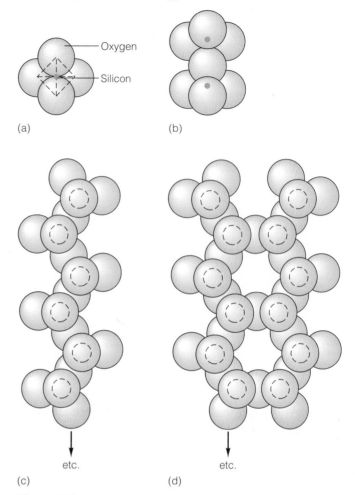

Figure 7.3 The silicon-oxygen tetrahedron (*a*) is the basic structural unit of all silicate minerals. It consists of a single silicon ion (charge = +4) surrounded by four oxygen ions (charge of each = −2). Connecting the centers of the oxygen ions forms a four-sided geometric figure known as a tetrahedron. Each of the four sides of the tetrahedron is an equilateral triangle. Variations in composition and complexity of silicate minerals are achieved by sharing of oxygen ions between adjacent tetrahedra. Examples shown are the (*b*) double tetrahedron, (*c*) single chain, and (*d*) double chain. Additional ions of other elements (not shown) attach themselves to complete a mineral's structure.

The effect of such a paired substitution can be seen in the chemical formulas of two varieties of plagioclase. This type of substitution is the rule in nature rather than the exception. It also explains the last part of our mineral definition, that the chemical composition of a mineral is fixed or varies within narrow limits. The narrow limits are determined by the amount of ionic substitution. Variations in composition resulting from ionic substitution, as well as the presence of minor amounts of impurities within the mineral, account for some differences in physical properties among individual specimens of the same mineral. For example, the sodium variety of plagioclase feldspar, known as albite, is always lighter in both color and weight than the calcium variety, anorthite.

Now that you are familiar with the principles of ionic substitution and electrical neutrality in minerals, return briefly to a

(a)

(b)

Figure 7.4 Feldspars, by far the most abundant rock-forming mineral group, are members of the silicate chemical mineral class. Pictured is the pink variety microcline (K-feldspar) (a) alone and (b) in combination with other silicate minerals in the rock granite.

Figure 7.5 An example of the growth of a quartz crystal within a natural rock cavity. Such an environment, where the growing mineral is unconfined, is the best situation for the development of the mineral's crystal shape.

consideration of the silicates. It was stated that the silicon tetrahedron consists of a combination of one silicon ion and four oxygen ions. The electrical charges on these ions are +4 and −2, respectively. If we add the charges on the four oxygen ions they total −8 (compared to the +4 of the silicon). What this means is that the silicon tetrahedron by itself cannot be a mineral because it has a surplus of −4 charges and is electrically unbalanced. To form a mineral, other ions with charges totaling +4 must be added to the simple tetrahedron. In nature, this can be accomplished by the addition of ions, such as Fe^{2+} or Mg^{2+} ions. The result is the mineral olivine, $(Mg,Fe)_2SiO_4$. As more complex silicate minerals form, these same rules of substitution and electrical neutrality must be maintained. Table 7.2 gives examples of minerals that are characterized by ionic substitution.

Mineral Identification

The properties of minerals are identifiable through certain field and laboratory tests even a novice can make. One of the tests is based on a scale of *hardness* devised in the eighteenth century by the German mineralogist Friedrich Mohs (fig. 7.6). This test compares the hardness of an unknown mineral sample to that of the minerals on the hardness scale by "scratching" the unknown specimen. Each of the ten minerals on the scale is assigned a hardness value, with talc (1) being the softest and diamond (10) the hardest. Commercial sets containing all of the scale minerals (with the exception of diamond) are available. Harder minerals will scratch softer minerals, and the hardness of the unknown can be bracketed between the hardness values of minerals on the scale. For example, the unknown specimen may be scratched by quartz (hardness = 7) but not by feldspar (hardness = 6). Therefore, the unknown specimen has a hardness value somewhere between 6 and 7. While not a definitive method of identification, this technique does narrow the possibilities significantly. Neither earth scientists nor interested amateurs commonly carry Mohs' scale sets about with them. Nevertheless, hardness approximations can be made by using some readily available materials such as a fingernail, a penny, a knife blade, or a piece of glass.

Minerals normally display a predictable range of colors, and many also exhibit a characteristic called *streak,* which is the color of a chalklike mark the mineral makes when rubbed on a hard, unglazed ceramic plate known as a *streak plate.* The streak of a particular mineral (actually a powdered sample of that mineral) may not be the same color as the specimen as a whole. For example, the brassy appearance of pyrite, or "fool's gold," is different from its dark streak. An advantage of using streak for identification purposes is that its color tends to be more consistent than the variety of colors often displayed by different specimens of the same mineral. A major drawback is that the streaks of the vast majority of minerals are white and, therefore, not distinctive. For those minerals with characteristic streaks, such as

Table 7.2

Selected Minerals Displaying Ionic Substitution

Mineral Name	Chemical Formula	Substituting Ions
Plagioclase	$(Ca,Na)(Al,Si)AlSi_2O_8$	$(Na^{1+},Ca^{2+}),(Al^{3+},Si^{4+})$
Olivine	$(Mg,Fe)_2SiO_4$	(Mg^{2+},Fe^{2+})
Rhodonite	$(Mn,Ca)SiO_3$	(Mn^{2+},Ca^{2+})
Nepheline	$(Na,K)AlSiO_4$	(Na^{1+},K^{1+})
Biotite mica	$K(Mg,Fe)_3(OH)_2AlSi_3O_{10}$	(Mg^{2+},Fe^{2+})
Chlorite	$(Mg,Fe,Al)_6(OH)_8(Al,Si)_4O_{10}$	$(Mg^{2+},Fe^{2+},Al^{3+}),(Al^{3+},Si^{4+})$

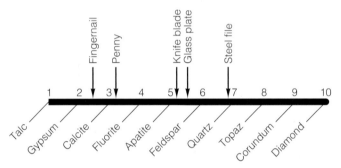

Figure 7.6 The Mohs hardness scale for minerals. The hardest of all natural substances is diamond, which is many times harder than corundum, the ninth mineral on the scale. A mineral that is higher on the scale "scratches" any mineral with a lower hardness value. Because hardness sets are usually not available in the field, hardnesses of some common items, which may be used in the identification process, are also shown on the scale.

the iron oxides hematite (reddish brown) and limonite (yellowish brown), however, this feature can be most useful. Related to color and streak is the physical property called *luster,* which is the way in which a mineral's surface reflects light. While some 20 different terms are used by mineralogists to describe the surface appearances of minerals, just the recognition that a mineral might be described as metallic, glassy, or dull can be useful in identification.

All minerals possess mass and therefore *density* (mass per unit volume). In the laboratory, precise measurements of this density compared to that of an equal volume of water, a property known as *specific gravity,* can be made. While such precise measurements would be impossible to make in the field, simply comparing the weight, or "heft," of a mineral to that of an approximately equal-sized piece of a common mineral such as quartz or feldspar may be useful in identification. Many metallic ore minerals, for instance, feel heavy.

Some minerals possess a characteristic known as *cleavage,* whereby they split smoothly in certain directions. Such directions are related to planes of weakness within the crystal structure. Because minerals most commonly occur as large clusters of many crystals, it is important when examining a specimen for cleavage to look at how a single crystal breaks. Often, breaking the mass of crystals simply separates them. When minerals break other than along smooth cleavage planes they are said to *fracture.* While fracture is usually considered to be an uneven, random breakage, some minerals exhibit peculiar types of fracture that may aid in identification. Large crystals of the mineral quartz, for instance, often exhibit smooth, curved breakage surfaces known as conchoidal fracture.

As noted, every mineral has its own particular crystal shape it assumes whenever sufficient space is available during the growth process. Because these shapes are geometric forms bounded by smooth surfaces, these too may be useful for identification. Two notes of caution must be sounded here, however. First, most minerals as they are found within rocks did not have sufficient room to grow and filled in whatever irregular spaces were available; thus they exhibit an irregular form. Second, even an expert may have difficulty differentiating smooth crystal faces ("growth" surfaces) from smooth cleavage ("breakage" surfaces). Often, the only way to be certain is to break the specimen, a technique that is obviously both destructive and costly. Figure 7.7a and b illustrate the crystal faces of quartz and the cleavage planes of calcite.

Some minerals can be identified by special properties. For example, the mineral magnetite, Fe_3O_4, is naturally magnetic and is attracted to a small magnet, making it easily identifiable. The mineral calcite, $CaCO_3$, "fizzes," or effervesces, when a drop of dilute hydrochloric acid (HCl) is placed on it. Calcite's close relative dolomite, $CaMg(CO_3)_2$, likewise effervesces in HCl, but only if the acid is placed on a powder formed by scratching the specimen.

ROCKS

The study of minerals is essential to the understanding of **rocks,** because rocks are aggregates of minerals that make up a significant portion of Earth's outer shell. Some rocks are *monomineralic,* consisting of grains or crystals of but a single mineral. Some sandstones consist wholly of quartz grains, and pure limestone is an accumulation of crystals or grains of calcite. Most rocks, however, are *polymineralic* in that they are physi-

Box 7.1 Further Consideration

Luminescence: Minerals That Glow in the Dark

Luminescence is a property possessed by certain minerals that causes them to emit light during and after being exposed to some external energy source, usually electromagnetic radiation. The radiant energy causes the electrons in the atoms that are normally at their lowest energy level (ground state) to be raised to a higher energy level (excited state). While excited, and as they return to the lower energy level with the removal of the external energy source, the electrons radiate energy. For most minerals, the energy given off is the same wavelength as that of the external source. For some, however, the radiant energy is of a longer wavelength than the external source. If they are exposed to electromagnetic radiation in the visible portion of the spectrum, they radiate some energy in a wavelength beyond the visible so that we do not see it. If exposed to radiation of a shorter wavelength than visible light, such as ultraviolet (UV), they radiate in the longer wavelength visible portion of the spectrum.

Minerals that give off light while being exposed to ultraviolet radiation are said to be fluorescent, from the mineral fluorite, which "glows" with a blue color due to the presence of small amounts of the rare earth elements. Such elements that are present as the impurities that cause fluorescence are called activators, and some minerals that do not normally fluoresce can be made to do so by adding activators to the crystal structure. The light of fluorescent minerals disappears when the UV source is withdrawn. A few minerals that continue to give off light even after the UV source is withdrawn are said to possess phosphorescence. Their color only fades gradually. If the external energy source is heat rather than electromagnetic radiation, the same type of visual effect is produced in certain minerals (and in the filament of an ordinary electric light bulb). This is called thermoluminescence. Some minerals, including fluorite and sphalerite, give off light when scratched or rubbed, a phenomenon known as triboluminescence. Many museums house special "darkroom" displays to illustrate the effects of luminescence.

Apply Your Knowledge

1. Besides fluorite, what are some other fluorescent minerals? What is the activator of each?
2. Cite some examples of phosphorescent minerals. Cite examples of thermoluminescent minerals.
3. Explain the thermoluminescent effect of an ordinary incandescent light bulb.

---

cal mixtures of two or more minerals. The common rock granite, often used for a memorial stone, is normally a mixture of quartz, potassium feldspar, biotite mica, and minor amounts of plagioclase feldspar.

You may better picture the place of rocks in nature by considering the following simple analogy. The 92 naturally occurring elements might be likened to the 26 letters of our alphabet. They are the fundamental building blocks. In isolation, most letters make little sense. Combining elements in different ways forms chemical compounds, including minerals, just as the combining of letters in meaningful ways forms words. Words by themselves may have little meaning until they occur in sentences. Similarly, minerals bound together form rocks. Just as it takes an educated person to read sentences, the educated earth scientist can interpret much of Earth's history by properly "reading" the rock record. Only 26 letters can be combined to form hundreds of thousands of words, and those words could form an indefinite number of sentences. Likewise, the number of possible rock types based on the potential combinations of more than 3,000 minerals staggers the imagination. Fortunately, for those involved with interpreting the rock record, only 7 groups (or mineral families) account for a very large percentage of the volume of rock (table 7.3). Note that of the 7 mineral families, 6 belong to the silicate, or "rock-former," class.

In reality, the possible variety of rocks is also limited by the nature of the rock-forming processes. Under various conditions of rock formation, only certain minerals are compatible with one another. For example, of the minerals shown in table 7.3, members of the pyroxene family would rarely occur in quartz-bearing rocks because quartz crystallizes under relatively low-temperature conditions while pyroxene crystallizes in a high-temperature environment.

An analogy for crystallization temperatures exists in the melting temperatures of butter and chocolate. Placed in similar pots and heated equally, butter melts before chocolate, and, upon cooling, chocolate hardens before butter. Some minerals crystallize at very high temperatures, while others are still hot liquids. When experimenter N. L. Bowen melted minerals and allowed them to cool under controlled laboratory conditions, he found that certain minerals crystallized before others. From his experiments, Bowen derived a crystallization-temperature series, which indicates the order of crystallization of igneous minerals and the common associations of minerals that form when liquid rock cools and hardens (fig. 7.8). Quartz and olivine, for example, occur at opposite ends of the crystallization spectrum, making their association in rocks unlikely, or at best, very limited. A corollary to this is the realization that those minerals that form at high temperatures are least stable in Earth's low-temperature surface environment and break down on contact with water, oxygen, and other atmospheric chemicals (the process of weathering covered later in this chapter). Thus, olivine and pyroxene are uncommon in the sedimentary rocks that form on or near Earth's surface, while quartz is a dominant constituent of these rocks.

Just which minerals will form from a cooling rock liquid is a complex matter. In a general way, those minerals with relatively simple internal structures and that require little rearrangement of the constituent ions form earlier than those with more complex ionic arrangements. Obviously, the availability

(a)

(b)

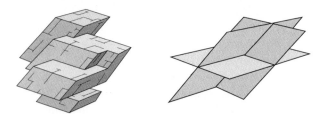

(c)

Figure 7.7 Crystal faces, fracture, and cleavage. (a) Smooth crystal faces on quartz. Note the irregular fracturing where this specimen was broken near its base. (b) Cleavage surfaces on a broken fragment of calcite. (c) Calcite always "cleaves" along three sets of planes oriented at constant angles to one another. All cleavage surfaces comprising one of the "sets" are parallel to one another (as are the top and the bottom of a box). Thus, calcite is said to possess three cleavage "directions." (c): (From R.D. Dallmeyer, *Physical Geology Laboratory Text and Manual*, 2d ed. Copyright © Kendall-Hunt Publishing Company, 1978. Used with permission.)

of necessary ions within the rock liquid also governs which minerals can or cannot form.

The Classification of Rocks

Just as they classify minerals, earth scientists also classify rocks. Their task here, however, is somewhat more difficult. Each mineral has a distinct chemical composition that automatically rel-

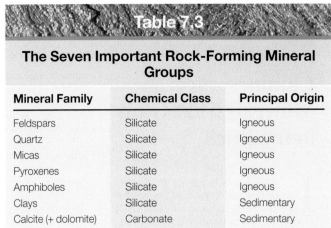

Table 7.3

The Seven Important Rock-Forming Mineral Groups

Mineral Family	Chemical Class	Principal Origin
Feldspars	Silicate	Igneous
Quartz	Silicate	Igneous
Micas	Silicate	Igneous
Pyroxenes	Silicate	Igneous
Amphiboles	Silicate	Igneous
Clays	Silicate	Sedimentary
Calcite (+ dolomite)	Carbonate	Sedimentary

egates it to a particular category. Rocks, consisting of physical mixtures of varying proportions of different minerals, cannot be sorted as easily on the basis of chemical composition. In fact, rocks with totally different mineral compositions and origins, such as granite and shale, might have very similar bulk chemical compositions.

Because earth scientists attempt to interpret Earth's history from clues provided from the rock record, they prefer to employ a classification that is based on rock origins. The traditional approach is such a "genetic" one. It is not perfect, but it has withstood the test of time. Under this scheme the rocks, with few exceptions, fall naturally into one of three groups. The **igneous** ("fire-formed") **rocks** are those that crystallize from magma. **Sedimentary rocks** form by the accumulation and consolidation of loose rock and mineral material and/or by the accumulations of fossils and salts. **Metamorphic** ("changed-form") **rocks** are formed by the alteration of preexisting rocks in the solid state by the application of heat and pressure and the introduction of chemical fluids. The relationships between these three rock types are shown by the rock cycle.

The Rock Cycle

The **rock cycle** is the traditional visual vehicle for portraying the origins of and the relationships between the three basic rock types (fig. 7.9). Note that each rock type is formed from a pre-existing rock type. For example, sedimentary rocks are formed by the weathering and erosion of igneous, metamorphic, or other sedimentary rocks. Metamorphic rocks are formed by the *metamorphism,* or solid state alteration, of sedimentary, igneous, or other metamorphic rocks. Igneous rocks are formed by the melting of sedimentary, metamorphic, or other igneous rocks. The rock cycle does not represent a truly "closed system," because some material is added from deep within Earth and some is added from the atmosphere and even outer space. The cycle does involve, however, the constant recycling of much the same material. Mineral grains that are now present in a sandstone might have been in a granite millions of years ago and millions of years from now might be in a metamorphic schist.

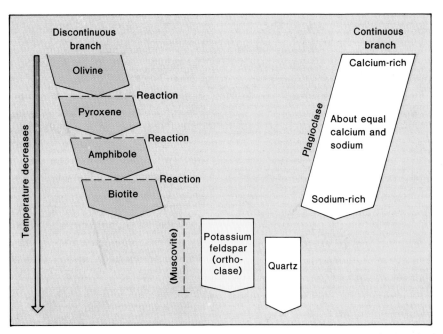

Figure 7.8 Bowen's reaction series. "Continuous" reactions (e.g., in plagioclase) involve continuous ionic substitution between the cooling magma and the growing mineral crystals. The mineral remains the same. In a "discontinuous" reaction (as from olivine to pyroxene) the earlier-formed mineral is destroyed as the new mineral forms.

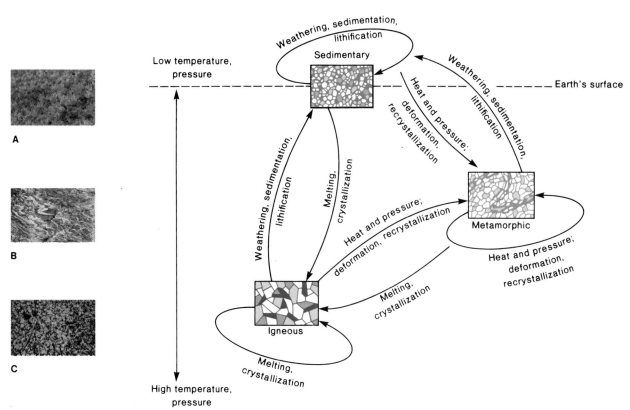

Figure 7.9 The rock cycle. While some new material is added from Earth's interior and some is exchanged with the atmosphere and the oceans, much of the same mineral material is refashioned and recycled again and again.

IGNEOUS ROCKS

Igneous rocks form by the crystallization of their constituent minerals out of a magma when it lies below Earth's surface or from **lava** on Earth's surface. The distinction between magma and lava is primarily one of location. All evidence indicates that there is no worldwide layer of magma within easy reach of Earth's surface, and that the source of magma must be in deep, isolated pockets produced by local melting. The melting of solid rock is caused by an increase in temperature under constant pressure, a decrease in pressure under constant temperature, or a combination of the two.

Heating the rock causes the constituent ions to vibrate more energetically in their lattice positions. If the vibrations become violent enough, the bonds between ions break, the lattice collapses, and the solid material changes to liquid. To some extent, high pressure counteracts the effect of elevated temperature by holding the ions in place and not allowing them to separate. Decreasing pressure, even without elevating the temperature, could allow the solid to melt if the rock was already at a sufficiently high temperature. Elevated temperatures within Earth might be generated by the decay of locally concentrated radioisotopes (described in chapter 14). Significant reductions in pressure could be caused when a mass of heated rock is raised by internal earth forces toward the surface and overlying rock layers are eroded away. Additionally, the presence of water, as in the pore spaces of sedimentary rocks, significantly lowers the melting points of rocks and minerals.

Once rock melts, the resulting magma becomes less dense than the surrounding solid rock and rises. It works its way upward through systems of cracks and crevasses, pushing aside, melting, and warping any overlying layers of rock. Some of the magma reaches the surface to be erupted as lava in **volcanic** (or **extrusive,** because it "comes out of") activity. The bulk of the magma, however, cools and crystallizes long before reaching the surface and, consequently, is emplaced as **plutonic** (or **intrusive**) rock. Such rock can only be exposed much later, when erosion removes the overlying rock layers. This difference in origin, which is reflected in mineral composition and certain physical characteristics, provides a convenient way to subdivide igneous rocks and, indeed, to provide names for individual igneous rock types.

Igneous Rock Textures

The most obvious physical characteristic of igneous rocks as a response to differences in origin is the size of the individual crystals in the rock. In general, the more slowly magma cools, the more time the mineral crystals have to grow and the larger they become. Conversely, rapid cooling leads to a matrix of many small crystals. Slow cooling in nature would most likely occur beneath Earth's surface, where heat loss is very slow because of the insulating effect of the overlying rocks. Lava erupted onto Earth's surface loses heat very rapidly to the atmosphere or, in oceanic eruptions, to the hydrosphere. Thin and/or shallow intrusive igneous bodies may also cool rapidly.

The size, shape, and arrangement of mineral grains or crystals in any rock is referred to as its **texture.** The mass of large (visible to the naked eye) crystals produced by the slow cooling of a magma is referred to as a **coarse,** or **phaneritic,** *texture* (fig. 7.10*a*). The matrix of tiny crystals formed by the rapid cooling of a lava or a shallow intrusive is a **fine,** or **aphanitic,** *texture* (fig. 7.10*b*). Under conditions of very rapid cooling, ions may not have time to arrange themselves into even the smallest crystals. The result is a *glassy texture.* Recall that a glass, though it has the rigidity of a solid, is not a true solid because it lacks an orderly internal arrangement of ions. The ions are simply "frozen" into random positions.

Other textures may develop under special circumstances, as when the upper portion of a lava flow hardens around escaping gas bubbles, leaving the rock with a **vesicular,** or **porous,** *texture.* Some vesicular igneous rocks are very light in weight because of the holes, or vesicles, in them. Instances of masses of the porous rock pumice floating on water have been reported. A magma often begins the crystallization process even as it rises toward the surface. This means that some crystals grow large slowly over a long period of time. When the mixture of large solid crystals and still liquid magma is erupted onto Earth's surface, the remaining liquid crystallizes rapidly as a "background" mass of fine crystals, resulting in a mixture of crystals of distinctly different sizes. Such a **porphyritic** *texture* normally indicates two stages of cooling: a slow early stage and a later rapid stage. The resulting rock is referred to as a *porphyry.* Its complete name is formed by adding the rock term that best describes the finer-grained fraction; for example, "basalt porphyry" or "andesite porphyry."

Sometimes, late in the crystallization of granitic magma, rock bodies with unusually large crystals may form. This very coarse-grained texture is a **pegmatitic** *texture,* and the resulting rock is called a pegmatite. While the bulk mineralogy of a pegmatite is quartz and K-feldspar, a number of unusual minerals, including the gemstones ruby and sapphire, can also be found in them. The very coarse texture is usually attributed to the high percentage of water at this stage of the crystallization process, which allows for easy migration of ions to the sites of mineral formation. The percentage of water is high at this stage because the crystal structures of most early-formed minerals do not allow for its inclusion.

Igneous Rock Composition

The second important factor in classifying igneous rocks is the mineral composition of the rock. Reference was made earlier to the laboratory experiments of N. L. Bowen, which substantiated the idea previously held from field experience that crystal formation is not a random process. Rather, the various minerals that crystallize from a magma do so in a specific order as the temperature of the magma decreases. Thus, the igneous rock basalt is composed only of minerals such as Ca-plagioclase and pyroxene that crystallize early at high temperatures. Granite, on the other hand, is formed of late-crystallizing, low-temperature minerals. Coincidentally, many of the early forming minerals,

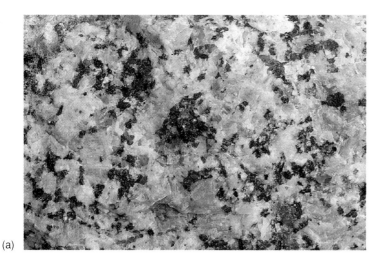

Figure 7.10 (a) Coarse-grained (phaneritic) igneous rock texture representing slow cooling and crystal growth at some depth beneath Earth's surface. (b) Fine-grained (aphanitic) texture developed by rapid cooling and solidification in the low-temperature environment of Earth's surface.

ty gives it a better chance of rising all the way to Earth's surface before it solidifies. You might liken the consistency of basaltic magma to a heavy motor oil and that of granitic magma to toothpaste. This "consistency," or resistance of the magma (or any fluid) to flow, is called *viscosity,* and we can characterize granitic magmas as being more viscous than basaltic magmas.

The upper part of figure 7.11 shows the approximate temperature range across which each of the important igneous rock-forming minerals crystallizes. Just as in the old expression "birds of a feather flock together," minerals that crystallize at relatively high temperatures (such as Ca-plagioclase, olivine, and pyroxene) can normally be found together in a rock, but are not found with low-temperature minerals (such as quartz, potassium feldspar, and muscovite), which form very different rock types. Because rocks are physical mixtures of minerals that can come together in varying proportions, unexpected mineral combinations may on occasion occur. Thus, the mineral percentages used to establish boundaries to separate one rock type from another (i.e., granite from diorite) are arbitrary and used by earth scientists for convenience.

An economically important aspect of igneous rock formation occurs during the latest stages of the process. When most of the common rock-forming minerals have already crystallized to form, for example, a granite, and even the pegmatites have formed, the residue of the original magma is a watery solution at several hundred degrees Celsius. Such a *hydrothermal,* or "hot water," *solution* contains an abundance of ions of rare metals that find no lattice positions to fill in common minerals such as quartz, feldspar, and mica. Instead, they are incorporated into and concentrated in rare and unusual minerals. The economic importance of these hydrothermal solutions and the minerals that crystallize from them is further discussed in chapter 13.

Volcanoes

Most igneous rocks form beneath the surface of the continents or beneath the sea, thus, away from our direct observation. Only when overlying rock layers are removed by erosion or the rocks are moved above sea level do we see them. Nature does, however, provide scientists with at least one type of occurrence where igneous rock in the process of formation is visible above ground, that is, the volcanic eruption.

A **volcano** is normally, but not exclusively, a cone-shaped topographic feature composed of accumulations of igneous rock and ash material that have been extruded through an opening, or *fissure,* in Earth's surface. As these features grow, material is extruded either from a central vent at the cone's summit, the *crater,* or from fractures on the flanks of the volcano. Because volcanoes can vary significantly in rock composition, history, and general appearance, earth scientists have subdivided them and named the different types. Two general approaches to volcano classification are commonly used: that based on eruptive history and that based on appearance (shape) and composition.

such as olivine, amphiboles, and pyroxenes, are rich in iron and magnesium (and called **ferromagnesian** *minerals*) and tend to be dark in color, while most late-forming minerals (quartz, K-feldspar, and muscovite) are light-colored. This means that igneous rock color can be used as a reasonable approximation to mineral composition. This is especially helpful in the identification of fine-grained rocks, the tiny crystals of which can only be identified under the microscope.

Figure 7.11 is a simplified classification scheme for the most common igneous rocks. The bottom portion of the diagram shows the names assigned to each combination of texture (vertical axis) and composition or color (horizontal axis). Although each of the basic rock types is given equal space in the diagram, the actual abundances in nature are far from equal. Granite and closely related rock types account for more than 90% of the total volume of intrusive igneous rocks, whereas basalt and its near relatives account for more than 90% of all extrusive igneous rocks. This distribution is not mere coincidence. Basaltic magma is much more fluid than granitic magma, and its greater mobili-

Crystallization temperature (°C)

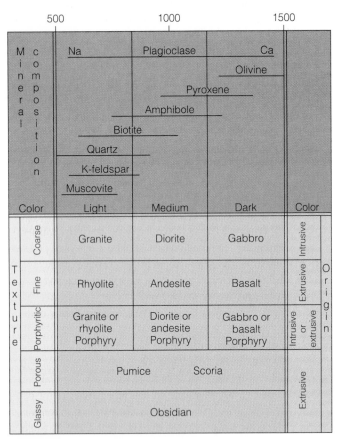

Figure 7.11 Classification scheme for the igneous rocks based on mineral composition and texture.

(a)

(b)

Figure 7.12 Photographs of Mount St. Helens in the Cascade Range of Washington state (*a*) before and (*b*) after its spectacular eruption of May 18, 1980.

The classification based on eruptive history recognizes three loosely defined categories: (1) those that were *active* and erupted in historic times; (2) those that have no historical record of eruption but are "fresh" looking with little sign of erosion and are thus regarded as only "sleeping," or *dormant;* and (3) those that are *extinct* because they are highly eroded and show no sign of any type of activity. Because this approach depends on the very short period of human experience compared to the enormity of geologic time, we have often been tragically mistaken in our assessment of a particular volcano's capacity for violent eruption. One such instance was the reawakening on May 18,1980 of Mount Saint Helens, one of the peaks of the Cascade Range in the state of Washington. In a violent eruption that literally "decapitated" the volcano, the surrounding landscape was altered with the tragic loss of 60 lives (fig. 7.12).

An even more devastating event shook the ancient Roman world almost 2,000 years ago: the eruption of Mount Vesuvius on the Bay of Naples in A.D. 79. Known to the Romans of that day as Monte Somma, this volcano had been quiet throughout the several hundred years of recorded human habitation in the area. Several years prior to its eruption, frequent small earthquakes in the area indicated that Somma was stirring. The A.D. 79 eruption was so violent that day was turned to night by thick clouds of volcanic ash. The port city of Pompeii was buried beneath many meters of rapidly falling ash, and the city of

Herculaneum was buried by massive mudflows generated by the eruption. Both cities were lost in time only to be unearthed by archeologists in the last century. The death toll at Pompeii and Herculaneum can never be known, but the impressions of many bodies preserved in the surrounding ash indicate that events overwhelmed the inhabitants, leaving them little chance to escape. Since that time, Vesuvius has been one of Earth's most active volcanoes, with dozens of recorded eruptions.

The classification based on topographic form, or appearance, recognizes four major and several minor types of features. The first is the *shield volcano,* so-called because it is rounded with a low topographic profile that resembles the shape of shields carried by warriors of old (fig. 7.13*a*). Such volcanoes are built up primarily by successive flows of basaltic lava. The slopes of the shield volcano are made gentle by the lava itself, which tends, while still in the fluid state, to flow down even the most gentle of slopes and to spread out. The foundation of the shield volcano is typically so broad, however, that even with its very gentle slopes, this feature can be built to an overall height of thousands of meters. Such is the case in Hawaii, where the volcanic shields have been built up from the seafloor, which is approximately 5,100 meters (17,000 ft) deep at that locality, to elevations in excess of 3,900 meters (13,000 ft) above sea level: a total height of more than 9,000 meters (30,000 ft). With its great height as measured from the seafloor and its size, the shield volcano Mauna Loa is the largest mountain on Earth. It is not, however, the largest in the Solar System. That distinction belongs to the shield

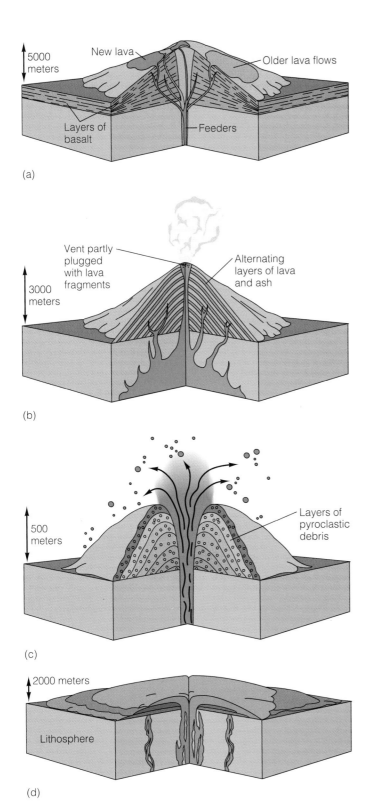

Figure 7.13 Types of volcanoes: (*a*) shield volcano, (*b*) strato-volcano, (*c*) cinder cone, and (*d*) flood basalts (basalt plateau).

volcanoes on Mars, including Olympus Mons, which was discussed in chapter 2.

A second feature, the *stratovolcano,* is built up of successive layers, or "strata," of alternating lava flows (usually of andesite rock) and solid volcanic particles (fig. 7.13*b*). The solid parti-

cles vary in size from fine dust to large blocks and are collectively called **pyroclastic** ("fire fragment") **debris.** Because andesitic lava is not as fluid as basaltic lava, and because the solid particles can form steeper slopes than lava, stratovolcanoes tend to be steeper-sided and to rise much more abruptly and spectacularly from their surroundings than do shield volcanoes. In general, the andesitic lava flows broaden the base of the volcanic cone and the pyroclastic debris builds it higher. Most of the world's most striking and beautiful volcanoes, such as Mount Fujiyama in Japan, Mount Kilimanjaro in Africa, Mount Mayon in the Philippines, and the peaks of the Cascade Range in the Pacific Northwest of the United States fall into this category.

Less spectacular than the others, but interesting in its own right, is a third type of volcano known as a *cinder cone* (fig. 7.13*c*). These are small, normally rising less than 450 meters (1,500 ft) above their surroundings, and often occur in groups. They consist almost exclusively of pyroclastic debris, which means that they are steep-sided, but their behavior is often modified late in their life cycles to include basaltic lava flows. Part of the fascination of cinder cones is that they appear to have very short life spans, on the order of 10 to 20 years. Earth scientists have, therefore, been able to study fully and to document completely their cycles of growth.

Other related volcanic features include maars, spatter cones, and calderas. A *maar,* which often contains a lake, is a broad volcanic crater of low relief formed by multiple explosions. *Spatter* **cones** are small basaltic cones, a few meters in height and formed from hardened accumulations of lava ejected as flying masses through fissures. *Calderas* are wide, deep depressions, many times larger than craters, that are the remnants of volcanic cones that erupted violently. The explosive eruption pulverizes much of the upper part of the cone and causes a good part of the remainder to collapse into the partially empty magma chamber below. The best example of such a feature in North America is Crater Lake in southern Oregon (see photo at the beginning of this chapter). Crater Lake formed by an explosive eruption of a volcano, now referred to by geologists as Mount Mazama, approximately 6,600 years ago. The result was a circular depression 8 kilometers (5 mi) in diameter and as much as 1,200 meters (4,000 ft) deep. Subsequently, rain and snow melt have filled the caldera to a depth of 600 meters (2,000 ft), forming the lake.

Basalt Plateaus

Very fluid basaltic lava can erupt out of many closely spaced vents, rather than from a single vent. When this happens, the lava spreads out to form a thin but widespread horizontal layer of basalt. Subsequent eruptions pile additional layers on top of the first until a sequence of many flows totaling a thousand meters or more in thickness has accumulated over the course of several million years. Such an accumulation of horizontal layers of basalt is called a **basalt plateau** (fig. 7.13*d*). In the United States, the most prominent such feature, the Columbia River plateau–Snake River plain, lies in the states of Washington, Oregon, and Idaho and includes much of the area drained by

Box 7.2 Investigating the Environment

The Long Valley Caldera: Will History Soon Repeat Itself?

The Long Valley caldera, an oval depression covering 600 square kilometers (170 sq. mi), is located on the edge of the Sierra Nevada in east central California (fig. 7.A). The caldera came into existence as the result of a violent eruption approximately 730,000 years ago. The eruption buried a 1,475-square-kilometer (570-sq.-mi) area with ash and pumice accumulations up to 1,350 meters (4,500 ft) thick and left a 3.2-kilometer-(2-mi-) deep depression. The thick blanket of volcanic ash is known to geologists today as the Bishop Tuff (fig. 7.B). Smaller-scale periods of eruption occurred again about 50,000 and 10,000 years ago, and volcanic activity has occurred as recently as 500 to 600 years ago at the nearby Mono Craters.

The most recent activity at Long Valley, however, has not been volcanic but seismic. In 1980, the town of Mammoth Lakes, California, a recreation center, began to experience earthquakes. This activity culminated on May 27, 1980, when four shocks of magnitude 6.0 were recorded. Over the next 2 years, the pattern of earthquake activity was so like that which preceded the Mount Saint Helens eruption that the director of the U.S. Geological Survey issued a notice of potential volcanic hazard for the region. Overall, the pattern of earthquakes, supplemented by other data, led seismologists to believe that the earthquakes were caused by the movement of magma toward the surface and that a mass of magma 4 kilometers (2.5 mi) across

lies less than 8 kilometers (5 mi) below the ground surface.

Two important issues are emerging from the work done at Long Valley. The first relates to the area as an important recreational resource. Warnings of impending volcanic eruptions can be devastating to a local economy based on tourism. The problem is compounded by the fact that the area's very remoteness would make a hurried evacuation,

should the need arise, a difficult proposition. On the positive side, regions with near-surface magma chambers such as Long Valley hold tremendous potential for geothermal ("earth heat") energy. Under the auspices of the U.S. Department of Energy, drilling was begun in 1989 on a 6,500-meter-(20,000-ft-) deep exploratory well within the caldera. The purpose of this well is to test the feasibility of tapping the magma's heat energy for electri-

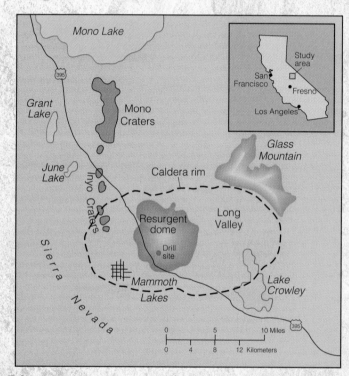

Figure 7.A Location map showing Long Valley caldera and energy exploration well.

---

these two river systems. In places the Columbia River plateau consists of a thickness of basaltic rock layers in excess of 1,500 meters (5,000 ft) that has been built up over the last 30 million years. Similar massive accumulations are found in the Parana River basin of Brazil and the Deccan Plateau of western India.

Plutonic Features

Those masses of magma, many of them granite in composition, that crystallized before reaching Earth's surface are found preserved in an array of sizes and shapes. All are referred to by the generic name **pluton,** after Pluto, the Roman god of the lower

world (fig. 7.14). Some plutons are intruded between and parallel to the surrounding rock layers . These are *concordant.* Others intrude across the surrounding rock layers and thus are considered *discordant.* If a pluton has a shape that resembles a tabletop, that is, relatively thin compared to its length and width, it is called a **tabular pluton.** Any other shape is a *nontabular pluton.* These two characteristics are used to describe and name the several types of plutons (table 7.4). For example, a tabular, discordant pluton is a **dike,** while a tabular, concordant pluton is a **sill.** A concordant pluton made nontabular by a pronounced thickening that "arches up" the overlying rock layers is a *laccolith.* A similar concordant feature, but with a distinctive downward cal power generation. While the results of this $8 million project won't be known for some time and the commercial use of this type of geothermal energy is still a number of years away, the prospects are certainly interesting to contemplate.

Apply Your Knowledge

1. What might be some problems in trying to tap the thermal energy of a near-surface magma chamber for electric power generation?

2. What social and economic problems might arise from a governmental announcement of an impending major volcanic eruption (consider both the short and the long term)?

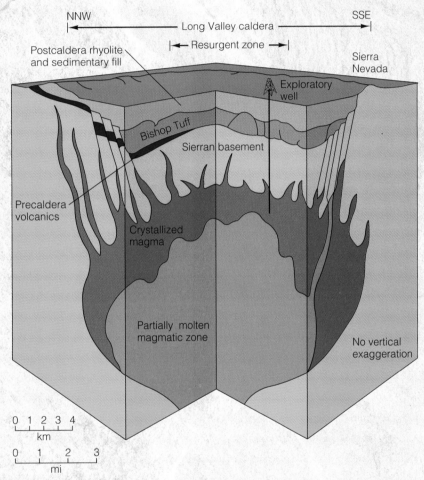

Figure 7.B Cross section showing subsurface geology through the Long Valley caldera.

"bow," is a *lopolith*. The largest pluton, by far, is the discordant, nontabular **batholith.** A batholith is usually composed of granite and/or related rock types and may occur over an area of hundreds of thousands of square kilometers. It must, however, by definition, have an exposed surface area of no less than 100 square kilometers (40 sq. mi).

A similar feature with less surface exposure, regardless of how large its underground dimensions might be, is called a *stock*. Elevated by forces operating within Earth, batholiths and stocks have been carved by the processes of weathering and erosion into some of the world's largest and most spectacular mountain ranges (fig. 7.15*a*). The classic example in the United States is

the Sierra Nevada of eastern California, a mountain system stretching some 650 kilometers (400 mi) in length and averaging 80 kilometers (50 mi) in width (fig. 7.15*b*). Similar bodies in North America are found in central Idaho, southern California, and coastal British Columbia.

WEATHERING AND SEDIMENTS

Most igneous and metamorphic, as well as some sedimentary, rocks form deep in Earth's interior in an environment of high temperatures and pressures. When ultimately exposed at Earth's surface by the wearing away of overlying rock layers,

Table 7.4

Summary of Igneous Plutons

Pluton	Tabular or Nontabular	Concordant or Discordant	Common Composition	Other Characteristics
Dike	Tabular	Discordant	Granite	
Sill	Tabular	Concordant	Gabbro	
Laccolith	Nontabular	Concordant	Granite	Thickened upward
Lopolith	Nontabular	Concordant	Gabbro	Thickened downward
Stock	Nontabular	Discordant	Granite	< 100 sq. kilometers
Batholith	Nontabular	Discordant	Granite	> 100 sq. kilometers

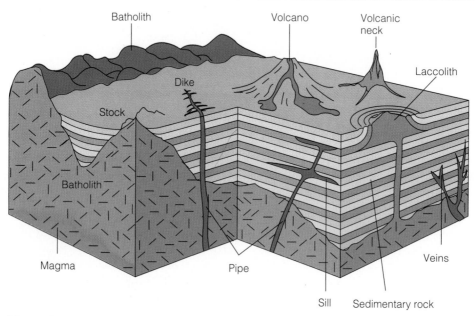

Figure 7.14 Igneous plutons.

these deeply formed rocks find themselves in a "hostile" low-temperature environment in which they are unstable. Additionally, they come in contact with such chemicals as oxygen, water, carbon dioxide, and various acids. Chemical reactions occur between these substances and the minerals in the rocks, leading to rock breakdown in a process known as **weathering** (fig. 7.16). This physical and chemical breakdown of rock material fulfills several important functions. Weathering prepares rock or any earth material for movement by converting large masses into smaller, more manageable particles that can be picked up and transported by air, water, and ice. With this subsequent removal of earth materials, weathering gradually wears down and lowers Earth's surface. Weathering is also partly responsible for the formation of soil on which not only plants but ultimately all life depends. Because of their importance as natural resources, soils will be considered in some detail in chapter 13.

Physical Weathering

Weathering is normally considered to occur in two fundamentally different yet related ways. The first is the purely mechanical breakdown of rock masses into smaller pieces without any change in chemical or mineralogical composition, a process also known as **physical weathering,** or *disintegration.* For example, a large piece of granite may be broken into smaller pieces of granite or separated into its component minerals (fig. 7.17a). The reduction in particle size not only allows for easier movement, but, more important, it exposes more rock and mineral surface area to attack by the chemical weathering agents. This mechanical change is brought about in several ways. The expansion of water as it freezes in cracks in the rock exerts tremendous pressure on the rock, often sufficient to break off large pieces (fig. 7.17b). Called *frost wedging,* this process is particularly important at high altitudes, where nighttime temperatures fall below freezing even during summer. Indeed, some landscapes above the treeline in mountain regions are literally covered by angular rock boulders that have been dislodged by this process (fig. 7.17c). Plant roots, as they work their way into rock cracks and grow, eventually pry loose rock fragments. Burrowing animals may also contribute to this overall process. In some regions, even alternate wetting and drying of mineral material or the expansion and contraction due to temperature changes may contribute to the process of disintegration.

Some igneous rocks, particularly granites, that formed at great depths undergo expansion due to relaxation of pressure when overlying rock layers are stripped away by erosion. This expansion leads to the development of fractures parallel to the ground surface, since this was the direction of least resistance

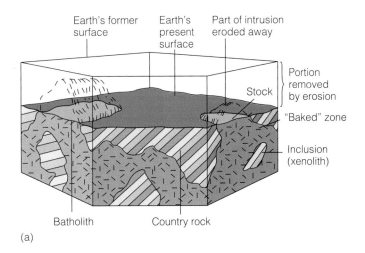

(a)

(b)

Figure 7.15 (a) Representation of how batholiths and stocks are exposed at Earth's surface by the removal of overlying rock layers by erosion. The inclusions (xenoliths) are pieces of "country rock" broken off and surrounded as the liquid magma works its way upward. (b) The Sierra Nevada of California, a batholith exposed and carved into a rugged mountain range by erosion. *Continued*

Figure 7.15 (c) A large inclusion (dark mass) within a light-colored granite (White Mountains, New Hampshire).

and, therefore, of easiest movement. Weathering then strips off these rock layers like peeling an onion in a process known as *exfoliation* (fig. 7.18). Mountainous masses with unique curved shapes that have been formed by this process are referred to as *exfoliation domes.* Yosemite National Park in the Sierra Nevada of California is especially noted for these features.

Chemical Weathering

More important than the mechanical breakdown of rock material are the chemical and mineralogical changes, known collectively as **chemical weathering,** or *decomposition,* that occur when rocks are exposed at Earth's surface (table 7.5). It is this chemical weathering, with its formation of new minerals, that finally brings the rock material into a state of chemical equilibrium with the surface environment. Such changes brought about by reactions with water (*hydrolysis*) and carbon dioxide (*carbonation*) are especially important in converting the most abundant igneous mineral feldspar into clay and quartz. These latter two minerals form the basis for most soils and for the abundant sedimentary rocks shale and sandstone. The red, tan, or brown color of most rocks exposed at Earth's surface is the result of the chemical reaction of iron-bearing minerals in the rocks with the oxygen and water in the atmosphere, a natural *oxidation,* or "rusting," process.

Climate is the overriding factor that governs the weathering process. In arid regions, where little water is available, physical weathering dominates over chemical weathering, soil development is curtailed, vegetation is sparse, and bedrock lies at or near the ground surface. All this produces a stark, "angular" terrain. In humid regions, with abundant moisture, the land becomes blanketed with a covering of soil and vegetation and develops a "smoother," more gently rolling appearance. Contrast, for example, the steep cliffs of the arid Grand Canyon region with the more gently rolling terrain of the humid Appalachians of north-central Pennsylvania (fig. 7.19). The underlying rocks are basically the same. It is largely the effect of climate that distinguishes the two areas.

(a)

(b)

Figure 7.16 Examples of weathering (a) Marble (composed of the mineral calcite) tombstone, highly susceptible to weathering (dissolution) in a humid climate. (b) "Sphaeroidally" weathered igneous boulder.

SEDIMENTARY ROCKS

Most earth material, once dislodged by weathering, begins another step in the endless rock cycle. It is picked up (eroded) and transported by water, wind, or ice until it finally is deposited and comes to rest. This loose material is called **sediment,** and it is most commonly deposited in nearly horizontal layers in the shallow waters at the edges of the continents, into which most large

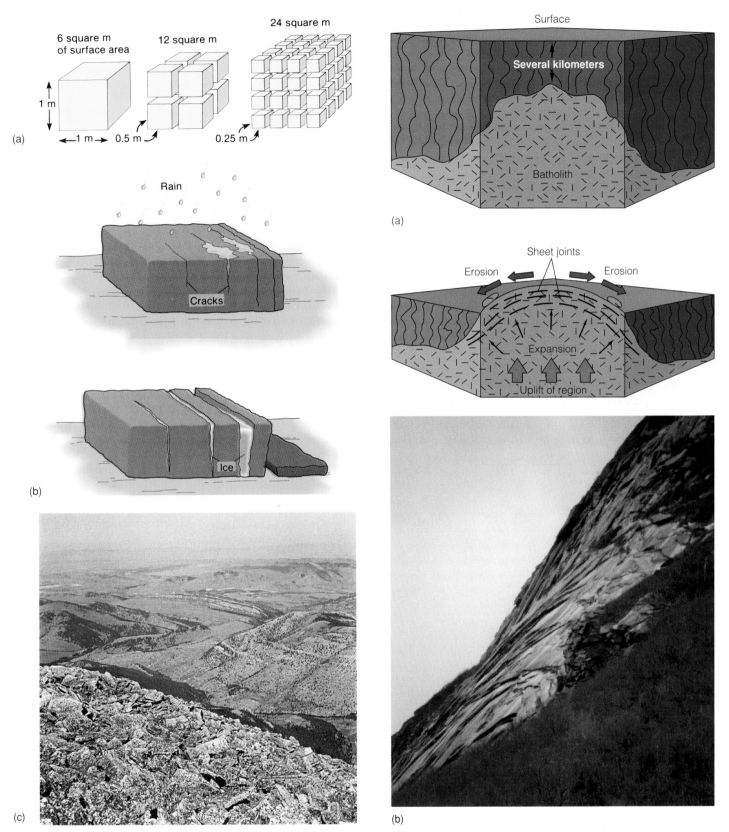

Figure 7.17 Physical (mechanical) weathering. (*a*) Physical weathering reduces particle size (without decreasing its overall volume), thereby increasing the amount of rock area exposed to chemical weathering. (*b*) Ice, expanding as it freezes in rock cracks, breaks off pieces of rock, a process known as frost wedging. (*c*) The results of frost wedging at high altitudes.

Figure 7.18 Exfoliation. (*a*) How pressure reduction due to erosion of overlying rock layers causes upward expansion in granite and fractures parallel to the ground surface. (*b*) Exfoliation at work in the White Mountains of New Hampshire.

Table 7.5

The Chemical Weathering of Common Rock-Forming Minerals

Original Mineral	Major Mineral Product	Other (mostly soluble) Products
K-feldspar	Clay minerals	SiO_2, K^{1+}
Plagioclase feldspar	Clay minerals	SiO_2, Na^{1+}, Ca^{2+}
Muscovite mica	Clay minerals	SiO_2, K^{1+}
Biotite mica	Clay minerals	SiO_2, K^{1+} Mg^{2+}, Fe oxides
Hornblende (amphibole)	Clay minerals	SiO_2, Na^{1+}, Ca^{2+}, Mg^{2+}, Fe oxides
Augite (pyroxene)	Clay minerals	SiO_2, Na^{1+}, Ca^{2+}, Mg^{2+}, Fe oxides
Quartz	Quartz	
Calcite	—	Ca^{2+}, Mg^{2+}, $(HCO_3)^{1-}$

(a)

(b)

Figure 7.19 (a) The Grand Canyon of the Colorado River in Arizona. (b) Pine Creek Gorge, the "Grand Canyon of Pennsylvania." Both canyons have been cut through horizontal sedimentary rock layers. The "smoother" topography of Pine Creek Gorge is attributable to the more intense chemical weathering and vegetation cover of "humid" Pennsylvania.

rivers ultimately drain. Successively younger layers deposited on top of older layers add to the sedimentary mass.

Because of the pressure of the overlying layers and the precipitation of mineral material within the body of sediment by circulating water, the loose sediment, over the course of millions of years, is converted into hard sedimentary rock. Because such rocks form at or near Earth's surface, they are the most common types found there and form a relatively thin veneer across approximately 75% of the continental surfaces. Bear in mind that although originally deposited in shallow seas, many of these rocks are eventually exposed when the seafloor is elevated or the seas drain off the continental surfaces, processes that will be discussed further in chapter 14.

Sedimentary Rock Types

The mineral material that makes up sedimentary rocks is transported and utilized in ways that give rise to three basic types of sedimentary rocks. The most common, by far, are those that form as the result of the movement by streams of rock and mineral particles of various sizes ranging from fine clay to large boulders. When deposited, buried, compacted, and hardened, or *lithified,* this material forms **clastic** ("fragmental") sedimentary rocks. This category includes such familiar rock types as conglomerate, sandstone, siltstone, and shale (fig. 7.20).

Some weathered material is not carried in the form of particles but rather as dissolved ions. These ions may be removed from solution by animals to construct their shells, as clams and oysters do, or by plants for their growth. The accumulated shell remains of many generations of animals, mostly living in the world's oceans, gives rise to the rock limestone, whose chief mineral constituent is calcite. The accumulation of dead plant material, mostly in swampy environments, forms coal. Both of these rocks are clastic in the sense that they consist of fragments. Since the fragments are of biological origin, however, they are more commonly considered as **bioclastic** rocks. Because of its importance as an economic resource, coal receives additonal consideration in chapter 13.

Box 7.3 Further Consideration

The Wentworth-Udden Scale: How Large Is a Sand Grain?

We all have an intuitive sense of the size of such common particles as sand grains and pebbles, and usually such a sense is sufficient for our purposes. The professional earth scientist, however, often must be more precise about the nature of sedimentary particles. Consequently, several particle size scales have been developed over the years. One of the oldest, but still the most widely used, is the Wentworth-Udden scale, first developed by J. A. Udden in 1898 and modified and expanded by C. K. Wentworth in 1922 (table 7.A). This scale is a geometric scale and as such has two distinct advantages over arithmetic scales, where each interval is of equal value. First, the boundaries for particle size classes differ from one another by a constant ratio, or multiplier, in this case one-half. For example, the upper limit of the "granule" is a diameter of 4 millimeters. The lower limit of the "granule" category is 2 millimeters. (4 x 1/2). Second, a geometric scale expands the

fine-grained end of the scale and compresses the coarse-grained end of the scale. The advantage of this is that small changes in grain diameter at the clay-silt end of the scale can be very significant, whereas even large changes in particle diameter at the boulder-pebble end of the scale are probably of little significance.

Although the boundaries for the sand category seem to contradict the constant ratio of one-half for determining such boundaries, they indeed do not. Such categories can be subdivided using the constant ratio as shown along the right hand portion of the chart.

Table 7.A

The Wentworth-Udden Scale

Particle Diameter (mm)	Name	Sand Particle Diameters (mm)
Greater than 256	Boulder	
256–64	Cobble	
64–4	Pebble	2–1 = very coarse sand
4–2	Granule	1–1/2 = coarse sand
		1/2–1/4 = medium sand
2–1/16	Sand	1/4–1/8 = fine sand
1/16–1/256	Silt	1/8–1/16 = very fine sand
Less than 1/256	Clay	

Apply Your Knowledge

1. Subdivide the "pebble" category as was done for the "sand" category and give appropriate names for each subdivision. Do the same thing for the "silt" category.

2. How does the expansion of the fine-grained end of a geometric scale and the compression of the coarse-grained end of the same scale differ from what would occur using an arithmetic scale?

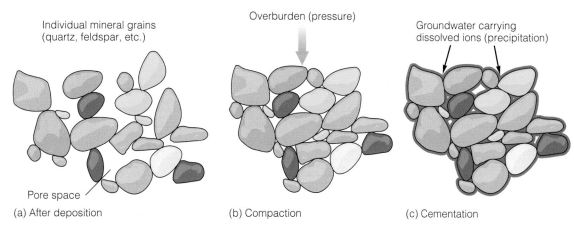

Individual mineral grains (quartz, feldspar, etc.)

Pore space

(a) After deposition

Overburden (pressure)

(b) Compaction

Groundwater carrying dissolved ions (precipitation)

(c) Cementation

Figure 7.20   (a) Mineral particles immediately after deposition. (b) Mineral grains squeezed closer together by the weight of the overlying sediment. (c) Deposition of mineral material by circulating water that cements the mineral grains together completing the lithification process.

A third category of sedimentary rocks is the **nonclastic,** or *crystalline,* rocks, which precipitate directly out of solution, usually as the water in which their constituent ions are dissolved evaporates. Because this process is similar to how minerals form from magma, the minerals in nonclastic rocks form a tight texture of interlocking crystals. This category includes some limestones and the chemically similar dolostone (consisting of the mineral dolomite); but it also encompasses an economically important subgroup known as the **evaporites.**

The evaporation process removes only pure water, so that when a restricted body of seawater in an arid climate slowly dries up, the dissolved salts become increasingly concentrated in the water that remains. Eventually the salts become so concentrated that they can no longer all remain in solution, and the less-soluble ones begin to precipitate out as horizontal rock layers on the floor of the shallow sea or lagoon. Three of the most important such salts are gypsum ($CaSO_4nH_2O$), widely used in plaster and plasterboard; halite ($NaCl$), or common table salt; and sylvite (KCl), an important source of potassium (K) for fertilizer. During the Permian period (approximately 250 million years ago), a large sea that covered much of western Texas and eastern New Mexico gradually dried up and, in the process, left a blanket of evaporite deposits hundreds of meters thick over an area of more than 250,000 square kilometers (100,000 sq. mi). Such thick deposits imply a delicately balanced situation. Once the process of salt precipitation began, just enough seawater was able to enter the area of restriction to replace that which was being lost, leading to continuous precipitation of salts.

Compared to igneous rocks, sedimentary rocks are compositionally very simple, with many being monomineralic. Three rock types, shale (clay minerals + quartz), sandstone (quartz), and limestone (calcite), probably account for more than 99% of the volume of all of the sedimentary rocks. Rocks such as dolostone, gypsum, and rock salt may have local economic importance, but their quantity is insignificant. Figure 7.21 is a general classification scheme for the sedimentary rocks based on particle size (for just the clastic rocks) and mineral composition.

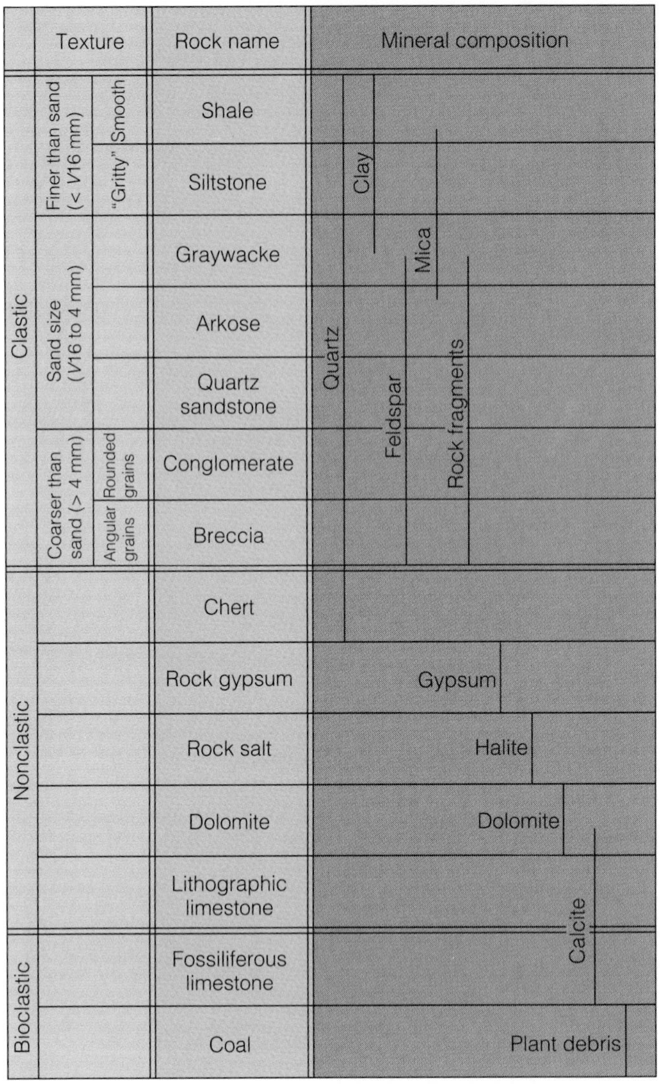

Texture			Rock name	Mineral composition
Clastic	Finer than sand (< 1/16 mm)	"Gritty" Smooth	Shale	Clay — Quartz
			Siltstone	
	Sand size (1/16 to 4 mm)		Graywacke	Mica — Feldspar — Rock fragments — Quartz
			Arkose	
			Quartz sandstone	
	Coarser than sand (> 4 mm)	Rounded grains	Conglomerate	
		Angular grains	Breccia	
Nonclastic			Chert	
			Rock gypsum	Gypsum
			Rock salt	Halite
			Dolomite	Dolomite
			Lithographic limestone	Calcite
Bioclastic			Fossiliferous limestone	Calcite
			Coal	Plant debris

Figure 7.21   General classification scheme for sedimentary rocks.

Clues to Earth's Past

Although simple in composition, sedimentary rocks contain a wealth of information that has allowed earth scientists to piece together much of what we know about Earth's history. The very fact that they form in almost horizontal layers tells us much about the nature of the formative processes and the environments in which they occur. *This horizontal layering, or stratification, is, in fact, the single most characteristic feature of sedimentary rocks.* **Fossils,** the direct evidence of past life, also represent significant information. Fossils most commonly are in the form of hard animal parts, such as shells, teeth, and bones, which can withstand the rigors of erosion and transport. The rare preservation of imprints of animal soft body tissues, footprints, burrows, and plant leaf impressions, however, can provide even more valuable information (fig. 7.22). The abundance of fossil material varies with rock type. Some limestone consists wholly or in large part of animal shell remains. Fine-grained limestones and shales provide the best media for the preservation of the various types of impressions. Sandstones often are lacking in fossil material because the constant pounding by waves in the beach environments where many form, destroys the fragile shell material.

Mud cracks, formed on tidal mud flats and in other environments subject to alternate wetting and drying, can also be preserved in the lithified rock. They develop when wet mud shrinks upon drying. During the next submergence of the surface, other sediment carried in by the water may fill in the cracks, preserving them (fig. 7.23). Other features commonly preserved in sedimentary rocks are the sand ripples (*ripple marks*) formed by the movement of sand particles on the surface of dunes and sandbars and the small-scale, inclined sand layers (*crossbeds*) within the bodies of the same features. Under certain circumstances, ripple marks and crossbeds tell earth scientists the direction of wind or water movement tens or hundreds of millions of years ago.

(a)

(b)

(c)

Figure 7.22   Examples of the variety of fossils. (*a*) The shells of marine animals preserved in limestone. (*b*) Leaf impressions in shale. (*c*) Petrified wood.

METAMORPHIC ROCKS

The third major group of rocks, the metamorphics, are those that have been altered while in the solid state in response to changes in temperature, pressure, and the chemical environment. The significance of "solid state" changes is that once melting of rock occurs, the mineral matter crosses the boundary into the igneous domain. Generally, the metamorphic processes are the opposite of the weathering processes. Rather than high-temperature, high-pressure rocks being exposed to the elements at Earth's surface, during metamorphism, low-temperature, low-pressure sedimentary rocks are made unstable by being buried deep in Earth, where temperatures and pressures are high. Although any rock can be metamorphosed, sedimentary rocks are the most easily changed.

Whereas elevated temperatures can bring about significant changes in sedimentary rocks (as the alteration of clay to mica and chlorite in hornfels), pressure is responsible for the most obvious changes. In some instances, as along fault zones (chapter 9), where rocks move against each other, the rock may be altered by crushing and grinding to form a new rock of angular fragments cemented together, known as a *breccia*.

More significantly, however, pressure affects all sedimentary rocks by first squeezing mineral grains closer together, thus eliminating open spaces and expelling the fluids that are present (fig. 7.24*a* and *b*). As pressure continues to increase, the mineral grains form a very tightly interlocking mosaic of crystals, much like a rock wall or fireplace, where irregularly sized and shaped rocks have been fitted tightly together. At still higher pressures, small mineral grains reform into fewer, but larger, grains in the process called recrystallization. During recrystallization, the mineral may remain unchanged, or a new mineral may form. In the latter case, the new mineral is of higher density than the old one, because recrystallization, like all of the responses to increased pressure, reduces the mineral matter to a smaller volume (fig. 7.24*c* and *d*). Pressure may also cause newly formed mineral grains, especially the elongated ones of the amphiboles and the flat ones of the micas, to be oriented so that their long or flat dimensions lie perpendicular to the direction of maximum pressure. This process imparts to the metamorphic rock a distinct alignment of linear minerals (*lineation*)

(a)

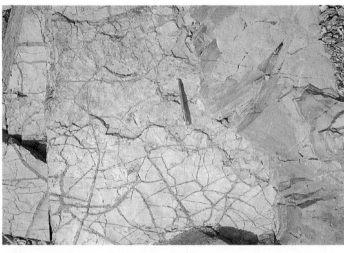

(b)

Figure 7.23 (a) Fresh mud cracks form as mud dries. (b) Mud cracks preserved in rocks that are more than one billion years old.

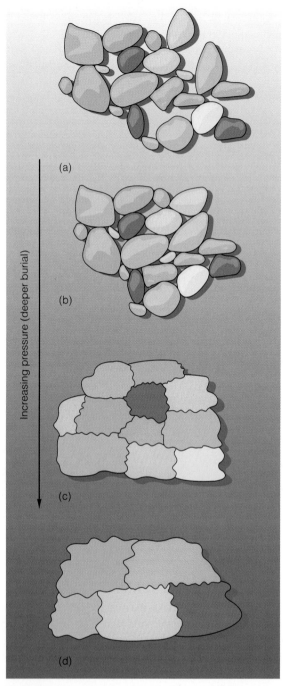

(a)

(b)

(c)

(d)

Increasing pressure (deeper burial)

Figure 7.24 Response of mineral grains to increasing pressure. (a) Original open sedimentary texture. (b) Pressure pushes grains closer together (eliminating open space) without altering their shape or size. (c) With all open space eliminated, grain shape changes to form interlocking network. (d) A few large grains form from many small ones (with or without a change in composition), thereby reducing overall volume. The final step is a process known as recrystallization.

or layering (*foliation*) in the case of flat minerals. Foliation is an important characteristic used in the classification of metamorphic rocks and is discussed more fully in that context.

A rock's "chemical environment" is largely determined by the presence of water carrying an abundance of dissolved ions, which come in contact and react with the changing rock. The source of such fluids is threefold: groundwater that fell as precipitation and works its way downward through soil, sediment, and rock; water that was trapped in the pore spaces of the sediment at the time of deposition, in a sense "fossil" seawater; and leftover, or residual, fluids associated with the final stages of igneous rock crystallization, and that, as previously stated, often represent an important source of metallic ore minerals.

Types of Metamorphism

The term *metamorphism* really encompasses a broad group of related processes that operate at different intensities under different conditions. These variations in processes and conditions give rise to a number of categories of metamorphism. For our purposes, however, only two such varieties need be considered: *contact metamorphism* and *regional metamorphism.* Contact metamorphism occurs in association with lava flows and shallow igneous intrusions. It involves the alteration of the rock that surrounds the intrusion (usually referred to as the *country rock*) by the high temperatures and chemical fluids associated with the intrusion. Thus, the effects are normally those produced by heat and chemical reaction. Pressure is often negligible. Alteration is localized in a "halo," or *metamorphic aureole,* surrounding the intrusion and extending only a few meters to a few tens of meters into the country rock. Just how far is normally a function of the difference in temperature between the intrusion and the country rock and the permeability of the country rock. The *permeability,* or ease of passage of fluids through a rock, is discussed in more detail in chapter 8. In general, as the distance from the intrusion increases, the intensity of metamorphism decreases and does so rapidly (fig. 7.25).

Regional metamorphism, as the name implies, occurs across vast areas and normally at great depth in association with the deep burial of sediments and massive igneous batholiths. While heat, pressure, and chemical fluids all play major roles in regional metamorphism, pressure becomes particularly important because of the great depths involved.

Metamorphic Rock Classification

Metamorphic rocks are initially separated and classified on the basis of whether or not they possess foliation (figs. 7.26 and 7.27). Those rocks that do are called **foliated**. A secondary subdivision is based on grain size, with increasing pressure leading to progressively larger grain size. If we examine the progression of rocks formed in response to increasing pressure, we would find that they possess large amounts of mica, because the flat crystals of the members of this mineral group accommodate themselves well to high pressure. Shale, the most common sedimentary rock, consists mostly of clay minerals. These minerals are converted to micas under even relatively low pressures. The result is a fine-grained rock in which all of the minute mica crystals are oriented with their flat surfaces parallel to one another. This parallelism provides directions of weakness in the rock along which the rock easily splits, or cleaves. This rock is a *slate,* and the quality whereby it splits is called *slaty cleavage,* a variety of foliation.

With increased pressure, the mica crystals become enlarged, and although still too small to be distinguished by the naked eye, they are large enough to reflect light, thereby giving a shiny appearance to the surfaces along which they split. A rock with this characteristic is called a *phyllite* and is easily recognizable by these shiny surfaces and its sometimes crenulated, or "wavy," appearance. At still higher pressures the mica crystals become large enough be distinguished with the unaided eye. The rock with such distinctive layering is called schist, and the foliation is *schistosity.* One other type of foliated rock is commonly found. Under the highest metamorphic temperature and pressure conditions, gneiss (pronounced "nice") forms. Gneiss is coarse-grained and has much the same mineralogy and appearance as granite, except that the minerals are segregated into distinct bands dominated alternately by light and dark minerals. This distinct banding is the variety of foliation known as *gneissosity.*

We have considered only the characteristic of foliation in naming the metamorphic rocks, and, indeed, this is the dominant characteristic. Metamorphic rocks are, however, as varied or more varied in their mineralogical composition than igneous rocks. While the basic rock name is determined by the type of foliation and by the texture (grain size), a more complete name can be formed by using compositional terms as modifying adjectives. For example, we might refer to a "granite gneiss" or a

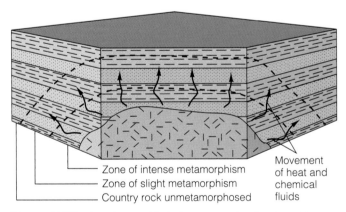

Figure 7.25 A metamorphic halo, or aureole, developed in sedimentary country rock in response to temperatures and chemical fluids associated with igneous intrusion. Total depth of metamorphic penetration may vary from a few meters to several tens of meters depending on intensity of heat and permeability of the country rock.

Box 7.4 Further Consideration

Granitization: Where Did All That Granite Come From?

A further consideration of metamorphism can not only add new information to the subject but also illustrate an important point about the nature of scientific inquiry and of scientific debate. Thus far, we have used the term *metamorphism* to describe all solid state changes that occur in rocks. Specialists, however, would distinguish between metamorphism, which involves no fluids (or fluids already within a rock mass), and *metasomatism,* in which fluids carrying dissolved ions enter the rock mass from outside. This concept of metasomatism is at the heart of a long-standing controversy dealing with the origin of granites. To this point we have accepted that granite is of igneous origin. Some geologists argue,

however, that some granites form by a specific type of metasomatism referred to as *granitization*. They envision fluids carrying dissolved ions passing through a rock mass. Chemical reactions between these ions and the minerals already present convert the original rock to one with the mineralogy and texture of a granite.

What evidence could be cited to substantiate this radically different concept? First, at the margins of some granite masses there is no sign of the alteration of the country rock, which should have been caused by the heat of liquid magma. Second, some sedimentary and/or metamorphic features present in the country rock can be traced *into* the "granite" mass before disappearing completely. Third, proponents of an igneous origin for granite have always been at a loss to resolve what is usually referred to as the "space problem." If massive batholiths were intruded into the country rock, they would have had to displace hundreds of cubic kilometers of material. Where did this material go?

On the other side of the argument, specialists point to the numerous cases in nature

where distinctive evidence for the forceful intrusion of hot, liquid magma can be cited. Which side is right and which is wrong? As with so many questions in the sciences, there is no clear-cut answer. Most earth scientists would probably acknowledge that both origins for granite are possible.

Apply Your Knowledge

1. Do you think that metamorphism or metasomatism is the more common process? Why?
2. What do you think might have happened to the rock material displaced by the intrusion of a granite batholith? Approximately how much rock material (in cubic kilometers) was displaced by the intrusion of the Sierra Nevada batholith? (Consult a map of California.)
3. If both magmatic and metasomatic origins are possible for a magma, which do you think is the more common? Why?

Texture (identifying feature)			Rock name	Metamorphism		Dominant mineral composition						Original rock
				Dominant kind	Degree							
Foliated	Fine grained	"Smooth" "Fractured"	Slate	Regional	Low grade	Clay						Shale
		"Shiney" "Layered"	Phyllite	Regional	Medium grade	Chlorite						Shale
	Coarse grained	"Banded" "Layered"	Schist	Regional			Mica	Quartz	Amphibole	Feldspar		Shale
			Gneiss	Regional	High grade							Shale or granite
Nonfoliated	Fine grained		Hornfels	Contact	Medium to high							Shale
	Coarse grained	No reaction with HCl	Quartzite	Contact or regional	Medium to high							Quartz sandstone
		Reaction with HCl	Marble	Contact or regional	Medium to high						Calcite	Limestone or dolomite

Figure 7.26   General classification scheme for the metamorphic rocks.

(a)   (b)

Figure 7.27   Metamorphic rock textures (*a*) nonfoliated (marble) and (*b*) foliated (slate).

"biotite schist" and thereby present a meaningful image of the rock under consideration.

The second broad group of metamorphic rocks is the **nonfoliated** rocks. Such rocks are either formed under low pressures, as when shale is "baked" around the margins of igneous intrusions to form the dark, fine-grained rock known as hornfels (which is easily mistaken for basalt), or where the minerals present are not flat and do not lend themselves to the development of good foliation. In the latter category are the important nonfoliated rocks marble and quartzite. Marble is metamorphosed limestone (calcite) or dolostone (dolomite), and quartzite is metamorphosed quartz sandstone. Marble has been used since antiquity for statues, monuments, and buildings because it is abundant, can be worked with relative ease, and is reasonably resistant to the elements.

Further Thoughts: Taking the Good with the Bad

There are negatives and positives to the mineral-and-rock-forming processes summarized by the rock cycle. The rock material, formed and reformed over billions of years, provides us the solid rock platforms on which we live. The same processes are still creating habitable surfaces in places like Hawaii and Iceland through volcanic eruptions, which add land area to those growing islands, and in deltaic regions, where the land is building outward into the sea. In other areas, similar processes form features of scenic wonder such as the volcanic peaks of the Cascade Range and the sculpted batholiths of the Sierra Nevada. Natural concentrating processes have formed mineral accumulations of economic significance that have been exploited by human beings for thousands of years.

Too often, however, human beings and other organisms have become caught up in those same volcanic eruptions, earthquakes, and floods with tragic results. The processes that formed economic concentrations of minerals may have left the soils and waters in certain regions with excessively high concentrations of certain elements or may have depleted the soils of vital plant nutrients. For the most part, however, human beings have learned to adapt to and live with this environment. When we cross the boundary into areas where we should not be or fail to care for what we have, nature has a way of catching our attention.

Significant Points

1. A mineral is a naturally occurring, inorganic element or compound with a definite internal arrangement and a chemical composition that is fixed or varies within narrow limits.
2. The silicates are the most abundant of all the mineral groupings. The basic silicate structural element is the silicon-oxygen tetrahedron.
3. Minerals form by crystallization from either magma or water.

4. A rock is an aggregate of mineral grains that makes up a significant portion of Earth's outer shell. The three principal groups are igneous, sedimentary, and metamorphic, and they are interrelated through the rock cycle.
5. Igneous rocks form by the crystallization of mineral material from magma (intrusive) or lava (extrusive).
6. The size, shape, and arrangement of the mineral grains in a rock form the rock's texture.

7. Early-formed igneous minerals tend to be dark in color, while late-formed minerals are light. There is rarely mixing of early and late minerals in a rock.

8. Basaltic magma is more fluid than granitic magma; therefore, basalts and related rocks dominate the volcanic (extrusive) rocks, while granites and their kin dominate the plutonic (intrusive) rocks.

9. Pegmatites and hydrothermal veins form late in the crystallization process of a granitic magma.

10. The more intense chemical weathering characteristic of humid regions leads to a topography that is smooth, whereas that of arid regions tends to be sharp and stark.

11. Sedimentary rocks form by the accumulation and consolidation of loose sediment at or near Earth's surface, where they are the most common rock type.

12. Sedimentary rocks are subdivided into clastic, bioclastic, and nonclastic on the basis of texture.

13. Metamorphic rocks form by the alteration of preexisting rocks in the solid state, largely through a process known as recrystallization.

14. Metamorphic rocks are foliated if they possess a noticeable layering. They are nonfoliated if such layering is not apparent.

15. Contact metamorphism usually occurs on a small scale around the margins of intrusive igneous rock bodies. The principal metamorphic agents are heat and chemical fluids. Regional metamorphism occurs on a large scale, usually at great depth and in association with batholiths. Heat and pressure are the dominant metamorphic agents.

Essential Terms

mineral 158	metamorphic rock 164	basalt plateau 169	nonclastic 178
crystallization 159	rock cycle 164	pluton 170	evaporites 178
magma 159	lava 166	batholith 171	fossil 178
crystal 160	volcanic (extrusive) 166	weathering 172	foliated 181
ionic substitution 160	plutonic (intrusive) 166	physical weathering 172	nonfoliated 183
rock 162	texture 166	chemical weathering 174	
igneous rock 164	volcano 167	sediment 174	
sedimentary rock 164	pyroclastic debris 169	clastic 176	

Review Questions

1. What is the basis for mineral classification? What are some of the categories?
2. How does a crystal differ from a crystal lattice?
3. What is meant by "ionic substitution" in minerals? Give at least one example.
4. What are some of the important characteristics used in mineral identification?
5. How does cleavage differ from fracture? How do cleavage surfaces differ from crystal faces?
6. What are the principal igneous rock textures? Under what conditions does each form?
7. What is a volcano? What varieties are recognized in terms of activity?
8. How do the three principal types of volcanoes differ from one another in terms of size, shape, and composition?
9. What is a caldera?
10. What are the important concordant plutons? What are the important discordant plutons? What is the largest of all plutons?
11. What is the difference between chemical and physical weathering?
12. What are some examples of clastic sedimentary rocks? of bioclastic rocks? of nonclastic sedimentary rocks?
13. What is a fossil? Give some examples.
14. What are some examples of foliated metamorphic rocks? of nonfoliated metamorphic rocks?

Challenges

1. Examine some of the rocks exposed in road cuts in the area you live. Are they generally igneous, sedimentary, or metamorphic? Photograph or sketch several typical exposures. Identify and describe the rock types, indicating particularly the characteristics that aided in your identification.

2. Make a collection of rocks and/or minerals that can be found near your home or that you have picked up on a trip.

3. Select several mineral specimens you have found or bought or were made available to you by your instructor. Examine each in terms of the identifying properties

described in this text. Note on a checklist which features or properties you can observe on the specimens and which ones you cannot observe.

4. Examine a single (or several) rock specimen you have found or is provided by your instructor. Describe it as completely as you can, particularly with reference to those characteristics that tell you something about the rock's origin and history.

5. Select a volcano that is located in your state (if none exists, select one from anywhere). With the use of reference materials available in your college or public library, write a complete history for that volcano. Pay particular attention to type, severity, and frequency of eruptions.

8

EXTERNAL FEATURES AND PROCESSES

Head of glacier on Mount McKinley, Alaska.

Chapter Outline

INTRODUCTION

During the course of an ordinary human life, Earth appears to change very little. Its surface features and processes seem constant to the casual observer. Hills continue to stand above the lowlands, streams steadily run through their channels, and shorelines play host to annual influxes of vacationers. Our perceptions notwithstanding, Earth's surface is constantly changing. Streams undercut their banks and carry the resulting sediment away. Rock and mineral particles dislodged from the hills move downslope, gradually lowering the surface of the land. On the beach, the sand migrates under the influence of wind and water. Other subtle changes slowly reshape the landscape as we go about the daily business of our lives.

From time to time, however, such catastrophic events as floods, hurricanes, volcanic eruptions, and earthquakes radically alter a locality, sometimes within a matter of minutes. After a nineteenth-century earthquake, the citizens of New Madrid, Missouri, had to relocate their town because the old site was inundated by a shifted Mississippi River. In 1986, a *lahar,* or rapidly moving volcanic mudflow, rushed off the summit of Colombia's Nevada del Ruiz volcano at 160 kilometers (100 mi) per hour toward the town of Armero. By the time the flow stopped, more than 20,000 people had been buried, and more than 41 square kilometers (16 sq. mi) of the Lagunilla River valley had become a flat, muddy plain (fig. 8.1).

GEOLOGIC WORK

All changes in the physical world require work. Throughout this chapter, we will emphasize over and over again that the landscape we see is the product of countless small changes compounded over thousands of years. Each surface agent that we will consider operates in its own way to produce unique landscape features and each leaves its telltale mark. The common theme among all of them, however, is that they do geologic work.

Energy and Change

Geologic work, or the movement of material, requires the expenditure of kinetic energy (chapter 2). In every instance of change, some force releases potential energy and causes it to be converted to kinetic energy. In the case of Armero's destruction, potential energy was stored in millions of tons of volcanic ash, mud, and glacial ice perched high on a mountain. This energy was released when the volcano's heat melted the ice, saturating the ash and mud until the entire mass was too heavy for its resting place. Pulled by gravity and laden with water, the almost liquified matter converted its potential energy to kinetic energy in its downslope race toward Armero.

Although gravity directly influenced the expenditure of energy that changed the Lagunilla River valley and the lives of the citizens of Armero, it was not the sole physical cause of the event. Energy from the sun, traversing the near vacuum of space in the form of electromagnetic radiation, was also involved. Solar energy differentially heats the surfaces of our planet (see chapter 4), causing wind and driving the hydrologic cycle (fig. 8.2). The massive water supply that was locked in glacial ice above Armero was the product of the hydrologic cycle and the weather conditions associated with high altitudes. In its rush downhill, the water in the mudflow worked to complete the cycle by moving back toward the sea, from which it had originally evaporated. Because sediment was moved by the water, the result was geologic work.

Figure 8.1   The town of Armero, Colombia, buried beneath a volcanic mudflow (lahar) generated by melted glacial ice during the eruption of Nevada del Ruiz, some 48 kilometers (30 mi) distant. The 30-meter-(100-foot-) thick flow killed all but 3,000 of the town's inhabitants.

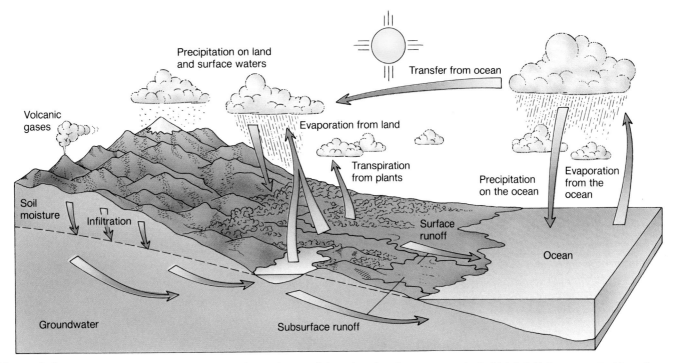

Figure 8.2 The hydrologic cycle. The diagram shows the possible paths taken by water, most of which is initially evaporated from the surfaces of the oceans.

Geologic Agents and Geologic Work

The external processes that constantly reshape Earth's surface involve the concentrated movements of air, water, and ice. These substances in motion are the primary **geologic agents:** streams, groundwater, glaciers, wind, waves, and currents. These six agents expend energy on a landscape under the guiding influences of atmospheric pressure and gravity.

All of the geologic agents do work by moving earth materials from one place to another. Their task has three basic components: (1) the picking up of rock material, or **erosion,** (2) the actual movement of rock material, or **transportation,** and (3) the ultimate laying to rest of rock material, or **deposition.** Each component differs from the others in its expenditure of energy. Consider the energy required to move a heavy box of books across a room. The work done in bending over and picking up the box differs from the work necessary to carry it across the room. Compare those two actions to dropping the box in a corner. Bending over and picking up the box uses more energy than carrying it, and dropping the box requires none at all. This is analogous to the work of the geologic agents. More energy is expended in erosion than in transportation, while deposition requires no energy. Deposition is the natural consequence of the earlier actions, since any material eventually comes to rest, even though that rest may be geologically temporary.

Erosional and Depositional Landforms

Earth's landscapes show the results of the work of geological agents in a series of individual features known as *landforms.* Each landform is shaped by either the erosional or deposition-al work of one of the six agents. A valley, for example, is a stream erosional feature, whereas a sinkhole is a groundwater erosional feature, and a moraine—a pile of rock debris—is a glacial depositional feature. The landforms are manifestations of the geologic agents, and studying them in conjunction with the agents reveals the interrelationship between process and product.

MASS WASTING: THE PRELIMINARY STEP

Before beginning a study of the six external geologic agents and their resultant landforms, you should consider a set of related natural processes known collectively as **mass wasting.** These normally occur in the alteration of landscapes prior to the work of the agents. Mass wasting is the downslope movement of earth material under the direct pull of gravity. Rock and mineral matter broken loose by one of the weathering processes (see chapter 7) may begin the downhill movement without the intervention of any of the geologic agents. A pebble broken loose from a conglomerate rock may roll down a slope for a short or a long distance before it temporarily comes to rest. A large mass of bedrock forming part of a cliff face may break loose and fall under the pull of gravity to the valley floor below. There are varieties of such movements, some very slow and others virtually instantaneous. Some movements are barely perceptible, while others are highly visible and even destructive. At times dry rock masses move, but often the movement requires some lubrication by water or ice. Generally, the steeper the slope the less lubrication required for movement (fig. 8.3).

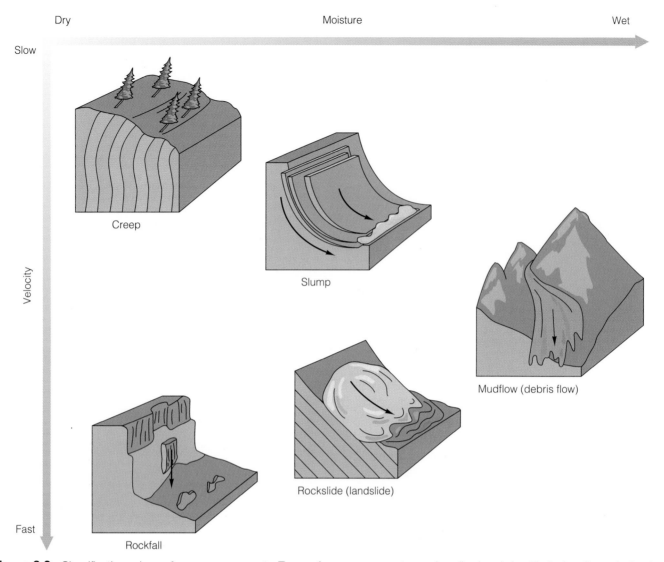

Dry   Moisture   Wet

Slow

Creep

Slump

Mudflow (debris flow)

Velocity

Rockslide (landslide)

Fast

Rockfall

Figure 8.3 Classification scheme for mass movements. Types of mass movements are described and classified primarily on the basis of two variables: moisture content of the moving mass and relative velocity of movement. No attempt is made in the diagram to assign actual water content or velocity values.

Creep

On all slopes, regardless of climate, individual soil, rock, or mineral particles work their way downslope. Such movement is usually very slow, with the results noticeable only after months or years of observation. This gradual movement is called **creep.** In many climates, the most common cause of creep is the alternate freezing and thawing of water in the soil. As water freezes, it expands (chapter 11), and this expansion in the soil lifts individual particles. When the water melts, it contracts and drops the lifted particles. Although the lifting is perpendicular to the ground's surface, the dropping is vertical under the direct pull of gravity toward Earth's center. On a sloping surface, the directions of lift and drop do not coincide. This process of alternate freezing and thawing with the consequent downslope displacement of mineral grains is a phenomenon known as *frost heaving* (fig. 8.4). The results of frost heaving and creep can be seen in curved tree trunks, displaced cemetery headstones, tilting utility poles, and crooked fence lines (fig. 8.5). In addition to slow-

ly moving material downslope, frost heaving also causes potholes in highways in regions with temperate climates.

Although it is slow, undramatic, and seemingly unimportant, creep is the most significant of all mass movements because it occurs in so many areas and affects the shallow surface zone where humans live and build structures. Designers of homes, barns, retaining walls, and highways must take into account this inevitable movement.

Rockslides and Rockfalls

Far more spectacular, but geologically much less important, are **rockslides** and **rockfalls.** Such movements are virtually instantaneous and involve the displacement of rock masses down very steep or vertical slopes. Rockslides and rockfalls are common and dangerous in mountainous areas. They are, nevertheless, limited in scope, and they normally do not affect large numbers of people (fig. 8.6).

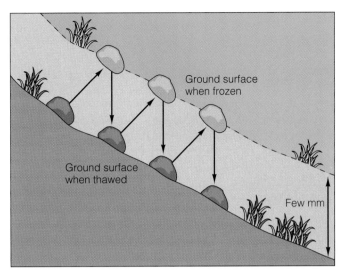

Figure 8.4 Slow progress of a rock particle downslope as a result of the alternate freezing (expansion) and thawing (contraction) of water in the pore spaces of the soil. This process is known as "frost heaving."

Figure 8.6 A rockslide blocks the highway at Boone, North Carolina.

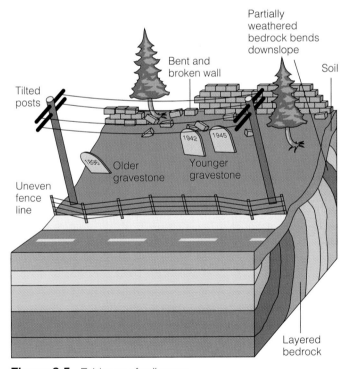

Figure 8.5 Evidence of soil creep.

Rock on a steep slope is inherently unstable. Often, large rock masses are held in place above a fracture surface or a sedimentary bedding plane only by friction. If something reduces the frictional resistance, the unstable mass above the surface may begin to move. Historically, the shaking and vibrations associated with earthquakes have been the major culprits in triggering rockslides and rockfalls. Such was the case on August 17, 1959, when a major earthquake in the vicinity of

Yellowstone National Park caused the movement of more than tens of millions of cubic meters of rock material from the steep walls of the Madison River canyon into the valley bottom, where it dammed the river and formed an 8-kilometer-(5-mi-) long lake. Eventually the U.S. Army Corps of Engineers dredged a channel through the natural dam to drain the lake and restore the normal stream drainage.

Heavy rainfall can also induce landslides by infiltrating fractures and lubricating the slide surface, thus reducing frictional resistance (fig. 8.7). In addition, water adds more weight to an already unstable slope. Streams, too, may initiate rockslides by undercutting steep slopes. Human activity has added to the instability of some slopes by road construction, mining, and the building of homes or other structures. On a somewhat smaller scale and commonly in hilly regions, masses of soil and bedrock slip along curved surfaces in a natural rotational movement. This results in the type of mass wasting known as *slump*. The upper portion of a slump is characterized by an exposed, steep, and curved slip surface, while the lower portion shows an irregular flow of material. Such movement may threaten highways or other structures near the base of the slope.

Mudflows

Residents of the hills and canyons of southern California are plagued by another variety of mass wasting, the **mudflow.** Mudflows are common in arid and semiarid regions or anywhere where rainfall is highly seasonal and torrential when it does occur. Southern California has very dry summers (see chapter 6). Brushfires strip the hillsides of their covering of vegetation. When the heavy rains come during the winter, the water picks up the unprotected soil to form masses of mud, which follow the natural stream channels through the hills and out onto the flatlands below. Widespread destruction of the expensive homes that line the canyons and cover the flats is as common as the change of seasons.

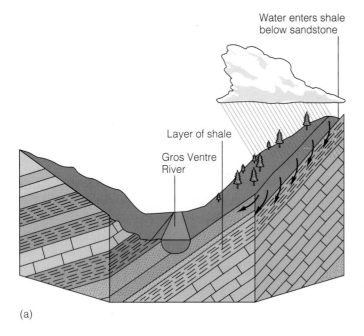

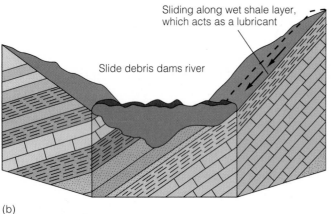

(a)

(b)

(c)

Figure 8.7   (a) Before and (b) after diagrammatic views of the 1925 landslide along the canyon of the Gros Ventre River in Wyoming. Movement was triggered by a period of heavy rainfall. (c) The Gros Ventre landslide "scar."

Subsidence

Mass wasting also occurs when humans or nature disturbs the rock balance underground. Such a disturbance causes **subsidence,** the vertical collapse or settling of the ground surface. A number of human activities have been responsible for subsidence. In some regions of Pennsylvania, the deep mining of coal and the ensuing subsidence has caused millions of dollars in damage to buildings, homes, and roads. Consider the mining of a 3-meter-(10-ft-) thick seam of coal in a mine 90 meters (300 ft) deep. If the roof collapses under the weight of the overlying layers of rock, 0.3 meter (1 ft) or more of vertical displacement may occur at the ground surface (fig. 8.8). A full 3 meters (10 ft) of displacement does not occur at the surface because as the overlying rock layers break up and settle, they expand and occupy a larger volume than they did prior to subsidence (fig. 8.9a).

In other areas, subsidence is caused by pumping groundwater and petroleum from loose sands (fig. 8.9b). Without the water or oil's support, the sand grains compact, reducing the overall rock volume and causing the ground surface to sink. Houston, Texas, Venice, Italy, and Mexico City are just a few of the major metropolitan areas where severe surface subsidence has resulted from extensive overpumping of groundwater to meet the demands of increasing populations and industries.

From the 1920s to the 1960s, the city of Long Beach, California, suffered as much as 8 meters (27 ft) of subsidence as a result of the extraction of oil and natural gas from the sands beneath the city. Because it lies along the coast, any drop in elevation subjects Long Beach to flooding. Over the years, the city spent millions of dollars to construct seawalls and levees to hold back the sea. In the early 1960s, subsidence in the Long Beach area was significantly reduced by the pumping of seawater under pressure into the sands from which oil had been extracted. The water became a substitute support, and oil companies profited because the pressurized water forced additional oil from the sands.

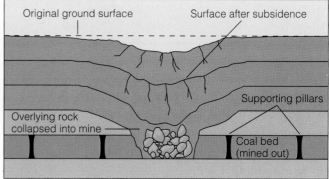

Figure 8.8   Subsidence caused when a coal seam is mined, leaving the overlying rock layers and the ground surface without sufficient support.

(a)

(b)

Figure 8.9 Subsidence (a) Surface collapse and sinkhole formation in Winter Park, Florida, due to excessive pumping of groundwater and resultant lowering of the water table, (b) damaged store in Belle Vernon, Pennsylvania. Subsidence was caused by underground coal mining.

STREAMS

Streams, the channelized flow of continental surface water, represent the most important of the geologic agents. They do more work than any other agent and perhaps more work than all other agents combined. Even in arid regions where few permanent streams flow, the bulk of sediment undergoes removal during those short periods when running water occupies the otherwise dry channels, known locally as "arroyos" or "washes" in the southwestern United States and as "wadies" in the Middle East and North Africa.

General Characteristics

Streams result from precipitation over the land's surface. Moisture that is not immediately evaporated, taken up by plants, or absorbed by the ground begins to run off the surface under the influence of gravity. For a short distance, this flow is in the form of a very thin sheet. Very soon, however, the sheet of water begins to divide, and the water becomes concentrated into small channels, or *rills*. As the water flows downslope, these small channels merge to form larger channels and streams form.

All streams flow downslope regardless of the geographic direction. They drop in elevation as they flow from their point of origin, *stream head,* to their end point, or *stream mouth.* This drop in elevation as a stream flows a certain distance is called the *stream gradient* and is measured in terms of meters dropped for each kilometer traveled, that is, meters per kilometer (or ft per mi). Normally, this drop in elevation is not constant over the stream's course; rather, the stream drops most steeply near its head and least steeply near its mouth. Consequently, stream flow is usually turbulent in the high ground where the stream originates and sluggish in the channel near the mouth, resulting in different expenditures of energy over the length of the stream. Energy is available for erosion on the steeper slope, but reduced energy leads to extensive deposition near the mouth. Figure 8.10 shows the *long,* or *longitudinal, profile* of a stream: literally a cross-sectional (side) view from its head to its mouth. Note that the profile is "concave upward," indicating diminished gradients as you follow the stream course toward its mouth.

The seemingly lazy flow and dominant deposition that occurs along the lower course of a stream can be misleading in terms of stream velocity. Because of the large volume of water that must be moved through the stream near its mouth, flow velocity is usually much higher than in the more turbulent portion of the stream near its head. The volume of water passing any point on a stream in a given interval of time is called the **discharge** and is normally measured in cubic meters per second (or cubic ft per sec).

Erosion

Streams do not flow indefinitely. At some point they end, either by flowing into another stream, a lake, or an ocean. In arid regions, they may simply dry up or infiltrate into the ground. When a stream reaches an ocean, its water is dispersed, and it is incapable of doing geologic work. Sea level is thus the lowest point to which a stream can erode its valley. This downward limit of erosion for all streams is called **ultimate base level.** Rare exceptions to sea level as base level for stream systems are isolated arid basins such as Death Valley in California and the Dead Sea in Israel and Jordan. These are areas cut off from the oceans and dropped below sea level by faulting (see chapter 9). Throughout its life and length, every stream strives to reach base level by eroding its channel and the surrounding landscape. Although the world's oceans serve as the ultimate base level for all stream systems, the downward erosion by individual streams may be controlled by the lakes or the larger streams into which they flow. Such base level control is considered to be *local,* or *temporary, base level.*

Erosion by streams involves a multifaceted expenditure of energy. On the smallest scale, the work entails the picking up of loose sediment delivered downslope either by mass wasting or by caving stream banks. As these mineral and rock particles move downstream, they bump against one another.

Box 8.1 Further Consideration

Stream Erosion on Mars

For many years the best-known physical attributes of the planet Mars were the straight, linear features known popularly as "canals." These were considered by many, including Sir Percival Lowell, the nineteenth century's most prominent astronomer, to be the marvelous engineering works of a long dead civilization. Closer views in photographs taken by the several *Mariner* and *Viking* space probes show no evidence of the existence of the canals, and they are generally regarded now as optical illusions. What the 1972 *Mariner 9* probe did reveal, however, was strong evidence for large-scale water erosion on the martian surface. Because of the extremely low temperatures found there, liquid water cannot exist on the surface of Mars today. In addition, the atmosphere of Mars is so thin and holds so little water vapor that rainfall cannot occur. The evidence for surface erosion by running water, therefore, requires the existence of very different climatic and environmental conditions at some time in the past.

Water-carved channels on Mars come in all sizes. The largest, the Valles Marineris (named for the *Mariner* probe), measures 4,500 kilometers (2,800 mi) long, has a maximum width of 600 kilometers (370 mi), and is up to 7 kilometers (4.4 mi) deep. If superimposed on the United States, it would stretch from New York to California. The principal origin was probably the rapid melting of large quantities of buried ice, which gave rise to a series of catastrophic floods. Such occurrences are known on Earth. During the Pleistocene ice age glacial ice dams burst, releasing the impounded waters of lakes in floods of almost unimaginable sizes.

Eroded channels of intermediate size are sinuous and range from as much as 56 kilometers (35 mi) to as little as 90 meters (288 ft) in width (fig. 8.A). They possess short, blunt tributaries forming dendritic patterns, as well as meandering channels, braided channels, islands, and terraces. The numerous small channels occur mostly on crater rims and have the appearance of channels and rills formed by runoff from rainfall and spring outflow on Earth.

The water that caused channel erosion on Mars probably was absorbed into the ground, where it is stored as ground ice, as in the permafrost regions of Canada and Alaska. Whatever their origin, the stream channels on Mars are remarkably similar to those on Earth and offer unique challenges to scientific interpretation.

Apply Your Knowledge

1. Describe the characteristics of the stream system shown in figure 8.A.
2. Which appears to be younger, the stream system or the impact craters?
3. What possible explanations can you offer for the melting of ice on Mars that led to stream channel erosion?

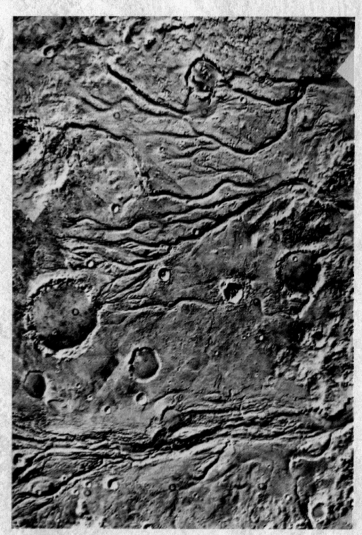

Figure 8.A   Dendritic valley system on Mars that has been cut deeply into the old cratered terrain. North is toward the top.

The bumping process gradually reduces the sizes of the particles in the downstream direction. In addition, any sharp edges or corners on the particles undergo a chipping that smoothes, or rounds, them. The particles also bump against the stream banks, dislodging other particles.

On the largest scale, the geologic work of streams cuts **valleys,** or linear depressions (V-shaped in profile), into the landscape. It is a standard geologic principle that valleys are the products of the streams that occupy them, although streams can also occupy valleys originally cut by other streams and/or that

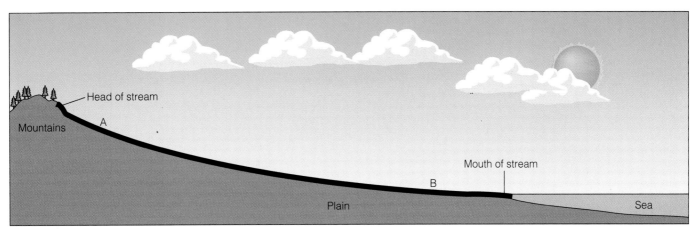

Figure 8.10 The long, or longitudinal, profile of a stream: literally a cross-sectional view from the head to the mouth. Note that the profile is "concave upward," indicating diminishing slopes (gradients) as you follow the stream from point A to point B.

have been extensively modified by glaciers. The valley is the principal stream erosional feature and is probably the most common of all landforms.

Valley Floor Deposition

Within stream valleys lie a variety of depositional landforms created by the loss of stream energy and the dropping of sediment. Streams transport and deposit material selectively. Depending on its velocity, a stream either carries or drops gravel, sand, silt, or clay. With greater energy, a stream can carry larger particles. When a stream loses its energy by slowing down, it deposits its materials, starting with the largest particles, followed by progressively smaller ones. Eventually, even the clay-sized sediments settle where the flow is very slow or the stream loses its identity. This action leads to the separation, or *sorting,* of stream sediment deposits by size. These relationships are illustrated by the Hjulstrom diagram (fig. 8.11).

Streams transport sediment of different sizes in different ways. Because of chemical weathering, some earth materials move through a stream system as dissolved ions invisible to us. More apparent, especially in times of high flow when the stream more actively erodes its banks, is the movement of silts and clays. They are transported by *suspension,* riding in the water column of the moving stream. Whenever these particles are present in abundance, the stream looks brown or muddy. Larger particles, sand, pebbles, and even boulders move only under high-energy flows during a stream's *flood stage.* Too large for suspension, these particles skip and bounce (a process known as *saltation*) or roll along the stream bottom during transport; but during times of low stream energy, they remain motionless on the channel bed (fig. 8.12).

The landforms created by the accumulations of sediments in a stream system include natural levees, backswamps, and point bars. **Natural levees** are low, parallel ridges on either side of a channel. They form as a result of repeated floods. After leaving the confinement of its channel, the stream quickly loses energy and begins to deposit its sediment. The largest particles (sand and silt) and the greatest volume of sediment settle first

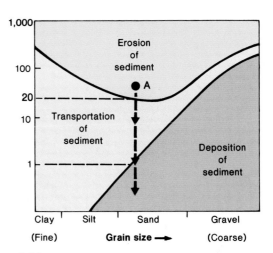

Figure 8.11 Modified version of the Hjulstrom diagram, which relates stream velocity to sediment particle size. To pick up (erode) a sand-sized particle such as *A,* stream velocity must exceed 20 centimeters/second. The same particle could be kept in motion until the velocity drops below 1 centimeter/second, at which time it would be deposited.

in the area adjacent to the channel, forming more massive accumulations than those farther out on the *floodplain,* the broad, flat area between the stream channel and the valley walls that is periodically subjected to flooding. The silts and clays, deposited some distance out on the floodplain, form a waterproof seal that prevents the water from draining into the soil. The result is the formation of *backswamps* between the natural levees and the distant valley walls. Figure 8.13 illustrates these relationships. In an attempt to control flooding and protect life and property, humans have constructed artificial levees. During ordinary meteorological conditions these levees are usually sufficient to contain a stream. When rainfall leads to excessively high water, as it did during the summer of 1993 along the upper Mississippi River, flooding can occur if the levees are breached or overtopped.

Natural streams never travel in straight lines for any significant distance. Instead, they follow curving, meandering paths.

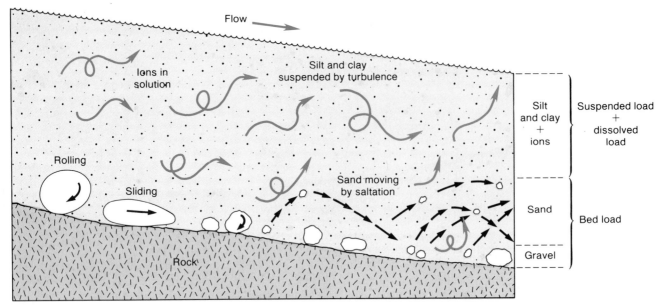

Figure 8.12 Transport of dissolved ions and sediment in a stream. The coarser part of the bed load remains at rest except for those periods of high stream flow (discharge) and consequent high energy.

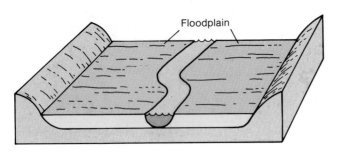

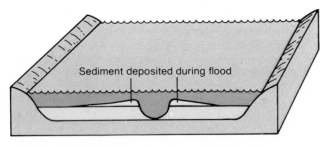

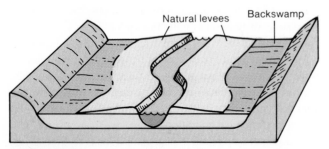

Figure 8.13 Water velocity drops as soon as the water leaves the confinement of its channel and spreads out across the floodplain. Drop in velocity causes immediate deposition (coarsest material first). The accumulation of large amounts of coarse sediment along both stream banks forms ridges called natural levees.

As the water flows around each bend, or *meander,* it travels faster on the outside of the curve than on the inside because of the greater distance to traverse. Greater velocity means more energy. The outside of the curve undergoes erosion, while the inside of the curve, where less energy is available for work, receives an accumulation of sediment. The deposit on the inside of the meander is called a *point bar* (fig. 8.14). Eventually, erosion of the outside bank lengthens the meander to the point where it may be cut off forming a shorter route for the water. Such an event normally occurs during flood stage when water flow is fast and high, the stream possesses extra energy, and a portion of the water leaves the stream channel. What remains is a horseshoe-shaped, detached segment of the former channel that fills with water to form an *oxbow lake.*

Deltas

When a stream enters a standing body of water (lakes, ponds, reservoirs, or oceans), it disperses and drops its load of sediment. If the shoreline is relatively free of wave and current activity, that sediment may accumulate at the mouth of the stream, forming a **delta.** As additional sediment is transported to the stream mouth, the delta extends farther out into the standing water body. Forced to travel through its own deposits, the stream lengthens and often divides into a series of smaller channels known as *distributaries,* so called because they distribute the flow of the stream. Each distributary seeks its own route across the surface of the delta to reach its base level. Figure 8.15 illustrates delta formation and structure, and figure 8.16 is a satellite view of the most actively building portion of the Mississippi River delta. More will be said about deltaic environments in chapter 12.

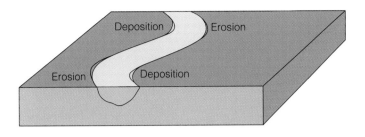

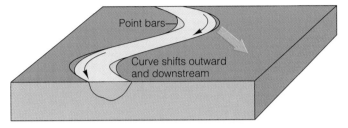

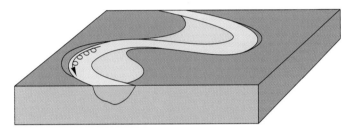

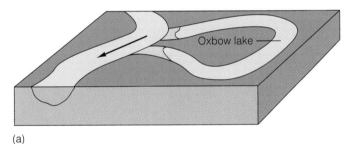

(a)

(b)

Figure 8.14 (*a*) Erosion occurs on the outside bank of each meander with deposition occurring on the inside bank. The deposits are called point bars. Continued moving and "stretching" of the meanders eventually leads to their being cut off and isolated (usually during flood stage) as oxbow lakes. (*b*) A floodplain in Florida showing meanders, oxbow lakes, and point bar deposits.

Stream Patterns

From head to mouth, a stream system of *tributaries* and main channel makes a pattern on the landscape that is usually only evident from above. Among the most common of these patterns are the dendritic, rectangular, radial, and parallel (fig. 8.17). In an area of horizontal sedimentary rocks or homogeneous igneous rocks, the stream appears to branch out randomly in all directions. This *dendritic stream pattern* evolves because no control, other than the everpresent pull of gravity, mandates a flow direction. The randomness of the downhill flow of tributaries into the main channel gives the aerial appearance of a treelike structure.

Where the surface rocks are not of equal resistance to weathering and erosion, streams erode through zones of weakness particularly in areas of alternate belts of weak and resistant rock units. The resulting stream system has a *trellis pattern*. By following the belts of weak rock the streams take the paths of least resistance and have less work to do in eroding their channels.

On volcanoes or other conical landforms, which form rapidly, streams adopt a *radial pattern*. Water falling on the higher elevations takes the shortest routes downhill. Because this occurs evenly on all sides of the volcano, when viewed from above the pattern is much like the spokes on a wheel. Similar to the radial pattern is the *parallel pattern,* where stream flow is determined by the general slope of the ground surface in a single direction. All streams will flow in that direction and thus be parallel to one another. An example of the occurrence of parallel patterns is along sections of the gently sloping coastal plains of the Atlantic and Gulf coasts.

GROUNDWATER

Like streams, groundwater comes from precipitation. Some of the rain and melting snow sinks into, or *infiltrates,* the soils and underlying rocks. Moving from space to space between and around mineral grains, this water is pulled slowly downward by gravity.

The Water Table and Other Features

In the uppermost ground layer that water traverses in its downward movement, the spaces between mineral grains are partly filled with air and partly occupied by the water. This is the *zone of aeration.* Eventually the water reaches a level where all the pore spaces are completely filled with water, the *zone of saturation* (fig. 8.18). The upper surface of this zone, or the boundary between the aerated zone and the saturated zone, is called the **water table.** The water occupying the pore spaces within the saturated zone is the **groundwater.** The depth of the water table is largely a function of climate. In arid regions (B climates, see chapter 6), it might be a hundred meters or more deep, whereas in more humid regions, it is very near the surface. The water table also rises and falls seasonally. During the wettest time of year in some locations it can actually lie at the ground surface. In fact, when we look at a lake, a marsh, or a large river, we are really

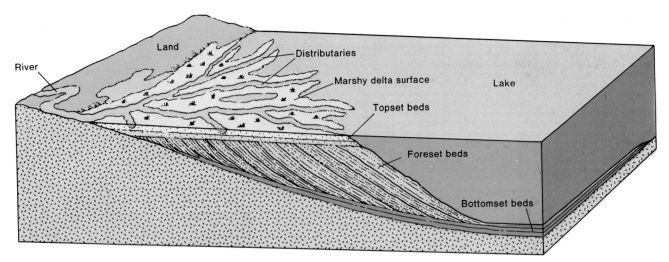

Figure 8.15 General form and internal structure of a delta.

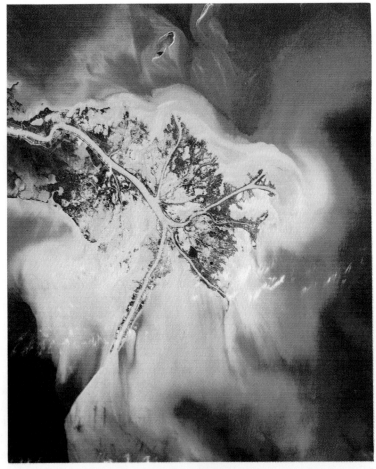

Figure 8.16 A *Landsat-1* view of a part of the Mississippi River delta from an altitude of 914 kilometers (568 mi). Barrier islands, distributaries, and interdistributary bays are clearly visible as is the light-colored plume of sediment in front of the delta.

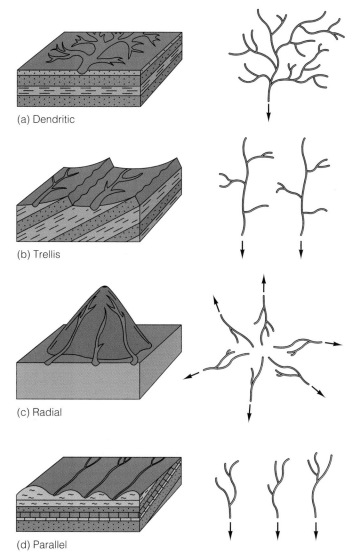

Figure 8.17 Some common stream patterns and their relationship to the underlying geologic structure. (*a*) Dendritic— uniform resistance to weathering and erosion of the rocks exposed at the surface. (*b*) Trellis—alternating weak and resistant rock units. (*c*) Radial—uniform slope in all directions from a central high point. (*d*) Parallel—streams flow down an inclined, or sloping, surface.

looking at the water table. The water table also intersects the ground surface on steep slopes or along fracture systems to form seepages of water called *springs* (fig. 8.19).

Within the saturated zone, groundwater moves slowly from opening to opening. Velocities from a few meters per day to a few meters per year would encompass most groundwater velocities, with the exception of streams flowing through caverns in areas of limestone bedrock. Just how rapidly groundwater moves is largely a function of rock *permeability,* that is, the extent to which the pore spaces are interconnected, allowing for passage of the water. Rock and sediment units, such as sand, gravel, and cavernous limestone, that allow relatively easy passage of water are called **aquifers** and are important economic resources such as the Dakota Sandstone discussed in Box 8.2. The amount of

water that can be stored in an aquifer is a function of the percentage of open space within that body, known as the *porosity.* Units with low permeability, which preclude the passage of water, are called **aquicludes.** A major difficulty in areas dependent on wells for their water supply arises from the very slow movement of groundwater. If groundwater is pumped from a well at a rate greater than it can be replaced by water moving through the aquifer, the water table surrounding the well drops. This cone-shaped depression of the water table centered on the well is called a *cone of depression.* If pumping continues unabated, the point of the cone will fall below the bottom of the well, and the well will go dry.

The movement from pore space to pore space within the soil, known as *percolation,* is also important for homes with septic systems. The percolation rate must be rapid enough to permit the movement of fluid wastes quickly away from the house, yet at the same time slow enough to permit the filtering of contaminants out of the water.

While some groundwater may eventually emerge at the ground surface as a spring, a large amount seeps into the channels of nearby streams. Such seepages provide the water that allows streams known as *perennial streams* to flow year round. Frequently, the water table falls below the beds of smaller streams, causing the stream channel to become dry and creating a flow called an *intermittent stream.* Intermittent stream channels carry water during the wetter times of the year but are dry during other times. A third type of flow regime is an *ephemeral stream* and is characteristic of streams in arid regions. Ephemeral stream channels carry water only during and shortly after a period of intense rainfall. In arid regions, such an event might occur several times a year or once every several years.

Erosion and Deposition

As an agent of erosion, groundwater performs most of its geologic work by *dissolution.* When it comes in contact with the soil and bedrock during its slow movement, groundwater dissolves some of the minerals and transports them in solution. Because it moves so slowly and through such tiny openings, groundwater rarely transports suspended silts and clays. Groundwater becomes acidic by absorbing carbon dioxide from the atmosphere and the soil, so its capacity to do geologic work is enhanced in areas of soluble limestone bedrock. So effective is this work that large masses of limestone may be removed, leaving subsurface voids known as **caverns.** Without its bedrock support, the ground surface above these voids occasionally collapses to form surface depressions called **sinkholes** (fig. 8.20). The occurrence of surface and subsurface features, including caverns and sinkholes is named **karst topography** after a limestone plateau in Yugoslavia, where such features are especially well developed.

In addition to the presence of limestone bedrock, three other conditions contribute to karst formation: (1) a humid climate , (2) fractures in the otherwise impermeable limestone to provide passageways for the downward movement of water, and (3) a major

Box 8.2 Further Consideration

Artesian Water Systems

An **artesian aquifer** water system consists of an inclined aquifer enclosed above and below by impermeable rock layers. If the aquifer is exposed at the ground surface (intake, or recharge,

area) at a sufficiently high elevation and a significant volume of water is taken in, pressure is generated within the aquifer that is much greater than atmospheric pressure. When the aquifer is penetrated by a well or when a natural system of fractures is present, the water rises through the opening, driven by the weight (pressure) of the water in the aquifer. Theoretically, the rise of the water should be to the same elevation as the location of the intake area. In practice, the water loses some

energy to friction during its passage through the aquifer and the rise is somewhat less. In general, the farther the water travels through the aquifer, the less it rises through the well or the fracture system (fig. 8.B). The surface defined by the level to which the water rises in all the associated wells or fracture systems is the piezometric, or potentiometric, surface.

If the hydraulic gradient (elevation difference between the intake area and the ground at the surface opening or well site) is

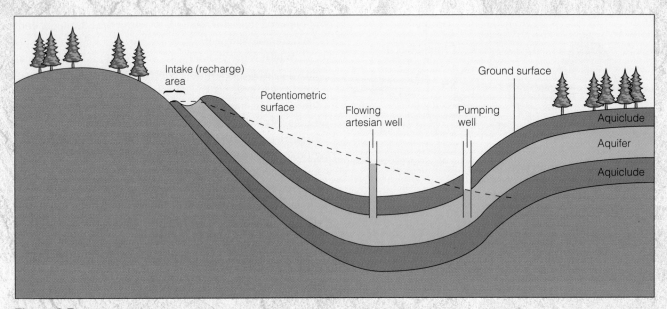

Figure 8.B   A simplified natural artesian system.

stream in the area, which provides a base-level control for the downward moving water.

As the groundwater progresses through the soils, sediments, and bedrock, some of its dissolved minerals precipitate out of solution. Near Earth's surface, this deposition occurs chiefly in two locales: (1) among loose sediment, the precipitating mineral matter serves as a cement that binds the individual grains and converts the sediment into rock and (2) as calcite-rich and clay-rich layers within the soils of arid and humid regions, respectively (see chapter 13).

When groundwater saturated with dissolved mineral material seeps out of fractures in the roof of a cave, it undergoes a reduction in pressure, loses some of its carbon dioxide, and becomes less acid, reducing its ability to retain calcium carbonate in solution. Some of this mineral material then precipitates to form delicate, iciclelike projections, called *stalactites,* from the roof of the cave (fig. 8.21). Continuing its downward movement, some of the water drops onto the cave floor, where it de-

posits more of the mineral matter. This accumulating material forms a cone-shaped feature known as a *stalagmite.* In many instances the stalactites and stalagmites merge to form *columns.*

GLACIERS

Great accumulations of snow produce a third geologic surface agent. When annual snowfall exceeds an area's climatic ability to melt, evaporate, and sublimate it, the accumulation can create icefields from which, in turn, glaciers form. A **glacier** is a thick mass of ice that moves as its weight responds to gravity and to pressures within.

Anatomy of a Glacier

Two kinds of glaciers transform landscapes. Those glaciers that move primarily in response to gravity are long, linear, and confined to stream valleys within mountainous regions. They are

sufficient, water rises to the ground surface and flows out as an **artesian spring** or an **artesian well.** If the gradient is less, water must be pumped from the well. The artesian system operates much like a municipal water system. Here, water flows through a network of pipes (the "aquifer") and into the plumbing systems of homes and businesses under the gradient produced by the high location of the water reservoir or water tower (fig. 8.C).

In the United States, an important artesian aquifer is the Dakota Sandstone, which underlies a large area of the Great Plains region from South Dakota to Texas. The intake area for the Dakota Sandstone is in South Dakota, where the upturned edges of the sandstone are exposed around the Black Hills. For thousands of years, water accumulated within and flowed slowly southward through the Dakota Sandstone. In the last 150 years, however, more than 10,000 wells have been drilled into this unit to tap the water for irrigated agriculture in this semiarid region. The result is that this water resource has literally been "mined," that is, removed at a far greater rate than nature can replace it. Consequently, both the volume of water remaining in the aquifer and the pressure within the aquifer have dropped drastically. Most of the wells still producing from the Dakota Sandstone must now be pumped.

Apply Your Knowledge

1. Discuss the significance to an artesian aquifer of (a) being confined between impermeable rock units and (b) being exposed somewhere at the ground surface.

2. Discuss the likelihood of the Dakota Sandstone aquifer recovering to its resource potential of 150 years ago.

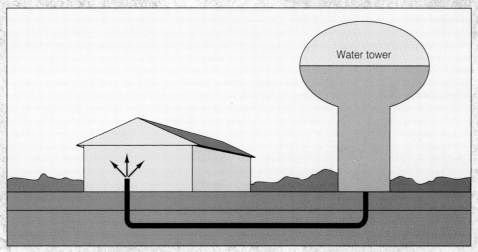

Figure 8.C   A simple municipal water system operating under the same principles as an artesian system.

variously referred to as *ice streams, alpine glaciers, mountain glaciers,* or *valley glaciers* (fig. 8.22a). Glaciers that move from areas of high pressure (thick ice) to low pressure (thin ice) are *ice sheets, ice caps,* or because they are often geographically very large, *continental glaciers* (fig. 8.22b). Such glaciers are formed at high latitudes and may have surface areas measured in millions of square kilometers and thicknesses in excess of 3 kilometers (2 mi).

New ice forms across the entire area of a glacier when fallen snow changes by melting and refreezing to a granular form called **firn.** Even as new ice forms, older ice is lost. That portion of a glacier where accumulation of new ice exceeds loss of old ice (usually at higher elevations or latitudes) is the *zone of accumulation.* That part of the glacier where loss of ice exceeds gain is the *zone of ablation.* Such ice loss, or *ablation,* may be by melting, sublimation (the direct change from the solid to the gaseous state), or *calving,* the breaking off of ice masses that ex-

tend into a body of water. The calving process is responsible for the formation of *icebergs,* or floating masses of ice, that have become detached from the glacier. If the addition of new ice exceeds that lost to ablation, the front of the glacier advances across the landscape. If ablation exceeds accumulation, the ice mass shrinks, and the ice front retreats. In both cases, ice movement within the glacier is forward; it is only the ice front that moves forward or backward. The line separating the zone of accumulation from the zone of ablation is the *firn limit.* Like groundwater, glaciers fluctuate with the seasons and move very slowly at rates generally between a few meters per day and a few meters per year.

The geologic work of glaciers is limited today because glaciers cover only about 10% of the land. In the recent geologic past, however, glaciers covered as much as 30% of Earth's land surface, and many areas still bear the imprint of glaciation in the form of distinctive landscape features.

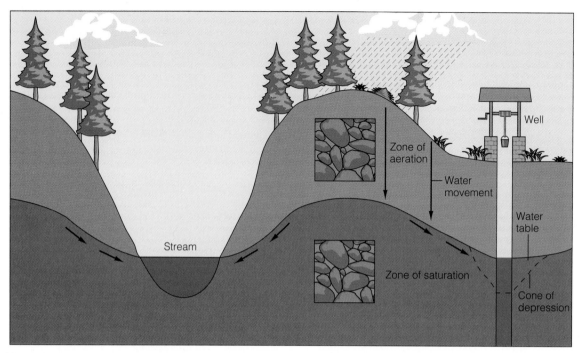

Figure 8.18   Saturated zone, unsaturated zone, and the water table, which separates them. Where the water table intersects a stream channel, groundwater feeds and sustains the stream. If this situation persists year round, the stream is a perennial one. If the water table periodically drops below the stream channel, the stream goes dry and is considered to be intermittent.

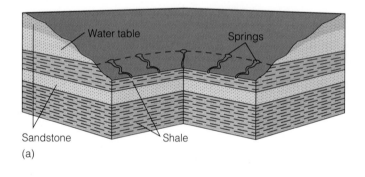

Figure 8.19   (a) A spring forms where water moving along the upper surface of an impermeable shale bed emerges at the ground surface. (b) A series of springs developed within a sequence of alternating sandstones and shales in southwestern Pennsylvania.

Glacial Erosion

Glaciers are very powerful agents of erosion. They grind and push their way downslope and across the ground surface, scraping up loose sediment and plucking fragments from masses of solid rock. In mountainous regions, where their erosive power is best displayed, they carve the higher elevations into a series of sharp peaks called *horns* and narrow ridges called *arêtes* (fig. 8.23). At their points of origin, they scour flat-floored, semicircular depressions called *cirques* that fill with water after the glacier melts to form small lakes, or *tarns*. Alpine glaciers also broaden and deepen the valleys they occupy. When the ice melts such valleys are left with a characteristic U-shaped cross profile, making them easy to identify.

The rate of downcutting by a tributary stream is governed by the level of the main stream into which it flows, an example of local base-level control. Consequently, at their junction a tributary stream and a main stream valley are at the same elevation. Glaciers behave much differently. They erode proportionately to their thickness, with large glaciers eroding more deeply than small glaciers. When the glacial ice melts and streams once again occupy the valleys, the smaller tributary streams flow through

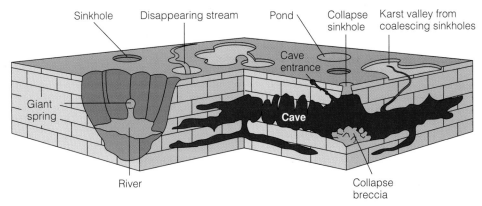

Figure 8.20 Development of groundwater (karst) features in an area of limestone bedrock.

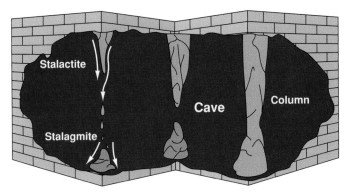

Figure 8.21 The formation of stalactites, stalagmites, and columns within a cavern. Arrows show movement of water carrying $CaCO_3$.

hanging valleys above the level of the main stream. To join the main stream, the tributaries must flow down steep rapids or over waterfalls. Bridalveil Falls and Yosemite Falls in Yosemite National Park, California, are classic examples of such features.

Glacial Deposition

Sediment picked up by scraping and material that falls onto the ice surface from overhanging valley walls is incorporated into the moving ice mass or carried on top of it. Because ice is a solid and can support the weight of material even when not moving, sediment is not deposited when the ice slows down, only when it melts. The sediment dumped by the melting ice is called **glacial till.** Because ice lacks the sorting capability of streams, till is a mixture of all sizes of particles, from fine clay to large boulders. (Some of these are known as *erratics* because they have been transported great distances by ice.) Landforms produced by the accumulation of till are called **moraines,** and they occur in a variety of shapes and sizes (fig. 8.24). Some are low, linear ridges formed at the end, or terminus, of the glacier. These *end,* or *terminal, moraines* are crescent-shaped like the lobate form of the ice front. Rock debris that falls from overhanging valley walls and accumulates along the sides of a valley glacier forms *lateral moraines.*

Where two valley glaciers come together, the lateral moraines merge to produce an elongated ridge of till on the ice surface and later in the center of the valley called a *medial moraine.* As glaciers melt back, they may pause for a period of time in their retreat so that a ridge of till accumulates along the ice front forming a *recessional moraine.* This type of moraine most commonly occurs as one of a group of such features paralleling the terminal moraine in front of it and marking the path of ice retreat. Should conditions permit the glacier to advance once again, the recessional moraines would be highly modified or erased. Where large ice sheets cover relatively flat areas, the till is deposited as a broad blanket with little topographic relief, known as a *ground moraine.* In general, deposition is a process more characteristic of continental glaciers than of valley glaciers. The deposition of till by continental glaciers creates a smoothness to the landscape in contrast to the sharp, erosional landforms of mountain glaciation.

Other landforms are also characteristic of regions of continental glaciation. Sediment carried by streams flowing within or beneath the ice may be deposited when ice movements block the channel. Such deposits, known as *eskers,* remain after the ice sheet has melted. These curving linear ridges mimic the meandering pattern of the stream channel in which they were deposited. An advance of an ice sheet across previously deposited till may shape those deposits into a series of elongated hills, or *drumlins,* which indicate the direction of ice movement.

Some of the sediment deposited along the ice front may be picked up by meltwater streams and carried down the valley away from the ice front. This transported sediment is glacial **outwash.** The meltwater sorts and deposits this material in a geologic action typical of streams. Occasionally, large chunks of ice may become detached from the main glacier, stagnate, become buried by outwash, and then melt. This causes the ground surface, now composed mainly of outwash, to subside. The depression created by the subsidence is a *kettle.* Filled by water, the kettle becomes a lake or pond. In glaciated areas, such water bodies occur by the thousands. Once covered by continental ice sheets, Minnesota is now the "land of 10,000 lakes" because of its vast number of kettle lakes. Figure 8.25 illustrates

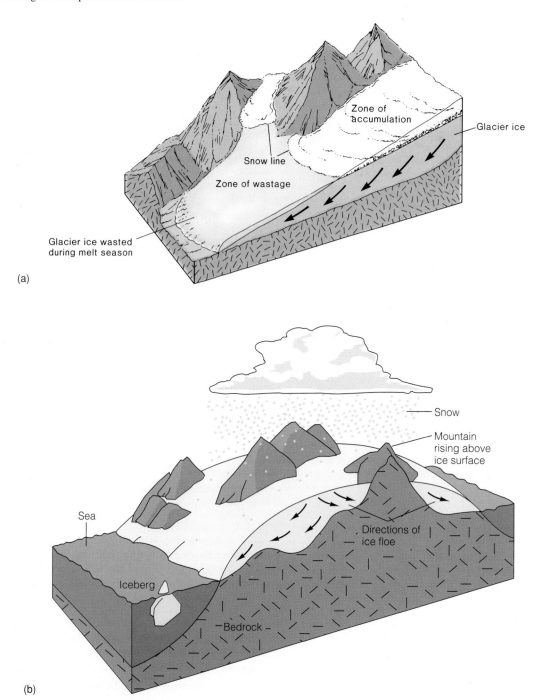

(a)

(b)

Figure 8.22 (a) Ice movement in response to gravity within a valley glacier. The zone of accumulation is the area of the glacier where more snow and ice accumulate each year than are lost by melting or other causes. In the zone of wastage (ablation), annual loss exceeds annual accumulation. The snow line separates the two zones. (b) Ice within an ice sheet (continental glacier) flows outward in all directions from the region of high pressure (thick ice) to the zones of lowest pressure (thin ice).

the principal erosional and depositional features associated with an ice sheet.

IND

The effect of wind as a geologic agent is often overrated. Although the wind is present everywhere, it is usually effective in performing geologic work only under certain circum-

stances. Wind is most effective when loose, dry, and fine-grained sediment is exposed at the ground surface. Particles larger than fine sand are normally too big to be picked up and transported by the wind's energy. Wet sedimentary particles tend to bind together, inhibiting the lifting potential of the wind. Because the vegetation common to humid climates shields sediment from the action of wind, this agent is most effective in arid and semiarid regions.

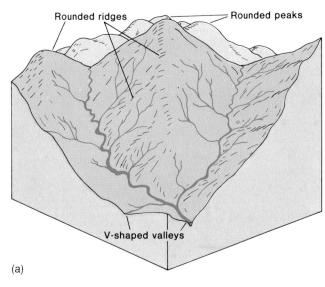

(a)

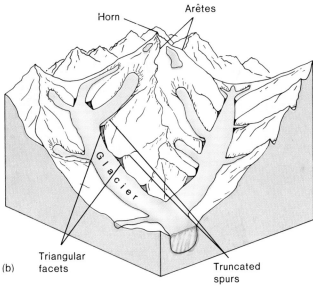

(b)

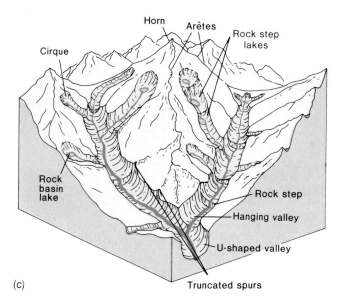

(c)

Figure 8.23 Modification by glacial erosion of a "smooth," stream-formed landscape to a "sharp," glacial landscape: (a) prior to glaciation, (b) during glaciation, and (c) after glaciation.

Wind Erosion

Erosion by wind occurs in two ways: lifting and carrying away particles is a process called **deflation,** and the sandblasting of rock by wind-borne sediments is a process called **abrasion.** In some areas, deflation creates broad, shallow depressions called *blowouts,* or *deflation basins.* These are particularly common in the American Great Plains region of Nebraska and eastern Colorado and in the High Plains of northeastern New Mexico and the Texas Panhandle. They serve as collection basins for water during periods of rainfall and as natural water holes for buffalo and other forms of wildlife that once populated these areas in vast numbers. Wind abrasion and deflation working in combination with weathering have also carved the spectacular rock arches in Arches National Monument in eastern Utah.

Wind Deposition

Like streams, the wind is a highly selective transporter of sediment. The size of the particle that can be moved is dependent on wind velocity. Because the air is "thin," however, it is only capable of lifting the finest particles to significant heights. Even fine sand particles can only be lifted a meter or two and are primarily transported by bouncing and skipping along the ground surface.

When wind velocity diminishes, sediment is progressively deposited, coarsest first and finest last, to make depositional landforms. The most common and best known of the wind-generated features are **dunes,** accumulations of sand that either are moving or have moved but are now stabilized by a protective covering of vegetation.

Dunes form wherever there is sufficient sand and where the wind is almost constantly blowing in the same direction. The wind moves the dune's sand grains by rolling and bouncing them on the upwind (windward) side of the dune toward the peak, or *crest,* of the dune, where they then tumble onto and down the downwind (leeward) side. The effect of all the individual sand grain movements is the transport of the entire dune (fig. 8.26). While dunes are often envisioned as desert features, they are equally common along coastlines (as along the southern shore of Lake Michigan and the eastern shore of Cape Cod, Massachusetts) and some stream courses, especially in the Great Plains region. Figure 8.27 illustrates the shapes of several common dune types, as well as the prevailing wind patterns necessary to form them.

Dunes, by their very nature, are mobile features, and their locations are temporary. Such mobility often comes in conflict with the desire of humans to build on or otherwise utilize the land. Consequently, efforts to stabilize and halt the movement of dunes, particularly in coastal areas that are important

Box 8.3 Investigating the Environment

The Great American Dust Bowl

The classic example of wind erosion in the United States and one of the best-known examples in history occurred in the 1930s across much of the Great Plains region. Dubbed the "Dust Bowl" by an enterprising reporter, the area affected stretched from Montana to Texas but was especially severe in the panhandle regions of Oklahoma and Texas. These were storied times. Indeed, one of the classics of literature, John Steinbeck's *The Grapes of Wrath*, chronicled the movement of one Oklahoma family as it worked to overcome their tribulations.

The storms of the Dust Bowl began in 1932 and lasted until the end of the decade. Those rolling down from the north were low, boiling, thick clouds: the infamous "black blizzards." During March of 1936, parts of Texas and Oklahoma recorded 22 days of dust storms (fig. 8.D). The greatest storm of the decade raged for 4 days during May of 1934, when a 2,400-kilometer-(1,500-mi-) long dust cloud, covering 3.5 million square kilometers (1.35 million sq. mi), stretched from Canada to Texas and from Montana to Ohio. As the storm moved eastward, day was turned to night in cities along the eastern seaboard. When it was over, an estimated 272 million metric tons (300 million tons) of soil had been blown away. The dust that formed each storm was so pervasive that it could not even be escaped indoors. It would sift through the finest openings around windows and doors, leaving a coating on every surface, including sleeping human beings. Piles of dust would accumulate unseen in attics until their weight would cause the ceilings to collapse.

What might have caused a once highly productive agricultural region to be turned into a part of the Great American desert? This portion of the Great Plains was, in a sense, a victim of its own success. The soils of these endless prairies, under natural conditions, were protected from erosion by the dense root system of the native grasses. When the land was opened by the federal government for homesteading in the late 1800s, farmers (the

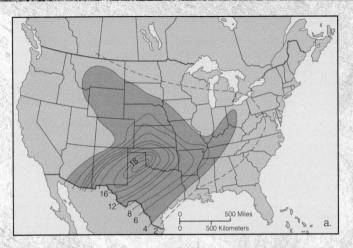

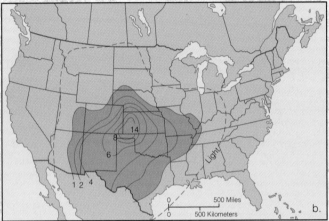

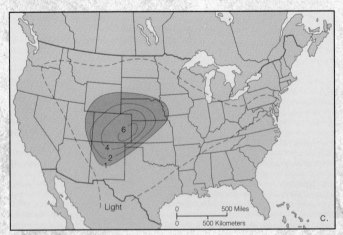

Figure 8.D   The sequence of maps shows the number of days with dust storms in the United States for the months of (a) March, (b) April, and (c) May 1936.

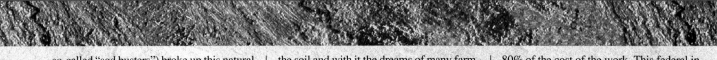

so-called "sod busters") broke up this natural sod armor with their plows, converting natural grasslands to wheat farms. During World War I, tractors came into large-scale use in this region for the first time, and in a matter of a few years the amount of prairie under cultivation increased dramatically.

As long as rainfall was plentiful, wheat grew in abundance and the region prospered. The good times, however, could not last in a region that has historically been subjected to periodic droughts. The year 1932 began a 6-year period in which rainfall averaged little more than half of the normal amount. As a result, the wheat crops failed. With no vegetation to bind and hold the soil particles, the winds that blow almost ceaselessly across the prairies were able to pick up and carry away the soil and with it the dreams of many farmers. With crop failures came farm mortgage foreclosures and unemployment.

While the Dust Bowl days of the 1930s are the best known, they have not been the only times of drought in this century in the Great Plains. Similar conditions occurred in the mid-1950s, the mid-1960s, and the mid-1970s (fig. 8.E). During the drought of the 1950s, Congress passed legislation creating the Great Plains Conservation Program. Under this program, individual farmers and ranchers planned, scheduled, and contracted with the U.S. Department of Agriculture for a conservation plan to be instituted over a 3- to 10-year period. Technical help was available from the Soil Conservation Service, and the federal government paid from 50% to 80% of the cost of the work. This federal investment has made a significant difference in terms of decreased erosion and increased agricultural income.

Apply Your Knowledge

1. Examine figure 8.E. Does there seem to be a cycle to the recurrence of periods of intense wind erosion in the Great Plains? What year was characterized by the greatest loss of soil acreage to wind erosion?

2. What human activities might lessen the frequency and/or degree of future "dust bowl" periods?

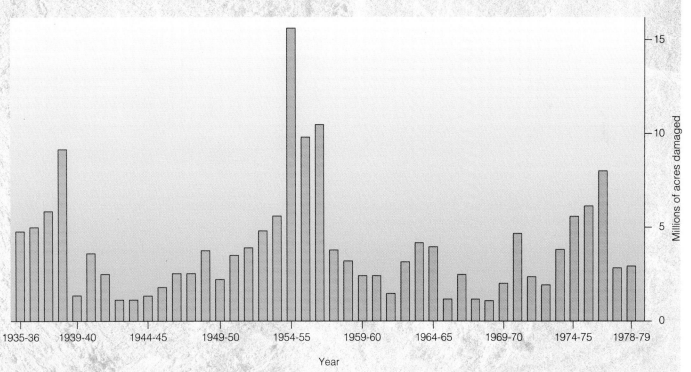

Figure 8.E Wind damage to croplands in the Great Plains by years. In addition to the damage in the mid- to late 1930s, note the high instances of soil erosion in the mid-1950s and late 1970s. (Source: Soil Conservation Service, *America's Soil and Water: Condition and Trends,* U.S. Department of Agriculture.)

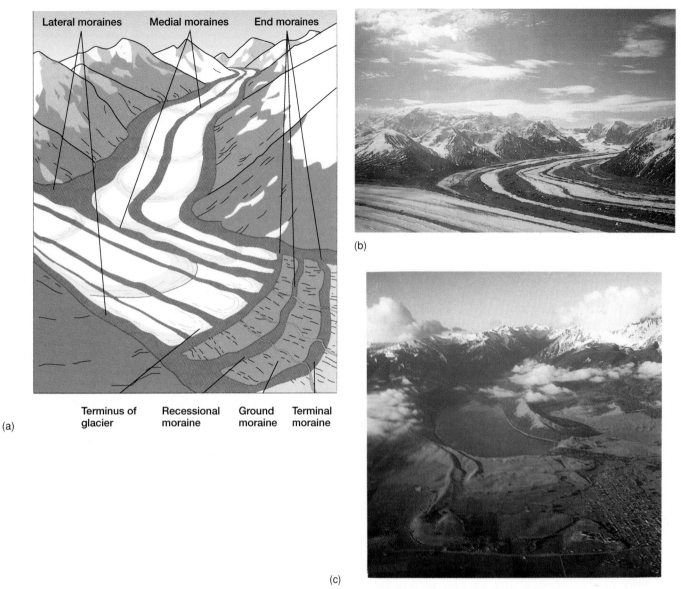

(a)

Lateral moraines   Medial moraines   End moraines

Terminus of   Recessional   Ground   Terminal
glacier   moraine   moraine   moraine

(b)

(c)

Figure 8.24 (*a*) Positions of the several types of moraines commonly associated with valley glaciers. (*b*) The lateral and medial moraines on a typical valley glacier. (*c*) Lateral, terminal, and recessional moraines remain as topographic features after the glacier has melted away. Note the general "sharpness" of the glacially eroded topography.

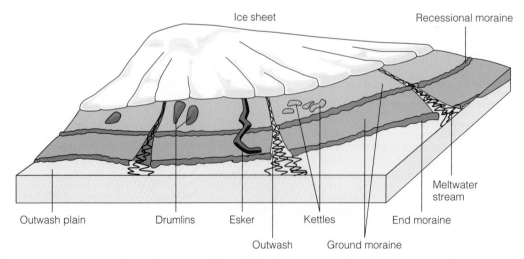

Ice sheet   Recessional moraine

                                            Meltwater
                                            stream

Outwash plain   Drumlins   Esker   Kettles   End moraine

           Outwash   Ground moraine

Figure 8.25 General assemblage and locations of characteristic features associated with continental glaciation. All are of depositional origin and the resulting topography is "smooth" and gently rolling.

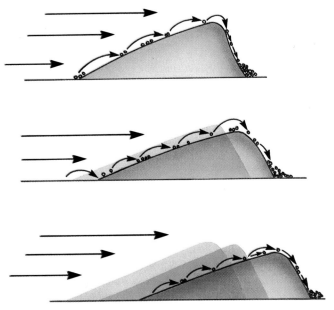

Figure 8.26 Dune formation and movement. The wind moves sand, grain by grain, up the windward side of the dune. The grains tumble down the leeward side and come to rest. The net movement is downwind.

recreational resources, are widespread and ongoing. These efforts commonly involve the planting of vegetation (usually some variety of grass) and/or the placement of sand fences to break up and slow the wind currents (fig. 8.28).

WAVES AND CURRENTS

Waves and currents are characteristic of standing bodies of water. As geologic agents, they differ fundamentally in their orientation to the energy that generates them. A **wave** is the circular movement of water particles within a body of water in response to the passage of energy. The rise and fall of wave action is essentially perpendicular to the direction of the passage of energy within the water. A **current,** however, is a movement of water within a larger body of water that is aligned with the energy that produces it. Additionally, in waves only the energy moves forward, with the water particles staying essentially in place. Currents, on the other hand, move the body of water from one place to another, much like a stream flowing on land.

Wave Formation and Behavior

Figure 8.29 illustrates the motion of individual water particles as the energy and wave form pass a particular location. Each water particle that is disturbed by the passage of the wave describes a circular path, returning to its original position before the arrival of the next wave. The diameters of the paths decrease with depth until at a water depth equal to *wave base,* or approximately one-half wavelength, such motion, for all practical purposes, ceases. Thus, waves are a surface phenomenon.

The immediate source of energy for most waves and currents is wind, which transfers its energy to the water by contact

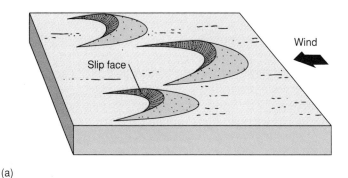

(a)

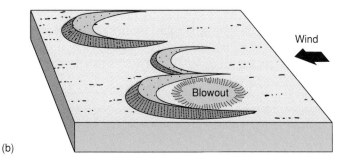

(b)

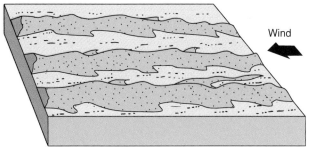

(c)

Figure 8.27 Types of dunes. (*a*) The barchan forms when the sand supply is limited and the wind direction is constant. (*b*) The parabolic dune requires abundant sand, constant wind, and anchoring vegetation. (*c*) Longitudinal dunes, or seifs, require abundant sand, a fairly constant wind to elongate the dune, and an occasional crosswind to bring in more sand.

as it blows across its surface. The extent of the water surface over which the wind blows is called the *fetch.* With increasing fetch, the wind transfers more of its energy to the water, amplifying the waves and currents. Waves begin as small ripples and increase in height and length with stronger wind, longer fetch, and greater wind duration.

Individual wave systems, or *wave trains,* fall into two categories based on the water depth. Those waves that occur in water deeper than one-half wavelength are said to be *deep-water waves.* Those waves that occur in water shallower than one-half wavelength are *shallow-water waves.* Because wavelengths associated with different wave trains are different, wave base is variable for any location.

On a shallowing, or *shoaling,* seafloor, a shallow-water wave begins to lose energy to friction with the bottom. The result is an increase in the wave height and a decrease in the wavelength as the wave begins to slow. The top of the wave, not slowed by friction, moves forward over the leading trough, where the water has no support. The crest then falls, or *breaks,* into the trough, creating the *surf zone* (fig. 8.30). Wherever the energy of the wave intersects the bottom, particularly in the surf zone, the wave performs geologic work by moving sediments and eroding rock. Transferring its energy to the bottom, the wave spends itself in a final rush up the beach, carrying sediment with it. The *swash* is the end of this landward movement of water. Robbed of its energy, the water rolls back down the beach under the pull of gravity as *backwash.* Like the swash, the backwash moves loose sediment. While swash carries sediment up the beach face in the direction of the incoming waves, backwash carries it directly down the beach face. The net result is a movement of sediment along the beach by each incoming wave, a process known as *beach drifting.*

Figure 8.28 Attempts to stabilize dunes along the New Jersey shore by the planting of dune grass.

Wave Erosion

In areas where the shoreline is irregular, projections of land into the body of water, called *headlands,* take the brunt of wave energy because of **wave refraction** or bending. The shoals in front of the headland slow part of the approaching wave. The deeper water to each side has no effect on the wave velocity, so the wave bends, concentrating its energy and erosive power on the sides of the headland (fig. 8.31). The same wave refraction leads to a dispersal of energy within the adjacent bays and consequently to deposition. Attacked by the continuing wave action, the headland is worn back, often forming flat, terracelike surfaces called *wave-cut platforms.* Eventually the shoreline is straightened, with low areas marking sites of former bays alternating with sea cliffs at the locations of former headlands (fig. 8.32). Isolated rock masses, not yet eroded, remain offshore as **sea stacks.** Sometimes the waves carve out **sea caves** and **arches** before the tip of the headland is isolated from the mainland (fig. 8.33). All these features are especially well developed along the Pacific coast of North America.

Wave Deposition

The depositional work of waves is manifested in a series of sand features known as **beaches** and *bars.* These are created by the longshore, onshore, and offshore movement of sediment. These elongate sand bodies move landward or seaward depending on the available wave energy. Strong waves move beach sands offshore, while gentle waves usually build up the beach. Typically, the stronger wave systems occur in winter, especially along the West Coast, and beaches are usually narrow at that time of year. When the calmer waves of summer return, the beach widens as sands are moved landward.

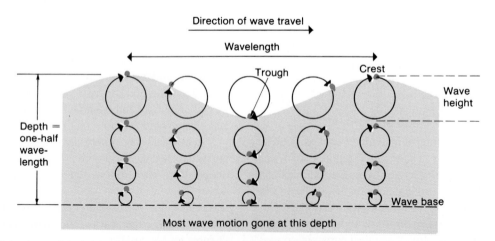

Figure 8.29 Each water particle that is disturbed by the passage of a wave describes a circular path, returning to its original position before the arrival of the next wave. The diameter of the path decreases with depth until at a water depth equal to one-half wavelength (wave base), such motion, for all practical purposes, ceases. Waves are a surface phenomenon.

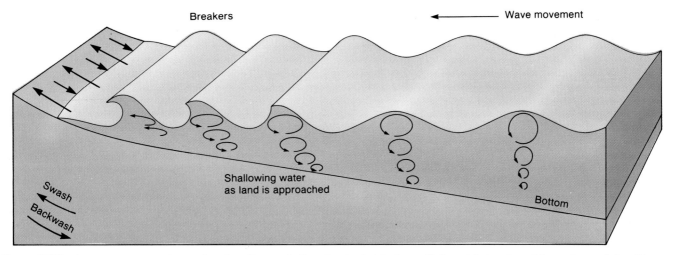

Figure 8.30 As a wave approaches a shoreline, the sea bottom begins to interfere with the orbital motion of the water particles. The orbital paths are distorted, the bottom of the wave is slowed relative to the top, the wave becomes unstable, and eventually it collapses (breaks). Water rushes up the beach face as swash and flows back down as backwash.

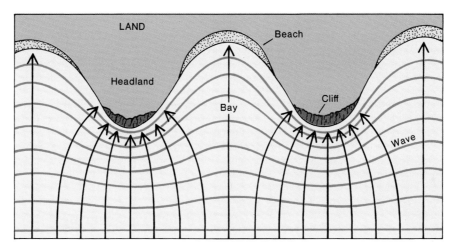

Figure 8.31 As they approach a shoreline, waves "feel bottom," are slowed, and refracted first in front of the projecting headlands. Energy associated with the wave (shown as lines, or "rays" perpendicular to the wave fronts) is concentrated on the headlands, causing erosion and retreat of the headlands.

Currents

Wave trains rarely approach the shore straight on. The refraction of waves in shallow water works to align the wave fronts closer to a position parallel to the shoreline. Even so, refraction is rarely complete and the wave fronts strike the shore at some angle, resulting in a pushing of the water in the direction away from the incoming waves (fig. 8.34). This action of waves in the surf zone creates a flow of water parallel to the shoreline known as a **longshore current.** Particles picked up by the longshore current move down the beach in a constant migration. Thus, while the beach seems to be unmoving, it is, in fact, constantly in motion. Because of seafloor and shoreline configurations at some locations, not all the water may be able to flow parallel to the shoreline. Some of it is directed perpendicularly to the shore-

line and back out to deep water. These are the notorious *rip currents,* or undertow, that have dragged many an unwary swimmer to danger and even death.

The movement of sediment within the surf zone by longshore currents is called **longshore transport.** Such transport occurs continuously along the submerged portions of beaches. Wherever the longshore current encounters an embayment in the shoreline, however, it carries its sediment across the opening to form a *spit,* an extension of the beach into the bay (fig. 8.35a). The spit ends in the deeper, open water of the bay because the sand cannot be built up high enough by the longshore current to stay above water. If the water at the mouth of the bay is relatively shallow, the spit may extend all the way across to seal off the bay and diminish or eliminate any further interchange between the bay and open water. Such a feature is called

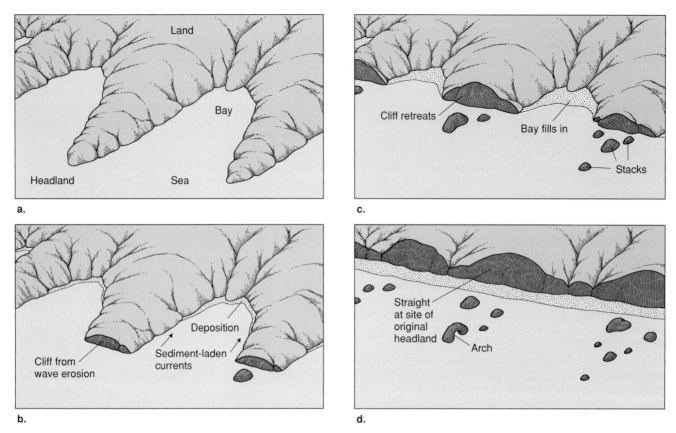

Figure 8.32 With time, the headlands along irregular shorelines are worn back. The shoreline becomes straightened but with sea cliffs marking the former headland positions.

Figure 8.33 Sea stacks and sea arches along the Oregon coast.

a *baymouth bar.* Once formed, the bar also traps sediment brought into the bay by streams. With time, the accumulated sediment converts the bay first to marsh and finally to land. Deposits of this type also may be built up as a connection between the mainland and an offshore island, forming a *tombolo* (fig. 8.35*b*).

Other types of currents associated with the world's oceans are discussed in chapter 11.

Beach Engineering

Natural objects such as sea stacks or any artificial offshore structure can rob the beach of the incoming wave energy necessary to sustain longshore transport. Sand piles up and builds seaward. Many communities have attempted to provide calm anchorages for pleasure boats or to protect their beaches from erosion by constructing engineering works of various kinds, often with unexpected and undesirable results. The most common of these structures are breakwaters, groins, and jetties.

Breakwaters perform the function their name suggests: they break up the energy of the incoming waves and create a quiet, protected zone between the breakwater and the shore. Often the area the breakwater is meant to protect becomes choked by accumulated sediments trapped without the energy for further transport. The usual solution to the problem of clogged waterways is dredging. Such is the case at Santa Barbara, California, where a large breakwater constructed to provide a haven for boats has caused a large spit to form. The spit developed in the harbor because the sediment-laden longshore current was deprived of some of its energy by the breakwater placed in its path.

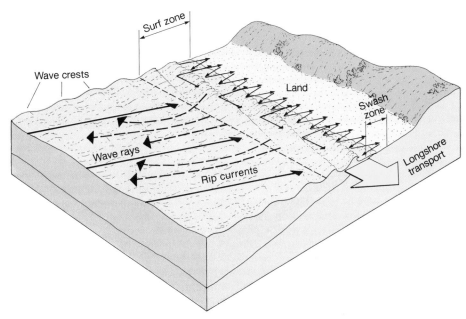

Figure 8.34 Formation of longshore currents and rip currents.

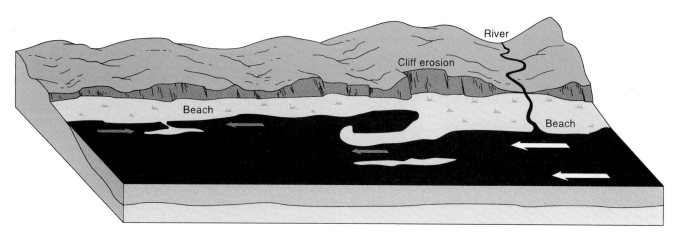

(a)

(b)

Figure 8.35 (a) Formation of a spit and a tombolo. (b) A tombolo connects an otherwise isolated sea stack to the mainland.

Jetties are walls built perpendicularly to the shore and usually constructed in pairs, one on either side of a channel or river mouth. Their purpose is to keep the passage open to navigation. Jetties divert the flow of the longshore currents to deeper water, thereby causing deposition of sediment away from the channel or river mouth (fig. 8.36a and c).

Groins are similar to jetties in that they are built perpendicularly to the shoreline, but their purposes are to capture sand in its migration and to prevent beach erosion. Sand piles up on the upcurrent side of the structure where the energy of the longshore current is reduced. As the current moves around the end of the groin and continues along the shore, it has less than its normal load of sediment and becomes capable of picking up, or eroding, more. This results in the loss of beach sand on the downcurrent side of the groin. To compensate for this loss, it is necessary to construct another groin, with the same result. This process then may be carried out for many kilometers along the beach front (fig. 8.36b and d). Expensive to construct, groins are nevertheless common where communities attempt to control the flow of beach sand initially altered by human activities. Miami Beach's waterfront hotels have beaches today in part because of the construction of groins but also because of another technique commonly used to maintain beaches: *beach nourishment*. With this technique sand is continuously applied to a beach to replenish that carried away by longshore transport. The source of the sand may be offshore dredging or it may be trucked in from some distance.

Seismic Sea Waves

A **seismic sea wave,** or **tsunami,** is a large-scale wave system generated by an earthquake or a volcanic eruption rather than by the wind. These waves have very long wave periods and travel at speeds of several hundreds of kilometers per hour. When they approach a beach, seismic sea waves can steepen to heights of tens of meters. Seismic sea waves have been responsible for thousands of deaths in and around the Pacific Ocean basin, where earthquakes and volcanic eruptions are common. Tsunamis generated by the eruption of the Indonesian volcano Krakatoa in 1883 caused an estimated 30,000 deaths as they inundated low-lying coastal villages. In July of 1993, tsunamis with wave heights exceeding 9 meters (30 ft) were responsible for most of the deaths associated with a severe earthquake that struck the northernmost of the Japanese islands. At low-lying villages and towns the waves washed people seaward and tossed boats onto the land. The work of seismic sea waves is short-lived but significant. They have left their mark on the geologic landscape, and their effects from millions of years ago have been studied for evidence of major geologic events. The origin of seismic sea waves is discussed in chapter 9.

Tides

The largest of all shallow wave systems are the **tides.** Driven by the gravitational attraction of the moon and the sun for Earth, the tides are bulges of water that encompass whole oceans, making their wavelengths far in excess of the water depth. The gravitational pull of the moon and the sun affect the continents, all water bodies, and the atmosphere, but its effect is most notice-

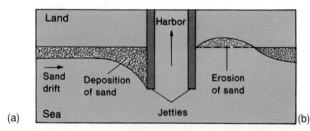

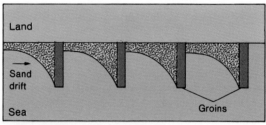

Figure 8.36 (a) and (c) Jetties designed to keep the mouth of a stream or channel free of sediment. (b) and (d) A system of groins placed to prevent shore erosion.

able on the waters of the oceans. The moon, though smaller than the sun, exerts the greater influence because of its proximity to Earth. The mean distance from Earth to the moon is only 384,404 kilometers (238,857 mi), whereas the mean distance from Earth to the sun is 149,600,000 kilometers (93,000,000 mi).

The tides are waves with exceptionally long periods (approximately 12 1/2 hours) between successive wave crests (the high tides). Figure 8.37 shows the effect of the moon's gravitational pull (the sun's effect is not considered here) on the oceans. As Earth rotates on its axis and the moon moves in its orbit around Earth, *high tide* is experienced at points on Earth's surface directly facing the moon because this portion of Earth is closest to the moon. High tide is experienced at the same time at points on the opposite side of Earth because here the moon's pull is weakest and the outward directed centrifugal effect associated with any rotating body becomes most noticeable. Between these two bulges of water are offsetting depressions of water level called *low tides*. When the moon's revolution and Earth's rotation are taken into account, the high tide can be expected at any location every 12 hours and 25 minutes (the moon rises 50 minutes later each day).

In the open ocean, the ebb and flow of the tides do little geologic work. Where the incoming tide is funneled through a narrowing inlet, however, the water movement is rapid and its energy is concentrated so that significant erosion and movement of sediment can occur. One location where the tide's influence is felt on a large scale is the Bay of Fundy between New Brunswick and Nova Scotia, Canada. There the *tidal range*, or difference in

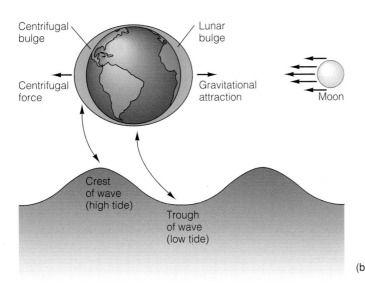

Figure 8.37 The tides are basically waves with exceptionally long periods (approximately 12 hours) between successive wave crests (the high tides). The diagram shows the effect of the moon's gravitational pull (the sun's effect is not considered here) on Earth's waters. As Earth rotates on its axis and the moon moves in its orbit around Earth, high tide is experienced at the point on Earth's surface directly facing the moon and at the same time at the point on the opposite side of Earth. The former location because the moon's pull is greatest here and the latter because the moon's pull is weakest and hence centrifugal effects are most noticeable.

(a)

(b)

Figure 8.38 (a) High and (b) low tides in the tidal creeks of a saltwater marsh along the South Carolina coast.

Box 8.4 Investigating the Environment

Venice: Where (More and More) Land and Water Meet

World attention became focused on Venice, Italy, on November 4, 1966, when a storm surge from the Adriatic Sea submerged more than 80% of the city. The water rose 1.9 meters (6 ft) above its normal high level. Saint Mark's Square lay beneath 1 meter (3.3 ft) of water. Total damage was estimated at $70 million. The November 1966 event was but one of an increasingly frequent series of floods that are causing the city's historic palaces, libraries, and museums to deteriorate at an alarming rate.

Venice was founded in the fourth century A.D. by refugees from the Italian mainland on a group of low-lying islands that occupy a part of a 10-kilometer-(6-mi-) wide lagoon (fig. 8.F). The daily tide, with a range of approximately 0.65 meter (2 ft), moves into and out of the lagoon through three inlets. This flow of water has always provided a natural flushing action for the lagoon and the city's complex of canals. In an early attempt to preserve the city's environment, the ruling magistrates of the sixteenth and seventeenth centuries had three rivers that fed into the lagoon diverted to reduce the amount of siltation. Over the years seawalls were built for protection from storms. Only in this century was the city linked to the mainland by railroad and highway bridges.

The problems of Venice represent an example of the interrelationship of the several earth sciences. The city's location in a shallow lagoon adjacent to the Po River delta sets the stage. Meteorological and climatic factors lead to the storm surges and submergences. Sharp drops in atmospheric pressure are capable of raising sea level by as much as 30.5 centimeters (12 in). Heavy, concentrated precipitation during the winter months can raise water levels from 10 to 20 centimeters (4 to 8 in). Southeastern winds, known as *siroccos*, push water northward up the narrowing Adriatic Sea toward Venice. Like all coastal cities, Venice is also affected by the general rise in sea level, estimated at 10 centimeters (4 in) during this century, due to glacial melting.

The increasing frequency and severity of flooding can generally be attributed to human causes. Because of the heavy pumping of groundwater in the first two-thirds of this century, the ground surface has subsided significantly. The situation has been alleviated somewhat by a 1969 ban on the drilling of new wells. Production of natural gas from wells in the Po River delta and the weight of new factories and other buildings may also have contributed to the subsidence. Filling of portions of the lagoon for urban expansion and the Marco Polo Airport and the sealing off of other sections for fish ponds has reduced the overall volume of the lagoon, leaving the incoming water nowhere to go but *up*. Deterioration of the seawalls and enlargement of the three natural inlets to accommodate shipping have also led to greater influxes of water. The net result of all of this is that average high-tide levels have increased some 35 centimeters (10 in) since the beginning of the century.

What can be done to save Venice for future generations? Plans have been drafted to construct additional barriers to the sea as well as inflatable storm gates that could be raised from positions on the floor of the lagoon to block the tidal inlets during times of very high water. Any solution that restricts the free flow of tidal water, of course, decreases the natural cleansing action of the tides.

Apply Your Knowledge

1. Will the problems of Venice become greater or worse as a result of the increasing greenhouse effect? Why?

2. What coastal regions of the United States might be particularly susceptible to rising sea level as a consequence of excessive building and filling of estuaries and lagoons?

3. Discuss the pros and cons of measures you would consider applying to control the flooding problem at Venice if you were in a position of responsibility there.

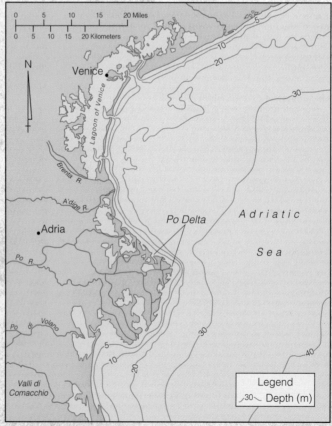

Figure 8.F The location of the Lagoon of Venice and the city of Venice (Venezia), Italy, at the northern end of the Adriatic Sea.

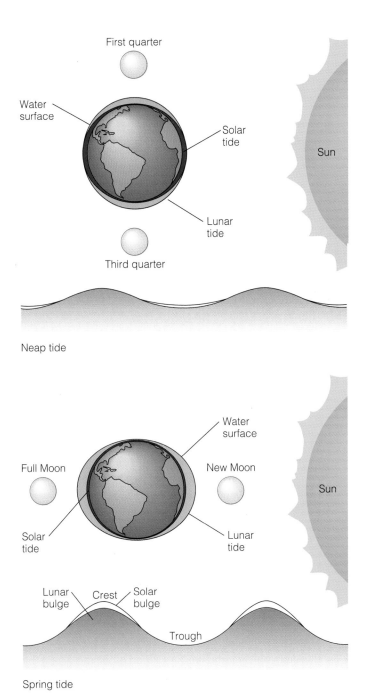

height between low and high tide, can exceed 12 meters (40 ft) and the incoming tide is observable as a low wall of advancing water known as a *tidal bore*. At the Rance Estuary on the Channel coast of France, the high tidal range is used to generate electricity. Water fills the reservoir behind the dam during high tide and then is sealed in behind the dam as the tide falls. At low tide, the water in the reservoir is released to run down through the dam and turn the turbines that generate electrical power.

Tides also influence marshlands, such as those found in New Jersey and South Carolina. The rising water flows up *tidal channels,* spreads out over the marsh, and gently moves clay and silt particles among the marsh grasses (fig. 8.38). This twice daily movement of water into and out of the marshes has the important effects of moving nutrients into the environment and carrying out pollutants. These marshes are important breeding grounds for many species of marine fish and invertebrates.

One particular tidal condition, though ineffective by itself as a geologic agent, can amplify the geologic work of waves during powerful storms. This occurs twice each month when Earth, the moon, and the sun are in a straight line so that the gravitational forces pull in the same direction. This leads to higher-than-average high tides and lower-than-average low tides, a condition known as a *spring tide* (fig. 8.39). Alternatively, when the sun and the moon are oriented perpendicularly to each other twice each month, tidal ranges are smaller than normal. This is the *neap tide*. When a hurricane approaches the land, its low pressure pulls the water up. If this uplift of water, combined with the push of the winds, coincides with a spring tide, the water may inundate the land to a great extent, and the waves do their work on sediments they would not ordinarily affect. Along low-lying coasts such as that of Bangladesh on the delta of the Ganges River, this combination of factors has historically led to massive influxes of water far inland across the islands of the delta, with devastating results. Death tolls have been measured in the hundreds of thousands. The most recent such occurrence, in 1985, took an estimated 20,000 lives.

Figure 8.39 When the sun and moon are oriented at right angles with respect to Earth (as when the moon is in the first and third quarters), their gravitational attractions are operating in different directions and partially cancel one another. The result is a neap tide, a lower than normal tidal range (i.e., lower than normal high tide and higher than normal low tide). When the sun and the moon pull in the same direction (during times of the full and new moons), gravitational attraction is reinforced, resulting in a greater than normal tidal range, or spring tide (i.e., higher than normal high tides and lower than normal low tides).

Further Thoughts: Humans as the Ultimate Geologic Agent

The natural surface processes described in this chapter have operated throughout most of Earth's history and presumably in much the same way as they do now. In recent years, however, humans have come to operate as a seventh geologic agent capable of modifying Earth's surface dramatically. Such modifications may be direct, by the application of explosives and heavy earth-moving equipment, or indirect, by the clearing of forests or the construction of breakwaters, causing an acceleration or an alteration of natural surface processes. As world population increases exponentially and underdeveloped countries adopt modern engineering techniques, the pace of conflict can only accelerate with consequences that one can predict will not be beneficial.

Significant Points

1. Earth's surface constantly changes. Any geologic change involves work, the movement of earth material, with an expenditure of energy.
2. Earth's surface is shaped by six geologic agents that do work by eroding, transporting, and depositing earth materials. These agents obtain their energy ultimately from solar radiation and their direction from gravity.
3. The prelude to action by surface agents is a gravity-driven process called mass wasting, which includes a variety of downslope movements.
4. Creep is commonly caused by the alternate freezing and thawing of water in the soil. Earthquakes and rainfall are the most common triggering mechanisms for rockslides and rockfalls.
5. Streams are the most important of the geologic agents, doing more work than all of the others combined. The stream-eroded valley is the most common of all landforms.
6. The top of the zone of saturation in the ground is the water table. The water occupying the saturated zone is the groundwater.
7. Rock and soil units that are sufficiently permeable to allow the passage of groundwater are called aquifers. Those that are impermeable to the passage of groundwater are called aquicludes.
8. Groundwater does most of its geologic work by dissolution.
9. A glacier is a thick body of ice that moves in response to gravity or to its own mass. The two principal types of glaciers are valley glaciers and continental glaciers.
10. If more ice is added to a glacier than is lost by ablation, the front of the glacier advances. If ablation exceeds accumulation, the ice front retreats.
11. Landforms associated with valley glaciers are predominantly erosional in origin, while those associated with continental glaciers are predominantly depositional.
12. Wind as a geologic agent is most effective where the ground surface consists of fine, loose sediment unbound by moisture or vegetation.
13. In waves only the energy moves forward, while the water particles move only in circular orbits. In currents both the energy and the water move in a forward direction. The immediate cause of most waves and currents is the wind.
14. Waves approaching a shore at an angle generate longshore currents that flow through the shallow surf zone parallel to the shoreline.
15. Seismic sea waves, or tsunami, have very long wavelengths and are generated when the ocean basin floor is disturbed, either by an earthquake or a volcanic eruption.
16. Tides are very long wavelength, shallow-water waves caused by the gravitational attraction of the moon and the sun for Earth.

Essential Terms

geologic work 188	subsidence 192	porosity 199	wave 209
geologic agent 189	stream 193	cavern 199	fetch 209
erosion 189	ultimate base level 193	karst topography 199	current 209
transportation 189	valley 194	glacier 200	wave refraction 210
deposition 189	delta 196	glacial till 203	beach 210
mass wasting 189	zone of saturation 197	moraine 203	longshore current 211
creep 190	water table 197	outwash 203	longshore transport 211
rockslide 190	groundwater 197	deflation 205	seismic sea wave (tsunami)
rockfall 190	spring 199	abrasion 205	214
mudflow 191	aquifer 199	dune 205	tide 214

Review Questions

1. What are the six geologic agents? How does each form?
2. What are the three principal components of geologic work?
3. Name and describe some of the important types of mass movement.
4. What are some of the common causes of subsidence?
5. What are some of the stream depositional landforms typically found within a valley?
6. How and where do deltas form?
7. What are some of the common stream patterns and why do they form?
8. What is the difference between perennial and intermittent streams?
9. What are some of the erosional and depositional features associated with karst topography?
10. What is the name of the sediment deposited directly by melting glacial ice? What landform does it produce? What is the name for stream-deposited glacial sediment?
11. What is the principal wind depositional landform? In what environments does it occur?
12. What are some of the principal depositional landforms produced by waves and currents?
13. How are longshore currents formed?
14. What are some of the common engineering structures found along a shoreline? What is the purpose of each?
15. How do spring tides differ from neap tides? When does each form?

Challenges

1. Look around your yard, your neighborhood, or your town. What examples of change or potential change brought about by the geologic agents can you identify?
2. Give some example from nature of potential energy being converted to kinetic energy. Describe the geologic work being done in each case.
3. Solar energy is utilized at Earth's surface in two ways: by heating and by evaporating water. How is each of the geologic agents caused or affected by these two processes?
4. Consider the six geologic agents. Which of them are (or have been) active in the region where you live? What are some of the landforms that you can identify and to which of the agents would you attribute each?
5. If you live near the ocean or a large lake, examine the shoreline for engineering structures. Can you determine the effect that each has had on the shoreline?

9

EARTHQUAKES, GEOLOGIC STRUCTURES, AND EARTH'S INTERIOR

The Teton Range in northwestern Wyoming is a classic example of a fault-block mountain range. Vertical movement along a normal fault has elevated the mountains thousands of meters above the adjacent down-faulted block.

Chapter Outline

INTRODUCTION

All of us are aware of atmospheric disturbances such as tornadoes and hurricanes that periodically sweep across Earth's surface, not only disrupting our daily routines but frequently putting life and property at risk. Events such as the "World Series" earthquake that rocked the San Francisco area in October of 1989 (chapter 1) likewise serve as periodic reminders that planet Earth is not only an unstable platform drifting through space but an often hazardous place to live. Earthquakes are not the only reminders of our precarious foothold on the planet today. An examination of the twisted and broken rocks exposed at many locations at or near Earth's surface verify that forces of unimaginable strength have been at work throughout all of geologic time.

In recent years, however, the passage of the energy associated with these forces through Earth has painted a picture of the planet's interior for scientists who have learned to read earthquake records. Without such information, we would know little about Earth's interior beyond what could be observed in deep mines or in well bores. Such knowledge has unlocked secrets related to the movement of continents and the formation of ocean basins discussed in chapter 10. Indeed, it has laid the groundwork for theories that explain the origin of many mineral resources, and the historical evolution and development of our planet.

EARTHQUAKES

More than one million earthquakes are recorded each year (an average of almost two per minute). Fortunately, most of these are minor tremors detectable only by sensitive instruments, but, in an average year, 20 are strong enough to cause severe damage and loss of life. Typically, at least one earthquake each year can be expected to be catastrophic.

Simply put, an **earthquake** is a trembling, shaking, or vibration of the ground surface caused by the passage of energy, in the form of waves, through the rocks of Earth's outer shell. The most common cause of the release of this energy and, thus, of the vibrations is the breaking and shifting of rocks in Earth's outer zones. This breaking and movement of rocks is a process known as **faulting.** Thus, the immediate cause of most earthquakes is faulting.

Hazards Associated with Earthquakes

In addition to the shaking of the actual earthquake, associated phenomena are the direct consequences of the earthquake and often account for more damage and loss of life than the earthquake itself. In urban areas, with their development and high population densities, fires, resulting from overturned appliances, broken electrical cables, and ruptured natural gas pipelines, represent a deadly hazard. Once started, the fires are difficult, or even impossible, to control because of broken water mains and streets clogged with rubble that restricts access to the affected areas by fire fighters and emergency crews. During the World Series earthquake, fire damage was restricted to a relatively small district because of a far-sighted system that employed seawater for fire-fighting purposes (fig. 9.1). The emergency procedures that prevented widespread destruction in 1989 were not in place during the 1906 San Francisco earthquake. Consequently, during that earlier earthquake, approximately 80% of the city was destroyed, mostly by fire. As in many earthquakes, fire fighters stood by helplessly until the fires burned themselves out.

Perhaps the most tragic example of devastation wrought by fire occurred as the result of a major earthquake that struck Japan

Figure 9.1   Fire, triggered by the October 1989 earthquake, caused significant damage in the Marina District of San Francisco.

in 1923. Open cooking fires easily engulfed the paper and wood structures and set up whirlwinds of fire driven by their own heat. The fires consumed oxygen, causing people to die by the thousands from suffocation. Even those who sought refuge in the rivers and shallow coastal waters did not survive. Hardest hit were the cities of Tokyo and Yokohama. In all, 150,000 people lost their lives, most from the fire storms, and approximately 80% of Tokyo and virtually all of Yokohama were destroyed— almost 400,000 buildings.

In mountainous regions, where slopes are already unstable, rockslides and other earth movements generated by earthquake vibrations represent another major hazard. This type of event was responsible for the greatest natural catastrophy in history: a series of earthquake-generated movements of unconsolidated wind-deposited material in northern China in 1556. Whole villages were buried beneath masses of sediment and an estimated 830,000 lives were lost. In more recent times, a massive avalanche of snow and rock debris, triggered by a major earthquake off the Pacific coast of South America, roared at high speed down a series of valleys in the Andes Mountains and buried the cities of Yungay and Ranrachirca, Peru. An estimated 70,000 people died. Compounding the tragedy was the memory of a similar, but smaller, event that had struck the same area only 8 years earlier, leaving more than 3,500 dead.

Along coastlines and within the ocean basins, yet another earthquake-induced event must often be dealt with; this is the tsunami, or seismic sea wave, described in chapter 8. These waves, generated by sudden motion when the seafloor is displaced by an earthquake, radiate outward from the point of the disturbance. They can do great damage along low-lying shorelines even thousands of kilometers distant (fig. 9.2).

The movement of portions of the seafloor, such as those that generate seismic sea waves, can also have profound effects if they occur along a coastline. During the so-called Good Friday earthquake, which rocked Alaska on March 27, 1964, some 207,000 square kilometers (80,000 sq. mi) of seafloor rose as much as 15 meters (50 ft) to form new land (fig. 9.3). In other areas, the coastal land sank below the water, carrying shoreline structures with it.

Smaller-scale features and changes of less consequence and less danger also accompany earthquakes. One such common feature involves the liquefaction of saturated soils and their eruption through ground cracks as features known as *sand spouts.*

Regardless of the damage done by fires, landslides, and seismic sea waves, the earthquake is first and foremost a series of vibrations. If severe enough, these vibrations do considerable damage to urban structures. Apart from the strength of the earthquake, which is considered later in this chapter, several factors govern how much damage an earthquake causes. An important and obvious factor is the distance a structure is located from the earthquake's point of occurrence. Earthquake energy is dissipated, and the resulting vibrations are reduced with distance from the point of occurrence. This is a general relationship, however, and subject to modification by other factors, including the nature of the rocks through which the earthquake waves pass. At the location of any structure in the path of the waves, the nature of

the rocks becomes even more important. Because soil and loose sediment amplify the earthquake waves much more than solid bedrock, buildings with foundations in such loose material are subject to much stronger vibrations and, consequently, much greater damage than those built on bedrock. Studies of damage to similar buildings after the 1906 San Francisco earthquake indicated that structures built on loose sediments suffered damage up to 12 times greater than those built on solid bedrock. (See box 9.1 for a discussion of earthquake engineering.)

Structural damage is also dependent on the duration of the earthquake. In most small earthquakes, the vibrations last but a few seconds. In larger earthquakes, however, structures may be subjected to vibrational swaying for up to several minutes (San Francisco, 1 minute; Alaska, 3 minutes). A building that resists damage when vibrating for a few seconds may collapse from the accumulated stress of several minutes of vibrations. Those structures that do not collapse outright may be so weakened that they must be condemned, or they may collapse during subsequent earthquake stresses. The last case may be used to introduce the point that an earthquake does not occur in isolation.

The Nature of an Earthquake

Any earthquake is preceded by a series of small, but measurable, tremors. Such *foreshocks* are not dangerous but have been, in recent years, the focus of one line of investigation into the prediction of the larger earthquakes with which they are associated (see box 9.2). Of more concern to the survivors of an earthquake are the tremors that invariably follow a major shock. These are *aftershocks,* some of which are almost as large and destructive as the main event. This is particularly true where buildings are already weakened. While some buildings may survive the main shock, the additional stresses of the aftershocks, occurring over the next few hours or days, are enough to bring them down.

Although our most important concern is the damage and loss of life associated with an earthquake, scientists are also concerned with the causes of earthquakes. While some small earthquakes result from the pushing and movement of rock layers by liquid magma as it works its way up toward Earth's surface, the large destructive earthquakes are the result of other stresses applied to Earth's outer layers. Such stresses are explored in more detail in the next chapter. For now, we will focus our attention on the consequences of these forces.

As stress is applied to rock, the rock initially stretches and deforms, much like a rubber band that is stretched at both ends. The more it is stretched, the more strain it is under. With continued stretching, more energy accumulates in the system. There is a limit to how much a rubber band can be stretched and how much energy it can accumulate. When this limit is reached, the rubber band breaks as the loose ends move rapidly apart and then snap back (rebound), releasing the stored energy (fig. 9.4). When the limit of the rocks' resistance to stretching is reached, they too break and move rapidly, creating a **fault,** a break in rocks of Earth's crust or mantle along which movement has taken place. In the process, the energy that has accumulated in the rocks for hundreds or thousands of years is suddenly released and travels

(a)

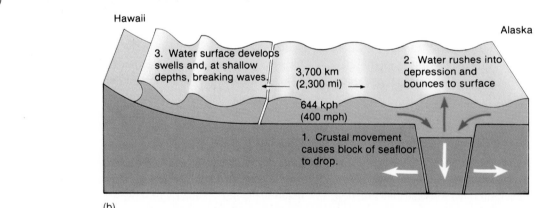

(b)

Figure 9.2 (a) Boats carried inland at Kodiak by seismic sea waves during the 1964 Alaskan earthquake. (b) Seismic sea waves, or tsunami, are very long wavelength waves generated by volcanic eruptions, earthquakes, or any sudden displacement of the seafloor.

outward through the surrounding rocks as high-speed waves. The passage of these waves, in turn, generates the series of vibrations that is the earthquake.

The actual location, or zone, at or below Earth's surface where faulting occurs is the **focus** (sometimes called the *hypocenter*) of the earthquake. If the focus is below the surface, then the point on the surface directly above the focus is termed the **epicenter** of the earthquake. The epicenter provides a geographic reference point for describing the location of an earthquake (fig. 9.5). The depth to the focus is a useful criterion for

the classification of earthquakes, which are often described as *shallow focus, intermediate focus,* or *deep focus* (table 9.1).

An interesting observation is that the frequency of earthquakes diminishes with depth. Shallow-focus earthquakes are the most common, while deep-focus earthquakes are the least common. Indeed, no earthquake foci have ever been recorded below a depth of approximately 700 kilometers (435 mi). At least two physical conditions might be cited to help explain this observation. First, rock distortion and faulting occur when the stresses applied to the rocks are unequal (i.e., stronger in one di-

Box 9.1 Investigating the Environment

Earthquake Engineering: "Bracing" for the Big One

For buildings constructed on the same type of foundation materials, an important factor in the different levels of earthquake damage sustained is the type of construction employed. In general, buildings are designed to withstand vertical stress, that is, the weight of the upper floors must be supported by the lower floors. It is more difficult to design a building to withstand horizontal stress: whiplash, shifting, and rotational motion that might be equal to the weight of the building itself. While we normally think of the "whiplash" effect as being intensified by length, this does not necessarily apply to the height of buildings.

The history of earthquake engineering is relatively short and not overly successful. Some of the early, taller buildings in San Francisco that were designed to withstand strong winds also survived the great earthquake of 1906, probably for the same basic engineering reasons. After 80% of the buildings in Tokyo were destroyed in the Great Kwanto earthquake of 1923, building codes were established that, among other things, required the use of reinforced concrete in certain buildings and limited building height to 160 meters (100 ft). Following the Long Beach earthquake of 1933, in which a number of schools were severely damaged, the California legislature passed the Field Act, requiring rigid building standards for all public buildings. In subsequent years, specialized seismographs were developed to study and understand ground motion, and "shakers" were attached to buildings to determine their critical frequencies of vibration. From these studies, a number of techniques are now employed in an attempt to minimize structural damage.

In *ductile-frame construction,* the framework of the building is designed to absorb energy by permanently bending and cracking *but not collapsing.* During the 1971 San Fernando earthquake in southern California, the Olive View Hospital, which employed ductile concrete columns, swayed and moved 46 centimeters (18 in) out of plumb, but it did not collapse. The columns had been wrapped in wire, which trapped the fractured concrete and prevented its disintegration. Columns on freeway overpasses had not been so treated, and their collapse was common. Likewise, the collapse of untreated supporting columns of the upper level of I-880 (the Nimitz Freeway) in Oakland, California, led to the greatest loss of life during the 1989 Loma Prieta earthquake.

Another technique used to a limited extent is *base isolation,* in which the building is mounted on a series of blocks, each consisting of alternate layers of rubber and steel. The blocks are designed to move and absorb the earthquake energy. This technique is not suitable for tall buildings, however, because they might become top-heavy and capsize. Reinforcement by *cross-braces* has been done in many buildings and is probably the most common engineering strategy to date. In *shear wall construction,* a building's walls consist of a number of smaller triangular steel or concrete components that are free to shift when subjected to earthquake stress. Friction between the moving components absorbs most of the energy. The walls bend and crack but do not collapse.

Despite the best knowledge of engineering principles and the finest and most costly construction materials, the bottom line as to whether a structure is truly "earthquake proof" is whether or not it survives a major earthquake.

Apply Your Knowledge

1. You are considering the construction of a 10- to 12-story office complex in the San Francisco area. What environmental factors would you take into account? Which engineering techniques would you consider using? (Don't forget the cost factor.)

2. Your company transfers you to a city in a known seismic area. What questions would you ask realtors as you looked for a home? What agencies and other sources might you contact to obtain this type of information?

(Source: Data from Charles C. Plummer and David McGeary, *Physical Geology,* 5th ed. Copyright© 1991 Wm. C. Brown Communications, Inc., Dubuque, Iowa.)

---

rection than in others), so the rock is "stretched." With increasing depth, rock pressures become more uniform and equal-sided, just as do water pressures. Second, rocks at increasing depths are hotter than at the surface, so hot that in some cases they melt to form magma. Even without melting, however, they may lose their brittleness and become more **plastic.** Thus, rather than snapping and breaking, they stretch and deform slowly.

Another important observation about earthquakes is that they are not randomly distributed about Earth's surface. For the most part, they are confined to certain distinct geographic zones (fig. 9.6). Most earthquake epicenters lie in a linear zone around the margin of the Pacific Ocean basin, the Circum-Pacific belt, which coincides with the major zone of volcanic activity that has led to the designation "Pacific ring of fire." Both earthquakes and volcanoes arise from the same underlying causes and are related to the position of boundaries between adjacent units of Earth's outer shell, called *plates,* which are examined more fully in chapter 10. A second major earthquake trend can be traced from southern Europe and Asia to the islands of Indonesia. This is the Mediterranean-Himalayan belt. The third major trend follows the axis of the submarine oceanic ridge system, which can be traced for tens of thousands of kilometers through all of the ocean basins. It is interesting to note that only shallow-focus earthquakes are recorded along the oceanic ridge system, where magma is forcing its way into the crustal rocks at or just below the surface of the ocean basin. Deep-focus earthquakes are known only in association with the deep-sea trenches of the ocean basins, where crustal plates are being forced as rigid slabs to great depths within Earth (figs. 9.7 and 9.8).

Measuring Earth Tremors

Earlier in this chapter we noted that earthquake strength varies and that variation is an important factor in the amount of damage associated with any particular earthquake. How, then, is

Figure 9.3   Portion of the seafloor at Cape Cleare, Alaska, elevated more than 8 meters (26 ft) by movement associated with the 1964 Alaskan earthquake.

earthquake strength assessed? The answer is in two ways: intensity and magnitude.

Intensity is measured by one of several scales based on the results of an earthquake: property damage and human experience. Information needed to determine earthquake intensity is gathered by field examination of damage, photographic assessment, personal interviews, and analysis of questionnaires. The most widely used intensity scale, the **Modified Mercalli scale,** assigns numerical values of strength from I to XII based on such qualitative assessments and perceptions (table 9.2). Results of such analyses are often portrayed for easier interpretation in the form of an *intensity map* (fig. 9.9). Note on the map that intensity values vary from place to place, and that, in a general way, the closer a location is to the earthquake epicenter, the greater is the earthquake intensity.

The second measure of earthquake strength, **magnitude,** is a calculation of the amount of energy released during the earthquake as determined from seismic records. Any earthquake releases only a certain amount of energy, so that for a particular earthquake there is only a single magnitude value. It matters not where an individual was during the earthquake, how strongly an individual perceived the earthquake, or whether that person felt it at all: there is only a single magnitude value for that earthquake.

Magnitude values are measured on the open-ended **Richter scale.** This scale is most familiar to all of us, because it is the one used by the media in news accounts to report and describe the occurrence of an earthquake. The Richter scale is a logarithmic rather than a linear scale, meaning that numerical increments on the scale are not of equal value but increase by some multiple. Each whole number increase in Richter magnitude actually represents an increase in energy released of 31.5 times the previous value (table 9.3). By convention, an earthquake with a Richter magnitude value of 7.0 or greater is considered to be a major earthquake, while one of 8.0 or greater is catastrophic. The 1906 San Francisco earthquake is estimated to have had a Richter magnitude of 8.2 (the Richter scale had not been devised at that time). In the same year, an earthquake in Peru is estimated to have been of magnitude 8.9. The 1964 Alaskan earthquake had a magnitude of approximately 8.4 and in 1993 the Japanese island of Hokkaido was shaken by a magnitude 7.8 earthquake that killed more than 160. No earthquake has ever been recorded with a magnitude value greater than 8.9, for the simple reason that it is unlikely that any rocks are strong enough to store the tremendous amounts of energy that must necessarily be released to cause earthquakes of this magnitude.

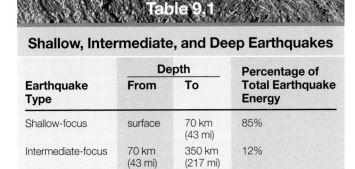

Table 9.1

Shallow, Intermediate, and Deep Earthquakes

Earthquake Type	Depth From	To	Percentage of Total Earthquake Energy
Shallow-focus	surface	70 km (43 mi)	85%
Intermediate-focus	70 km (43 mi)	350 km (217 mi)	12%
Deep-focus	350 km (217 mi)	700 km (435 mi)	3%

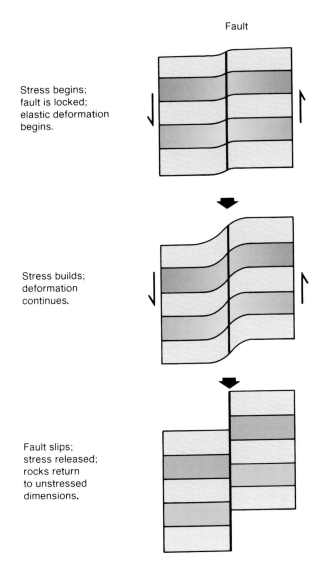

Fault

Stress begins; fault is locked; elastic deformation begins.

Stress builds; deformation continues.

Fault slips; stress released; rocks return to unstressed dimensions.

Figure 9.4 The elastic rebound theory.

Table 9.2

Damage and Perception for Each Value on the Modified Mercalli Scale of Earthquake Intensity

I. Not felt by humans except under very unusual circumstances.

II. Possibly felt by persons at rest on the upper stories of buildings. Some suspended objects such as lights might swing.

III. Noticeable indoors, especially on higher stories of buildings. Might be mistaken for the vibrations of passing trucks.

IV. Some sleepers may be awakened. During the daytime, felt indoors by most, outside by a few. Stationary automobiles are rocked noticeably. Sensation of heavy impact as a truck striking the building. Dishes rattle, walls creak.

V. Felt by most people. Many sleepers awakened. Dishes and windows may be broken and plaster walls cracked. Unstable objects may be overturned and poles and trees disturbed. Pendulum clocks stop.

VI. Felt by all. Many frightened people leave buildings. Some furniture moved. Plaster and poorly constructed walls cracked. Some chimneys damaged.

VII. Most people leave buildings. Chimneys broken. Noticed by people in vehicles. Damage to poorly constructed (adobe) houses; some cracks develop in buildings of ordinary construction.

VIII. Heavy damage to buildings of poor or even ordinary construction. Some collapse. Even well-designed buildings sustain some damage. Disturbance of people in moving automobiles. Heavy furniture overturned. Chimney, columns, and monuments collapse. Mud spouts and changes in the level of the water table.

IX. Well-designed buildings thrown out of alignment. Some damage to specially constructed buildings. Poorly constructed buildings are destroyed, while ordinary buildings are heavily damaged. Conspicuous ground cracks develop and underground pipelines are broken.

X. Most masonry and frame structures are destroyed. All structures are severely damaged. Extensive ground cracking. Railroad rails are bent. Landslides are common along stream banks and steep slopes, with water slopping over the banks.

XI. Few masonry structures remain standing. Extensive bending of rails and collapsing of bridges. Broad open cracks in the ground. Underground pipelines completely out of service. Slumps and landslides are common.

XII. Destruction is total. Waves are visible on the ground surface. Sight and level lines are distorted. Objects are thrown into the air.

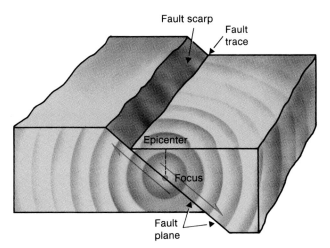

Figure 9.5 Illustration of earthquake focus and epicenter.

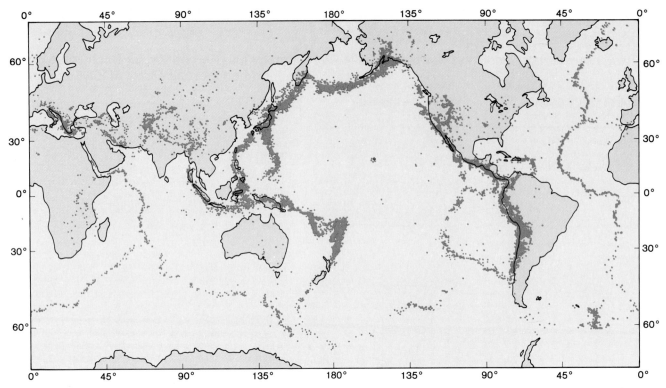

Figure 9.6 Worldwide distribution of earthquake foci between 1961 and 1967.

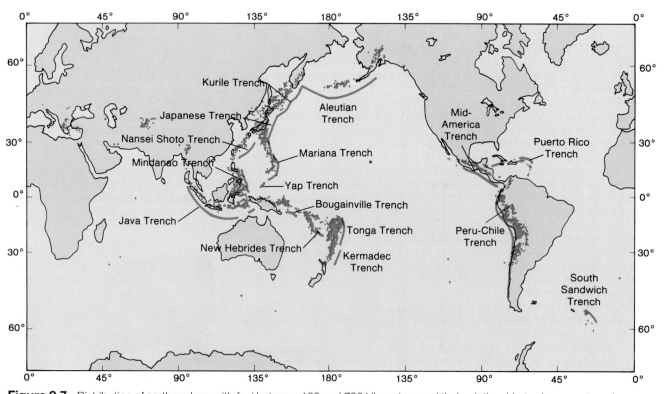

Figure 9.7 Distribution of earthquakes with foci between 100 and 700 kilometers and their relationship to deep-sea trenches.

Table 9.3

Earthquake Magnitude Values

Earthquake Magnitude	Frequency (per year)	Conventional Explosives Equivalent	Area Felt (sq. km)	Distance (km)	Approximate Mercalli Intensity
1.0	10,000,000	122 gm (6 oz)	—	—	none
2.0	1,000,000	5.85 kg (13 lb)	—	—	I-II
3.0	200,000	178.65 kg (397 lb)	—	—	II-III
4.0	30,000	5,400 kg (6 tons)	1,950	25	IV-V
5.0	3,500	179,000 kg (199 tons)	7,770	50	V-VI
6.0	500	5.64×10^6 kg (6,270 tons)	38.850	115	VII-VIII
7.0	50	1.79×10^8 kg (199,000 tons)	130,000	200	IX-X
8.0	5	5.64×10^9 kg (6.27×10^6 tons)	520,000	400	XI-XII
9.0	≤1	1.8×10^{11} kg (2.0×10^8 tons)	2,072,000	800	XII

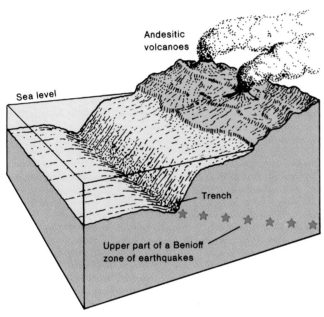

Figure 9.8 The topography and structure of the seafloor along a deep-sea trench. The dipping trend of the earthquake foci outlines a feature known as the Benioff zone and delineates a downward moving tectonic plate (see chapter 10 for a fuller explanation).

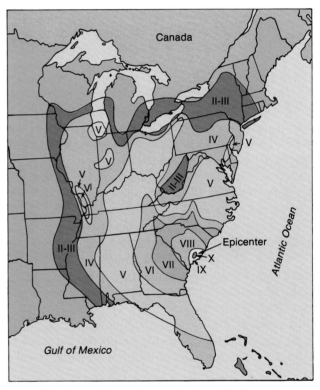

Figure 9.9 Seismic intensity map for the 1886 Charleston, South Carolina, earthquake. (Source: Data from U.S. Geological Survey.)

Earthquake Waves

When rocks break during the process of faulting, they release energy that has been accumulating for many years. The energy travels outward from the point of the disturbance in the form of waves, which vibrate and displace the rocks through which they travel. Two categories of waves are created by the disturbance, and their formation and movement might be compared to what occurs when a pebble is tossed into a quiet pond. The most obvious result of the impact of the pebble is a series of ripples that spread outward across the surface of the pond from the point of the disturbance. In the same manner, energy travels from the earthquake focus to Earth's surface and then outward on the surface from that point. The ground is thrown into a series of waves by its passage; these are **surface waves.**

If the earthquake is strong enough, such ground wave motion is actually visible to observers near the epicenter. Some of the fans standing in the parking lot outside Candlestick Park reported seeing cars rise and fall in a sequence across the lot during the World Series earthquake. During a series of earthquakes that rocked a large area centered on the small southeastern Missouri town of New Madrid in late 1811 and early 1812, whole forests were uprooted and the Mississippi River flowed backward for a short time and had its course permanently altered.

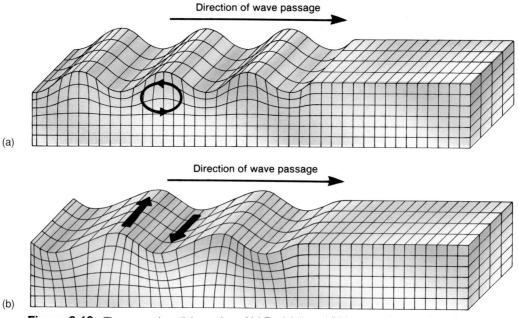

Direction of wave passage

(a)

Direction of wave passage

(b)

Figure 9.10   The ground particle motion of (a) Rayleigh and (b) Love surface earthquake waves.

Surface waves are actually of two types (fig. 9.10). The first, called *Love waves,* cause the ground to rock back and forth in a side-to-side motion. The second variety, known as *Rayleigh waves,* causes ground particles to move in elliptical orbits similar to the movement of water particles with the passage of an ocean wave (chapter 8). In Rayleigh waves, however, orbital motion is vertical and in a counterclockwise direction with respect to the direction of wave movement as opposed to the clockwise motion that occurs in water waves.

A listening device placed on the bottom of the pond into which a pebble is thrown detects the sound caused by the pebble's impact. This is because the sound travels downward through the water in wave form much as it does through air. The passage of sound energy disturbs and displaces the water particles. In the same way, earthquake energy in wave form travels downward through the solid Earth, disturbing and displacing rock and mineral particles. Because these waves travel through the "body" of Earth rather than around its surface, they are called **body waves.**

Body waves are also of two types, differentiated because they displace the rock particles through which they pass in different ways. The first of these are **primary waves, or P-waves,** so-called because they are the fastest of all waves and arrive first at any recording station. "P" might also stand for "push-pull," which describes the compressional motion by which they form. Such motion might best be visualized by imagining a flexible spring (or a child's "Slinky™" toy) with one end attached to a wall or a doorknob and the other end in your hand (fig. 9.11*a*). Stretching the spring horizontally and giving it a sharp push inward causes some of the coils of the spring to compress together in a sequence. To compensate for the area of compression, coils in the adjacent portions of the spring stretch, or expand. The zone of compressed coils moves down the length of the spring (followed by the expanded zone) until it reaches the wall. Each push generates another series of compressions and expan-

sions (also called *rarefactions*). The coils of the spring are not, of course, moving from your hand to the wall, only the energy imparted to the spring makes this journey. The coils travel only a short distance back and forth as the energy passes each part of the spring. As the P-waves pass through the coils (or through rock particles in Earth's crust), the back-and-forth movements, parallel to the direction in which the waves are moving, are identical to sound waves. Thus, like sound waves, they are capable of being transmitted through any medium: gas, liquid, or solid.

The second type of body wave is the **secondary wave,** or **S-wave,** so-called because it travels more slowly than the P-wave and, consequently, arrives at any recording station "secondarily," or after, the P-wave. "S" might also stand for "shake" (or, more technically, *shear*), which describes the motion of these waves. The S-waves act similarly to a rope attached at one end to a stationary point and shaken (fig. 9.11*b*). The individual particles that the rope is composed of do not travel from hand to wall, only the energy travels this course. At most, the rope particles move a short distance up and down as each wave passes that part of the rope. This type of up-and-down particle motion perpendicular to the direction that the waves are moving is possible only in bodies that are "rigid," that is, in solids. In gases and liquids, the energy is dissipated and the wave motion dies.

The Movement of Waves through Earth

The action of S-waves provides an important key to interpreting the structure of Earth's interior. Solid layers within Earth allow the passage of both P-waves and S-waves. Any liquid layer within Earth, however, only transmits P-waves. (It is not possible for a gaseous layer to exist within Earth because of the high pressures found there and the compressibility of gas.)

Observations of the loss of S-wave energy within Earth has led *seismologists,* those earth scientists who specialize in the

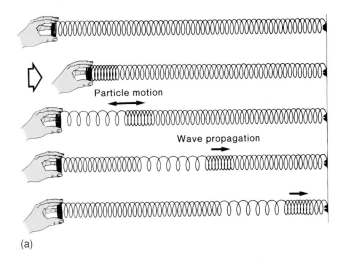

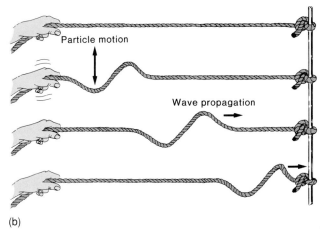

Figure 9.11 (*a*) A toy "Slinky™" (actually a long, coiled spring) illustrates the back-and-forth (push-pull) motion of the passage of a P-wave. (*b*) A rope attached to a wall and given a shake shows the up-and-down particle motion associated with S-waves.

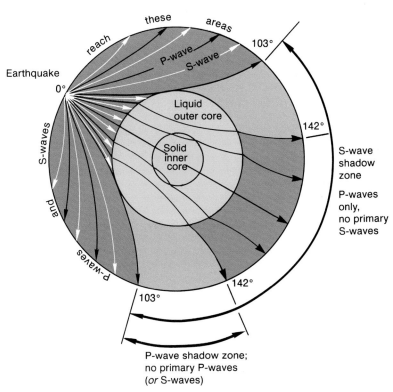

Figure 9.12 S-wave shadow zone on the side of Earth away from the earthquake epicenter results from the liquid composition of the outer core, which does not allow for the passage of such waves. The P-wave shadow zone results from the fact that P-waves are slowed and refracted as they pass through this zone.

study of earthquakes, to conclude that at some depth lies a major zone that is liquid rather than solid. They have identified zones on Earth's surface some distance from any earthquake's epicenter where the S-wave energy from that earthquake is not received. Such zones are called **shadow zones,** and their occurrence is shown in figure 9.12. The explanation for the shadow zones is that a deep liquid layer has caused any S-wave energy that impinges on it to die out. A further examination of figure 9.12 shows that numerous small refractions of both P-waves and S-waves impart to them generally curved paths as they travel downward through Earth. This bending brings them back to the surface at locations closer to the epicenter than if they had followed straight line paths through Earth. Because P-waves are slowed dramatically at the upper boundary of Earth's liquid zone, the bending is accentuated. This leads to the formation of a shadow zone for P-wave transmission as well as that for S-waves.

Seismic wave velocities vary as they travel through different kinds of rock units. As they pass through the different rock layers, their changing velocities cause them to be refracted to varying degrees. Wave velocities and the amount of refraction are governed by rock characteristics, such as rigidity, density, and elasticity. These, in turn, are controlled by temperature, pressure, and mineralogical composition.

Instrumentation

Earth scientists use a network of instruments to measure and record movements of the ground surface during the passage of seismic energy. The primary instrument for recording these movements is the **seismograph,** or *seismometer.* Although modern seismographs are complicated and sophisticated instruments, the basic principle on which they operate is simple (fig. 9.13*a*). A weight suspended by a slender spring from a rigid support firmly anchored in bedrock holds a pen or a light source. In front of the pen, and in contact with it, is a clock-driven, rotating drum covered with paper (photosensitive paper in the case of a light source). As the drum rotates in front of the weight, the pen traces a straight line on the paper. When earthquake waves reach the seismic station, the ground, the support, and the rotating drum begin to shake. Everything vibrates except the suspended weight, because the slender spring cannot transfer enough energy to overcome the weight's inertia and cause it to move.

The simple act of carrying a cup of coffee illustrates this energy transfer. If you attempt to walk with a cup firmly gripped in your hand, you will probably spill a good portion of your coffee by the time you reach your destination. Much of the energy associated with the movement of your body travels through your

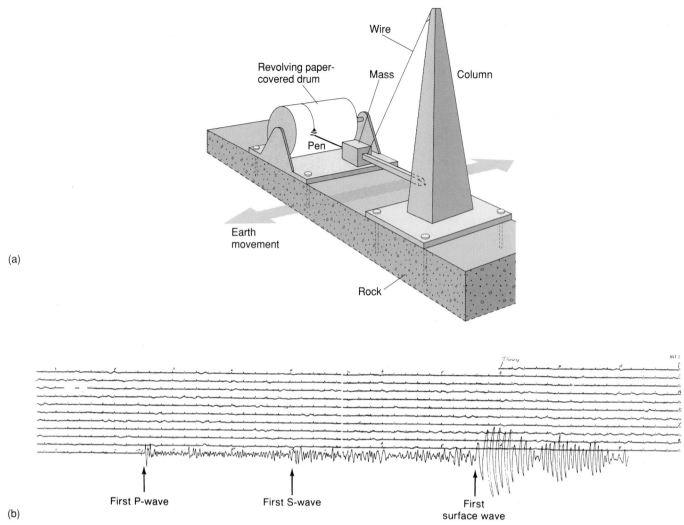

Figure 9.13 (a) A seismograph for recording horizontal ground motion. Movement of the paper-covered drum as the ground moves is recorded as a series of lines in front of the stationary mass. The same station would also have a seismograph for recording vertical ground displacements. (b) A typical seismogram. The small "tic" marks represent time in minutes. The longer the time lag between the arrival of the first P-wave and the arrival of the first S-wave, the greater is the distance from the seismic recording station to the earthquake epicenter. (Source: Seismogram courtesy of University of California, Berkeley.)

hand and to the cup, causing the coffee to vibrate violently and splash out. If you carry the cup with only your thumb and fore-finger, however, you decrease the area of contact between your hand and the cup, transferring less energy to the cup and reducing the movement of the coffee.

When earthquake waves vibrate a seismograph, the rotating drum moves in front of the stationary weight that holds the pen, and the normal straight line being drawn becomes a series of small to large deflections on the paper. These deflections constitute a **seismogram,** the written record of the passage of the earthquake waves (fig. 9.13b). The first deflection represents the arrival of the first P-wave, that which took the most direct route from the focus to this station. This is followed by other small deflections, which indicate the subsequent arrivals of additional P-waves, which took longer routes. The arrival of the first S-wave is followed by more deflections, indicating the arrival of additional P-wave and S-wave energy. Time markings on seismograms allow scientists to determine precisely when the first P-wave and the first S-wave arrived at a station. This time dif-

ference is extremely important in determining the location of the earthquake epicenter. The largest deflection in the line marks the arrival of the first surface wave. The large deflection is a testimony to the impact of surface waves. In fact, their large-scale vibrations are responsible for most earthquake damage.

Depending on a number of factors, the seismogram deflections range from small to large. One important factor in this equation is the direction from which the waves emanate. Because this is variable, most modern seismic stations have at least three seismographs. One is oriented to record east-west displacements, another records north-south displacements, and the third records vertical displacements of the ground.

Where Did the Earthquake Occur?

Shortly after the World Series earthquake, seismologists pinpointed the epicenter near the town of Santa Cruz and the Loma Prieta peak in the Santa Cruz Mountains, south of San Francisco. How did they arrive at their conclusion so rapidly?

Box 9.2 Investigating the Environment

Earthquake Prediction: Where, When, and How Big?

Of the ten natural disasters that have taken the highest tolls in human life, six have been earthquakes. As world population grows, the press of people into earthquake-prone areas becomes ever greater. These two perspectives, one historical and the other current, make the accurate prediction, and even prevention, of earthquakes an increasingly desirable scientific goal.

Four nations, China, Russia, Japan, and the United States, are active to varying degrees in the field of earthquake prediction. To date, China can be cited as having both the best and the worst examples of predictions. On February 4, 1975, a magnitude 7.3 earthquake destroyed 90% of the city of Haicheng in northeast China as well as many surrounding towns. Fortunately, 9 1/2 hours earlier the government had ordered several million people living in this region to leave their homes and move temporarily into open areas. This governmental directive, resulting from the prediction of an imminent earthquake, probably saved tens or even hundreds of thousands of lives. Eighteen months later, on July 28, 1976, an earthquake in excess of magnitude 8.0 leveled most of the city of Tangshan, 150 kilometers (93 mi) south of Beijing. Loss of life was in the hundreds of thousands. While Chinese scientists had predicted that a major earthquake would occur in this region sometime within a 2-month period, this lack of immediacy was virtually the same as no warning, so no preventive actions were taken.

Scientists have identified a number of events that could be related to the imminent occurrence of an earthquake. These are referred to as *precursor phenomena*. Among those most actively investigated are the occurrence and frequency of foreshocks (as at Haicheng), changes in ground level (either swelling or tilting), changes in the degree to which the ground conducts or retards (resistivity) the flow of electrical currents, rock density changes, increases in the amount of the radioactive gas radon in well water, anomalous animal behavior (such as snakes leaving their holes in winter and restlessness of livestock), and even changes in atmospheric pressure or the alignment of the planets. In addition to looking for and monitoring precursor events, studies are also being done on the statistical probability of earthquakes of certain magnitudes occurring along specific faults (or segments of faults) within a given period of time. One outgrowth of this last approach has been the recognition of *seismic gaps:* segments of active faults where the buildup of stress in the rocks has not been reduced by the almost continuous slight displacements known as *creep*. Stress continues to build until the rocks can no longer withstand it and all of the accumulated energy is released suddenly to cause a large earthquake.

No type of precursor is universally accepted as the most fruitful avenue for further research. Likewise, no model detailing the events leading to an earthquake that can account for all such phenomena is accepted by everyone. One such model, however, the *dilatancy model,* probably has more adherents and accounts for more of the observed phenomena than all of the others. The dilatancy model holds that rocks within the fault zone continue to accumulate stress until the stress is relieved by the fracturing of the rock along a network of countless microscopic cracks. This cracking results in a phenomenon known as *dilatancy hardening* and, at least temporarily, relieves some of the accumulated stress. This period of dilatancy hardening can be recognized by the fact that P-waves from minor seismic events or those artificially induced move through the rocks in the vicinity of the fault zone with lower than normal velocities. (A more common method of expressing this change in seismic velocities is as a reduction in the ratio of P-wave to S-wave velocities.) Over a period of time, even as stress continues to be applied to the rock, groundwater moves in to fill the fracture system and seismic wave velocities gradually increase. When the velocities reach their normal values, the hardening effect is canceled and an earthquake can be expected at any time. What studies have been made of these seismic velocity changes also suggest that the longer the anomaly, the larger the earthquake will be.

An earthquake prediction would ideally include three factors: time, location, and magnitude. As desirable as such an end result would be, however, success will not be simple, nor will it come without a price.

Apply Your Knowledge

1. Discuss the significance of being able to determine in advance *how big* an earthquake will be.

2. Discuss the economic and social implications of the prediction of an imminent earthquake for a major metropolitan area. Consider how long people might have to stay at home and how long they might have to avoid the freeways and public buildings.

3. Under what conditions might you recommend the evacuation of a metropolitan area? Consider the time factor.

4. Discuss the consequences of an erroneous prediction in terms of economics, disrupted lives, and future credibility. Who would be responsible (morally and legally) for these consequences?

Because seismologists have compiled and tabled the seismic velocity data from thousands of earthquakes, they now know the time intervals required for P-waves and S-waves to travel given distances. Examples from this data in table 9.4 shows that a P-wave taking the most direct route requires 4 minutes and 6 seconds to travel from the focus to a seismic station 2,000 kilometers (1,243 mi) away. An S-wave leaving the focus at the same time and following the same route requires 7 minutes and 25 seconds to reach the same station. The P-wave obviously travels much faster than the S-wave. This situation is analogous to two automobiles leaving the same location at the same time but traveling at different speeds. The faster automobile reaches any given destination before the slower one. Furthermore, the farther the destination is from the starting point, the greater the time interval between the two arrivals. In the same manner, the P-wave pulls farther ahead of the S-wave as they travel farther from the focus. Therefore, the difference in arrival times (S − P) of the two types of waves at any recording station can be used as a measure of the distance from focus to that station.

Armed with the type of information given in table 9.4, seismologists can easily pinpoint an epicenter through the cooperative efforts of at least three seismic stations (fig. 9.14).

Table 9.4

Sample Seismic Velocity Data

Distance from Source (km)	Travel Time				Interval between P and S arrivals (S–P)	
	P-wave		S-wave			
	Minutes	Seconds	Minutes	Seconds	Minutes	Seconds
2,000	4	06	7	25	3	19
4,000	6	58	12	36	5	38
6,000	9	21	16	56	7	35
8,000	11	23	20	45	9	22
10,000	12	57	23	56	10	59
11,000	13	39	25	18	11	39

(Source: Data from S. Judson and M. Kauffman, *Physical Geology*, 8th ed., p.185 Copyright © 1989 Prentice Hall.)

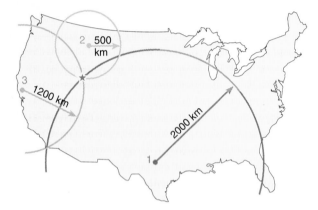

Figure 9.14 The epicenter of an earthquake is located by drawing circles centered on each of three seismic stations. The radius of each circle is the distance that the station is from the epicenter as determined from S-wave minus P-wave velocity data. The intersection of the three circles is the location of the epicenter.

Assume that a seismograph at station 1 records the data described in the previous paragraph. Seismologists at that station, by consulting their table of travel times, know that the earthquake epicenter lies 2,000 kilometers (1,240 mi) from their station. To locate the epicenter, they first draw a circle on a map, centered on station 1, with a radius equal to 2,000 kilometers. The earthquake epicenter must lie somewhere on the circumference of that circle, because those are all of the points on Earth's surface that are exactly 2,000 kilometers from station 1. By itself, the circle is relatively useless, but in conjunction with circles drawn on the basis of information provided by seismologists at other stations, it becomes the first step in the locating process.

Seismologists at station 2, a site closer to the epicenter than station 1, provide travel times that allow the plotting of a second circle, one with a radius of 500 kilometers (300 mi) and centered on station 2. The two circles intersect at two points. The epicenter must be one of these two points, because these

are the only two points on Earth's surface that satisfy the conditions: 2,000 kilometers from station 1 and 500 kilometers from station 2. Determining which of these points is the epicenter requires data from a third station, station 3. If the distance to the epicenter from station 3 was 1,200 kilometers (745 mi), a circle of that radius and centered on station 3 is drawn. This third circle should intersect one of the two earlier determined points or come very close to it. This is the earthquake epicenter: the only point on Earth's surface that is 2,000 kilometers from station 1, 500 kilometers from station 2, and 1,200 kilometers from station 3.

The fact that the circles often do not meet precisely at a point is the result of several influences. The most important of these is that distances from an epicenter to any seismic station are measured across Earth's surface, while the waves take a "shortcut" from the focus through Earth (fig. 9.15). In addition, seismic wave velocites increase with depth, so waves traveling more deeply in Earth travel faster than those following shallower paths. For example, as the distance from an epicenter to a station doubles, the (S − P) does not double but is less than that amount (table 9.4).

GEOLOGIC STRUCTURES

That Earth undergoes an extraordinarily high number of tremors and earthquakes each year was earlier cited as evidence that Earth is a restless body. Earth scientists can cite additional evidence to support the contention that Earth has always been restless. For example, rocks containing the fossil remains of animals that once lived in the sea have been raised to such heights that they now make up portions of some of our highest mountains. In addition, layers of coal, known to have been formed at or above sea level, have been pushed downward to depths of hundreds or thousands of meters below sea level. Surely, extraordinary forces must have been at work over long periods to bring about such changes. As compelling as these examples are, geologists can point to even more com-

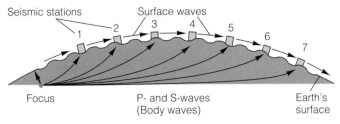

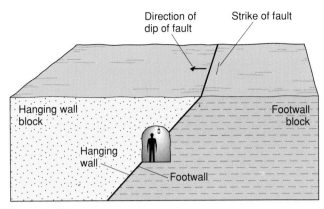

Figure 9.15 The relationship of seismic wave paths to earthquake focus and receiving stations. The deeper the focus and the more distant the stations, the shorter the paths of the P-wave and the S-wave compared to the surface waves.

Figure 9.16 The hanging wall block (that which would be above an observer's head) and the footwall block (that on which the observer would stand) in relation to the fault plane.

monplace rock features that tell much the same story. These features, which were produced when strong, deforming forces were applied to the rocks, may be grouped together as the **geologic structures.**

The geologic structures can be broadly separated into two categories, principally on the basis of how the rocks respond to stresses: they break if they behave as brittle solids, and if they respond plastically, they bend. Structures that form when rocks break can be further sudivided into two types: *faults* and *joints*. If rocks respond plastically, they form *folds*. The process by which all the geologic structures are formed is called **diastrophism.**

Faults

Faults are of several types. When the movement is predominantly in a horizontal, or sideways, direction, they are referred to as **lateral faults** (or as *strike-slip, transcurrent,* or *transform faults*). This group includes the famous San Andreas Fault in California, movement along which was responsible for the 1906 San Francisco earthquake as well as the 1989 Loma Prieta earthquake. During the 1906 event, a portion of the San Andreas Fault underwent an instantaneous horizontal slippage of 6.4 meters (21 ft), a remarkably large displacement for a single instance of movement.

When the movement along a fault is predominantly vertical, geologists normally employ a pair of terms that have been adapted from coal miners: **hanging wall** and **footwall** (fig. 9.16). The diagonal line on the front of the block represents a fault plane, which reaches from Earth's surface to some undetermined depth. If you could walk down the fault plane and into Earth's interior, you would have to place your feet on the right-hand side (or block of rock), while the left-hand side would "hang" overhead. Because of this relationship, the right-hand block is referred to as the footwall, while the left-hand block is the hanging wall.

Figure 9.17 depicts the several types of faults with displacements recognized by earth scientists. In the instance where the hanging wall has moved downward *relative* to the position of the footwall, the resulting fault is called a **normal fault** (an unfortunate application of terminology, because this type of fault is no more or less "normal" than any other type). Understand the importance of the term *relative* displacement. As far as the *ac-tual* movement is concerned, there are really five distinct possibilities.

1. The hanging wall may have indeed moved down while the footwall remained stationary.
2. The footwall may have moved up while the hanging wall remained stationary.
3. The footwall may have moved up while the hanging wall moved down.
4. Both blocks may have moved up, with the footwall moving farther.
5. Both blocks may have moved down, with the hanging wall moving farther.

During faulting, and for a period of time after, the upthrown block may stand topographically higher than the hanging wall block, with the steep surface separating the two called a *fault scarp*. In time, however, the processes of weathering and erosion wear down the elevated block and restore a uniform surface. The evidence of past movement remains, however, in the broken and displaced rock layers visible on the front of the block. Normal faults form when rocks of Earth's crust are subjected to a stretching force (referred to by earth scientists as *tension*). Where blocks of Earth's crust are raised or lowered between pairs of faults, *horsts* and *grabens*, respectively, are recognized.

If the hanging wall block is forced upward relative to the position of the footwall block (the reverse situation from what caused the normal fault), the result is a *reverse, or thrust, fault*. If the fault plane is inclined at a low angle to the horizontal, an *overthrust fault* is formed. Portions of the southern Appalachians as well as the northern Rockies have been moved tens of kilometers from their original positions by overthrusting. Both reverse faults and overthrusts occur when the rocks are subjected to squeezing, or *compression*. Figure 9.18 shows the front (fault scarp) and back sides of a major overthrust in northern Montana.

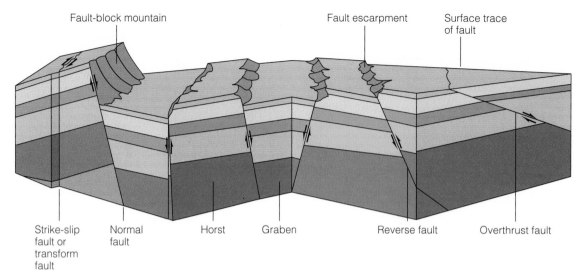

Figure 9.17   Various types of faults.

Joints

The second major type of structure resulting from rock breakage is the **joint,** a break in the rock along which *no movement* has taken place. Joints are, by far, the most common type of structural feature found in rocks. They form usually from the release of stress rather than from its direct application. That is, when deeply (or evenly shallowly) buried rocks are brought to or near Earth's surface, by uplift and erosion of overlying rocks, the lower-pressure environment allows for a certain amount of expansion. When expansion occurs, cracks develop in the rocks. Such cracks, or joints, normally occur in parallel sets. They may be spaced centimeters or meters apart, but they are almost always there. Joints may also form in rocks by contraction rather than expansion, as when igneous rocks cool.

Folds

Rocks may respond to deforming stresses by behaving plastically. This flowlike behavior produces a group of structural features known as **folds.** Although earth scientists recognize a variety of folds, there are but two fundamental types: anticlines and synclines. Both result from the application of strong compressive stresses. **Anticlines** form when rocks are bent, or arched, upward (fig. 9.19). The name derives from the fact that the rock layers on either side of the fold center, or *axis,* are "inclined" away from or "against" (*anti-*) each other and the fold axis. In an anticline, the axial area is also referred to as the *crest.* The very action of pushing the rocks upward carries older rocks nearer to the surface, where they may be exposed by erosion. Indeed, if you examine the ages of rocks in the eroded core of an anticline, you find the oldest exposed layer along the axis, with progressively younger rocks exposed in either direction away from the axis.

The opposite situation occurs in a **syncline,** or downfold, where younger rocks, originally near the surface, are pushed downward to greater depth. As a result, younger

(a)

(b)

Figure 9.18   (*a*) Front (facing the fault scarp) and (*b*) backside of an overthrust fault in the Sun River Canyon area of Montana. Massive limestone units have been pushed up and over much younger shale beds for a distance of several kilometers.

Box 9.3 Further Consideration

Sideling Hill: Not Your Ordinary Rest Stop

The expenditure of public money to build a museum dedicated to the geology exposed in a road cut is anything but a common occurrence. Not only was it done in Maryland at Sideling Hill, but it has been an uncommon success. The deep cut made for the passage of I-68 through Sideling Hill in the western mountainous portion of Maryland revealed one of the finest exposures of a major geologic feature in the northeastern United States: an almost perfectly developed, tightly folded syncline (fig. 9.A). Within the syncline are exposed a sequence of sandstones, siltstones, shales, conglomerates, and coals that were laid down as sedimentary deposits some 340 million years ago, when this area was alternately dry land and shallow seas. The synclinal structure was formed approximately 240 million years ago, the result of intense compression caused by the collision of Africa and North America during the Alleghenian Orogeny.

Located 9.7 kilometers (6 mi) west of Hancock, Maryland, this spectacular cut is visible on a clear day from as far away as South Mountain, 60 kilometers (37 mi) to the east. It took 16 months to excavate the cut, which was completed in August of 1985, and required the removal of 3.4 million cubic meters (4.5 million cubic yd) of rock. The cut is 216 meters (720 ft) wide at the top, 60 meters (200 ft) wide at the bottom, and 102 meters (340 ft) deep.

Work began on the Sideling Hill Exhibit Center in 1990 and was completed 1 year later. The center, with its tourist information, rest area, and walkway across I-68, was officially opened to the public on August 2, 1991, by Governor William Donald Schaefer. It includes 20 exhibits, which explain the geologic setting and the geologic history of the Sideling Hill area. Some of the exhibits are outdoors, some are "hands-on," and there is a block model of the cut for the visually impaired. In its first year, more than a half million people visited the exhibit center.

Apply Your Knowledge

1. What benefits (economic, social, and educational) can you cite in support of the establishment of a facility such as that at Sideling Hill, Maryland?

2. Would you support the expenditure of public funds for the construction of similar geological museums in your state? Why or why not?

Figure 9.A Aerial view showing Sideling Hill road cut and the Sideling Hill Exhibit Center.

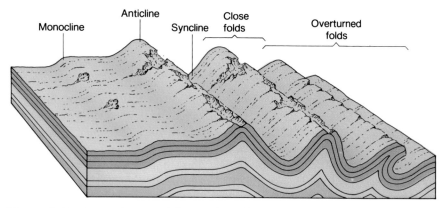

Figure 9.19 Simple folds: anticlines and synclines.

rocks are found near the axis, or *trough,* of a syncline, with increasingly older rocks exposed outward on the fold flanks (fig. 9.20). Synclines are so-named because the rock layers on either side of the axis are "inclined" toward or "with" (*syn-*) one another.

Anticlines and synclines may be highly varied in scale, ranging from microscopic to whole mountains or mountain ranges. For example, the individual ridges of the Allegheny Mountains and the Front Range of the Colorado Rockies consist of single, broad, anticlinal folds.

EARTH'S INTERIOR

Most of Earth, of course, is not directly observable. As indicated, we must rely on the techniques of remote sensing and extrapolation to determine the composition and structure of Earth's interior. For many years, scientists have relied on two principal sources of information in attempting to piece together a picture of Earth's interior: evidence from meteorites and seismic studies of Earth's interior.

Evidence from Meteorites

Scientists have derived a theory of Earth's composition from some of the millions of meteoroids that impinge on Earth from space each year. Most of these are consumed by the frictional heat generated by their passage through the atmosphere, which produces the brilliant moving light displays called meteors. Some, however, survive this ordeal and reach Earth's surface, where scientists collect and study them as meteorites (so-called once they are found on Earth's surface). Years of collecting and study have led to classifications based on the composition of the meteorites. Two basic types occur: those that are metallic (an iron-nickel mixture) and those that are "rocky" or "stony." It has been suggested that meteorites represent the remains of a disintegrated planet that once occupied an orbit between Mars and Jupiter (the present site of the asteroid belt). Such a planet, it is reasoned, could have been similar to Earth. Meteorites, then, would be samples of the interior of some planet that approximated the composition of Earth's deep, unsampled interior. This analogy seems reasonable when used in conjunction with other data and, indeed, helps us to explain some of the other data.

Seismic Studies of Earth's Interior

A second remote source of data for interpreting Earth's interior is the study of seismic body waves, because, as we have seen, their velocities are determined by the physical characteristics of the materials through which they pass. Figure 9.21 shows that Earth is divisible into four major zones. The first of these, the **inner core,** occupies approximately the inner 1,600 kilometers (1,000 mi) of Earth's radius and is thought to consist of a mixture of iron and nickel, called *kamacite.* This high-density metallic material probably worked its way toward Earth's center as lighter material became segregated toward the surface during the formative stage in Earth's history. Because of the very high

Figure 9.20   The Tremont Syncline along I-81 in Pennsylvania.

pressures that must exist at this depth, as well as P-wave characteristics, the kamacite in the inner core is considered to be in a solid state.

From the boundary with the inner core to approximately 3,200 kilometers (2,000 mi) out from Earth's center is the **outer core.** Although somewhat less dense, this zone is thought to consist of the same mixture that constitutes the inner core. This is the deep layer, as discussed earlier in this chapter, where the energy associated with S-waves becomes lost and the velocity of P-waves is drastically reduced. To the seismologist, this means that the material of the outer core is a liquid rather than a solid.

The liquid composition of Earth's outer core has further important implications for scientists. Because liquids flow, it is quite likely that heat-driven currents of metallic material exist within this zone. Such movement would constitute an electric current, and an important principle of physics is that electric currents generate and surround themselves with magnetic fields. Thus, the outer core is a likely source for Earth's magnetic field. This idea is explored in some detail in chapter 10.

The next zone out from the center, the **mantle,** occupies approximately the outer half of Earth. With the exception of isolated pockets of magma in its upper portion, it is assumed, because of its seismic behavior, to be in the solid state. Unlike the metallic core, the mantle is considered by scientists to be a rocky layer. While opinion differs on its exact composition, which almost certainly varies significantly from place to place,

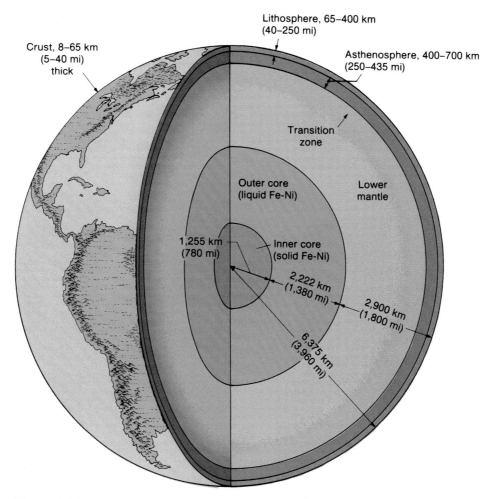

Figure 9.21   Cutaway view showing the major zones of Earth's interior.

a gross approximation might be that of the igneous rock peridotite, a mixture of the iron- and magnesium-rich minerals olivine and pyroxene. A variety of *peridotite* known as *kimberlite* is usually associated with the diamonds of South Africa and elsewhere. Both the fragmented nature of the kimberlite rock and the high-pressure origin of diamonds suggest that this rock was forced up from some depth in the mantle.

As kimberlite (or any other material) moves upward from the mantle, it must pass through a last zone, the **crust,** Earth's outer rocky shell. On the scale of Earth, the crust is an extremely thin layer, varying from a minimum of about 4 kilometers (2.5 mi) thick beneath the ocean basins to a maximum of about 65 kilometers (40 mi) thick beneath the continents. These thicknesses have been determined by a number of seismic studies. Although small in volume and thickness, the crust is of immense importance to us. Because of its relative accessibility, it is the zone about which we know the most.

Figure 9.22 portrays both the composition and the dynamics of the crust. Two dominant types of crustal material are recognized. The first is essentially granite in composition and underlies only the continents (including the shallow submerged seafloor around their margins). Because of its location and its composition, this material is referred to as either the *continental crust* or the *granitic crust.* Because the minerals that compose a granite are rich in silicon (Si) and aluminum (Al), this material has also been referred to as *sial,* or *sialic crust,* but this terminology is not in vogue in newer texts. The thickness of granitic crust is highly varied; it is absent beneath the ocean basins and in direct proportion to elevation in the continents. That is, it is thickest beneath high, mountainous regions and thinnest beneath low-lying plains. This distribution leads to comparisons of continents to icebergs, where the ice mass hidden beneath the water is in direct proportion to that visible above the water. As the exposed ice melts and its surface is lowered, the bottom of the iceberg rises to compensate for this loss of mass.

As mountains wear down by erosion, the base of the granitic crust is thought to rise to keep this system in balance, so that the continents are in a sense "floating and bobbing" on the underlying mantle. This system of balance with respect to the rising and lowering of continents is referred to as *isostasy.* The varying thickness of the granitic crust as determined through seismic studies, its high position, and its low average specific gravity (2.7) compared to the materials of the other earth zones, appear to substantiate a view long held by earth scientists that the continents represent a light "scum" that has accumulated on Earth's surface throughout the planet's development.

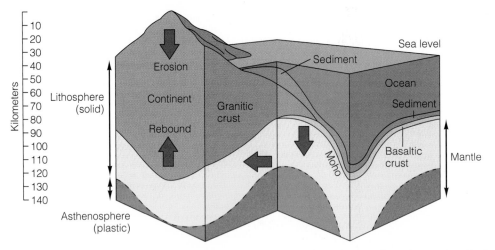

Figure 9.22 Expanded view of Earth's crust and uppermost mantle. As the continent is worn down by erosion, the base of the continental crust rises to maintain balance. Material moves laterally slowly to replace it.

The ocean basins are underlain by a very different material, rocks rich in the ferromagnesian minerals best represented by volcanic basalt and its plutonic counterpart, gabbro. This is the second dominant type of crustal material. Because of its location and its composition, this material is often referred to as *oceanic crust* or *basaltic crust*. Its composition, dominated by silicate minerals rich in magnesium and iron, has led in the past to the designation *sima,* an abbreviation of silicon (Si) and "mafic," a term used to describe dark minerals and rocks.

The basaltic crust is not as restricted as the granitic crust, and some seismic evidence suggests that it forms a continuous layer stretching beneath the continents as well as the ocean basins. If it does lie below the granitic crust in continental areas, it is because of its greater specific gravity (3.0). Unlike the granitic crust, the basaltic crust maintains an approximately uniform thickness between 4 and 5 kilometers (2.48 to 3.1 mi).

Because seismic wave velocities are functions of the physical properties of the rock layers through which they travel, seismic waves undergo a marked increase in velocity as they pass downward from the crust into the mantle, where specific gravity increases markedly (fig. 9.23). Such an abrupt velocity increase represents an important boundary that was first recognized in the early part of this century by the Croatian seismologist A. Mohorovicic. Earth scientists now refer to this seismic representation of the crust-mantle boundary as the *Mohorovicic Discontinuity,* the *M-discontinuity,* or simply, the *Moho.*

Seismologists have found that the downward increase in seismic wave velocity through the crust (including the sharp increase at the Moho) continues through the uppermost part of the mantle. At a depth of approximately 80 to 100 kilometers (50 to 60 mi), however, seismic wave velocities not only stop increasing, they actually slow somewhat. This depth represents the top of a zone that extends downward to a depth of approximately 320 kilometers (200 mi). Because of its seismic characteristics, this is referred to as a *low-velocity layer.*

Material in this layer is apparently still solid but not brittle and rigid. Rather it is likened to plastic, which may be distorted in shape and caused to flow over long periods of time. Thus, occasional reference is made to a "plastic layer" in the upper part of the mantle. What could cause such plastic behavior in seemingly solid rock? The best explanation is that the material within this layer is probably very hot—not hot enough to melt, but close enough to its melting point that some of its solid characteristics are altered. Because of the loss of solid characteristics, this layer is referred to as the **asthenosphere** (from the Greek *asthenes,* meaning "weak"). The more rigid material above the asthenosphere, consisting of the continents, the ocean basins, and the uppermost layer of the mantle, acts as a single unit in most geologic processes. This unit is called the **lithosphere** (from the Greek *lithos,* meaning "rock"). As important as the discovery of the Moho was, it was really the recognition of the distinction between the asthenosphere and the lithosphere that has helped to answer questions that had for so long seemed unanswerable.

For example, what is the source of magma for volcanic eruptions? If the material of the asthenosphere is indeed very close to its melting point, then a decrease in pressure or the application of a small amount of additional heat would be enough to melt the rock and form magma. Indeed, plots of successive earthquake foci have tracked the upward movement of magma from the upper part of the asthenosphere just prior to its eruption at the surface. If material is, in fact, moving vertically at or near the surface (as in the "floating and bobbing" of the continents), there must be a "cushion" at depth that has sufficient flexibility to allow for such movement. A plastic layer would provide the necessary flexibility. It could also account for the continuing "rebound" of Earth's crust following the melting away of the continental glaciers of North America and Europe. The formation of massive ice sheets in excess of 3.25 kilometers (2 mi) thick placed an enormous strain on Earth's crust, literally pushing it downward. The melting of the ice, a relatively rapid geological event, left the crust out of isostatic balance. Since the

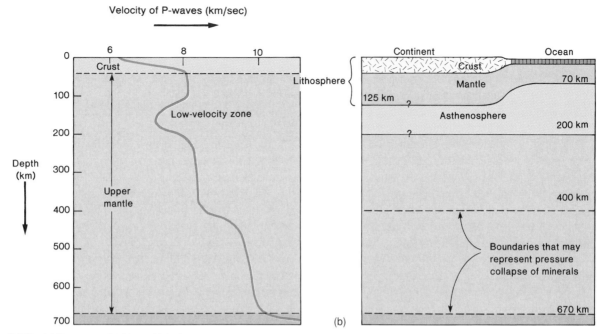

Figure 9.23 (a) P-wave velocities within the crust and upper mantle. Note the progressive increase in velocity through the crust and the pronounced velocity decrease in the upper mantle, the "low-velocity zone." (b) Interpretation of the nature of the crust and upper mantle based on the seismic wave data.

melting (approximately 8,000 to 10,000 years ago), the crust has been slowly recovering to its former position. Evidence for such rebound, particularly in high-latitude regions, is in the form of elevated beach ridges and a continuing decrease in coastal tide gauge readings.

Lastly, a plastic layer, capable of flow, would provide a plausible mechanism for the movement of large segments of lithosphere. They could be carried along as rafts. Long a shortcoming of the old theory that the continents had somehow drifted from ancient sites to their present positions, the "continental drift" theory now holds that the continents move as integral parts of larger, "floating" lithospheric units, called plates, whose movements can now be logically explained. This new theory, the topic of the next chapter, is called **plate tectonics.**

MAJOR FEATURES OF EARTH'S SURFACE

Whether we view Earth from space or simply look at a world map, it is readily apparent that Earth's surface is a mosaic of two fundamental types of features: continents and ocean basins. Before examining continents and ocean basins in more detail, we first examine their similarities and differences.

Similarities and Differences

The continents account for approximately 30% of Earth's surface area and average 800 meters (2,670 ft), or 0.8 kilometers (0.5 mi), above mean sea level. The ocean basins average −3,740 meters (−12,464 ft), or 3.62 kilometers (2.25 mi), deep. These are rather obvious geographic differences. Earth scientists are also interested in two fundamental geologic differences that are not so readily apparent. First, as we have seen, the foundations of the

continents consist principally of the rock granite, whereas basalt forms the floors of the ocean basins. Second, the continents contain the oldest rocks yet dated by geologists, granites and gniesses from Greenland and Canada that are at least 3.8 billion years old. The ocean basins, on the other hand, contain no rocks older than 180 million (0.18 billion) years (at least none that have been discovered to date).

This last relationship represents a complete reversal in geologic thinking from 30 years ago. Until that time, it was assumed that the oceans were primordial and unchanging and that the continents were constantly being altered, their rocks destroyed and regenerated. Ideas related to the concept of seafloor spreading, which is examined in chapter 10, now tell us that new ocean basin material is constantly being formed along features known as oceanic ridges and destroyed in subduction zones at oceanic trenches. None has the time or opportunity to age to the same degree as continental rock material, which, because of its lower density, always remains "floating" on the surface.

In addition to these general characteristics that apply to entire ocean basins and continents, it is also instructive to examine each in terms of the large-scale features that characterize them. (We examined the smaller landforms of the continents in chapter 8, and those of the ocean basins are considered in chapter 12.)

The Continents

An examination of any of the continents for such variables as elevation, rock type and structure, and rock age and history reveals three fundamental types of landforms: plains, plateaus, and mountain systems.

Plains are areas characterized by low elevations and low *relief* (the difference between high points and low points). The underlying rocks are usually sedimentary in origin, horizontal (or near so) in altitude, and show few signs of having been disturbed by tectonic forces. Prominent examples in North America are the coastal plains, which are especially well developed along the Atlantic and Gulf coasts, the central lowland region from western Ohio to eastern Iowa, and the Great Plains, stretching from central Canada to West Texas.

Plateaus are similar to plains in that they are underlain by rock layers generally horizontal and predominantly sedimentary (although layered basaltic lava flows comprise the structure of the Columbia River plateau of Washington and Oregon). Elevations of plateaus generally are higher than those of plains, because most plateaus have been elevated vertically by tectonic forces, sometimes by hundreds of meters. Such elevation often leaves the margins of the plateau bounded by steep, clifflike surfaces known as *escarpments*. Because of the elevations of most plateaus, the processes of weathering and erosion carve out steep canyons and gorges (such as the Grand Canyon of the Colorado River through the Colorado Plateau in the southwestern United States) or at least dissect the plateau into a "rolling" region of moderate to high relief. In addition to the Columbia River and Colorado plateaus, the Appalachian Plateau, stretching from New York State to Alabama, is a prominent feature of continental proportions.

Mountain chains, or *systems,* in the sense used here, are features that are thousands of kilometers long and hundreds of kilometers wide. Topographic characteristics include high elevations, high relief, mostly steeply sloping surface areas, and very limited summit areas. The rocks of mountain systems may have been folded, faulted, tilted, intruded by magma, metamorphosed, or even extruded as ash and lava. While most of us can readily visualize mountains in the geographic sense, to the earth scientist, it is also important to visualize mountains in the geologic sense, that is, how they formed and what has happened to them. To the geologist, mountains are more than just towering peaks such as those of the Rocky Mountains or Sierra Nevada. Older mountain systems like the Appalachians have been worn down by millions of years of erosion. The broad area of igneous and metamorphic rock occupying most of eastern Canada is today an area of low relief. The rock types and structures, however, tell a story of mountain systems to rival the Rockies and Appalachians that existed there billions of years ago.

The Ocean Basins

Like the continents, the ocean basins can be divided into three broad realms: continental margins, the deep ocean basin floors, and the oceanic ridge systems. Because these features are dealt with in detail in the section on oceanography, brief descriptions, without formal definitions, will suffice here.

The **continental margins** form transition zones from the continents to the deep ocean basins. These zones are highly variable in width, ranging from broad, shelflike areas hundreds of kilometers wide off the eastern coast of North America to narrow belts only tens of kilometers wide off the tectonically active western coast. Not only are the margins topographically transitional but geologically as well, with the granitic continental material thinning to zero at approximately the halfway point of this transition.

If any part of the ocean basin approaches our early perception of it as a deep, flat, featureless surface, it is the deep ocean basin floor. Broad tracts have been rendered flat by millions of years of deposition of sediment. Even so, the basin floor is interrupted by a myriad of relief features, both large and small, which are examined in chapter 12.

The **oceanic ridges** form a submarine mountain system with an aggregate length of 65,000 kilometers (40,000 mi) and an average width of 2,000 kilometers (1,250 mi). This system can be traced around Earth, into and out of the ocean basins. In some of the basins, as in the Atlantic, the ridge system occupies a central position, while in others, such as the Pacific, the system is offset to one side. In some places, the higher portions of the system break the water surface to form isolated islands or island groups. The origin of the oceanic ridge system is volcanic, and, as was noted earlier in this chapter, it is the site of numerous shallow-focus earthquakes. This and other information have led earth scientists to conclude that the oceanic ridges are the sites of formation of new oceanic crust.

Further Thoughts: Adapt to What Cannot Be Controlled or Changed

Earthquakes have played a significant role throughout all of human history. It is suspected by historians and archeologists that they may even have contributed to the decline and demise of whole civilizations. Scientists now understand the causes and mechanisms of earthquakes, but while major efforts are made to predict their occurrence, no reliable system has as yet been developed. The prospect of being able eventually to control them is remote. Perhaps the best that can be achieved is to learn to live with earthquakes in a sort of uneasy truce. Avoid them if possible, but where you cannot, exercise good judgment by following the advice of professional earth scientists and engineers.

Significant Points

1. Planet Earth is a restless body undergoing constant change, as attested to by the more than one million earthquakes recorded each year.

2. An earthquake is the sudden movement or vibration of Earth caused by the passage of energy in wave form released by faulting. The energy accumulates in the

rocks as the result of the application of stress over many years.

3. Structural damage resulting from an earthquake is a function of earthquake strength and duration, distance from the epicenter, and type of foundation material.

4. The location of energy release (faulting) is the focus of the earthquake. The point on Earth's surface directly above the focus is the epicenter.

5. Earthquakes are classified as shallow, intermediate, or deep, based on their depth of focus. This is also their order of abundance.

6. Earthquakes are, for the most part, confined to distinct geographic zones: the Circum-Pacific belt, the Mediterranean-Himalayan belt, and the oceanic ridge system.

7. Earthquake intensity is measured by the Modified Mercalli scale and is a qualitative assessment based on property damage and human perception which varies from place to place. Magnitude, as designated by values on the Richter scale, is a measure of the amount of energy released. Each earthquake is assigned only a single magnitude number.

8. Energy is transmitted by surface waves and body waves, the latter of two types: P-waves and S-waves.

9. P-waves are the fastest of all earthquake waves, they are compressional in origin, and they travel through solids, liquids, and gases. S-waves are shear waves and are transmitted only through solids.

10. The creation of a "shadow zone" for both P-waves and S-waves is taken to mean that Earth has a liquid outer core.

11. The basic instrument for recording the passage of seismic waves is the seismograph, which operates on the principle of a stationary mass. The "written" record of an earthquake is the seismogram.

12. The epicenter of an earthquake can be determined by plotting the travel time differences for P-waves and S-waves (S − P) to any three seismic stations.

13. Faults and joints develop when rocks respond as brittle solids to earth stresses, and folds result from plastic rock behavior. All are grouped together as geologic structures.

14. Fault movements may be dominantly horizontal or vertical in nature. The former gives rise to lateral faults, while the latter forms normal or reverse faults.

15. Normal faults result from tensional stresses; reverse faults from compressional stresses.

16. Joints involve no rock movement but form from the release of stress or from the cooling of igneous rock.

17. Folds are of two types: anticlines and synclines. Both are of compressional origin.

18. The nature of Earth's interior has been deduced largely from the analysis of seismic waves and the study of meteorites.

19. Four major zones are recognized within Earth: inner core, outer core, mantle, and crust.

20. Crustal material is of two types: granitic, associated with the continents, and basaltic, associated with the ocean basins.

21. The crust and uppermost mantle act as a single unit divided into a number of segments called plates.

22. Below the lithosphere is a plastic layer of low seismic velocity known as the asthenosphere. Movement of heat-driven currents within the asthenosphere is thought to be responsible for plate movement and a number of other geologic phenomena.

*E*ssential Terms

earthquake 222	surface wave 229	hanging wall 235	asthenosphere 240
faulting 222	body wave 230	normal fault 235	lithosphere 240
fault 223	P-wave 230	reverse fault 235	plate tectonics 241
focus 224	S-wave 230	joint 236	plains 241
epicenter 224	shadow zone 231	fold 236	plateau 242
plastic 225	seismograph 231	anticline 236	mountain chain 242
intensity 226	seismogram 232	syncline 236	continental margin 242
Modified Mercalli scale 226	geologic structure 235	inner core 238	oceanic ridge 242
magnitude 226	diastrophism 235	outer core 238	
Richter scale 226	lateral fault 235	mantle 238	
	footwall 235	crust 239	

*R*eview Questions

1. What is an earthquake and what is its origin?
2. What are some of the potentially deadly side effects of an earthquake?
3. What are the factors that determine the extent of earthquake damage?
4. Where do most earthquakes occur?
5. What is the potential for damage of foreshocks? of aftershocks?
6. What is the difference between the focus and the epicenter?

7. What is the relationship of earthquake frequency to depth of focus?
8. Why are deep-focus earthquakes only associated with deep-sea trenches?
9. How does intensity differ from magnitude?
10. How do surface waves differ from body waves? How do P-waves differ from S-waves?
11. What is the origin of the shadow zones?
12. How do seismologists determine the epicenter of an earthquake? What are some potential sources of error in a such a determination?
13. How does the occurrence of the various types of faults and folds relate to whether the applied stresses were tensional or compressional?
14. How does a normal fault differ from a reverse fault?
15. How does an anticline differ from a syncline?
16. How might the composition of meteorites tell us something about the composition and structure of Earth's interior?
17. How do we know that Earth's inner core is solid and its outer core is liquid?
18. What are the differences between the continental crust and the oceanic crust?
19. Why is the asthenosphere referred to as a "low-velocity layer," and what is the geologic significance of this?
20. What are the principal differences between the continents and the ocean basins?
21. What are the principal subdivisions of the continents? of the ocean basins?

Challenges

1. Consider the area in which you live. Are you aware of or can you find any reference (oral or written) to a history of or potential for earthquakes? Consider your immediate neighborhood. What are some of the potential earthquake hazards you might encounter there? How would you rate your home with regard to its potential to withstand a moderate to severe earthquake?
2. Suppose you were given the task of assessing the intensity of an earthquake that occurred in or near the area where you live. How would you go about it? What lines of evidence would you pursue?
3. How would you construct a simple seismograph? What materials would you need? What difficulties would you encounter in terms of potential sources of error? (Hint: consider other sources of vibrational energy.)
4. Examine some exposures of bedrock in the area where you live (in road cuts, stream valleys, etc.). Photograph or sketch any geologic structures (faults, joints, or folds) that you can identify.
5. Examine a world map that includes the major relief features of the continents and the ocean basins. What other mountain systems, besides the Rockies and the Appalachians, can you identify? Can you discern any major plateaus? Where are some extensive low-lying plains? What are some of the names applied to major segments of the oceanic ridge system?

10

PLATE TECTONICS: A UNIFIED THEORY OF EARTH

Aerial photograph of the San Andreas fault as it crosses the Carrizo Plain between San Francisco and Los Angeles.

Chapter Outline

INTRODUCTION

If the ultimate goal of the earth sciences is a holistic view of the planet and its processes, then the single most unifying theory currently available is one based on the interactions between Earth's interior and its outer shell. This theory, now called *plate tectonics,* began as a series of *hypotheses,* or topics for investigation, which became progressively interrelated as scientists in different fields gathered observational data. The evolutionary history of this unified view of Earth traces a series of purposeful and accidental discoveries that a few insightful scientists used to develop a new model of the planet. The transformation of this perspective from hypothesis to theory is a good example of how science often works. Not just the product of researchers in laboratory coats, scientific discovery and theory often begin with straightforward observations of the obvious, such as observations you can make.

The initial tenet of plate tectonic theory was simple: the continents are changing units on a mutable lithosphere. They move, break up, collide, and even form new continents. The movement creates oceans where breakup has occurred, mountain ranges where collisions take place, and numerous volcanoes and earthquakes wherever the lithosphere releases heat and relieves stress. All the movements depend on activity within Earth's asthenosphere, a zone in which heat generated through radioactivity, pressure, and friction cause convectional movements that press against the bottom of the lithosphere.

Casual Observations, Unsubstantiated Conclusions

When you look at a map of the world, you might notice a linearity in the mountain systems. The lines of mountains, often called ranges or ridges, occur on every continent. Among some of these linear systems are extensive tracts of volcanoes, also "lined up," making volcanic chains and, in some instances, volcanic islands. What could cause such linear systems to form?

The question is compounded by another seeming coincidence. Earth scientists have long noted that not only the mountains and volcanoes exhibit this trend, but earthquakes, too, generally follow patterns in narrow zones across Earth's surface (fig. 10.1).

The location and origin of our planet's surface features, such as mountains, have long generated numerous hypotheses about the evolution of Earth's surface, but many of these have been based more on imagination than on accurate observations and sampling. At the beginning of the twentieth century, for example, the prevailing idea was that mountains formed when a cooling Earth contracted and wrinkled. Many held that view even though no scientific observations supported it.

Scientific Observations

After World War II, scientists adapted a number of military technologies to earth science research and, armed with instruments appropriate for their objective studies, began to make accurate observations of Earth, which were not possible before the war. Earth scientists soon had sound observational data on which to base their hypotheses about continental movements.

Some of the major scientific advances of the postwar explorations occurred in 1958, the International Geophysical Year (IGY), during which the concerted efforts of several cooperating countries led to the discovery of mountain systems hidden in the depths of the oceans, systems that are also linear. The technology that made their discovery possible was an adaptation of sonar, the sound reflection device used to detect submarines. Sound vibrations, which pass through water just as they pass through air and solids, can be bounced off the ocean floor to reveal its **bathymetry,** that is, its depth and configuration. Signals sent to the bottom are timed, and any differences in successive return times indicate varying distances to the ocean floor. A shorter return time for signals means shallower water, whereas longer returns indicate deeper water (fig. 10.2).

THE BEGINNING OF A UNIFIED THEORY

The discovery of the undersea volcanic mountains led a Princeton University scientist, Dr. Harry Hess, to speculate that the oceans are actually the birthplace of oceanic crust. He believed that hot material from the mantle wells up to form new crust and creates the large linear volcanic systems called oceanic ridges (fig. 10.3). Hess's speculations were preceded by those of a British scientist, Arthur Holmes, who had suggested that rising cells of hot mantle material upwelled to "stretch," or thin, the ocean floor and move the continents apart, but he believed that the ocean floor and the continents were similar in composition. Because of the information IGY scientists had provided, Hess knew that differences existed in the kinds of rock that dominated the two types of crust: *the basaltic ocean floor differs from the predominantly granitic continental rocks.* In Hess's view, the oceanic ridges on the ocean floor were volcanic areas where the crust formed. Although Hess had little proof for his speculations, he helped to renew scientific interest in an older notion, one proposed early in this century by the German meteorologist Alfred Wegener.

Wegener's Idea of Continental Drift

If you look at a world map for nothing more than the shapes of the continents, you will notice that some of the continental units have coastlines that would, if merged, fit into one another like the interlocking pieces of a puzzle. This "fit" caught the attention of some thinkers shortly after the first crude maps of coastlines were published. The idea of connected continents somehow separating did not generate research in the earth sciences, however, until Wegener examined a growing body of evidence that suggested that formerly linked continents had moved from their previous positions.

When Wegener suggested in his 1915 work, *The Origin of the Continents and Oceans,* that the continents had moved from their initial locations, his fellow scientists ridiculed the idea. As a theorist, Wegener worked with limited information, but he saw

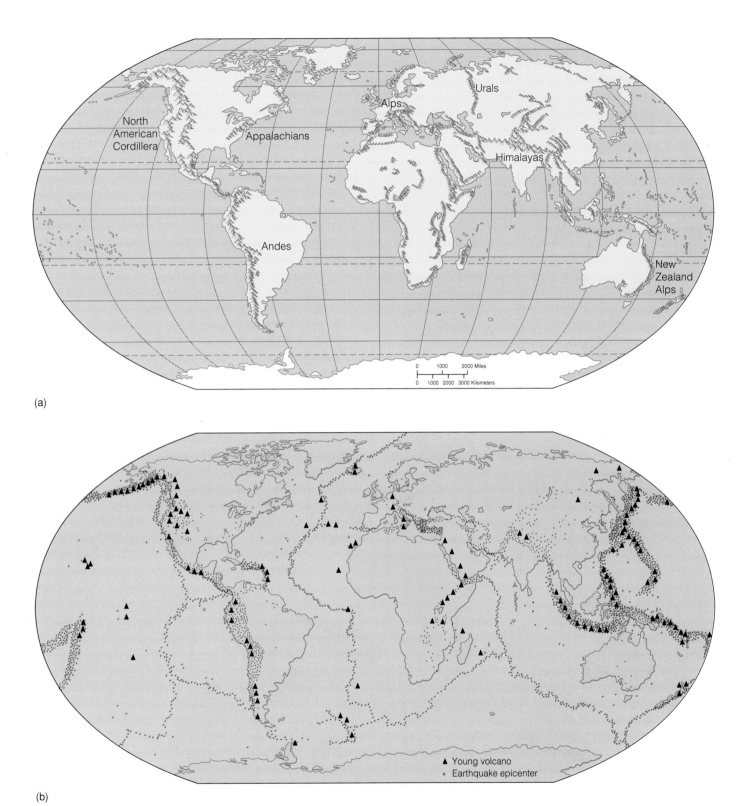

(a)

(b)

Figure 10.1   (a) The locations and linearity of the world's major mountain systems. (Sources: Data from Environmental Data and Information Service, National Oceanic and Atmospheric Administration, and the U.S. Coast and Geodetic Survey.) (b) Plot of earthquake epicenters and active volcanoes shows the linear arrangements of these systems. (Source: Echo-sounder record courtesy of Peter A. Rona, NOAA.)

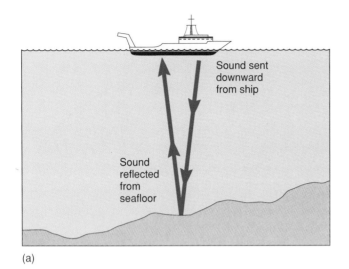

(a)

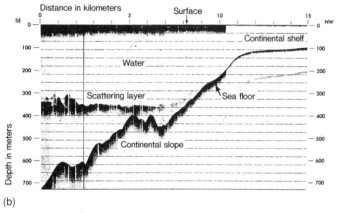

(b)

Figure 10.2 (*a*) Surveying underwater topography (bathymetry) by means of shipborne sonar. (*b*) Bathymetry as portrayed on a typical echo-sounder record. The "scattering layer" may be due to a water temperature anomaly.

convincing evidence for what he called "continental drift." To Wegener, the world was a changing place, and the changes were much larger than most scientists had imagined.

Wegener's Proofs for Continental Drift

The impetus for Wegener's work initially came from a commonsense examination of the coastlines on either side of the Atlantic. You probably have noticed that the western edge of Africa and the eastern edge of South America make the two continents look remarkably like two separated puzzle pieces. Unlike the casual observer, Wegener followed his geologic impressions with fact finding.

He found that the Atlantic continental shelf edges of Africa and South America match even more closely than the present shorelines, which are subject to erosional and depositional changes as well as to changes in worldwide sea level. The edge of each continental shelf (chapter 12), called the *shelf break*, represented the "true" continental boundary in his view. Wegener used such boundaries to reassemble the continents cartographically into one united landmass, the supercontinent he called **Pangaea,** or "all lands" (fig. 10.4).

Fossils played a key role in Wegener's research. He knew, for example, that fossilized palm trees occur in the rocks of Spitsbergen, an Arctic Ocean island covered by ice. The island houses fossils of temperate plants, too. Relatives of oak, maple, and beech trees also once grew in what must have been dramatically different climatic conditions on Spitsbergen. For Wegener, the presence of the fossils was evidence that the island had moved from tropic through temperate to its current boreal climate.

Wegener also used occurrences of similar fossils on the opposite sides of oceans to support his claim. Two fossils in particular seemed to confirm his theory: one, the fossil of a small dinosaur known as *Mesosaurus* (fig. 10.5), and the other,

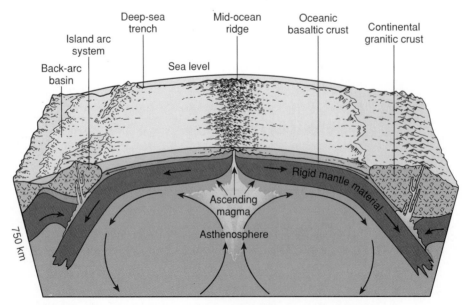

Figure 10.3 Magma rising from the asthenosphere forms new oceanic crust at the oceanic ridges.

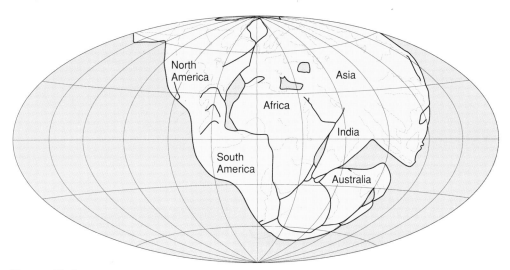

Figure 10.4 Wegener's Pangaea.

generally represented by the leaves of the plant *Glossopteris* (fig. 10.6).

The occurrence of similar fossils on different continents is significant because most animals and plants live in specific climates and habitats. The *Mesosaurus,* for example, appears to have been a shallow-water swimmer, inhabiting lakes and estuaries similar to Chesapeake and Hudson bays. Its life-style and habitat are inferred from its body structure and the types of rock in which it was fossilized. *Mesosaurus* apparently did not have the capability to cross a wide, deep ocean like the Atlantic, so its range would have been limited. Although many animals do have wide ranges, the distribution of a species over Earth is usually limited by geologic and climatic barriers, such as ocean basins and deserts. *Mesosaurus* appears to have encountered no such barriers between the two continents, evidence, perhaps, that the continents were once linked and not separated by an ocean (fig. 10.7).

Plants are especially sensitive to climatic change. The redwoods of Oregon and California, for example, do not grow along the beaches of Florida, and palm trees do not grow in the Adirondacks. *Glossopteris* and related plants (collectively called the *Glossopteris* flora) grew in temperate, humid areas, probably in swampland, as long ago as the Permian period, a geologic interval that began 285 million years ago and lasted 40 million years. Also, plants of similar age and distribution called the *Gangamopteris* flora flourished before the supercontinent rifted into various landmasses until after the breakup. Today, these plant fossils occur in rocks found in areas with climates ranging from tropical to polar. These two floras have left fossils of 58 species of seed plants as evidence of continental separation and climatic change.

Regardless of the proof that both the shapes of continents and fossils provided, Wegener had difficulty convincing others of his theory of continental drift because he had proposed no credible mechanism, or process, that might have moved the landmasses. Without such a mechanism, the theory languished in the annals of science as an imaginative, but foolish, proposal.

Continental drift received a further setback as late as the 1950s with the discovery of the undersea ridges. If the continents had drifted, some argued, in their movements they would have destroyed these large mountain systems.

Other Early Proofs

Even the scientific efforts of Australian geologists who accepted the theory of continental drift and the breakup of a southern supercontinent, called **Gondwanaland** by Edward Suess, did not convince the scientific community at large, particularly Northern Hemisphere scientists. Another Southern Hemisphere geologist, Alex du Toit, also proposed the breakup of a supercontinent. Du Toit's evidence, which influenced Wegener, lay in similarities he observed in the rocks and fossils of South Africa and Brazil and from a pattern of glacial till that encompasses parts of different continents (fig. 10.8). Like Wegener, du Toit had no acceptable mechanism for the continental splitting, that is, no Earth process that could move landmasses and change the shape of the crust.

CONFIRMATION OF THE THEORY

Regardless of its quality, the work of Wegener, du Toit, and others during the early decades of this century remained in the background of mainstream earth studies until the late 1950s. Then, an explosion of information about Earth's crust became available through the concerted efforts of many scientists. In fact, investigating hypotheses about the dynamic nature of Earth's crust became an international cause for financing earth studies.

The Deep-Sea Drilling Project

One scientific breakthrough that earth scientists needed to accept the existence of ancient supercontinents and continental breakup came with the initial voyage of a research ship capable

(a)

(a)

(b)

Figure 10.5   (a) Fossil remains of the small, freshwater dinosaur *Mesosaurus*. (b) Artist's drawing of a *Mesosaurus*.

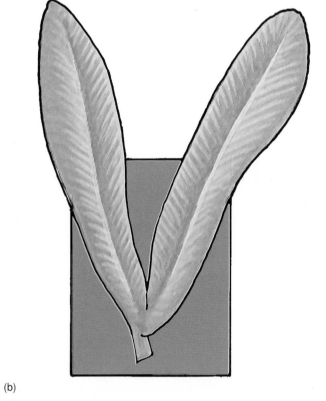

(b)

Figure 10.6   (a) Fossil remains of the land plant *Glossopteris*. (b) Artist's drawing of the *Glossopteris* leaves.

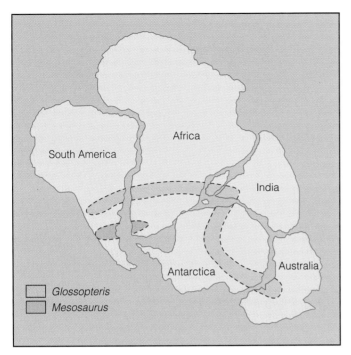

Figure 10.7 Distribution of *Mesosaurus* and *Glossopteris* fossils on a reconstructed Pangaea.

of drilling in deep water to recover cores of sediments and rocks. The Joint Oceanographic Institutions for Deep Earth Sampling (JOIDES) launched the *Glomar Challenger,* its primary research vessel, in March 1968 to begin a series of explorations of the ocean floor called the *Deep-Sea Drilling Project* (fig. 10.9).

Through their initial voyages, the ship's scientific crews provided two new bits of indirect, but convincing, evidence that the continents had moved. The first evidence was that the ocean floor was geologically young, everywhere less than 180 million years old. By comparison, some continental rocks approach 4 billion years in age. (Zircon crystals found in 1984 in the Acasta gneisses of Canada's Northwest Territories were radiometrically age-dated at 3.96 billion years.)

The second bit of evidence came from the relative dating of different areas of the ocean floor. The rocks and sediments that scientists cored near the axis, or center, of the oceanic ridge system appeared to be younger than those at increasing distances from the ridge: that is, the ocean floor became increasingly older away from the ridge axis. Apparently, the geologically young ocean floor was almost new at the ridge axis (fig. 10.10).

Seafloor Spreading: The Missing Mechanism

Harry Hess's proposal of crustal birth seemed to offer the mechanism that explained the ages of ocean floors and the movement of the continents. The apparent continental drift was the result of **seafloor spreading,** as it was termed in 1961 by R. S. Dietz, a colleague of Hess's. As new crustal material wells up from the mantle, the older material is pushed aside, constantly renewing the igneous rock at the axis by replacement.

The seafloor spreading that moves the continents begins when a section of the crust succumbs to pressure from beneath, pressure exerted by a rising segment of a partially molten upper mantle. The uplift breaks, or "rifts," the crust, allowing volcanic flows and mountains to form in fault-bordered structures called *grabens* and *half-grabens* (fig. 10.11). The rifts occur on both seafloors and continents. When the latter occurs, a large continent can separate into two or more smaller landmasses, allowing seawater to fill the intervening, low-lying gap. A series of such rifts divided Wegener's Pangaea into today's continents. One of the larger rifts opened the Atlantic by separating the Americas from Africa and Europe, and another created the Indian Ocean. Aware of the consequences of rifting, contemporary earth scientists now study both continental and oceanic rift valleys to determine their dynamic nature, the forces operating within the depths of the crust, and the history of crustal movements responsible for distributing the landmasses.

Two Rifts You Can Visit

One of the most studied crustal rifts occurs on Iceland, a land of significant volcanic and geothermal activity. The rift of the Mid-Atlantic Ridge runs through the Reykjanes Peninsula from southwest to northeast toward the Krafla shield volcano (fig. 10.12). Earth scientists, using lasers for precision in yearly surveys, have carefully measured the rate at which the rifting widens Iceland: their studies reveal that Iceland is growing longitudinally about 1 centimeter per year. At its current rate of growth, Iceland will be 10 kilometers (6.2 mi) wider in a million years!

The growth of Iceland has been occurring simultaneously with the formation of the Atlantic Ocean. A look back in time, therefore, reveals a world without the Atlantic and Iceland. Iceland formed through a series of volcanic eruptions that raised the oceanic ridge above sea level. In 1963, an example of Iceland's formation manifested itself in the eruption of Surtsey. Now a volcanic island just south of Iceland, Surtsey is a part of the oceanic ridge that stands above sea level. Continued volcanism in its vicinity could eventually produce another island similar to Iceland, one that will grow by the slow accumulation of basaltic lavas.

In eastern Africa, the crustal rift is a two-armed split in the continent (fig. 10.13). The eastern arm of the rift has associated volcanism, including the famous Mount Kilimanjaro of Tanzania. The western arm is filled with large lakes, including Lake Tanganyika. If the rifting continues for millions of years, seawater will eventually inundate the rift valley forming a new sea similar to the Red Sea.

The notion of rifting and a spreading seafloor appears to answer a number of questions, particularly those surrounding the occurrences of similar fossils on different continents and the varying ages of seafloor rocks. It does not, however, explain how the continents move without breaking into disparate pieces, all unconnected and grossly deformed.

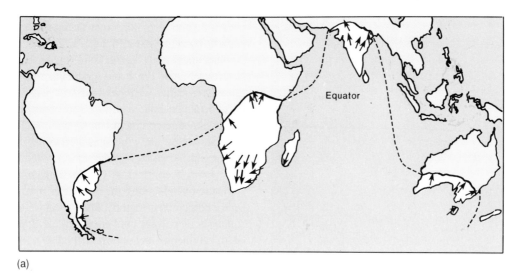

(a)

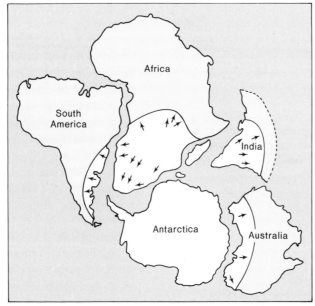

(b)

Figure 10.8 (a) Location of evidence of late Paleozoic glaciation in the lands of the Southern Hemisphere. Arrows show direction of ice movement. Patterns in several instances indicate movement of ice from the oceans onto the surfaces of the continents, an impossible situation. (b) Glaciation pattern on the southern continents in their premovement locations. Glaciers moved outward from a central location on the reconstructed "supercontinent." (Source: Data from Arthur Holmes, *Principles of Physical Geology,* 2d ed. Copyright 1965 Ronald Press.)

CRUSTAL UNITS

If massive, rigid continents move, would they not constantly break into random pieces even within their interiors? The theory that accounts for the continental movement without much catastrophic destruction is the product of research by Daniel P. Mckenzie and Robert L. Parker.

The Plates

Mckenzie and Parker postulated that the lithosphere consists of a series of movable, interlocking segments. They named these segments "plates." These plates are of varying thicknesses: some carry oceanic crust solely, while others underlie both oceanic and continental crust. The approximate thicknesses have been

determined more recently through *seismic tomography,* which gives a three-dimensional view of Earth's interior. The process mimics the medical CAT scan used to look into and image the brain, but it derives its image from an array of seismic stations spread over a wide area. The lithosphere has also been imaged through *electromagnetic induction* methods that use electric conductivity, which is influenced by temperature, pore fluids, and the presence of conducting materials.

Plate Interactions: Divergence, Convergence, and Lateral Sliding

The plates can interact in one of three ways (fig. 10.14). Some plates separate from each other or *diverge,* such as those on either side of the Reykjanes Peninsula and the East Africa rift.

Figure 10.9 The *Glomar Challenger*. This vessel had six computer-controlled engines that held it precisely on station over a drilling site. After 15 years of service, the *Challenger* was decommissioned and replaced by the JOIDES *Resolution*.

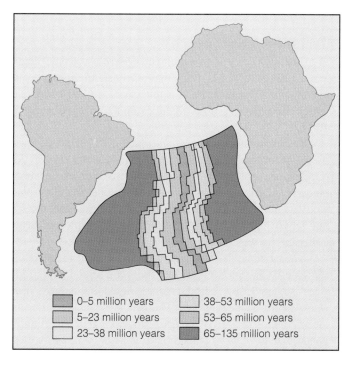

0–5 million years

5–23 million years

23–38 million years

38–53 million years

53–65 million years

65–135 million years

Figure 10.10 Plot of ages of oceanic crust in the southern Atlantic. The crustal rocks, cored by such ships as the *Glomar Challenger* and its successor, the JOIDES *Resolution,* have been dated through radiometric methods. Note the symmetrical distribution of ages about the central youngest rocks. (Source: After W.C. Pitman III, R.L. Larson, and E.M. Herron, Geological Society of America.)

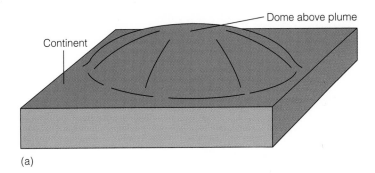

(a)

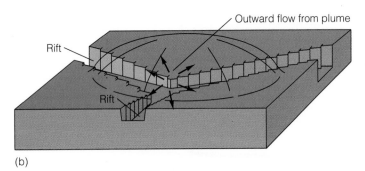

(b)

Figure 10.11 The rifting of a continent. (*a*) Pressure from the upward movement of magma forms an area of upwarping within the continent. (*b*) Continued pressure generates a series of grabens and half-grabens, down-faulted blocks, which eventually fill with lava flows and volcanoes. (*c*) Movement of continental fragments and intrusion of oceanic crustal material (basalt and gabbro) form a small basin subject to flooding. Continued movement further enlarges (perhaps greatly) this embryonic ocean basin.

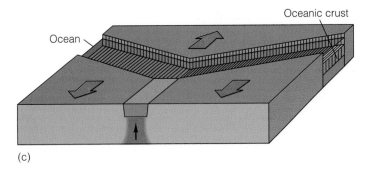

(c)

Divergence also occurs along the length of the oceanic ridge system. Other plates collide, or *converge,* driving massive sections of crust together like the two plates that meet on the western side of South America. Convergence can produce both mountains and deep trenches. Finally, some plates *slide* past each other, which is the geologic condition along the southwestern coastline of the United States. This last type of interaction, generally known as transform faulting, produces numerous earthquakes that relieve the stress built up by passing plates.

PLATE TECTONICS: A UNIFYING THEORY

In each instance, the movement of the crustal plates "builds" a geologic feature. This building led to another refinement of the initial continental drift theory, which is now known as plate tectonics for the geologic "architecture" generated by the several plate movements. In plate collisions, mountains (Himalayas), volcanic chains (Cascades), volcanic islands (Aleutians), and deep-sea trenches (Marianas) form. Finally, in the lateral sliding that moves one plate past its neighbor, faults act as the slip surfaces along which sections of rock pass others in the spastic earth movements we know as faulting (San Andreas Fault).

Tectonics and Linearity

The linearity evident in mountain systems, volcanic chains, and earthquake zones now makes sense in light of plate movements. The linearity is a reflection of plate boundaries, where divergence, convergence, and sliding occur (fig. 10.15).

To outline the positions of the plates, McKenzie and Parker used plots of moderate and large earthquakes for a 10-year period. The locations demonstrated the linear nature of most earthquake zones. Since they assumed that the movement of the plates would generate earthquakes, both men suggested that the seismic activity outlined the plate boundaries, where most of the tectonic activity would occur.

The movements of the plates seemed evident wherever they separated. Charles Richter, for whom the seismic magnitude scale is named, and Beno Gutenberg noted that the axis of the Mid-Atlantic Ridge is the site of numerous earthquakes. Because the orientation of the rifts along a central axis generally runs perpendicular to seafloor spreading, scientists could more easily plot the trends for some of the plates, such as those of the South American and African plates, which figure 10.10 shows in successive stages of separation.

Plate 1 Atlantic Ocean floor. (Source: World Ocean Floor map by Bruce C. Heezen and Marie Tharp, 1977. Copyright © 1977 Marie Tharp. Reproduced by permission of Marie Tharp, 1 Washington Ave., South Nyack, N.Y. 10960.)

Plate 2 Indian Ocean floor. (Source: World Ocean Floor map by Bruce C. Heezen and Marie Tharp, 1977. Copyright © 1977 Marie Tharp, 1 Washington Ave., South Nyack, N.Y. 10960).

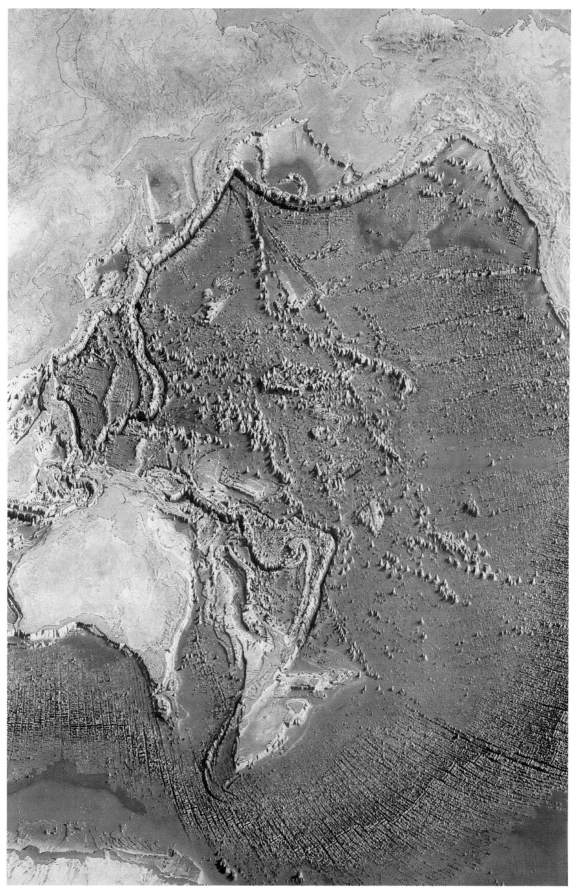

Plate 3 Pacific Ocean floor. (Source: World Ocean Floor map by Bruce C. Heezen and Marie Tharp, 1977. Copyright © 1977 by Marie Tharp. Reproduced by permission of Marie Tharp, 1 Washington Ave., South Nyack, N.Y. 10960.)

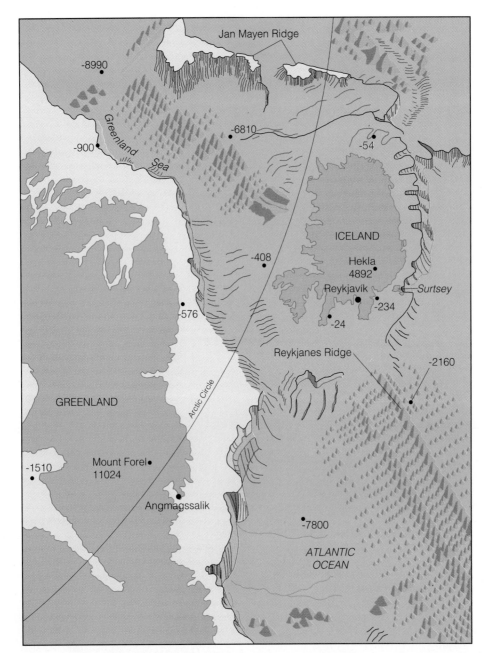

Figure 10.12   Iceland is located on the Reykjanes Ridge, a part of the Mid-Atlantic Ridge system. The small, recently formed volcanic island of Surtsey lies off the southeast coast.

Past, Present, and Future Plate Movements

Other plates, however, did not appear to move so predictably or their boundaries were questionable. Because the interactions of the plates cause earthquakes and volcanism, knowing their directions provides scientists with information relevant to the prediction of regions with a potential for disasters. Present directions also provide clues to both past and future plate movements. For scientists, however, predictions are an inadequate substitute for the truth. To determine the past movements of the plates, and particularly the movements of the continents, required the approach of a new scientific discipline: *paleomagnetic studies*.

PALEOMAGNETISM AND PLATE MOVEMENTS

Evidence for the continental movements came in a roundabout way from the study of Earth's magnetic field and the magnetism captured in rocks. After molten mineral matter cools and crystallizes to form igneous rock, the rock continues to lose heat. Whenever a rock cools to a temperature known as the *Curie point*, its magnetic properties reflect the prevailing magnetic field of the planet. This ancient magnetism is called **paleomagnetism** (from the Greek *palaio*, meaning "ancient").

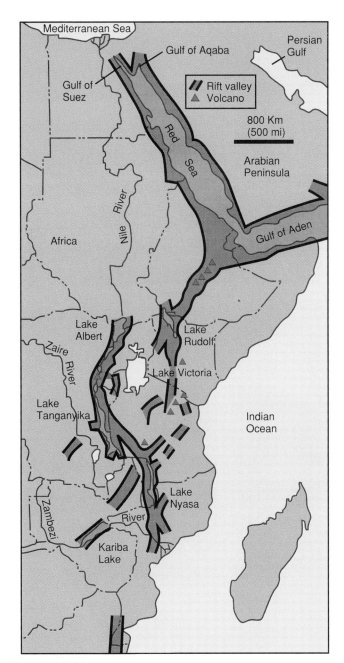

Figure 10.13 The East Africa rift is actually a bifurcated rift with volcanism in one "arm" and no volcanism in the other.

The Magnetic Recording Mechanism

The oxide magnetite (Curie point = 578° C or 1,072° F), a common mineral found in many rocks, aligns itself on the atomic level to the prevailing magnetism of Earth. The alignment can occur during the formation of plutonic and volcanic rocks, and it can also result from the orientation that settling flakes of magnetite may take when they are washed into the sediments of quiet waters, such as lakes.

Earth acts like a giant *dipole* magnet, with the lines of force exiting the Southern Hemisphere, where the south magnetic pole is located, and entering at the north magnetic pole (fig. 10.16). The lines are invisible, but they are the force that directs a compass needle toward the north magnetic pole, which is lo-

cated on Bathurst Island in northern Canada. Thus, following a compass needle does not lead you to the North Pole on the axis of planetary rotation, but it does take you to magnetic north. The magnetic force lines encompass the planet as an asymmetrical *magnetosphere.* (It is this magnetosphere that directs the solar wind to generate the northern lights, while the sun's corpuscular radiation shapes the field lines into an asymmetrical form.)

A hardened lava in the vicinity of the equator should exhibit an almost horizontal magnetic alignment, while one in British Columbia should reveal a dipping alignment. The direction of dip depends on the relative locations of the magnetic poles and the hardening molten mineral material. Exceptions to this general rule on magnetic alignment do occur, however, and these exceptions are highly significant.

In numerous instances throughout the world, older rocks have shown magnetic alignments, or *declinations,* that are *not* parallel to the lines of force in Earth's magnetosphere. Using magnetometers to study the magnetism captured in rocks of various types and ages, scientists discovered a magnetic mystery. Not only do some rocks exhibit an alignment oriented away from the present magnetosphere, but they also point magnetically to the Southern Hemisphere.

Anomalous Magnetic Readings and Magnetic Reversals

A number of earth scientists took up the challenge of the strange magnetic readings that some rocks yield. Among them were three Americans, Brent Dalrymple, Allan Cox, and Richard Doell, who asked themselves whether or not the magnetism captured by the rock was the result of special minerals that would adopt a magnetic alignment opposite that of the prevailing field. They soon realized that the minerals were not in themselves responsible for the differences in magnetic alignments. Instead, they recognized that the geomagnetic field had reversed itself: *The north magnetic pole had actually lain in the Southern Hemisphere* (fig. 10.17).

By dating the volcanic rocks through radiometric methods (see chapter 14), these American scientists and others around the world set out to determine the time of the most recent magnetic reversal. The north magnetic pole, according to the best estimates, was actually in the Southern Hemisphere about 800,000 years ago. At that time it appears to have reversed itself, or flipped, into the Northern Hemisphere. The time of the "flip," which includes the present, has been called the Brunhes Normal epoch. Throughout geologic time, each reversal has ushered in a new geomagnetic epoch, which the rocks have recorded (fig. 10.18).

Remember that the development of plate tectonic theory required the work of numerous scientists in different fields. The work on paleomagnetic reversals was just one more step in the transformation of hypothesis to theory. It provided the framework that allowed scientists to tie a number of discreet facts together in a unified theory. The geomagnetic reversal scale made possible another discovery, which supported the seafloor

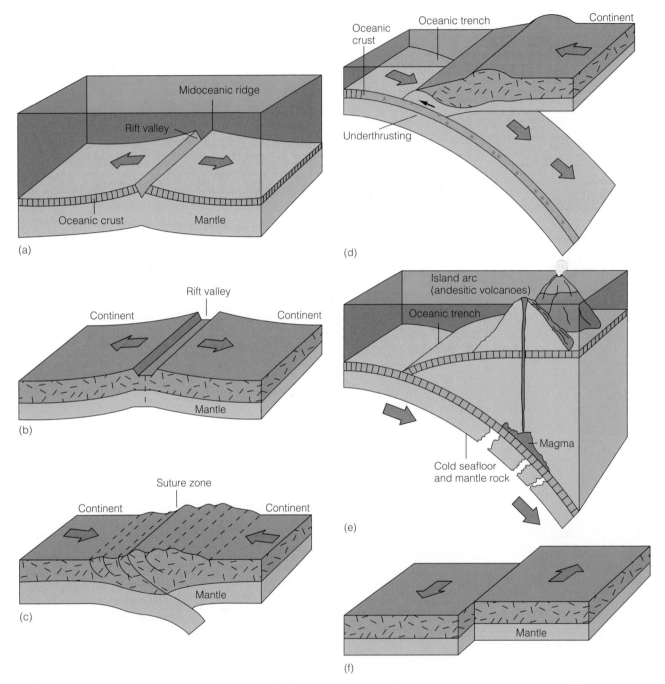

Figure 10.14 Plate boundaries: (a) divergent-oceanic, (b) divergent-continental, (c) continent-continent convergent, (d) ocean-continent convergent, (e) ocean-ocean convergent, and (f) transform fault. (Source: After W. Hamilton, U.S. Geological Society.)

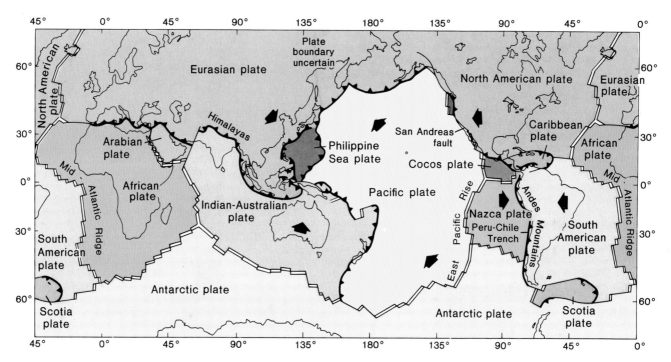

Figure 10.15   The crustal plates and their relative movements. The plate movements have been determined on a large scale, but "microplates" may operate in such areas as southern California and Central America.

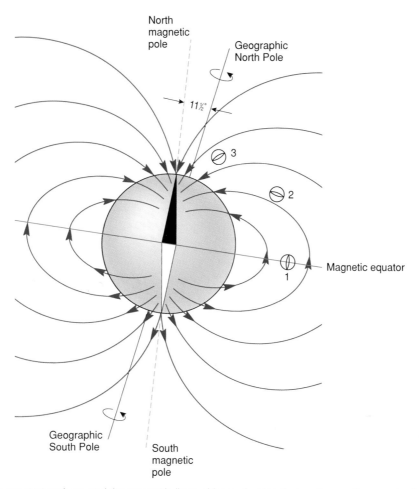

Figure 10.16   Earth's magnetosphere and the magnetic lines of force. A volcanic eruption on the equator (location *1*) produces lava, the iron-bearing minerals of which align themselves parallel to the lines of magnetic force of the dipole magnetic field. The dip of the lines of magnetic force, and thus of these minute "compass needles," increases toward the poles, as can be seen at locations *2* and *3*.

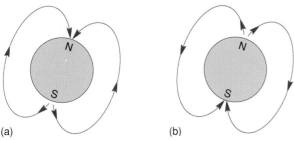

Figure 10.17 (a) The current (normal) geomagnetic field and (b) a reversed field.

spreading theory, and helped to trace the paleodirections of the plates and their continents.

Polar Reversals as Evidence for Seafloor Spreading

When a British scientist, Drummond Matthews, made a magnetic map of an oceanic ridge, the Juan de Fuca Ridge, by trailing a magnetometer behind a ship, he discovered something unexpected. The magnetic map of the ocean floor was mystifying to Matthews and others because it recorded "stripes" of apparently nonmagnetized rock (fig. 10.19). The "nonmagnetized" rock, lying next to the magnetized rock, seemed inexplicable, so the magnetic readings for these apparently nonmagnetized rocks were called "magnetic anomalies," and they remained a mystery for about 10 years.

Typical of the interrelationships found among the various disciplines of the earth sciences, those between geomagnetic research and the budding studies of seafloor spreading led to an important discovery. When Allan Cox and others defined the magnetic reversals, they laid the groundwork for the resolution of the irregular magnetism Drummond Matthews had found. Taking the paleomagnetic evidence of reversals and the model of seafloor spreading devised by Harry Hess, a British researcher, Fred Vine, determined that the stripes the magnetometer had recorded were actually evidence for seafloor spreading and magnetic reversals.

According to Vine, as the newly formed crust captured the prevailing geomagnetic field, it was moved aside, giving way to more crust, which recorded a different magnetism. The mapping stripes were actually magnetizations aligned to the two hemispheres. Those that align to a north magnetic pole in the Northern Hemisphere are *normally* aligned, whereas those pointing to a north magnetic pole in the Southern Hemisphere are *reversely* magnetized (fig. 10.20).

The credence Vine's explanation of the magnetic anomalies lent to the seafloor spreading hypothesis was only part of the benefit derived from his work. Vine had also provided scientists with a mechanism for determining the relative motions of the continents.

Still called anomalies in spite of his explanation, the magnetic stripes of different ocean floor rocks established the pathways for the movements of the crustal plates. The sepa-

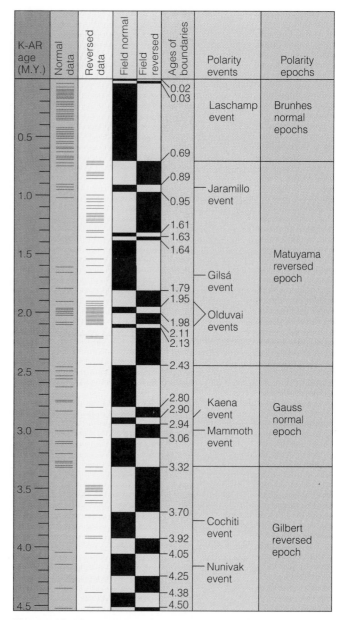

Figure 10.18 The geomagnetic polarity time scale for the past 4.5 million years.

ration of Wegener's Pangaea into the continents we know today can be traced backward through time to recreate the supercontinent before its breakup during the Mesozoic era, or the Age of the Dinosaurs (see fig. 10.10). The colors in figure 10.10 are parallel on either side of the spreading center, or ridge axis, because the rocks emerged simultaneously from the central rift and, as they cooled, captured the prevailing geomagnetic field.

THE MOVEMENT OF INDIA: A TECTONIC HISTORY

The complexity of plate movements is nowhere more evident than in the path of the Indo-Australian plate, which carries the two landmasses for which it is named. Since the breakup

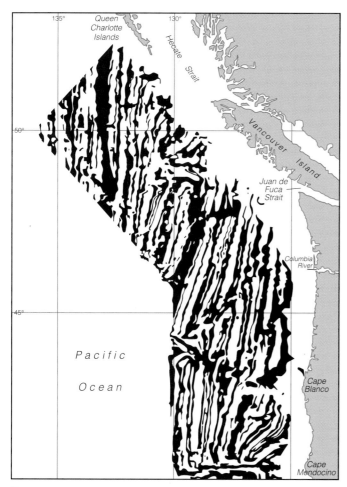

Figure 10.19 The magnetic anomalies of the Juan de Fuca Ridge. The alternating black and white lines represent the magnetic readings for strips of ocean floor rock. The "zebra" pattern yields a set of magnetic and nonmagnetic stripes. (Source: After A.D. Raff and R.G. Mason, "A Magnetic Survey of the West Coast of North America, 40N to 52½N," *Bulletin of the Geological Society of America,* vol. 72, 1260.)

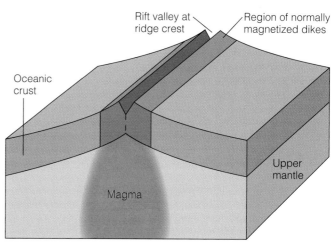

(a) Time of normal magnetism

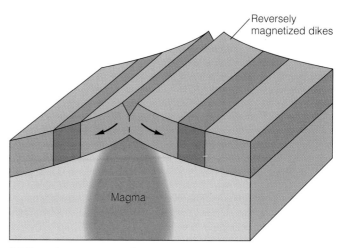

(b) Time of reverse magnetism

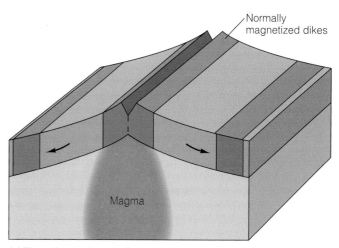

(c) Time of normal magnetism

Figure 10.20 The formation of magnetic "anomalies" by the process of seafloor spreading. Oceanic crust is continuously moved away from the oceanic ridge axis as new crustal material is emplaced.

of Pangaea, India has rotated and moved at varying rates on an eventual collision course with southern Asia (fig. 10.21).

Beginning about 70 million years ago, the subcontinent moved northward relatively fast for about 15 million years. Rapid movement in tectonic terms is about 15 centimeters (5.85 in) per year. During the ensuing 15-million-year period, its movement slowed as India probably encountered an island arc in its path. From about 55 to 40 million years ago, according to the Australian earth scientist C. McA. Powell, the subcontinent rotated counterclockwise toward Africa. For some unknown reason, India resumed its northward movement about 40 million years ago and, at the same time, changed the direction of its rotation. During this renewal of movement, the northwestern boundary of the Indo-Australian plate progressed northward at about 3 centimeters (1.2 in) per year. Some 20 million years ago, India's convergence with Asia was expressed in a squeezing of Iran and Afghanistan, two smaller continental masses that lay to the Northwest between India and more massive continental material. Thus, sporadic seafloor spreading over millions of years produced the current configuration of the Indian Ocean.

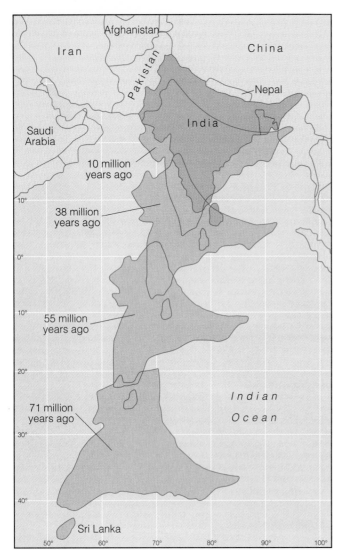

Figure 10.21 The movement of the Indian subcontinent over the past 70 million years based on the magnetic anomalies in the Indian Ocean floor.

Turning an Ocean Basin into a Mountain Range

The convergence of India and Asia continues even today as the subcontinent builds the Himalayas higher and causes earthquakes throughout Asia. The earthquake and mountain-building activities, however, are just part of the 70-million-year history of India. The path of the Indo-Australian plate carried India 2,000 kilometers (1,240 mi) and over the site of an ancient tropical sea, the **Tethys,** which the subcontinent crushed into the high mountain range that includes the world's two highest peaks: Mount Everest and K2.

To produce the mountains required shortening and buckling Earth's crust. Forming the Himalayas, therefore, probably resulted in 300 to 500 kilometers (186 to 310 mi) of *crustal shortening.* In Pakistan, the energy of India's movement shortened the crust by producing a belt of folded mountains. You can visualize this process by placing a sheet of paper on a table, then sliding its two ends toward each other. The "fold" in the paper is analogous to the shortening and mountain-building process.

As the Indo-Australian plate continues its northward trek, it continues to push itself beneath the southern end of Asia in a process of underthrusting called **subduction** (from the Latin for "lead under"). If the plate were fully oceanic (basaltic), it would tend to sink into the mantle. Because India is continental, however, it has a tendency to "float on" rather than sink into the asthenosphere. The convergence that is forcing northern India downward, therefore, has created a continuously thickening continental area and raised the height of the Himalayas. Continental thicknesses average about 35 kilometers (21.7 mi), but the Tibetan Plateau is now approximately twice that thickness.

Additional Evidence

Evidence for India's movement lies in both the earthquakes that shake China and in a series of ocean floor and continental features. India's northward movement has generated earthquakes of devastating power in China. One of the most catastrophic of these occurred in 1976, when the city of Tangshan was hit by an earthquake with a magnitude of 7.6 (see box 9.2). Because the Indo-Australian plate continues to move, other, similarly catastrophic earthquakes will occur within the region. This release of kinetic energy is not the only proof for the plate's movement. There are also features both on the ocean floor and on the subcontinent of India that provide evidence for tectonic activity.

The Indian Ocean, like the Pacific and the Atlantic, has a curving oceanic ridge that trends from south to northwest; then its rift valley exits the Indian Ocean and enters the Gulf of Aden, the Red Sea, and the Dead Sea. On the eastern side of this spreading center is an oceanic plateau that traces the subcontinent's movement. The basaltic Chagos-Laccadive Plateau is a series of volcanic piles that formed over a magma chamber as the plate moved northward. Its slightly curved shape may reflect India's counterclockwise rotation and mark an ancient transform fault. The Deccan Traps (see chapter 7), a high basaltic plateau on India, also appears to be related to the plate movement. The basaltic flows spread over the land through fissures in response to the rifting of Pangaea (see chapter 14). Similar flows of basaltic magma occurred in the Americas during the initial rifting of the supercontinent.

TRANSFORM BOUNDARIES

Crustal units can also slide past each other in a movement that has dire consequences for inhabitants along the interplate boundaries. In 1906, such sliding devastated San Francisco, and a similar action caused the October 1989 earthquake.

In fact, one of the best examples of lateral sliding between plates occurs in California, site of the famous San Andreas Fault, the division between the North American and Pacific plates. The San Andreas is the main fault in a system of faults that begins where Baja California separates from the mainland of Mexico, runs past Los Angeles, and exits the continent through San Francisco. It is a deep split between two plates, running somewhat vertically into the crust. Its vertical extent was highlighted during the 1989 earthquake, when scientists traced the focus of the Loma Prieta earthquake to a depth of 16 kilometers (10 mi).

The San Andreas belongs to a class of faults known as *transform faults*. These faults, first identified in 1965 by Tuzo Wilson, are the locations of lateral movements between adjacent plates. Most of the transform faults lie along the oceanic ridges, running perpendicular to the ridge axis, but some do not have a conspicuous relationship with a spreading center. One of those faults not readily associated with an oceanic ridge is the San Andreas. The Pacific plate moves northwestward relative to the North American plate along the San Andreas Fault, producing earthquakes, subsidiary faults, and mountains (fig. 10.22).

Seafloor Features and Plate Tectonics

A number of seafloor features associated with plate tectonics have been identified through detailed mapping beginning with and subsequent to the IGY.

Nemataths

On the ocean floor itself, a series of volcanoes that mirror each other on both sides of a spreading center make a **nematath** (fig. 10.23). Retracing the stages of development of a nematath leads to a reconstruction of the original ocean floor.

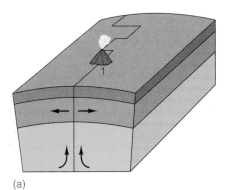
(a)

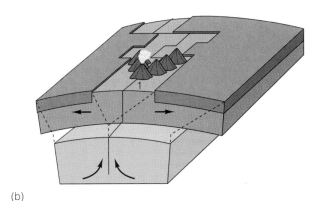
(b)

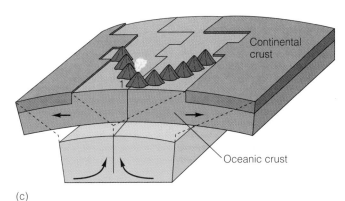
Continental crust

Oceanic crust

(c)

Figure 10.23 A nematath. Volcanoes become progressively older as they are carried away from their site of formation along the oceanic ridge (location *1*).

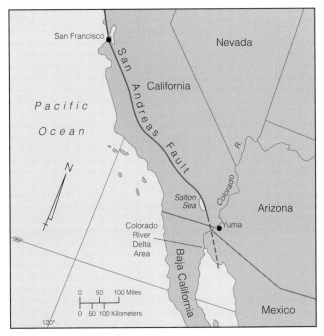

Figure 10.22 The San Andreas Fault system.

Bathymetry

Because the upward pressure beneath the spreading center is greatest, the crust is upwarped most where the ridge buildup occurs. Away from the rift, the seafloor begins to subside at a rate generalized by John G. Sclater.

Three factors influence the subsidence of ocean floor features:

1. The movement away from the concentrated upward pressure beneath the ridge axis reduces the upwarping.
2. The great weight of volcanic and plutonic rocks piled onto a young and thin oceanic crust also initiates a downwarping.
3. The oceanic rock is the recipient of sediments that originate as the hard parts of life forms in the pelagic

waters. The sediments accumulate slowly, with maximum rates rarely exceeding 21 meters (69 ft) per million years; yet, over many millions of years they can add up to a considerable weight on the crust.

Trenches

The convergence of oceanic crust and continental crust usually leads to the destruction of the thinner and denser oceanic material through subduction. The subduction zone is associated with frequent earthquakes along deep **trenches** (fig. 10.24). The sinking crust begins to break up and partially melt beneath the lighter and thicker continent. The partially melted crust shudders at all levels, and its earthquakes follow a pattern of increasing depths for their foci beneath the adjacent continent. A good example of such activities occurs along the western coast of South America in the vicinity of the Peru-Chile Trench (fig. 10.25). The trench exceeds a depth of 6.5 kilometers (4 mi) and is the site of the shallowest earthquakes. Farther inland beneath South America and the Andes Mountains the foci are deeper.

Trenches vary in bathymetry, and the Peru-Chile Trench shoals in some places, notably where an oceanic plateau like the Juan Fernandez Ridge lies off the coast. Their significance lies in the role they play in seafloor destruction. It is subduction in trench areas that eliminates "old" ocean floor.

Island Arcs

In some instances, particularly on the western and northern edges of the Pacific plate, a subduction zone is bounded on the landward side by a series of volcanic islands in an arc. The **island arcs** of the Pacific include the Tonga-Kermadec Islands, the Japanese Islands, and the Aleutians. The islands are subject to numerous earthquakes and eruptions, such as the most recent volcanic activity in the Aleutians. Generally, the island arcs are

built from the andesitic lavas that result from the partial melting of oceanic rock and sediments in the subduction zones. Such lavas are intermediate in composition between rhyolite and basalt.

Seamounts

Another product of plate activity is a **seamount,** which is volcanic pile that does not reach the ocean surface. Most seamounts occur in bunches, such as those off the American East Coast, and some make up linear systems. A good example of a linear seamount system is the Emperor Seamounts of the western Pacific. At one time some of these seamounts may have been islands, but the inevitable subsidence of the oceanic crust caused all to sink below sea level.

The linearity of the Emperor Seamounts begins at the end of the Hawaiian Island chain, just northwest of Midway Island (fig. 10.26). The Hawaiian system is also linear, but both it and the Emperor Seamounts lie on the Pacific plate far from the plate's edges, where volcanism and earthquake activity is most likely. Why, then, are these volcanic piles where they are?

Hot Spot Volcanoes

Just as the Chagos-Laccadive plateau appears to mark the path of the Indo-Australian plate and the subcontinent of India, so the Hawaiian Islands and the Emperor Seamounts may plot the direction of the Pacific plate's movements. Since no rift valley is associated with these features in either ocean, however, there must be another mechanism responsible for the string of volcanoes.

The Kilauea volcano on the island of Hawaii has erupted almost continuously for the past 40 years. After a few inactive years, the volcano releases great quantities of lava through fis-

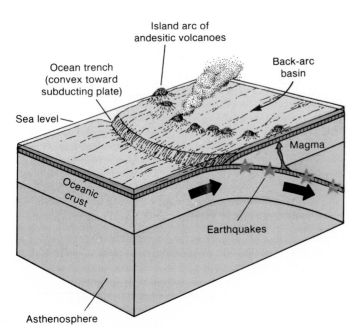

Figure 10.24 The generalized scheme of any island arc-trench system. On the landward side of the island arc lies a back-arc basin, which receives sediments from the continent.

Box 10.1 Investigating the Environment

Listening to an Erupting Volcano

On January 19, 1993, scientists using SOFAR channel (see chapter 11) hydrophones located near Oahu, Hawaii detected intense acoustical activity centered on a site about 200 kilometers (124 mi) southwest of the western side of the Rivera fracture zone (figs. 10.A and 10.B). Acoustic signals are now being used to identify submarine volcanic activity throughout the Pacific. Hydrophones located at the Laboratoire de Géophysique in Tahiti recently discovered a new volcano near the Eltanin fracture zone (54°S, 140°W).

Apply Your Knowledge

1. Would volcanic activity be common or rare on the ocean floor?

2. Why would some areas of the ocean floor be more likely sites of volcanic activity than others?

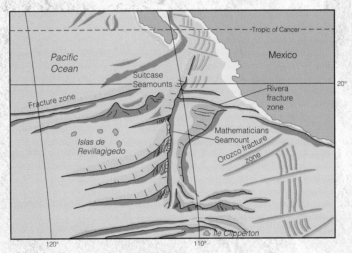

Figure 10.A The volcanic activity detected by hydroacoustic instruments is associated with a chain of seamounts.

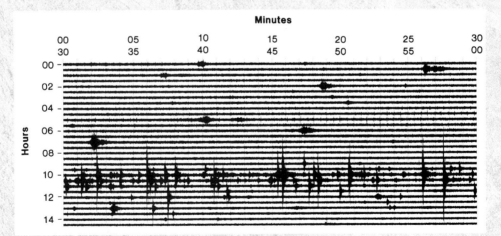

Figure 10.B Eruption signals from suspected submarine volcanic activity in the Revilla Gigedo Islands region at about 18.75° N, 110.8° W. These signals were recorded on January 19, 1993, by a SOFAR channel hydrophone (Kaneohe hydrophone 13SP) near Oahu at a distance of about 4,400 kilometers (2,728 mi) from the suspected eruption site. The intense episodes of unusual activity lasted for about 1 hour.

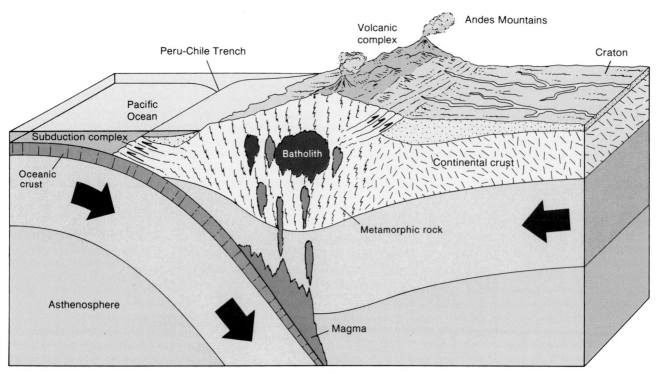

Figure 10.25   The Peru-Chile Trench is part of the grand architecture of plate movements.

sures along its flanks, building the shield ever larger. Kilauea's sister volcano, Mauna Loa, the largest mountain on Earth, has also erupted recently. Both lie at the southeastern end of the Hawaiian Island chain. Yet, on Midway Island, one of the most northwesterly islands of the chain, no volcanism occurs, and none has occurred for more than 20 million years.

The explanation for both the linearity and the volcanism lies in stationary **hot spots,** magma chambers located in the asthenosphere. As a plate passes over the hot spot, the rising magma produces a sequence of volcanoes in a line that traces the crustal movement. The great depth of the magma chamber can be inferred from its stationary nature: the movement of the plate seems to have no effect on the hot spot's location. Because the volcanoes sit on the plate, they move with it, passing away from their source of lava. Midway, a relatively low island, is more than 25 million years old, and its volcanism ceased when it moved off the hot spot responsible for its formation. Prior to the formation of the Hawaiian Island chain, the Pacific plate must have moved more north than west as evidenced by the orientation of the Emperor Seamounts. The date for the directional change in plate motion would coincide, therefore, with the oldest features of the Hawaiian Island chain and the youngest of the Emperor Seamounts.

Guyots and Coral Atolls

As the volcanoes move off the hot spot locality, they begin to subside and to be attacked by the processes of weathering and erosion. The islands in the northwestern part of the Hawaiian

chain are much lower and more eroded than those in the southeastern part. Unseen beneath the water are former islands that have subsided through the zone of wave action in these tropical waters. Coral reefs formed on the fringes of the volcano tops when the islands were close to sea level. As the volcanoes subsided, these coral reefs formed islands called **atolls** (see chapter 12). With further subsidence the underlying volcanoes carried their carbonate caps too deep for coral life, making relatively flat-topped, subsided features, now called **guyots** (fig. 10.27). Wherever subsided volcanoes formed in waters too cold for coral development, no carbonate cap tops the guyots.

TERRANES AND THE GROWTH OF CONTINENTS

One of the consequences of plate movements is the welding of different crustal pieces called **terranes.** Geologists have identified a series of adjacent, but unrelated, rock units from the northwestern states to Alaska (fig. 10.28). The types of rocks, fossils, and ages for these units are so radically different that their only explanation must lie in diverse origins.

Plate movements frequently combine pieces of crustal rocks and add to the size and volume of the continents. Western North America is largely a product of such crustal welding, but other areas, such as Newfoundland, have also grown through the addition of terranes. The probable sequence for the accretion of the terranes begins in the movement of the continent itself. As North America moved against the ancestral Pacific Ocean, it collided

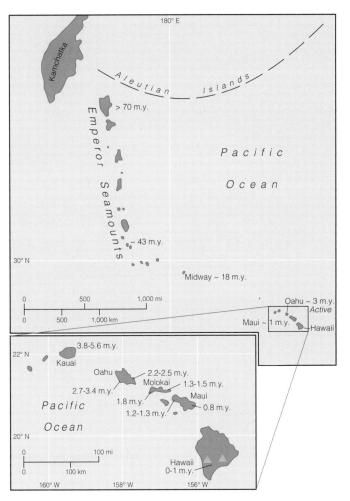

Figure 10.26 The Emperor Seamounts and the Hawaiian Island chain. The most recent volcanism occurs on Hawaii and on an undersea volcano called Loihi. As the plate continues to move toward the northwest, Kilauea and Mauna Loa will move off the "hot spot," leaving Loihi to be built above sea level as the next island in the chain. (Source: After I. McDougall, 1964, *Geological Society of America Bulletin*.)

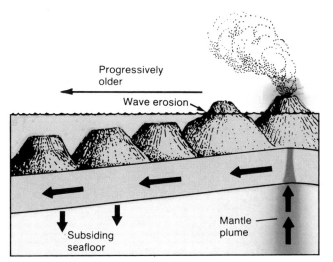

Figure 10.27 Movement of a plate across a mantle plume produces a linear succession of volcanic cones. The upper portions of the cones may be planed off by wave erosion. When combined with the continued subsidence of the seafloor, this erosion forms a flat-topped, submerged feature known as a guyot. In tropical climates, coral development takes place when the volcanic cone is still above, at, or just below sea level.

Box 10.2 Further Consideration

Could Seafloor Spreading Initiate an El Niño Event?

In box 6.2 you learned about the El Niño/Southern Oscillation (ENSO) and associated changes in world weather patterns. It now appears that seafloor spreading activity may be related to the warming of ocean water that is known as El Niño. Scientists aboard the *RV Melville* recently discovered 1,133 seamounts and volcanic cones covering 88,700 square kilometers (55,000 sq. mi) in the vicinity of Easter Island (fig. 10.C). According to Dr. Ken Macdonald of the University of California at Santa Barbara, two or three of the volcanoes could be active simultaneously.

The southeastern Pacific location of these volcanoes suggests that they may have some effect on the ENSO phenomenon. When a volcano erupts, it pumps heat into the surrounding environment. Numerous erupting volcanoes may sufficiently warm ocean water to initiate a change in the Southern Oscillation and to begin an El Niño event.

Apply Your Knowledge

1. What oceanic ridge is associated with the Easter Island volcanoes?
2. How could scientists keep track of underwater volcanic activity?
3. How could scientists relate an ENSO event and volcanic activity?

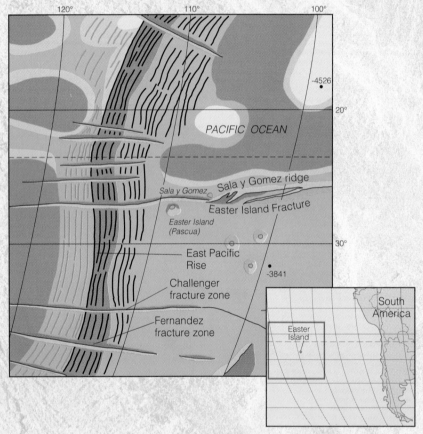

Figure 10.C   Scientists have discovered seamounts and volcanic cones in the vicinity of Easter Island.

with a series of island arcs, each of which lay on the landward side of a subduction zone. The oceanic rocks of western Idaho, which now lie hundreds of kilometers inland, represent one of the ancient boundaries between the ocean and the continent. Similar collisions have added to the size of other continents. Therefore, locating the boundaries of the terranes also means finding the ancient coast of a continent.

PLATE TECTONICS AND MINERAL RESOURCES

One of the most studied of terranes lies in the Mediterranean Sea on the island of Cyprus. The Troodos Massif is an *ophiolite,* or a section of ocean crust that was emplaced onto the island in a process called *obduction.* Evidence for its identity as a fragment of the ocean floor lies in both rock type and form: some of the igneous rocks of the massif are basaltic lavas of submarine origin.

Since Roman times, the Troodos Massif has been a mining site for copper and other metals. The occurrence of the metals seems to be associated with the origin of the rock itself. Wherever tectonic activity is associated with magma chambers, metallic salts are precipitated out of hot fluids to make **hydrothermal deposits.** Seawater circulating through cavities near the magma chamber dissolves these minerals and carries the ions away from the hot rock, to deposit them later.

Knowing where ancient sites of convergence and divergence occurred gives clues to the presence of mineral resources. The reverse can also be helpful. Alfred Wegener used, for example, the similarity between the diamond mines of Brazil and those of South Africa to provide evidence for his continental drift theory.

The discovery of deep-sea volcanic fissures and vents, where water in excess of 300° C (572° F) boils up through chimneys of sulfide deposits, confirmed the role of tectonics in metallic deposit formation (fig. 10. 29). The vents are active for a few hundred years at most, but during their short lives, they act as release valves for heated water, which circulates through the ocean's volcanic floor. The hot water picks up dissolved mineral matter and redeposits it in the vicinity of the vent. Some of the metals may circulate throughout the bottom water of the ocean to precipitate elsewhere in the form of oval or spherical *nodules* and encrustations rich in manganese, iron, copper, and other metals.

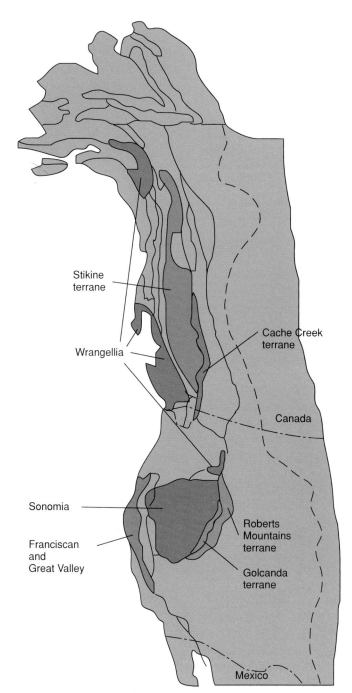

Figure 10.28 The terranes of western North America.

Figure 10.29   Black smoker on the deep-sea floor.

Further Thoughts: A Paradigm

The rejection of the hypothesis of continental drift by his contemporaries did not deter Alfred Wegener from pursuing a scientific explanation for Earth's features. Unfortunately, much of his work received the same response from the following generation, and our current scientific explanations for the shapes and locations of the continents and oceans did not gain acceptance until the second half of this century. Wegener's experience with rejection was not an exception in the history of science. Almost every advance in scientific understanding has met similar rejection because each upset the intellectual status quo. Once a model of Earth gains acceptance, it offers a secure basis on which humans can organize the diversity of the planet. Wegener laid the groundwork for a **paradigm,** a model so significant that it overturned the organizing principles of a previous scientific explanation. Because the plate tectonic model of Earth does not explain all features and processes, it may eventually give way to another paradigm. When that paradigm is offered, will it encounter the same scientific inertia that Wegener's idea of continental drift met?

Significant Points

1. Plate tectonics is currently the most unifying geophysical theory of the planet. It is based on the interaction between Earth's interior and its outer shell.
2. A series of hypotheses formulated by scientists of diverse fields combined interrelated observational data to create a new model of the planet.
3. The lithosphere is composed of movable interlocking segments called plates. Some plates carry oceanic crust solely, while others underlie both oceanic and continental crust.
4. The basic principle of plate tectonic theory is simple: the lithospheric plates move, break up, collide, and slide past one another to form new continents and oceans.
5. The result of plate movement is the creation of oceans where continents rift. Mountain ranges, volcanic islands, and deep-sea trenches result where collisions take place.
6. Plate movement is a direct result of convection currents within Earth's asthenosphere, exerting pressure against the bottom of the lithosphere.
7. Plates interact in one of three ways: divergence, convergence, and lateral sliding.
8. Mountain ranges, volcanic chains, and earthquake zones occur as linear systems in narrow zones across Earth's surface. The linearity is a reflection of plate boundaries.
9. The idea of continental drift, continents moving from their initial positions, was suggested by Wegener, a result of his observation that the Atlantic continental shelf edges of Africa and South America matched closely.
10. Each continental shelf, or shelf break, represents the "true" continental boundary, since shorelines are subject to erosional and depositional changes. Wegener used such boundaries to reassemble the continents cartographically into one united landmass, the supercontinent he called Pangaea.
11. The occurrence of similar fossils on different continents is significant since the distribution of a species is limited by geologic and climatic barriers.
12. The oceanic floor is geologically young in comparison to continental rock. The youngest rocks and sediments are

near the axis of the oceanic ridge systems, increasing in age with distance from the ridge axis.

13. The various ages of the seafloor and the movement of continents are a direct result of seafloor spreading; as new crustal material wells up from the mantle through rifts, older material is pushed aside, forming large, linear volcanic systems called oceanic ridges.

14. Continental rifting occurs as the continent begins upwarping because of rising pressure from below. A series of grabens and half-grabens form and eventually fill with lava flows and volcanoes. The rift allows seawater to enter, creating two or more smaller landmasses separated by a growing body of water.

15. As molten mineral matter cools and crystallizes to form igneous rock, its magnetic properties align parallel to Earth's magnetic field.

16. Dating of volcanic rocks along the seafloor through radiometric methods has provided evidence that magnetic north has shifted repeatedly between the Northern and Southern Hemispheres.

17. Magnetic anomalies form during the reversal of the magnetic north and south poles.

18. Where oceanic crust and continental crust converge, the thinner and denser oceanic material is subducted beneath the lighter and thicker continental crust. The result of this convergence is volcanic activity and earthquakes.

19. Nemataths, island arcs, and seamounts are features of the ocean floor that result from plate activity.

20. Stationary hot spots are magma chambers that extend downward to the asthenosphere. As a plate passes over the hot spot, the rising magma produces a chain of volcanoes that record plate movement.

21. The accretion of terranes along the western part of North America is the result of the North American plate's collision with island arcs as it moved in its northwesterly direction.

22. Whenever tectonic activity is associated with magma chambers, metallic salts are precipitated out of hot fluids to form hydrothermal deposits.

Essential Terms

Pangaea   248
Gondwanaland   249
seafloor spreading   251
paleomagnetism   258

Tethys   264
subduction   264
nematath   265
trench   266

island arc   266
seamount   266
hot spot   268
atoll   268

guyot   268
terrane   268
hydrothermal deposit   271
paradigm   272

Review Questions

1. In what ways do the plates interact with one another? What are the results of these interactions?
2. What are the proofs Alfred Wegener used to establish his continental drift theory?
3. How have earth scientists used fossils to reassemble continental masses cartographically?
4. What is the name of the southern supercontinent?
5. How does the age of oceanic rock serve as evidence for seafloor spreading?
6. Where can you go to visit a rift valley on dry land?
7. What are magnetic anomalies? How do they record the movements of oceanic crust?
8. What is a nematath?
9. What are the processes that destroy oceanic crust? Where do these processes occur?
10. Name examples of the following: divergent plate boundary, convergent plate boundary, transform plate boundary.
11. How do guyots and atolls form?
12. How is plate divergence associated with metallic mineral deposits?
13. Why is the San Andreas Fault the site of numerous earthquakes?

Challenges

1. How far is your home from a plate boundary? Would you expect to experience earthquake or volcanic activity in your area because of divergence, convergence, or transform faulting?
2. How could the redistribution of continents affect climate? (See chapter 6 for clues.)
3. Using the plate motion map in figure 10.15, determine the possible tectonic future of Athens, Greece; Los Angeles, California; Tokyo, Japan; and Dar es Salaam, Tanzania. What lies in the tectonic future of your area?
4. Are island arcs always associated with subduction?
5. How could the evolution of life be affected by seafloor spreading?

Ocean Waters and Environments

*T*he ocean waters *and* their basins are the most important controls on Earth's surface systems. Seafloor spreading is the starting point for most magmatic processes because it results in the movements of lithospheric plates that redistribute and reshape landmasses and ocean basins. Extensive ocean currents move heat, affecting climates, storms, and the hydrologic cycle. Gas exchanges between atmosphere and oceans can diminish or enhance greenhouse warming, and salt exchanges between seawater and life can multiply sediments and increase coral reefs.

The complexity of ocean systems is a reflection of interconnectedness. Thus, marine geology is inseparably related to marine chemistry, and both are tied to atmospheric processes, continental evolution, and life.

At the End of Part 5, You Will be Able to

1. relate the nature and significance of water;
2. identify and integrate the controls on seawater and its circulation;
3. illustrate air-sea interactions;
4. interpret ocean environments and their relationships to life.

from William Shakespeare's The Tempest, V, i

I'll deliver all;
And promise you calm seas, auspicious gales,
And sail so expeditious that shall catch
Your royal fleet far off.

11
OCEAN WATERS

Tabletop coral, Celebes Sea, off the coast of East Malaysia.

Chapter Outline

*I*NTRODUCTION

More than 1.3 billion cubic kilometers (312 million cubic mi) of water fill the ocean basins to overflowing, inundating the continental shelves and intruding upon low-lying coastal areas. In fact, only 29.1% of Earth's surface lies above the average level of seawater, as the *hypsographic* (from Greek *hypsos* , meaning "height," and *graphein,* meaning "to write") *curve* in figure 11.1 shows. The oceans contain 98% of the world's water and anchor the hydrologic cycle. They are largely responsible for the latitudinal distribution of solar heat, the amount of water vapor available for weather systems, and the storage and exchange of gases, such as CO_2. The ocean waters particularly in the large hurricanes they generate, also extend their influence to coastal geology and human structures. Along the everchanging shorelines, ocean currents and waves move unconsolidated sediments, erode cliffs, and undercut buildings, roads, and bridges.

With their rich supply of dissolved gases and nutrients, the oceans harbor vast amounts of life. Marine photosynthesizers and sulfide-reducing bacteria serve as the base of the food chain, making the oceans a supplier of protein for both marine and terrestrial organisms. Birds, terrestrial mammals, reptiles, and amphibians along coasts and island shorelines act as foraging marauders who invade the seawater for their sustenance. Even humans derive millions of tons of protein from sea life.

The Sink

Geologically and chemically, the oceans are the ultimate sink because they are so deep. Gravity insures that much of what is above water is destined to be washed into some sea or ocean through the continuous processes of weathering, erosion, and transport. Although lakes and floodplains may trap sediments for millions of years, they serve merely as temporary stopovers. A few dry inland basins (e.g., Death Valley) are lower than sea level or restricted by surrounding highlands, allowing local sediments to collect; but eventually, tectonic activity opens pathways to the sea or uplifts an area to initiate erosion and sediment transport. The pollutants in our streams, the soils on our farms, plus the upper bedrock of the continents are all seaward bound. The process may take tens, or even hundreds of millions of years, but it is inevitable. And yet, because Earth is dynamic, the oceans are themselves temporary repositories and parts of larger chemical and physical cyclic systems on Earth.

Oceanographic Studies

The complexity of ocean studies is manifested by the different kinds of scientific activities in which oceanographers engage. Marine geophysicists, geologists, and coastal geomorphologists research and explain the composition, evolution, and processes of the seafloor and basin boundaries, and they work to locate, exploit, and protect marine resources. Physical oceanographers study air-sea interactions, surface and deep-water dynamics, and ocean temperatures and salinities. Chemical oceanographers analyze the surface and deep water for distributions of the more than 70 elements that influence marine life and sediments. Almost half of those dedicated to understanding the oceans are marine biologists, who study the varied flora and fauna that reside above, on, in, and under the water column. Finally, ocean engineers design ships, drilling and research platforms, and coastal structures, such as seawalls, docks, and jetties. As you can see, oceanography is among the most interdisciplinary of subjects.

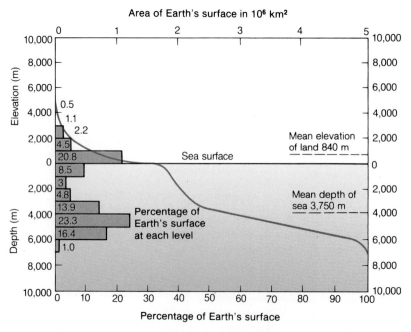

Figure 11.1 A hypsographic curve. The cumulative curve indicates the percentage of Earth's surface above any elevation on the curve. (Source: From Sverdrup/Johnson/ Fleming, *The Oceans,* © 1942, © renewed 1970, p. 18. Reprinted by permission of Prentice-Hall, Inc., Englewood Cliffs, New Jersey.)

THE NATURE AND SIGNIFICANCE OF WATER

Water is the most important compound on our world. It is also the most characteristic compound of the planet, leading to the term *watery planet* because the oceans cover most (71%) of Earth's surface.

The Significance of Water's Liquid State

Earthbound observations and data from spacecraft such as *Voyagers I* and *II* reveal that *no other planet in our Solar System maintains a liquid water ocean.* Water may have once flowed elsewhere in the Solar System, such as on Mars (see chapter 8), but ice (a solid) and water vapor (a gas) are the forms water takes outside Earth. Evidence of erosion by water on Mars shows that planet to have lost its liquid surface water billions of years ago. At 450° C (842° F), the surface of Earth's sister planet, Venus, is too hot for liquid water, and the other inner planet, Mercury, is a barren, cratered wasteland baked by a sun only 53 million kilometers (33 million mi) away. The Jovian planets are mostly gaseous balls, but Jupiter's moon Ganymede may house a liquid ocean mantle, and the rings of Saturn, Jupiter, and Uranus appear to be composed primarily of ice particles. Our planet is unique because of its liquid water.

A delicate orbital balance keeps Earth just the right distance from the sun for its water to exist in a liquid state. If our planet were about 16 million kilometers (10 million mi) closer to the sun, it would be too hot for liquid water to occur on its surface. If it were 16 million kilometers farther from the sun, all the surface water would be ice. This significant balance between freezing and boiling water has existed for more than 3 billion years, resulting in the continued occurrence of oceans, rivers, lakes, and groundwater.

The long-term presence of liquid water has influenced the origin, composition, and evolution of life. Life most likely originated in association with water, and water is necessary for its preservation. Organisms are composed chiefly of water, which comprises as much as 80% of a living cell. In dry climates, life collects at oases and on the windward sides of mountains. In humid climates, such as those in the eastern half of North America, the land is ubiquitously covered with life whose presence is made possible by water.

The Nature of Water

What is this substance that makes up almost 97% of the oceans? **Water** is a compound of hydrogen and oxygen that exists through the sharing of electrons (chapter 2). Although atoms are electrically neutral and balanced, they can be thought of as being incomplete whenever their most remote energy levels do not hold their full potential of electrons. It is their incompleteness that enables many atoms to react or to bind into molecules. Oxygen, an element with eight electrons, has an incomplete outer energy level, where six of its eight electrons are located. Hydrogen, the simplest of all elements, has only one electron in its single energy level, which can potentially hold two. In the

sharing of electrons, the hydrogen and oxygen both receive and lend electrons from their orbiting energy shells, and they make a covalent bond. The arrangement known as H_2O, therefore, completes the respective outer energy levels of the three atoms (fig. 11.2). The resultant molecule is unstable at best, breaking up and reforming as many as a million times each second and accounting for water's liquid nature.

There are many kinds of water that result from combinations of different isotopes (see chapter 2) of hydrogen and oxygen, and they possess different properties. The most common form of hydrogen contains a single, positively charged proton and a negatively charged electron (^1H). The balance of charges makes the atom electrically neutral. Other forms of hydrogen exist, however. Deuterium (^2H) is hydrogen with a neutron in its nucleus, and tritium (^3H) has two neutrons. Oxygen has isotopes with eight (^{16}O), nine (^{17}O), and ten (^{18}O) neutrons in their nuclei. The most common type of water is $^1H_2{}^{16}O$, but other forms, such as $^2H_2{}^{16}O$, or deuterium oxide, do occur in nature. Deuterium oxide is called *heavy water.* It does not behave the same as common water, and it does not quench your thirst. Nuclear reactors use heavy water as a moderator to slow the velocity of neutrons, resulting in an increase in fission.

The arrangement of atoms in the molecule gives water a special dipole (*di,* meaning "two") property much like a bar magnet's north and south poles. The hydrogen end of the water molecule exposes two protons, while the oxygen end is dominated by electrons. The isolated charges result from the pull of eight protons in the oxygen nucleus. For the electrons in the molecule, these more numerous protons exert a stronger tug than the two isolated hydrogen nuclei. This arrangement results in a

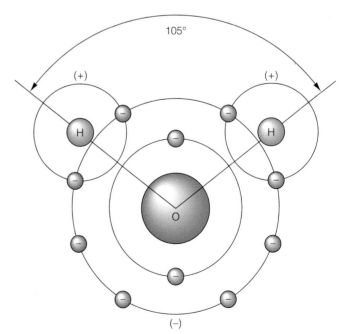

Figure 11.2 Arrangement of H and O ions in a covalent bond to form the water (H_2O) molecule. The molecule has a net positive charge on the H end and a negative charge on the O end because of its asymmetry.

positive hydrogen end and a negative oxygen end of the molecule. The positive end of one water molecule binds to the negative end of another and so on, making a glass of pure water virtually a single molecule. The lining up of opposite charges also makes pure water incapable of conducting an electric current.

Hydration

The dipole water molecule can act electrically to dissociate weakly bound molecules like sodium chloride, or common salt, which water separates into its component Na^{1+} and Cl^{1-} ions. In a process known as **hydration,** water molecules surround the component elements by binding to them in molecular form as the oxygen end of water molecules isolate the sodium and the hydrogen ends surround the chlorine (fig. 11.3). The significance of this process lies in the production of *dissolved salts* and water which is *saline,* or salty.

Seawater

Seawater, therefore, is composed of both water and dissolved salts. The salt water of the oceans contains the dissolved ions of more than 70 elements the most abundant of which are listed in table 11.1. A swimmer who swallows seawater drinks potassium, magnesium, silicon, and common salt, among other dissolved substances. More than 3% of the oceans is composed of these invisible salts, but these salts do not add to the volume of the ocean because they are contained between water molecules.

You can run a small experiment to demonstrate the relationship between volume and dissolved salt. Mark the water level, then add salt to a glass of water until a few of the salt crystals fall to the bottom of the glass without dissolving. You now have a *saturated solution,* or one in which there is no additional capacity for hydrating salt ions. A check of the water level against the mark should show no increase in the volume of water after you add the salt.

Residence Time

Elements pass into and out of the oceans because of physical, chemical, and biochemical reactions. Sea shells predominantly consist of calcium carbonate extracted from the water by certain enzymes in organisms. Coral reefs are the product of a similar

Table 11.1

The Most Common Elements in Seawater and Their Residence Times (where appropriate)

Element	Concentration (mg/l)	Known Residence Time (yr)
Oxygen	857,000	
Hydrogen	108,000	
Chlorine	19,400	
Sodium	10,800	2.6×10^8
Magnesium	1,350	4.5×10^7
Sulfur	885	
Calcium	413	8.0×10^6
Potassium	387	1.1×10^7
Bromine	68	
Carbon	28	
Strontium	8.1	1.9×10^7
Boron	4.5	
Silicon	3	8×10^3
Fluorine	1.4	
Nitrogen	0.5	

extraction. Thus, some ions do not reside permanently in the water but, rather, may become incorporated into marine sediments and rocks upon the deaths of the extracting organisms. Some of these materials may be returned to the water through dissolution, but many undergo burial and inclusion in the rock cycle. The time an element spends in the oceans is called its **residence time** (table 11.1).

SALINITY

Oceanographers measure dissolved salts in parts per thousand (‰, read "parts per mille"). So, in a liter of seawater, the saltiness might yield a reading of 34.5‰, meaning that 34.5 grams (or 3.45%) are dissolved substances. In 1902, an international commission defined saltiness, or **salinity,** as the amount of solid matter in grams contained in a kilogram (1,000 g) of water after any carbonate has been converted to an oxide, any bromine and iodine have been replaced by chlorine, and any organic matter has been oxidized. To find salinity oceanographers can use a *salinometer,* which is an instrument that determines salinity on the basis of electrical conductivity. Or they titrate (determine the quantity of dissolved salts by causing a chemical reaction) a water sample with silver nitrate. This process precipitates bromides, iodides, and chlorides, resulting in the *chlorinity,* or the total amount of chlorine. They then convert chlorinity to salinity through the following formula:

$$salinity = 1.808 \times chlorinity$$

The salinity of seawater differs with latitude and depth, location in open ocean or restricted water bodies, climate, stream

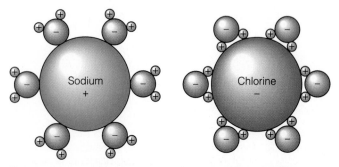

Figure 11.3 Water molecules surrounding and separating ions of sodium (Na) and chlorine (Cl) during the process of dissolution.

runoff, and ice formation and melting. Except for locally influenced waters along coastlines, the relative *constituent proportion of the various dissolved substances in seawater,* however, *remains the same.*

Controls on Surface Salinities

Oceanic surface salinities change with the seasons and with variations in precipitation amounts and evaporation rates. In equatorial zones, the surface salinity of the open ocean is relatively low because of the dilution from rain, whereas in the subtropics, the salinity of the surface water is higher because of evaporation. The July surface salinities shown in figure 11.4 reveal generally lower salinities in high latitudes than in low latitudes. Increased rainfall, runoff, and ice melt dilute the higher-latitude waters. The low equatorial readings that derive from excessive rainfall are shifted slightly to the north during the Northern Hemisphere summer and to the south during its winter.

Latitudinal salinity patterns established by climates can be offset by local seasonal influences in coastal waters. In restricted lagoons and bays, such as Laguna Madre along the Texas coast, for example, fluctuating salinities reflect wet and dry periods (fig. 11.5).

Salinity Profiles

The salinity of open ocean water varies with depth. The vertical record of this change is a **salinity profile.** The starting points for salinity profiles of vertical water columns are the surfaces of the seas and oceans. Wherever the profile reveals a rapid change in salinity with depth, it marks the **halocline** (from *halo,* meaning halite; and *cline,* meaning "transition"). A clearly defined halocline does not form throughout the oceans. In the equatorial regions, for example, the surface water salinity averages less than that for the subtropics. High-latitude surface salinities are actually lower than those in deep water, so a salinity profile reveals a slight increase in salt content down a high-latitude water column. Ultimately, mixing at depths overcomes local effects resulting in deep water of a relatively constant salinity. Figure 11.6 reveals the salinity characteristics for generalized water columns.

TEMPERATURE

Ocean temperatures vary. Vacationers on beaches in the Bahamas experience different water temperatures from those who attempt to swim along the coastal waters of Oregon, Washington, and British Columbia. These temperature differences result from controls not much different from those that affect atmospheric temperatures.

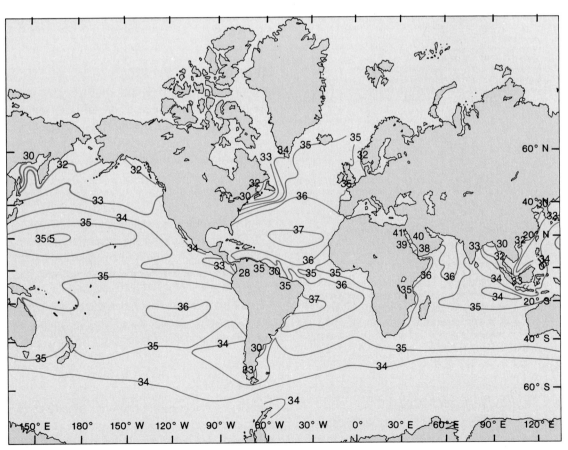

Figure 11.4 Average July surface water salinities in parts per thousand (‰).

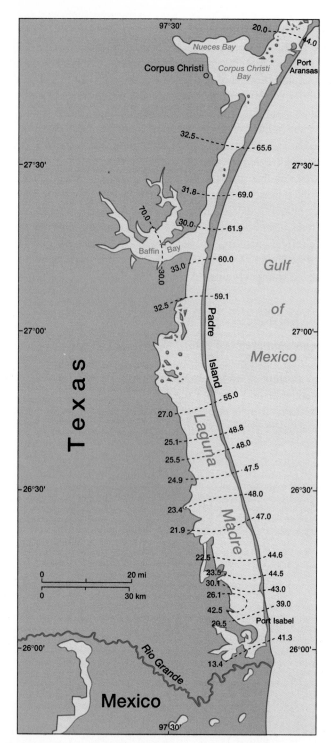

Figure 11.5 The salinity variations of Laguna Madre, a large lagoon along the south Texas coast. The extreme values at each end of the short lines indicate the annual range at each location: higher during the dry season and lower during the wet season. (After G.A. Rusnick, *Recent Sediments Northwest Gulf of Mexico*, ed. by Shephard and Moore, p. 155. Copyright 1960 American Association of Petroleum Geologists, Tulsa, Oklahoma. Reprinted by permission.)

Controls on Ocean Temperatures

Latitude is the chief control on surface temperature. Ocean waters in high latitudes, for example, receive less solar energy per unit area than low-latitude waters. Seasonal changes in the

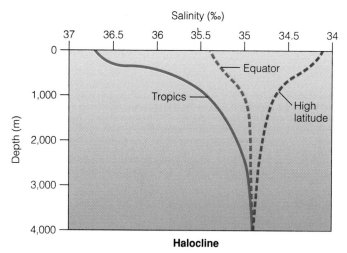

Figure 11.6 Haloclines at various latitudes. Note that the most pronounced change occurs in the tropical seas because of high evaporation rates but low precipitation.

amount of solar energy available also affect changes in temperature. Another control on temperature is depth, because deep waters are not influenced by the warming sun.

Typically, low-latitude surface water is warm and high-latitude surface water is cold, but these terms are only relative. Figure 11.7 shows average isotherms for surface waters during August, revealing very low temperatures in polar regions. Precise temperature readings by oceanographers and merchant marine sailors have enabled scientists to map not only the surface but also the deep-water temperatures and to draw temperature profiles.

Temperature Profiles

The resulting vertical profiles show varying patterns for different latitudes and indicate a zone of rapid temperature change with depth, called the **thermocline.** In equatorial and tropical waters, the thermocline is well developed because the surface temperatures are relatively high (fig. 11.8). A change to colder temperatures occurs rapidly in the top kilometer of these two water columns. In high-latitude surface waters, however, temperatures can fall below freezing (0° C, 32° F), so the change in temperature down the water column is not great.

The deep waters of the oceans are not universally cold. Along oceanic rifts with active volcanism, water that circulates near shallow magma chambers becomes very hot. Some of this water reaches 500° C (932° F) and exits the ocean floor through vents. Plumes of hot water then spread out over the ocean floor in localized areas, gradually cooling by contact and mixing with the ambient cold water of the deep ocean.

DENSITY

Density is weight per unit volume, usually given in grams per cubic centimeter. Our ordinary experiences give us a good sense of density. A block of wood, for example, is not as heavy as a similarly sized mass of iron. Of course, the atomic mass number

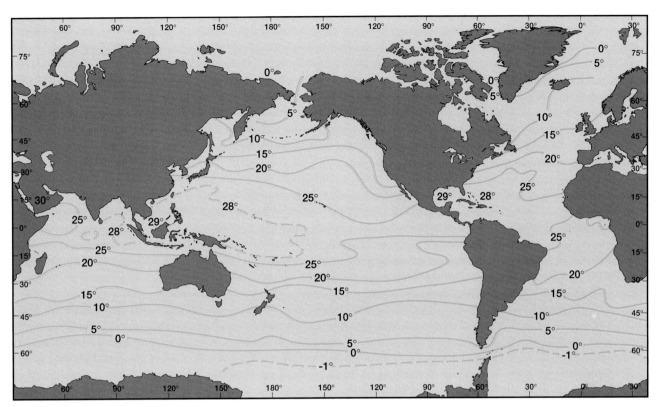

Figure 11.7 Mean oceanic surface water temperature during August (° C).

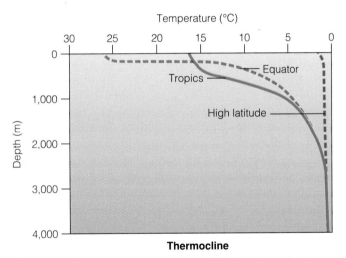

Figure 11.8 Thermoclines at various latitudes. Note that the sharpest change occurs in equatorial seas because of intense solar heating of the surface waters.

(chapter 2) of iron is higher than that of any constituent atom in the wood. Water, as you now know, can be made of different oxygen and hydrogen isotopes, some of which are heavier than others. But the most significant controls on the density of seawater are salt and temperature.

Salinity and Density

At 4° C (39° F), pure water weighs 1 gram per cubic centimeter. Salt water is heavier than pure water, and seawater averages about 1.025 grams per cubic centimeter. Wherever rains or in-

flows of river water dilute the salinity of the oceans, the local surface water becomes less salty and the density decreases. Many large river systems, such as the Amazon, for example, flow *onto,* rather than into, the ocean, forming a *freshwater lens,* which is a body of lighter, lens-shaped water that slowly mixes with the seawater (fig. 11.9).

Temperature and Density

Like salinity, the temperature of the oceans varies with latitude and depth. Surface waters in equatorial regions can exceed 30° C (86° F), whereas those of high latitudes can fall below 0° C (32° F). The water in the depths of the oceans can fall below 0° C. That the water is not frozen on the ocean bottoms is the result of three factors.

1. It is salt water, *which does not freeze at 0° C.*
2. Ice floats because it is less dense than water.
3. Even though water is *not very compressible,* it is under great pressure on the ocean bottom, as much as a thousand times the average sea-level pressure. The increase in pressure with greater depth causes a slight adiabatic warming.

Pure water is densest at 4° C (39° F). Although salt water is denser on average than fresh water, temperature exerts an even greater control than salinity on water density. Water warmer than 4° C is less dense, as one might expect because of more rapidly vibrating molecules. But 4° C marks a *pivotal point* in the density of pure water. Between that temperature and the initial

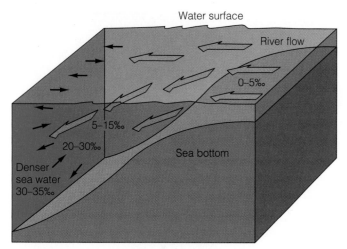

Figure 11.9 A freshwater lens rides above denser seawater for some distance until natural mixing processes bring about homogeneity.

freezing point, water decreases in density with the formation of partial and complete hexagonal ice crystals that add volume without adding mass (fig. 11.10). Relatively stable ice crystals make a latticework, which incorporates empty spaces because the angle between the hydrogen atoms in the water molecule increases to form hexagonal crystals.

As a consequence of the density differences between ice and water, ice floats. *Ice floes* form directly on the sea's surface (fig. 11.11*a*). *Icebergs,* which are *calved* (or broken) pieces of glaciers, drift with currents, becoming hazards to shipping in high latitudes (fig. 11.11*b*). It was a collision with an iceberg that sank the *Titanic,* the world's largest passenger liner, on her maiden voyage in 1912. Even larger ice accumulations can occur. The Arctic Ocean is covered by a number of moving *ice islands* of considerable size, and in the South Atlantic some tablelike icebergs attain an area equal to the size of Rhode Island.

The density of ice also influences organisms. Life on the ocean bottom is effectively protected by this density difference. If ice were denser than liquid water, the oceans would freeze from the bottom up, and marine organisms would have no secure habitat. A layer of ice formed from seawater insulates the water below from subfreezing atmospheric conditions. In addition, frozen seawater contains very little of the salt from its original liquid state. The absence of salt in icebergs has even led some to speculate about tugging them to desert countries like Saudi Arabia for a large freshwater supply. Any melting in transit would be offset by the great volume of ice.

The Sigma-t Diagram

Oceanographers use a special graph to plot the combined effects of salinity and temperature. This graph is called the **sigma-t** or **T-S, diagram** (s = salinity, and t = temperature), and it enables scientists to establish density characteristics for particular water masses (fig. 11.12). Earlier, you learned that pure water at 4° C has a density of 1 gram per cubic centimeter. The addition of salt

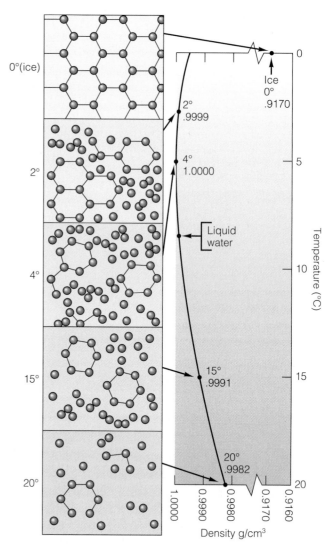

Figure 11.10 The relationship of temperature to water density. At 20° C, water molecules are separated with few attached to one another. As temperature decreases to 4° C the molecules are slowed and packed tightly together. This is the densest configuration. Below 2° C, the water molecules attach themselves together in a crystalline pattern to form solid ice. Because there is a specific spacing between ions in the crystal lattice they separate from one another as they link up, thus lowering the density.

or a change in temperature alters density. Seawater is always salt water, and it has, therefore, a slightly higher density than pure water, varying from about 1.025 to about 1.028 grams per cubic centimeter. On the sigma-t diagram, 1 gram is subtracted from each of the readings, which is then multiplied by 1,000, making, for example, 1.0255 grams per cubic centimeter read as sigma-t 25.5 at one atmosphere of pressure. Figure 11.12 reveals that warm, less-saline water is less dense than cold, saline water.

The vertical density profiles for both the equatorial and subtropical water columns show a rapid change in density with depth, called the **pycnocline** (fig. 11.13). Temperature rather than salinity has a greater effect on water density. The warm surface waters of equatorial and tropical oceans are relatively light, making the pycnoclines mimic the pattern of the thermoclines for these regions, as a comparison of figures 11.8 and 11.13 shows.

(a)

(b)

Figure 11.11 (a) Sea ice in an early stage of formation. (b) An iceberg in the polar sea.

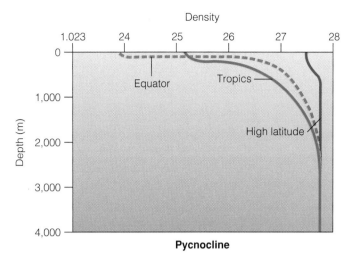

Figure 11.13 Pycnoclines at various latitudes. The most pronounced density changes occur in the equatorial seas.

GENERALIZED SURFACE PROFILE

By examining the surface water characteristics between 80°N and 60°S, which are plotted in figure 11.14, you can see the relationships among temperature, salinity, and density. Where temperatures are high and salinities are low, they produce a lowered density. In the subtropics, the warm temperatures of the water are balanced by the higher salinities derived from excessive evaporation, yielding higher densities.

THERMOHALINE CIRCULATION

The combined effects of temperature and salinity generate vertical currents in the oceans. Dense water falls below less-dense water. Thus, the cold surface waters of high-latitude seas sink because of their relatively high density. Similarly, water from climatically dry regions in mid-latitudes sinks because of increased salinity. The movement of ocean water driven by density differences is called **thermohaline circulation.**

Water Masses

Oceanographers and sailors have taken measurements of salinity and temperature for many years, and the compilation of their data has enabled scientists to characterize **water masses,** which are large, three-dimensional bodies of water with identifiable boundaries.

In the Mediterranean, evaporation is high and rainfall is seasonal. From its volume of 4.2 million cubic kilometers (1 million cubic mi) of seawater, the Mediterranean loses 1,235 cubic kilometers (296 cubic mi) through evaporation annually. The surface water that does not evaporate becomes increasingly more saline and, obviously, more dense. As the surface water sinks under its own weight to make bottom currents, it exits the sea at an estimated flow rate of 1,750,000 cubic meters (61,775,000 cubic ft) per second through the Straits of Gibraltar, mostly beneath incoming, less-dense Atlantic surface water that replenishes the sea (fig. 11.15). Once in the ocean, the great

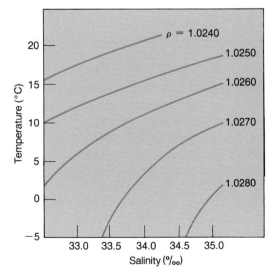

Figure 11.12 A sigma-t diagram showing variation in water density with temperature and salinity.

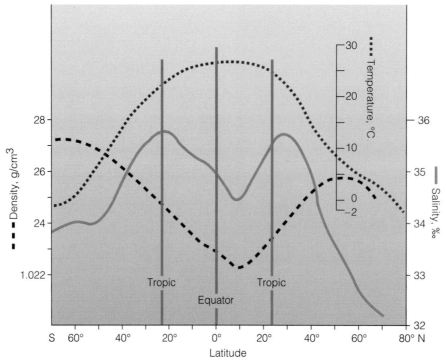

Figure 11.14 Latitudinal variation of temperature, salinity, and density of surface waters.

volume of Mediterranean water becomes an intermediately deep layer of water in the Atlantic, spreading laterally and eventually mixing with water sinking in the Norwegian Sea.

Because of its temperature, water from the Norwegian Sea sinks even lower than Mediterranean water and becomes the *North Atlantic Deep Water (NADW)*. At the bottom of the ocean lies a cold layer, *Antarctic Bottom Water* (AABW), derived largely from the Southern Ocean, that surface body of water that encircles Antarctica. AABW is very cold (-0.4° C, 31.3° F) and very dense (sigma -t over 27.8).

Downwelling

The formation of the AABW presents a good lesson on the complexity of the ocean systems. As the prevailing westerlies blow across the Southern Ocean, they drive different bodies of surface water together. Because water cannot form mountains at such a confluence, some of the converging water sinks. At the surface, the *Southern Ocean* averages 0.5° C (33° F) with a salinity of 34.68‰. This *downwelling water* merges with sinking shelf water that averages −1.9° C (28.58° F) and 34.62‰ to make the AABW water mass at a depth of 4 kilometers (2.48 mi). The AABW has a characteristic temperature of −0.4° C (31° F) and a salinity of 34.66‰, and it is among the densest of all water masses.

The autonomous characteristics of various water masses like the AABW make the ocean more or less *stratified*. The less-dense surface waters are underlain by moving layers of *intermediate, deep,* and *bottom waters*. Just like air masses, each of these water masses derives from a particular source region (fig.

11.16). A water mass can vary within its boundaries and, depending on neighboring densities, can extend upward or downward through other masses (fig. 11.17).

Upwelling

It should be obvious that the downward movements of water masses must be offset by upward moving water. Wherever surface water bodies are separated by prevailing winds, they diverge. Because a liquid such as water cannot have a permanent "hole," water from below rises to take the place of the displaced water. The result is oceanic **upwelling,** a mechanism that returns sunken water to the surface. Upwelling occurs both in the open ocean, where wind-driven currents diverge, and along coastlines, where prevailing winds blow seaward (fig. 11.18).

Along the western coast of South America, an important upwelling brings cold, nutrient-filled water to the surface to replace water driven westward by the trade winds (fig. 11.19). The nutrient-rich water that upwells has traveled from as far away as the Atlantic, where it sank to become deep water, carrying the elemental remains of organisms with it into the Pacific. Bolstered by the influx of phosphates and nitrates, compounds that are essential to plant growth, the upwelling waters off the west coast of South America support a rich flora, comprised of algae, which, in turn, feed an abundance of fish. The Peruvian fishery is, as a result, one of the largest in the world.

The vertical movements of water turn the ocean over about once every 2,000 years. That means that ocean-dumped pollutants, such as radioactive wastes, will resurface to plague future generations of sea life and humans. Possibly, dilution from mix-

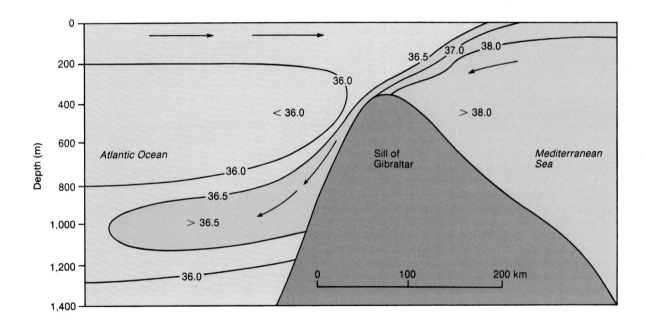

Figure 11.15 Cross section shows movement of dense Mediterranean water over the sill of Gibraltar and into the deep Atlantic Ocean basin. (Source: John A. Knauss, *Introduction to Physical Geography,* ©1978, p.168. Adapted by permission of Prentice-Hall, Englewood Cliffs, New Jersey.)

Box 11.1 Further Consideration

Did the Mediterranean Sea Dry Up?

In December 1972, *Scientific American* published an article by Kenneth J. Hsü in which the author reported evidence that the Mediterranean Sea had once dried up. The evidence was part of a multiple working hypothesis, one that adopted the findings of numerous scientists working in several research disciplines. Hsü's multiple evidence derived from deep-sea drilling by the crew on board the *Glomar Challenger,* from the studies of microfossils, from accounts of seismologists, and from other sources. Although other scientists had worked on the data from Leg 13 of the *Glomar Challenger* voyages, only Maria Cita and Bill Ryan, who were members of the scientific cruise, agreed with Hsü on the validity of a dessicated Mediterranean.

In particular, Hsü and his co-workers found layers of halite below the marine sediments of the western Mediterranean's Balearic Basin. The halite, which was associated with anhydrite ($CaSO_4 \cdot 2H_2O$), appears to have been produced by the evaporation of relatively shallow seawater. That it was found under 3 kilometers (approximately 2 mi) of Mediterranean water seemed to indicate that the sea had once been very shallow.

Corroborating evidence from drill sites on land came in the form of newly discovered canyons that lie buried beneath current stream systems, such as the Rhone in Europe and the Nile in Egypt. These canyons indicate a downcutting that reached below present sea level by thousands of meters. Because streams attain their erosive power through gravity, the base level of these streams must have been lower during the downcutting. Beneath deep water south of Greece is the large Mediterranean Ridge. This ridge also shows a downcutting that could not have occurred under deep-water conditions. Apparently, Atlantic seawater rushed in to fill the dried-up basins partly through the Mediterranean Ridge, eroding and cleaving the ridge (fig. 11.A)

The time of this desert Mediterranean has been estimated at 6 million years ago, and coinciding with this estimate is a change in the fauna that inhabited the region (chapter 14). A dry Mediterranean allowed the passage of animal life previously blocked by a deep sea. With fossil evidence supporting the theory, Hsü speculated that the Mediterranean opened and closed repeatedly at the Straits of Gibraltar. Each time the basin was flooded, a spectacular waterfall more than 1,000 times the volume of Niagara Falls spilled into the Mediterranean.

Apply Your Knowledge

1. What happens when you allow a glass of salt water to evaporate?

2. Is there any mechanism other than erosion by running water that could explain the cleft in the Mediterranean Ridge?

10 km

West   East

Figure 11.A   Seismic profiling showed a deeply eroded valley in the Mediterranean Ridge that formed as Atlantic water flowed into the Mediterranean basins. (Source: From Hsu, Kenneth J., *The Mediterranean Was a Desert.* Copyright © 1983 by Princeton University Press. Reproduced by permission of Princeton University Press.)

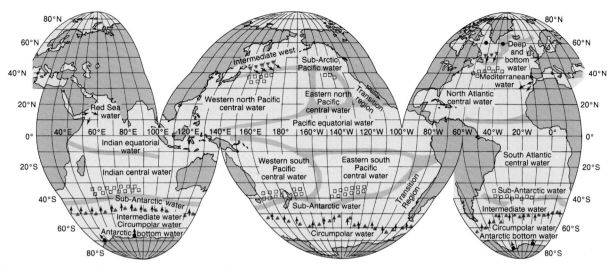

Figure 11.16 Major water masses and their origins.

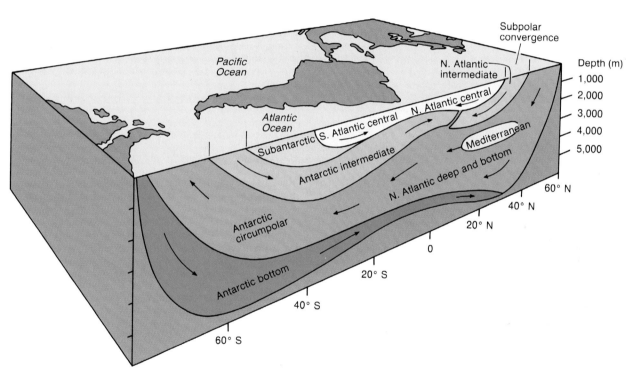

Figure 11.17 Stratification of water masses in the South Atlantic.

ing will minimize the effects, but the same hazards will still exist, especially any exposure to those toxic wastes with long half-lives.

HORIZONTAL CIRCULATION

Even though thermohaline circulation is important to the movement of seawater, the overriding influence on the surface is wind. The greatest volume of downwelling seawater results from converging currents driven by prevailing winds. The winds, in fact, also produce a planetwide surface circulation. In your study of meteorology earlier in this text you learned that six major wind systems scour the surface of the planet. These winds obey the

dictates of the Coriolis force by veering to the right in the Northern Hemisphere and to the left in the Southern, and their latitudinal boundaries migrate north and south with the seasons (fig. 11.20). The winds can also fade and intensify, or, as they do under the influence of changing pressure patterns, even reverse themselves locally like those that respond to monsoonal changes over the Arabian Sea.

Currents and Gyres

The general circulation of the atmosphere transfers some of its energy to the surface waters, pushing them along in relatively narrow bands called *currents*. Generally, these surface currents

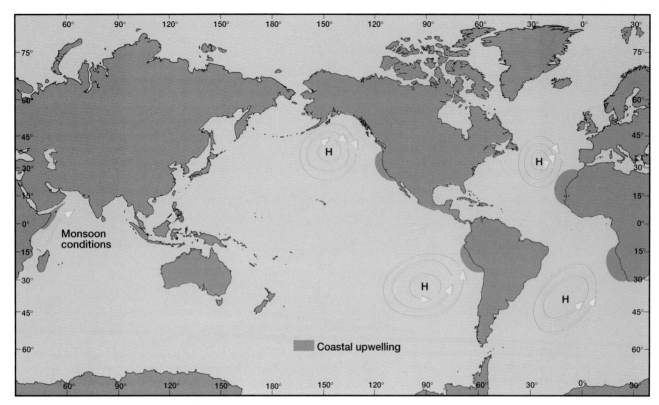

Figure 11.18 Principal areas of upwelling along coasts.

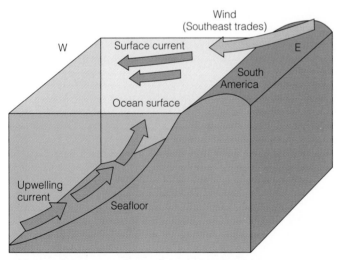

Figure 11.19 Upwelling along the western coast of South America.

veer to the left in the Southern Hemisphere and to the right in the Northern, though there are some exceptions in high latitudes. They also maintain identifiable characteristics. As the currents turn throughout the oceans, they appear, on the largest scale, to run in **gyres,** or large, somewhat circular clockwise and counterclockwise patterns that almost mirror the movements of the atmosphere (fig. 11.20). The water flow is not exactly the same as the winds because the ocean currents can veer up to 45° from the direction of the winds. Since water is a more viscous fluid

than air and its inertia makes it travel more slowly than winds, it is subject to an amplification of the Coriolis force.

Because of the configurations of the continents and the interruptions to flow caused by large islands and capes, the currents do not always conform to an ideal, smooth pattern. More like the surface winds influenced by friction than like the high-altitude jet stream, the surface waters of the oceans give only a general semblance of the "ideal" movements. No ideal basin boundaries exist, so the currents often diverge from the expected pathway and meander through the surrounding surface water.

Geostrophic Flow

As Northern Hemisphere currents flow, they trend toward the right, and water pealing off equatorial and subtropical surface currents mounds up. The oceanic mound acts just like a hill on land with water falling down its slopes, offsetting the Coriolis force and establishing a continual flow along the mound's base. This continual movement establishes the gyres, and even in the absence of winds, it would continue under the influence of Earth's spinning to make a turning current called a *geostrophic flow* (fig. 11.21).

The Ekman Spiral

As the wind blows across the ocean surface, it pushes against the more viscous water, which resists the flow with a greater inertia and veers, in the Northern Hemisphere, 45° to the right.

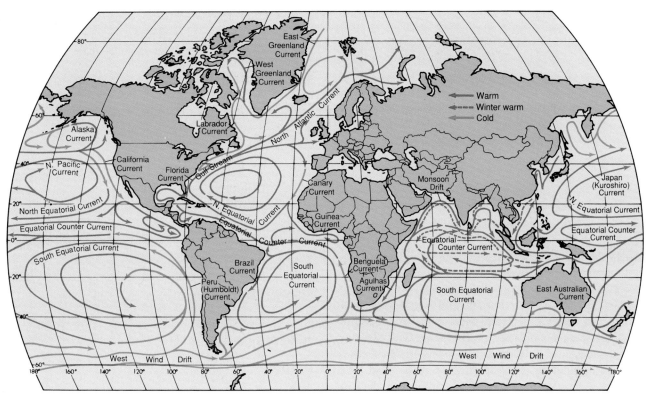

Figure 11.20   Major surface currents of the world's oceans.

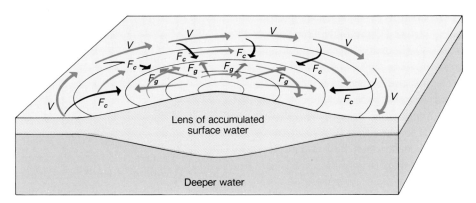

F_g = Gravity force
F_c = Coriolos force

Figure 11.21   Geostrophic flow around a mound of surface water.

Below the "pushed" surface water the progressively deeper water moves more slowly and even more to the right. Eventually, the water runs opposite the surface flow and 225° (180° plus the 45° of deflection from the driving wind) to the wind that initiated the surface movements (fig. 11.22). This deflection that increases with depth is termed the **Ekman Spiral.**

Types of Surface Currents

Oceanographers classify surface currents according to their relative temperatures. Usually, the designation accounts for movements across latitudinal boundaries, so that surface flows traveling away from or along the equator are termed *warm currents,* and those traveling from high latitude to lower latitude or longitudinally at high latitudes are called *cold currents* (see chapter 6). Figure 11.20 shows the warm currents in red and the cold currents in blue. The latitudinal position of the currents is not as important in this classification as is the direction of travel. Thus, the Alaska Current, which flows northward along the coast of northern British Columbia toward Alaska, is a warm current because it flows from warmer (lower) to colder (higher) latitudes. The Canary Current, which runs southward off the western coast of North Africa, is a cold current, carrying cooler water equatorward.

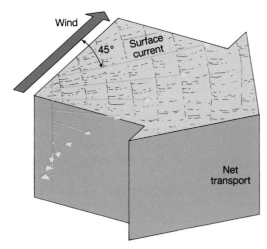

Figure 11.22   The Ekman Spiral.

Typically, warm surface currents are fast, narrow, and deep, and cold currents are slow, wide, and shallow. These are, of course, comparative terms. At 35°N the **Gulf Stream,** which is the most intense warm current in the Atlantic Ocean, averages 150 centimeters per second (3.35 mph). That may not seem fast, but at that rate, the current carries more than 50 million cubic meters (1.75 billion cubic ft) per second across its approximate 75-kilometer (46.6-mi) width and 2-kilometer (1.24-mi) depth. By comparison, off the West Coast, the Pacific's cold California Current, a broad sheet of water that travels little more than 10 kilometers (6.2 mi) per day, transports about 10 million cubic meters (350 million cubic ft) per second across its almost 1,000-kilometer (62-mi) width and 500-meter (1,640-ft) depth.

Oceanographers also classify surface currents as either *western* or *eastern boundary currents.* With the exception of high latitudes, warm currents run along the western sides of the oceans; cold currents, along the eastern. There is also an intensification of western currents that makes them flow faster in narrow bands against the continental margins.

The Western Intensification

The intensification of western boundary currents is the product of several influences. First, the trade winds move the North and South Equatorial currents across the oceans to collide with landmasses such as South America. At Recife, Brazil, the continent juts into a wedge and divides the equatorial flow of the Southern Hemisphere into two directions: southwest and northwest. The latter component moves water across the equator. In the Northern Hemisphere, some of the water flows into the Caribbean Sea through the southern Antilles and the rest, because it veers right, moves northward along the windward islands. The water that enters the Caribbean eventually moves into the Gulf of Mexico and exits between Cuba and Florida, as the *Florida Current.*

The Florida Current presents a good lesson on the *Bernoulli theorem,* which states that channeling a moving fluid increases its rate of flow. The narrow strait between Cuba and Florida channels the water, and the flow increases just as it joins the windward *Antilles Current.* At about the latitude where the two currents merge, the trade winds give way to the prevailing westerlies, and a vortex of winds at the convergence, plus the shape of North America, give impetus to the Gulf Stream. The large subtropic gyre called the *Sargasso Sea* also adds a geostrophic flow and helps intensify the current through the addition of water in a continuous flow.

At the equator the Coriolis force is zero, so the equatorial currents have a greater tendency to move longitudinally than do higher-latitude return currents on the northern end of the Sargasso Sea. The Sargasso Sea, therefore, tends to pinch the Florida Current and the southern end of the Gulf Stream toward the continent, with a consequence of channeling the flow. A channeled flow is a faster flow, and the Gulf Stream reaches its peak rate at about 35°N as it is squeezed against the southwestward-moving *Labrador Current* and shelf waters.

The Gulf Stream

The most studied current in the oceans is the Gulf Stream, the giant river of water described in the preceding section. As a climatic control, the current transfers heat from low latitude toward boreal regions and separates northern colder water from warmer water to the south. In this latter regard, the Gulf Stream acts like the high-altitude Jet Stream, which also serves as a fast-moving boundary between warm and cold air. Recent studies have shown that the Gulf Stream does not reach the shores of Ireland and England as long thought. Instead, the stream becomes very complex in the northern Atlantic, and the *North Atlantic Current* transfers heat to the *Norwegian Current.* Then, north of Scandinavia, the *North Cape Current* prevents sea ice from forming along the coast to the port of Murmansk, Russia.

The Gulf Stream does, however, influence the weather of the North Sea and eastern Atlantic countries. Denmark is a rainy land, and Britain has occasionally heavier rains when the current warms the western Atlantic more than usual. The heat that the Gulf Stream transfers to other North Atlantic currents also makes southern Ireland's climate milder than one would expect for its high latitude (see chapter 6).

One of the most intriguing aspects of the Gulf Stream is its meandering. As the current moves northward, it twists itself into loops and cutoffs very much like a terrestrial old-age stream that leaves oxbow lakes on an alluvial plain. As fluid expressions, the loops and cutoffs of the Gulf Stream do not create oxbows. Instead, the meanders isolate bodies of water from either the Labrador Current or from the neighboring Sargasso Sea. These isolated, spinning bodies of water are called **eddies,** or rings (fig. 11.23). As the loop closes on Sargasso Sea water from the south, it pinches off on the northern side of the current, making a *warm-core ring.* The opposite happens as the Gulf Stream loops around cold water from the Labrador Current. Thus, *warm eddies spin off north of the current, and cold eddies spin off to the south* (fig. 11.24).

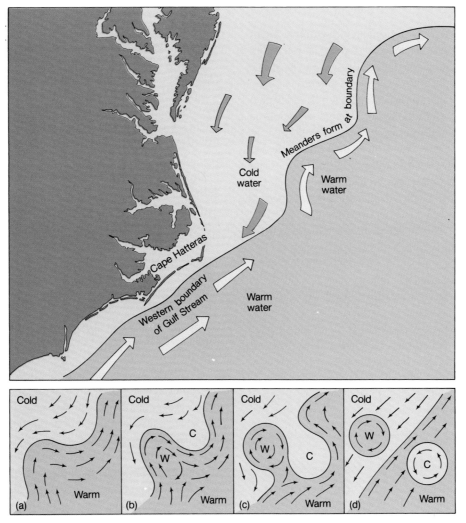

Figure 11.23   Eddies, or rings, form along the western boundary of the warm Gulf Stream Current.

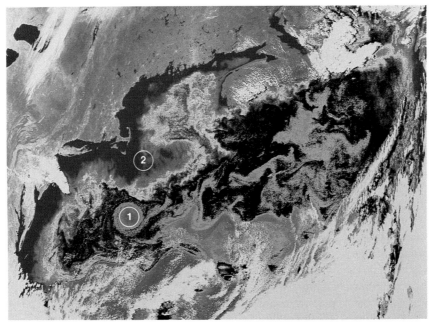

Figure 11.24   A warm core ring (*1*) is spun off of the Gulf Stream south of Georges Bank in the western Atlantic.

The Return Water

Just as upwelling balances downwelling, return currents resupply surface areas that lose water and keep the ocean water from piling up against and spilling onto the continents. The largest of the return currents is the Pacific's *Cromwell Current*.

The Cromwell Current is an equatorial *countercurrent* that transports an estimated 25 million cubic meters (875 million cubic ft) per second toward the eastern Pacific. Reports from tuna fishers in the early 1950s revealed that their lines drifted to the east below the surface while the prevailing Equatorial Current flowed west. The fastest speeds (150 cm/sec, 4.88 ft/sec) of this current occur at the 10-meter (32.8-ft) depth, so the surface current is superimposed on an opposite, shallow undercurrent. At the 350-meter (1,150-ft) depth, the current moves more slowly, indicating that it is truly a surface phenomenon. The Cromwell Current, which results from the wind's piling up water in the western Pacific, is a good example of water flowing down a water hill, and it acts to return water driven away from the eastern Pacific.

The return of water across an ocean or along the sides of gyres shows that the ocean currents describe a closed system. The system operates within the basinal confines established by seafloor spreading, moving continents, and, as might have occurred during the last glacial advance, the buildup of ice that creates temporary barriers to flow.

AIR-SEA INTERACTIONS

Although the currents make a closed system, the ocean and the atmosphere cross their mutual boundary through an exchange of gases vital to several of our planet's systems. Chief among the interchanges is the exchange of water vapor, the point of departure for the hydrologic cycle. About 334,000 cubic kilometers (80,160 cubic mi) of water evaporate from the ocean surfaces each year to enter the atmosphere as a significant gas. This water vapor captures heat that radiates from the surfaces of both land and water. It also adds energy to storm systems through the latent heat of vaporization and supplies the atmosphere with water for precipitation. As a consequence of this, oceans have weather and climate.

Another exchange of gas occurs when the O_2 produced by floating photosynthesizers leaks from the surface waters into the atmosphere. Since light is essential for plants and since darkness prevails below 1,000 meters (3,280 ft), photosynthesis is mostly a surface phenomenon. Plants also require phosphates and nitrates for growth. Wherever these nutrients are plentiful, such as within upwelling waters, the oceans house great abundances of Thallophyta, the plant subkingdom that includes the algae, dinoflagellates, bacteria, and fungi. Because of their relative abundance, the algae, particularly the **diatoms,** are the most important oxygen producers in the sea, with population densities exceeding 200,000 per liter (800,000/gal) in the North Pacific (fig. 11.25). Some higher plants, members of the subkingdom Embryophyta, also produce oxygen. One of these latter is the plant *Zostera,* the eel grass that grows in shallow coastal waters. Typically, oxygen production from photosynthesis varies with sun angle, which is a control on light intensity and duration. It also varies with the availability of nutrients.

The exchange of CO_2 at the ocean-atmosphere interface has led to a controversy about the role the oceans play in moderating the greenhouse effect (chapter 1).

Water has the ability to store dissolved gases, but the temperature and pressure of water controls its storage capacity. An analogy lies in the common experience all of us have had with carbonated beverages (soda pop, beer). Such drinks can go "flat" after their containers are opened and their temperatures rise. Streams of rising bubbles attest to the loss of dissolved gas, and warmer temperatures speed up the rate of loss. Because cold water and water under high pressure can hold more dissolved gas than low-pressure warm water, the heated surface waters of the Bahamas hold less CO_2 than the cold waters off Antarctica. The vertical profiles of oxygen and carbon dioxide show opposite trends as respiration removes oxygen from the water and produces carbon dioxide and photosynthesis does just the opposite (fig. 11.26).

With the increase in atmospheric CO_2 from the burning of fossil fuels (coal, natural gas, and oil), some scientists have taken different sides of an important issue: whether or not the oceans can act as a sink for the additional carbon dioxide. If the oceans have the capacity to accept the gas, then the greenhouse effect may be stabilized or even reduced. But if they are incapable of siphoning off the gas from the atmosphere, then worldwide atmospheric temperatures will increase. The warmer temperatures of the atmosphere will also mean warmer surface waters and a further diminished capacity to hold dissolved gas, compounding the problem.

Warmer surface waters lying under the influence of a warmer atmosphere would have an increased evaporation potential. Cloud cover might increase as a result, and world temperatures may be affected by blocked, reflected, and diffracted sunlight.

LIGHT IN WATER

The peculiar color of our blue planet derives from the dominant substance on its surface: water. In general, the oceans are blue for the same reason that the skies are blue. The shortwave end of the spectrum is scattered more than the longwave end, so blue light, as well as ultraviolet light, scatters most. The red end of the spectrum, which comprises longer waves than the blue end, is absorbed more readily, particularly by the water molecule (fig. 11.27).

To say that the ocean is "blue," however, is to miss the great variety of color reflected by the oceans. The deep ocean does appear a deep blue, but subtropical and tropical coastal waters have colors varying from light green to dark green, and cold Arctic waters look "oily" (fig. 11.28).

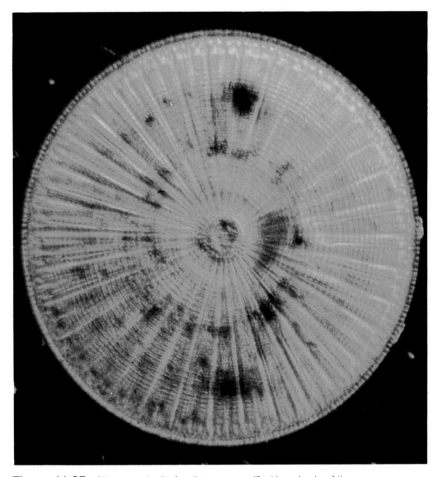

Figure 11.25 Siliceous shell of a diatom magnified hundreds of times.

The color of water is in part influenced by the angle of the sun's rays and by the roughness of the surface. The low angle of incidence in high latitudes means less penetration (or more reflectivity). At low latitudes, high angles of incidence over calm seas mean greater penetration and less reflectivity.

Another influence on water color is the amount of suspended matter. In coastal waters, plumes of sediment can make the water green or light brown by reflecting or absorbing light. In some areas, extraordinarily high concentrations of floating plants and animals also influence water color.

SOUND IN WATER

One of the most intriguing discoveries made by marine biologists is that whales can communicate with one another over great expanses of ocean. Their "songs" carry hundreds of kilometers in the modern oceans, and some speculate that, prior to the advent of noise pollution by mechanized ships and the sounds of machinery, those songs could carry across whole oceans.

Water is an efficient transmitter of sound. In water, sound travels more than four times faster than it does in air, but variations occur because of differences in temperature, salinity, and pressure. By analogy, sound bends at density boundaries just as light bends in water, a phenomenon that makes a utensil or a pencil look broken when half of it is submerged in a glass of water.

Submarine crews use this bending of sound whenever they attempt to hide their vessels. They know to look for the thermocline since it represents a zone of rapid density change. Because surface ships use sonar to locate submarines, the best place to hide is wherever sound bends radically because it changes speed (fig. 11.29). For every degree Celsius that seawater increases in temperature, there is a corresponding rate increase of 4.5 meters (14.75 ft) per second for sound. The overall effect within a well-developed thermocline is to establish an apparent, but false, location of the submarine as the sonar signal returns to the surface ship.

Sound can also be trapped in a "channel" by bending. The SOFAR (SOund Fixing And Ranging) channel is the product of sound bending as it moves through the water. The faster moving sound has a tendency to bend toward lower-velocity zones (fig. 11.30). Thus, sound from above and below a low-velocity zone bends to become trapped and channeled. Because little energy is lost in this low-velocity zone, sound can travel great distances, potentially thousands of kilometers. It is this channel that once enabled the whales to project their mating calls over entire oceans. Today, rescue operations often include the use of the SOFAR channel to locate those lost at sea.

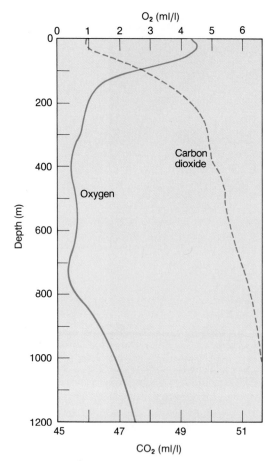

Figure 11.26 Oxygen and carbon dioxide profiles.

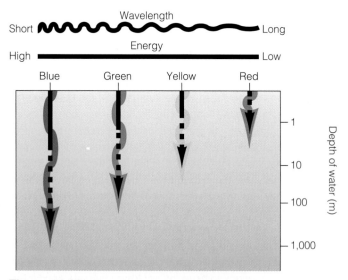

Figure 11.27 Representative penetration depths in clear water for the several wavelengths (colors) of the light spectrum.

Sound is an important tool in imaging the ocean bottom. The newly designed SEABEAM system uses sound to paint an accurate picture of the oceanic bathymetry, revealing seafloor features as small as 100 meters (328 ft).

Figure 11.28 Water color change in the Bahamas.

OCEAN WATER AND BOTTOM ZONES

Oceanographers classify the waters of the deep ocean and the continental shelves into zones based on location and the penetration of light. The most landward of these encompasses the water over the continental shelf and is called the **neritic zone** (fig. 11.31). This relatively shallow zone extends from the high water mark (high tide) to the edge of the continental shelf, so that at its deepest it reaches about 200 meters (656 ft). The shoreline between high and low tide is the **littoral zone.** The littoral lies between the **supralittoral,** the beach and dune area, and the **sublittoral,** the sea bottom immediately below the low water (low tide) mark.

The water beyond the continental shelf is classified as pelagic water. The upper 200 meters (656 ft) of the open ocean is the **epipelagic** (*epi,* meaning "on") **zone.** This zone is well-lighted and, as such, is also called the *euphotic* (*eu,* meaning "true"; *phot,* meaning "light"). The epipelagic is a layer with high primary productivity. Diatoms and other photosynthesizers have ample light for growth in this zone. From 200 meters (656 ft) to 1,000 meters (3,280 ft), the light attenuates rapidly, making the **mesopelagic zone,** the middle zone, one of "twilight." In this semidark, *disphotic* region, there are fish with *bioluminescent* spots very much like the chemical light of fireflies. Light does not penetrate beneath the mesopelagic, making all the water below 1,000 meters dark, or *aphotic* (*a,* meaning "without"). The uppermost lightless zone is the **bathypelagic,** which extends to 4,000 meters (2.5 mi). Along the bottom of the ocean is the **abyssopelagic,** the water column above the abyssal plains. As you learned in chapter 10, the deepest ocean water lies in the trenches. This water overlies the **hadal zone,** after the Greek underworld, "Hades." One other ocean zone has been identified, the **bathyal zone** of the continental slope, running from shelf to continental rise. It's believed that many of the organisms that inhabit the aphotic zones evolved from inhabitants of bathyal waters.

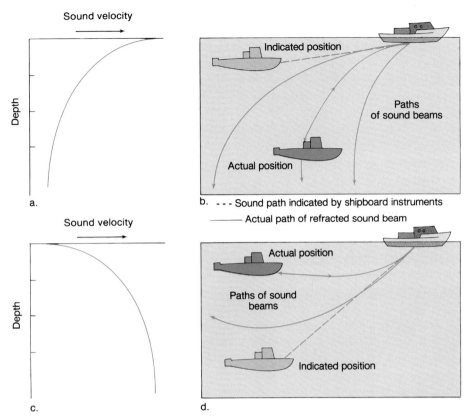

Figure 11.29 Hiding in the thermocline. (*a*) Sound velocity decreases with depth (tropical waters). (*b*) Refraction causes indication of target at shallow depth. (*c*) Sound velocity increases with depth (polar waters). (*d*) Refraction causes indication of target as deep.

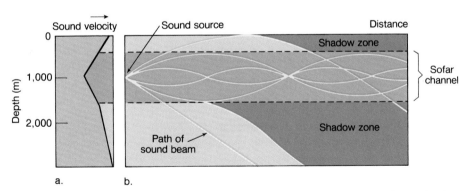

Figure 11.30 The SOFAR channel. (*a*) Sound velocity in seawater is at minimum in a zone centered on a depth of 1,000 meters (3,280 ft) because of changes in temperature, salinity, and pressure. (*b*) Refraction causes some sound waves to be trapped within the channel and other waves to bypass certain areas (shadow zones) through which the sound never passes. (Source: From *Fundamentals of Acoustics,* 2d ed., by Lawrence E. Kinsler and Austin R. Frey. Copyright © 1962 John Wiley & Sons, Inc., NY. Reprinted by permission of John Wiley & Sons, Inc.)

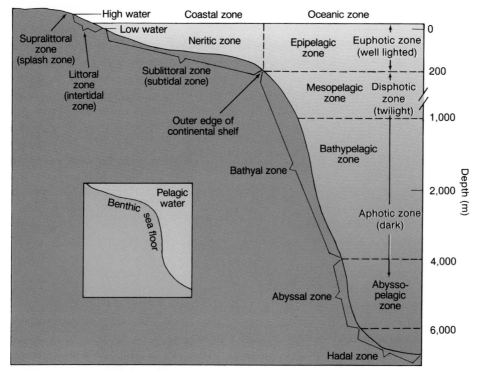

Figure 11.31   The ocean zones.

Further Thoughts: The Ocean-Atmosphere System

Earth's oceans are geographically connected because they cover 71% of the planet's surface. This interconnection allows the exchange of water among water bodies, which enables one ocean to influence another. In the Norwegian Sea, for example, cold water sinks and begins a flow back toward the Altantic as deep water. Although water that downwells initially carries nutrients to greater depths, it eventually upwells, taking those nutrients to the surface, where they once again provide plants—and eventually animals via the food chain—with material for growth. The constant circulation of the ocean waters beneath the "ocean" of atmosphere also leads to exchanges of matter, such as the incorporation of

dissolved carbon dioxide in seawater or the incorporation of water vapor in the atmosphere.

Knowing the interactions between atmosphere and oceans provides us with a warning. These highly interconnected systems can spread far beyond our local boundaries whatever we choose to dump into either air or water. Oceans affect human life in many ways, including the diet of inland peoples. The pollutants that spew from our factory smoke stacks or that run off farms into streams headed for the oceans may eventually end up on factory workers' or farmers' tables as part of their imported seafood.

Significant Points

1. The compound called water forms when two hydrogen atoms combine with an oxygen atom through the mutual sharing of electrons.
2. Salinity profiles reveal a change in salinity with depth. Wherever the profile reveals a rapid change in salinity, the halocline occurs.
3. The salinity of seawater differs with latitude and depth, location in open ocean or restricted water bodies, climate, stream runoff, and ice formation and melting.
4. Temperatures of the oceans also vary with latitude and depth. The zone of rapid temperature change with depth is called the thermocline.
5. That water is not frozen on the ocean bottom is the result

of three factors: it is salt water, which does not freeze at 0° C; ice floats; and it is under great pressure.

6. The T-S diagram is a graph used to plot the combined effects of salinity and temperature to establish density characteristics for particular water masses.
7. The combined effects of temperature and salinity generate vertical currents in the oceans. This vertical movement of ocean water driven by density differences is called thermohaline circulation.
8. Downwelling seawater is the product of convergent currents driven by prevailing winds; similarly, upwelling seawater is the product of divergent currents also driven by prevailing winds.

9. The general circulation of the atmosphere transfers some of its energy to the surface waters, pushing them along in relatively narrow bands called currents.

10. Generally, surface currents veer to the left in the Southern Hemisphere and to the right in the Northern, with some exceptions in high latitudes.

11. Surface currents are classified as warm or cold according to their relative temperatures. The latitudinal position of the currents is not as important in this classification as is the direction of travel.

12. Typically, warm surface currents are fast, narrow, and deep, and cold currents are slow, wide, and shallow. With the exception of high latitudes, warm currents run along the western sides of oceans; cold currents, along the eastern.

13. The Bernoulli theorem states that channeling a moving fluid increases its rate of flow.

14. Just as upwelling balances downwelling, return currents resupply surface areas that lose water and keep ocean water from piling up against and spilling onto coastal plains on the continents.

15. The return of water across an ocean or along the sides of gyres shows that ocean currents describe a closed system, whereby the system operates within the basinal confines established by seafloor spreading, moving continents, and the buildup of ice, which may have created temporary barriers to flow during the last glacial advance.

16. The ocean and the atmosphere do cross their mutual boundary through an exchange of gases. Three important gases that interchange are water vapor, oxygen, and carbon dioxide.

17. Water has the ability to store dissolved gas, but the temperature and pressure of water controls its storage capacity.

18. Cold water and water under high pressure can hold more dissolved CO_2 than low-pressure warm water.

19. Oceans cover 71% of Earth's surface, leading to the term *watery planet.*

20. The color of water is influenced by several factors: the angle of the sun's rays; the roughness of the surface; the amount of suspended matter; and, in shallow water, the colors of the bottom.

21. Since water is an efficient transmitter of energy, sound travels more than four times faster in water than it does in air. Variations in the rate at which sound travels in water occur because of differences in temperature, salinity, and pressure.

22. Sound can also be trapped in a "channel" by bending toward a zone of less resistance. Faster-moving sound has a tendency to bend toward lower-velocity zones, becoming trapped in the SOFAR channel.

23. Sound in the SOFAR channel can travel great distances because little energy is lost in this low-velocity zone.

24. Waters of the deep ocean and the continental shelves are classified into zones based on location, depth, and penetration of light.

25. The most landward zone encompasses the water over the continental shelf and is called the neritic.

26. The water beyond the continental shelf is classified as pelagic water.

Essential Terms

water 279	thermocline 282	downwelling 286	diatom 294
hydration 280	density 282	upwelling 286	neritic zone 296
seawater 280	sigma-t (or T-S) diagram 284	gyres 290	littoral zone 296
saturated solution 280	pycnocline 284	geostrophic flow 291	epipelagic zone 296
residence time 280	thermohaline circulation 285	Ekman Spiral 291	mesopelagic zone 296
salinity 280	water mass 285	Gulf Stream 292	hadal zone 296
salinity profile 281		eddy 292	bathyal zone 296
halocline 281		warm-core ring 292	

Review Questions

1. How do water molecules react with common salt?
2. What controls the salinity of pelagic surface waters?
3. Why do submarine captains look for the thermocline?
4. What prevents the bottom waters from freezing?
5. Why is Mediterranean water denser than the surface waters of the Atlantic?
6. What distinguishes the cold from the warm currents?
7. What is the role of ocean waters in the control of global carbon dioxide?
8. Why is the ocean blue? What controls ocean colors?
9. How and why do the oxygen and carbon dioxide profiles differ?

Challenges

1. Would you put hazardous materials in the oceans? Why?
2. If an oil spill occurs just north of the Yucatan Peninsula in Mexico, will the coastline of Texas be affected?
3. Using an atlas, find the southern boundary of sea ice in the Northern Hemisphere. Why does this boundary assume its particular shape? How does this boundary affect shipping?

4. What factors control the direction and intensity of the west wind drift?
5. How do the oceans interact with the atmosphere to influence climate?

12

MARINE AND NEAR-SHORE ENVIRONMENTS

Tidal marsh at Wood End Light, Cape Cod National Seashore.

Chapter Outline

INTRODUCTION

Wherever plants and animals reside in the seas, they both affect and are affected by the physical, chemical, and geologic systems of the oceans. In fact, these interrelationships are an example of the interdependence between the inanimate and animate worlds. Even humans, who are temporary visitors to the deep, but permanent inhabitants of coastal regions, can affect and are affected by the oceans.

An extreme example of humans affecting the oceans occurred during the 1991 Gulf War. The intentional spilling of more than three million barrels of Kuwaiti oil into the Persian Gulf brought havoc to the marine environment and demonstrated how catastrophic human actions can be (fig. 12.1). The sea birds, mammals, and fishes of the Gulf were subjected the world's greatest oil spill, one that eventually stretched along the entire western coastline and jeopardized the lives of sea life and humans from Kuwait to Qatar. By contrast, an event such as Hurricane Hugo's storm surge, with its devastation of the Charleston, South Carolina, area (chapter 1) demonstrates how the oceans affect humans.

These interactions, however, have not always been negative. Although many countries have increased their consumption of seafood, adding an important source of protein to the human diet, some seafood companies have acted to protect such animals as dolphins from becoming victims of commercial fishing methods. For many years, dolphins have been caught in drift seine netting used to surround and capture fish. An internationally signed agreement, called Marpol V, restricts this killing of dolphins. Within the past 10 years a worldwide public awareness has led to greater protection of the oceans and marine life largely through the efforts of those who have objectively observed, described, and modeled the marine environments.

Defining the interrelationships among marine and terrestrial environments and between marine environments and life is a holistic task that requires information obtained through the efforts of many scientists. Among these are different kinds of oceanographers whose work benefits both human and nonhuman life. Whereas ocean engineers and physical oceanographers work to control the coastal environment for millions of human inhabitants, marine biologists monitor the ecological effects of human activities and artificial structures. Because of a high population density along the coastlines of many countries, engineers must design seawalls, direct sediment dredging operations in ship channels, and protect and enhance the beauty of resort areas where seasonal influxes of tourists impinge upon the stability of local ecologies. (fig. 12.2).

Chemical oceanographers and marine geologists also analyze interactions between life and environment in the seas. The chemistry of seawater is, in part, controlled by organisms that remove dissolved substances to construct biological "hard parts" like shells. Marine sedimentation is also a product of life processes that produce those hard parts, which, upon the deaths of the organisms, settle to the ocean floor. That oceanographers devote so much effort to understanding these interrelationships among environments and life is an acknowledgment of the complexity in the oceans.

Marine environments exhibit a diversity of compositions, features, and processes. They extend from the shallow intrusions

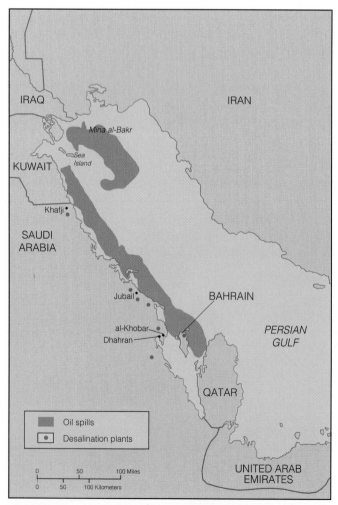

Figure 12.1 The extent of surface pollution from the deliberate release of crude oil during the 1991 Persian Gulf War.

Figure 12.2 Dredge clearing ship channel in the Bahamas.

of water in salt marshes and estuaries across the gently dipping continental shelf, over the continental slope and rise, onto the abyssal plains, and into the trenches (fig. 12.3). Many of these environments are hidden from human inspection unless inquisitive scientists expend money and energy in the risky endeavor of underwater exploration. Peering into the environments of the deep ocean requires oceanographers to use submersible ships such as the *Alvin,* which can carry three crew members, or robot submersibles such as *Jason.* Both types of vessels enable humans to see, for example, the hydrothermal deep-sea geysers that vent heated water and steam from a volcanically active oceanic ridge system (chapter 11). Describing the marine environments has also meant using *side-looking sonar,* instrumentation that bounces sound energy off the ocean floor from an oblique angle to get detailed images on a scale that cameras cannot provide in the darkness of the ocean bottom (fig. 12.4).

Salt Marshes: The Most Landward Extent of the Sea

One of the most short-lived marine environments is the **salt marsh,** a low-lying, constantly wet area of silt, sand, and organic material at the boundary between land and sea, found mostly in temperate regions. Poised in a delicate balance, the salt marsh is a fragile environment, one easily disrupted by human activities and by changes in *sea level,* which is the relative position of the surface of the ocean with respect to the land.

Because the ocean's surface reached its approximate current level only within the past 5,000 years, *present salt marshes are geologically young.* They will disappear—or move—whenever sea-level changes either inundate them completely or leave them as higher, drier land. The constantly wet condition of the marshes derives from the daily inflow of the tides through **tidal channels,** which act like streams in reverse during the tidal flow. As the tide moves into the marsh, it exceeds the capacity of the meandering channel system and floods the adjacent flats. Since the tides are a daily phenomenon and the marshes occur in temperate climates where evaporation cannot keep pace with wetting, *salt marshes do not become dry,* even in the warmest weather.

Salt marshes are more likely to occur on a tectonically passive continental margin characterized by gentle gradients rather than on an active one with steep gradients, where the movement of land from earthquake activity changes local sea level. Consequently, the Atlantic and Gulf of Mexico coasts, which have a more stable land-sea margin, have more area in salt marsh than does the Pacific coast of North America. However, valley marshes are not uncommon along the coasts of California, Oregon, Washington, and British Columbia, where streams enter the ocean through narrow valleys and their mouths come under the influence of the tides.

The marshes along the eastern seaboard of the United States have suffered from the intrusion of urbanization (fig. 12.5). Builders of housing and recreational complexes constantly encroach upon marsh areas. Along the coast of both Carolinas and Georgia, for example, recreational and retirement communities are growing rapidly, and a transient population of tourists places strains on the stability of the salt marsh environment.

Typically, a salt marsh is a geologically quiet environment, receiving the influx of the tides across its flats of fine particles and *cordgrass (Spartina)* and channeling the ebb and flow in tidal channels where the larger sand and silt sediments collect. The *tidal range* is the difference between the heights of the water

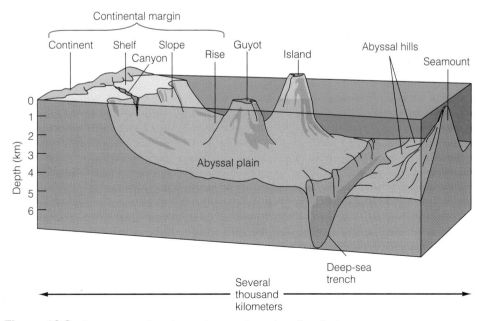

Figure 12.3 A summary of marine environments and seafloor features.

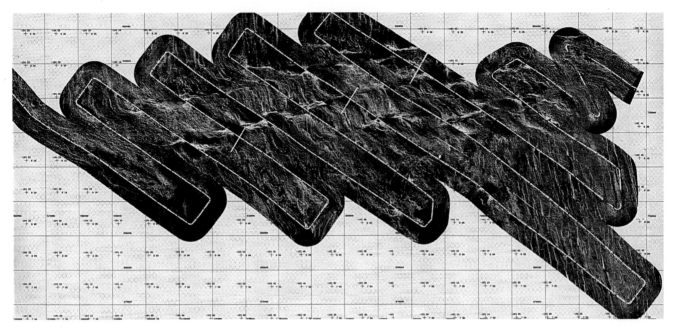

Figure 12.4 Side-looking (side-scan) sonar image obtained over the Siqueiros transform fault. (From D.J. Fornari, D.G. Gallo, M.H. Edwards, J.A. Madsen, M.R. Perfit, and A.N. Shor, "Structure and Topography of the Siqueiros Transform Fault System: Evidence for the Development of Intra-Transform Spreading Centers," *Marine Geophysical Researches,* 11:263–99, 1989. © Kluwer Academic Publishers.)

Figure 12.5 Housing development in the wetlands of Barnegat Bay, New Jersey.

Figure 12.6 Tidal marsh and channels near Jacksonville, Florida.

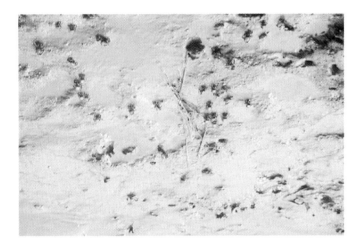

Figure 12.7 Crabs on a surface exposed at low tide.

surface during high and low tides. As the water rises, it flows up this and similar tidal channels and then spreads out over the salt marsh (fig. 12.6). The result of this inflow makes the mean depth of water greater during low tide, when the water lies only in the channels, than during high tide, when the water makes a shallow layer over a wide area. Additionally, water that fans out over the marsh slows as it encounters cordgrass and other plants which act as baffles to dampen the energy of the tidal current. The reduced flow rate causes suspended sediments and organic debris to settle through the relatively quiet water. This is one way in which life directly affects the shallow marine environment.

The daily ebb and flow of shallow water make a tidal marsh a nutrient-rich euphotic (well-lighted) shallow water zone that enhances photosynthesis and provides food for many organisms. Among the most numerous inhabitants of the salt marsh are crabs, which are relatives of lobsters and crayfish. With their five pairs of legs, these *crustaceans* manuever easily over a mud-supported surface of silts and sands that adheres to shoes and makes human passage difficult (fig. 12.7). The presence of crabs also indicates the abundance of *detrital food,* or fragments of other organisms, including parts of local plants that settle once the water ebbs. Crabs intensively burrow the marsh and, along with other burrowers like worms, *bioturbate,* or overturn, the

sediments. As the marsh develops, constant bioturbation eliminates most traces of its initial sedimentary *laminae,* the thin layers of small particles laid down by successive tides.

ESTUARIES

Many salt marshes border **estuaries,** partially open bodies of water that lie between the fresh water on the continents and the salt water of the oceans. Bays, lagoons, deltas, and river mouths all exhibit estuarine characteristics. Like the marshes, estuaries are also geologically ephemeral features, with the persistence of even the largest measured in thousands of years, a relatively short geological span. Estuaries disappear when river sediments fill them or changing sea level either isolates them or drowns them in the relatively deeper water. Seas have inundated continents throughout Earth's history and, during certain periods like the Devonian (395 to 345 million years ago) and Cretaceous (136 to 65 million years ago), have covered more than 50% of continental areas. Usually, these **epicontinental** ("on the continent") **seas** were shallow, with depths often no greater than 50 meters (164 ft). Both Hudson Bay and the North Sea are examples of recent **transgressions,** or intrusions, of seawater onto continental areas, and both have led to the development of estuaries along

Figure 12.8 A fiord, a glacially deepened estuary, along the coast of Alaska.

their boundaries. In fact, *all modern estuaries are related to the most recent rise in sea level,* which resulted from the melting of vast continental glaciers. After the retreat of the ice from its last advance over the Northern Hemisphere's landmasses, sea level rose rapidly, reaching an elevation close to present values between 6,000 and 5,000 years ago.

Essentially, the position and size of estuaries is dependent on sea level, local topography, and the configurations of river mouths or distributary channels. Whenever sea level rises in response to an influx of water from ice melt or in response to land subsidence, seawater encroaches on river mouths, drowning the lower part of the river, typically making a shallow estuary. Where streams empty into the lagoonal water behind barrier islands, shallow, elongate estuaries run parallel to the coast. If an alpine glacier retreats from a coastal valley cut below sea level, a deep estuary called a *fiord* develops (fig. 12.8). The deepest estuary along the eastern coastline of the continental United States ironically lies at the mouth of the York River, which is actually part of a larger, shallow estuarine system known as Chesapeake Bay, often described as a *drowned river system* (fig. 12.9).

A special kind of circulation occurs when the fresh waters discharging from rivers mix with the saltier seawater arriving through the open end of the estuary (fig. 12.10). This **estuarine circulation** results when the less-dense fresh water becomes *brackish,* or slightly salty, as it acquires some salts through mixing with the seawater. Both the fresh and the brackish water flow seaward over denser salt water, which flows landward along the estuary bottom. The landward flow traps suspended sediments that settle out of the discharging river mouths and enhances the filling of the estuary by blocking the exit of sediments. The flow also supplies plants with nutrients that flow off the land.

Estuarine circulation dominates the northeastern coastline of the United States and the southeastern coastline of Canada.

CONTINENTAL SHELVES

The shoreline so familiar to tourists and coastal residents is neither the edge of the continent nor the beginning of the oceanic basin. Instead, water overfills the basins and spills onto the continents, drowning a narrow area called the **continental shelf.** The inundated continental borders make up about 7.5% of the total ocean surface, but because they are not deep, shelf waters comprise only 0.15% of the total volume of seawater (fig. 12.11).

The structural tendency of continental shelves is a seaward *dip* or incline. This dip resulted from two processes. After the breakup of Pangaea, the trailing edges of continents, like the eastern coast of North America, began to subside in response to a decrease in tectonic activity along their trailing edges. As the continents moved away from the spreading center, they also moved away from the upward push of rising plumes of magma. In addition, many continental margins have been the sites of sedimentary deposition. The thicknesses of sediments deposited over the course of millions of years can be measured in kilometers, and their great weight has compacted the sediments and caused subsidence. On average the continental shelves dip 0.1°, but they exhibit a variety of bathymetric profiles.

For example, on some shelves lying in tropical waters, extensive organically produced calcium carbonate buildups called coral reefs have altered the original bathymetry. Nowhere is this more evident than along the northeastern coast of Australia, where the Great Barrier Reef has covered the basement rock with

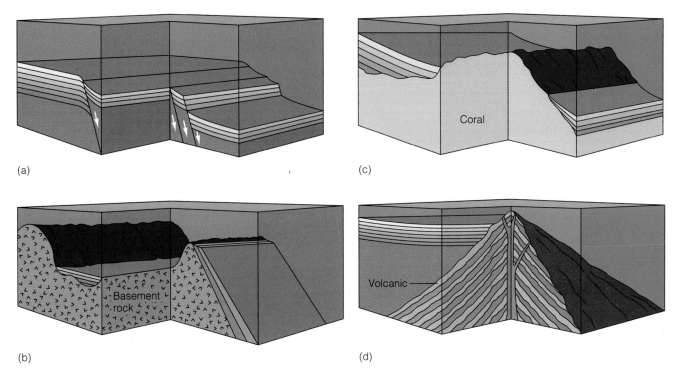

Figure 12.12 Possible origins for continental shelves: (*a*) faulting, (*b*) folding, or downwarping, (*c*) coral reefs, (*d*) volcanic. In most cases, some structure (reef, volcanic cone, etc.) acts to trap terrigenous sediment, leading to infilling of basins and a gently sloping surface.

Ganges-Brahmaputra system in India and the Huang Ho (Yellow) River in China, the foremost transporters of sediment, carry more sediment than the other seven rivers in the table combined. The ultimate destinations of the sediments from these two rivers are somewhat different. The Ganges-Brahmaputra flows onto the continental shelf at the Bay of Bengal, where the sediments migrate downslope and onto the floor of the Indian Ocean (fig. 12.13*a*). The Huang Ho empties into a marginal basin behind an island arc and two peninsulas, where there is no access to the deep ocean. If an access through these barriers were available for the sediments of the Huang Ho, they would spill into the trench system that borders the island arc (fig. 12.13*b*).

The sediments delivered to the continental shelves have been shaped by wind and water into both subaerial and subaqueous features. Where rivers deposit their silts and sands into the longshore transport system on relatively broad shelves, sediments can form a number of unconsolidated sedimentary features associated with coasts.

Sandy islands along the inner continental shelves develop as barriers to incoming wave energy from the open sea. These **barrier islands** have complex histories associated with local and worldwide sea-level fluctuations, sediment supply, and longshore transport (fig. 12.14). Sands move parallel to and along the seaward side of barrier islands in the longshore current, and they also move perpendicularly to the barrier island as wave energy increases or decreases (chapter 8). During periods of low-wave energy, sands migrate beachward. An increase in wave energy usually moves the sands seaward by gouging the unconsolidated particles on the beach face, causing a subsequent migration in offshore bars. Large storm waves can also move sands,

often those piled into dunes, across the narrow width of the barrier island to form fanlike deposits called *overwash deltas* in the adjacent lagoon (fig. 12.15).

Unconsolidated barrier islands can be altered by human activities. Whenever shipping lanes or boating inlets are needed to allow passage through the barrier system, dredging moves sands from their natural, if temporary, resting place, and changes the movement of water. With freedom to circulate into the lagoon behind the barrier island, water picks up and moves sediment, piling it elsewhere, often as an obstruction to boating and shipping and usually creating the need for further dredging. The construction of jetties, groins, and seawalls also changes the natural movement of unconsolidated material, and structures designed to protect a coastal community on a barrier island sometimes have an opposite, and detrimental, effect (see chapter 8).

One place where human activities have altered the movement of sands and changed the environment is Ocean City, Maryland, and Assateague Island. A jetty built along the inlet at the southern end of Ocean City has affected the movement of sands in the longshore current. Sands that used to replenish the northern end of Assateague Island now migrate seaward and farther to the south. As a consequence, the northern end of the island has been starved of its sediment supply and has changed its location and shape (fig. 12.16).

Barrier islands are also the sites of *dune complexes.* Because they are mounds of sands piled asymmetrically in arcing layers by the wind, beach dunes change easily with alterations in wind speed and direction, the movement of storm waves, and human trespassing. As *eolian* (wind-driven) deposits, they contain relatively uniform particle sizes, and they exhibit layering in cross

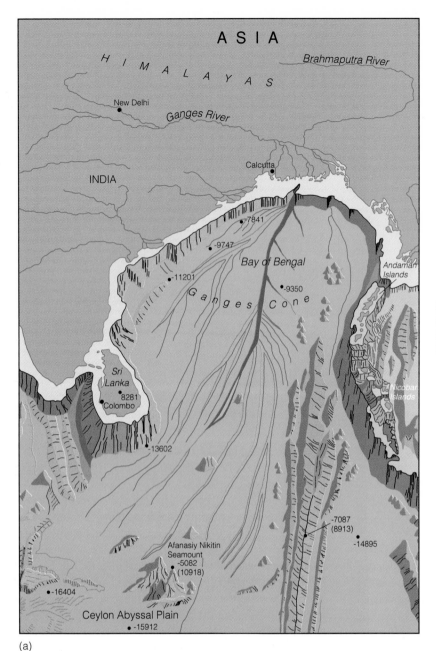

(a)

Figure 12.13 Depositional centers for the (a) Ganges-Brahmaputra system in the Bay of Bengal (Indian Ocean) and the (b) Huang Ho (Yellow) River system in the Yellow Sea.

section. Dunes can be relatively stabilized by vegetation. The initial plant colonizers, however, have only the most easily disturbed surfaces on which to grow, and even small animals can dislodge the roots. To protect the easily eroded dunes in some areas, "snow" fences have been erected, and signs prohibiting walking on the dunes or dune buggies have been posted (fig. 12.17). Plants can proliferate and eventually build soil fertile enough for woody bushes and, eventually, trees. The dislodging of the sands in a dune also occurs during severe weather. The storm waves of Hurricane Gloria in 1985, for example, removed 1.2 meters (4 ft) of sand from the beach at Ocean City, Maryland (fig. 12.18).

Marginal Seas

Because sea level has risen about 130 meters (425 ft) since the last glacial advance, the continental shelves house a number of marginal seas. Most, like the North Sea, are drowned continental shelf areas, but bordering the shelves in tectonically active areas is a number of marginal seas with complex origins. Some, such as the Sea of Japan, are thought to be the product of seafloor spreading.

Eustatically Formed Marginal Sea

The Atlantic shelf of Europe underlies shallow marginal seas, the Baltic and the North seas (fig. 12.19). Both of these seas are submerged parts of the continent, and both were overlain by ice

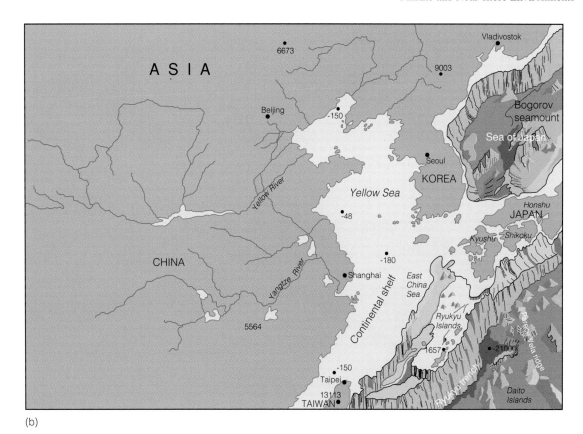

(b)

Figure 12.13 continued

during the last glacial advance. After the recession of ice, world-wide changes in sea level known as *eustatic changes* resulted in a transgression of shelves. The sediment of these seas is topped by a layer of glacial rock debris, and a few of the islands of Denmark, for example, such as Fyn (or Funen), are dominated by glacial till. The water between the North and Baltic seas is shallow enough for bridges to connect islands and mainland. Sweden and Denmark are now connected by such a bridge, and the Danish peninsular connection with Germany and the continent is joined by a bridge to the island of Fyn.

On the opposite side of the Atlantic lies a marginal water body called the Gulf of Maine (fig. 12.20). This shallow sea has been the site of a commercial fishery since colonial days, and it is still an important fishery. The southern end of the Gulf of Maine is bordered by Cape Cod, a glacially constructed feature that typifies in composition the kind of rock debris that lies over most of the floor of this water body.

Tectonically Formed Marginal Sea

The origin of the Sea of Japan is still a partial mystery for earth scientists. It differs from a eustatically formed marginal sea in several respects. First, its basement depth (the water depth plus the sedimentary layers) is considerably greater than pelagic ocean floor of similar age. Second, the water depth at places within the Sea of Japan is greater by an order of magnitude than marginal seas on the continental shelves. Third, it overlies a subducting section of the Pacific ocean floor. All of this indicates that

tectonic activity played a significant role in the formation of this marginal sea.

DELTAS

Deltas form at the mouths of rivers along coasts where average wave energy is rather small and where subsidence of sediments under their own weight is matched by the influx and buildup of new river deposits. As accumulations of sediments at river mouths, deltas build the land seaward in a process known as *progradation*. The buildup of sediments can form a variety of shapes. The name of this accumulation form derives from the Greek letter Δ because the most prominent delta in ancient western civilization, the Nile Delta, has an approximate triangular outline (fig. 12.21). Other shapes are possible, however, and the Mississippi Delta has been described as a birdfoot delta (fig. 8.16).

Sediments deposited in deltas can accumulate to make a number of related sedimentary deposits, including migrating channel sands, fan-shaped sand and silt deposits beside levees, mud layers within bays, bars of sands at stream mouths, silt deposits at the sea's edge, and muddy, subaqueous deposits.

Distributaries are streams that carry water from the alluvial valley through the delta and into the ocean. They are branches that receive water from the main channel, and they extend through the delta as channels in the very sediment they deposit. Distributaries can be numerous and ephemeral because rivers

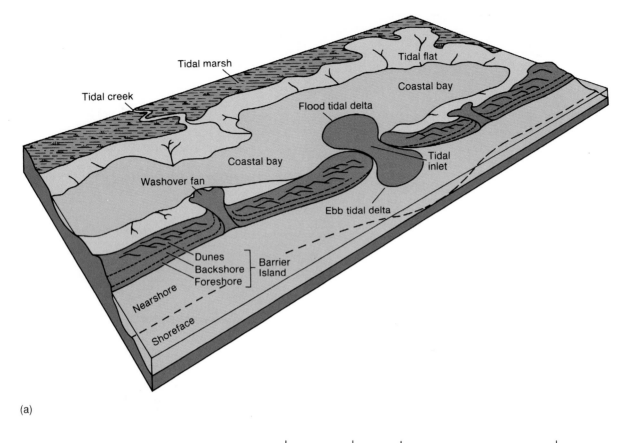

(a)

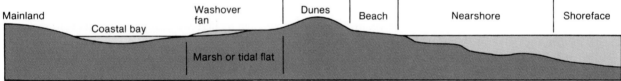

(b)

Figure 12.14   (a) Aerial view of a typical barrier island and (b) cross section of a typical barrier island. (Source: From Blatt/Middleton/Murray, *Origin of Sedimentary Rocks,* 2d ed., © 1980, p. 660. Adapted by permission of Prentice-Hall, Englewood Cliffs, New Jersey.)

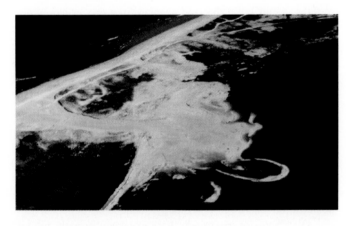

Figure 12.15   An overwash delta, or fan, at North Beach, Cape Cod, Massachusetts.

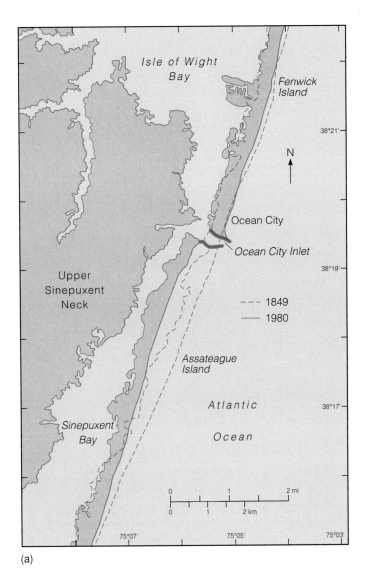

(a)

(b)

Figure 12.16 (*a*) Map and (*b*) photograph show offset of north end of Assateague Island as a result of the construction of jetties at the end of neighboring Fenwick Island. The southward-moving longshore currents are starved of sediment as they flow along Assateague, and, consequently, erosion has been considerable. Total offset has been more than 500 meters (1,640 ft).

Figure 12.17 Sand dunes along the New Jersey shore have been posted and planted to retard erosion.

Figure 12.18 Sand moved by storm waves from Hurricane Gloria in 1985 filled the parking lot at Ocean City, Maryland.

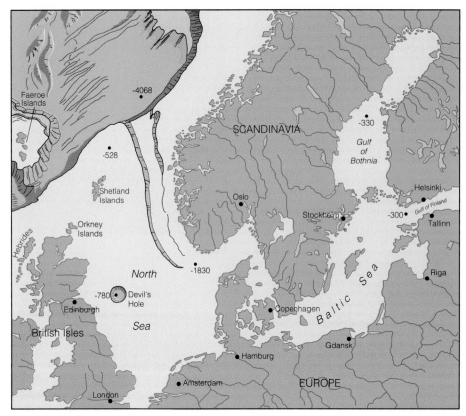

Figure 12.19 The floors of the Baltic and North seas.

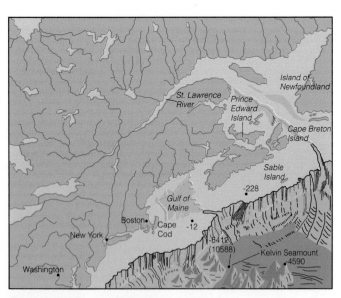

Figure 12.20 The Gulf of Maine, bordered on the south by Cape Cod.

will seek the most direct and least resistant path to the ocean, the ultimate base level. Whenever the channels receive an excess of water, they flood, breaching the natural levees and depositing sandy, fan-shaped *splays* of sediment more or less perpendicular to the channel (fig. 12.22).

As the river lays down its deposits during floods, the distributaries encounter sediments that block and alter their paths. The changing pattern of these streams leads to a varied development along the delta's seaward edge. In the case of the Mississippi Delta, seven different "lobes" of the delta have prograded the coastline over the continental shelf during the last 7,000 years. The typical pattern of development takes five steps.

In the first step, a distributary bifurcates. The two newly formed channels establish an approximate parabolic shape open to the sea, creating an *interdistributary bay* with marine sedimentation characterized by muds and burrowing organisms. The second step requires a flood that breaches the channel boundaries or natural levees, allowing silts and sands to migrate into the bay between the two channels. A third step sees the buildup of terrigenous sedimentation in the interdistributary bay, the water body between the distributary channels. An eventual shift in the channel chokes the flow of silt and sand in the fourth step. Fifth in the cycle of steps, is subsidence of the bay, producing a marine environment.

Figure 12.21 Satellite photo of the delta of the Nile River. Dark color indicates vegetation.

A delta progrades differentially seaward. While abandoned areas subside under their own weight, other accumulations build toward the shelf edge. At the leading edges of the delta lie deposits of sands called *distributary mouth bars.* Seaward of the bars silts form a convex upward feature called a delta front. At depth and farther seaward lies a fine-grained, muddy accumulation known as a **prodelta** (fig. 12.23).

A delta is a rapidly changing environment that exhibits both marine and terrestrial features. It is one of the most ephemeral of earth features, and its form and development depend on the flow of fresh water through distributary channels, changes in sea level, the amount and type of sediment carried by the distributaries, and the intensity of wave energy. The largest of deltas can stretch across the shelves to the shelf break.

CORAL REEFS

A predominantly biogenic (derived from life) subenvironment on the continental shelves and on stable platforms is the coral reef. Reefs form slowly by the layered accumulations of calcium carbonate, mostly generated by sack-shaped animals called coral polyps. Because they live symbiotically with algae known as zooxanthellae, coral reefs form in warm, shallow, clear water with a slightly elevated pH. Mixed into this environment are algae, numerous fish, sponges, echinoderms, molluscs, and arthropods. Reefs occur in all sizes and in a number of shapes, but all exhibit certain common features, including *reef flat, reef crest,* and *reef slope* (fig. 12.24). The slope is a

Figure 12.22 Principal features developed along a distributary channel of a delta.

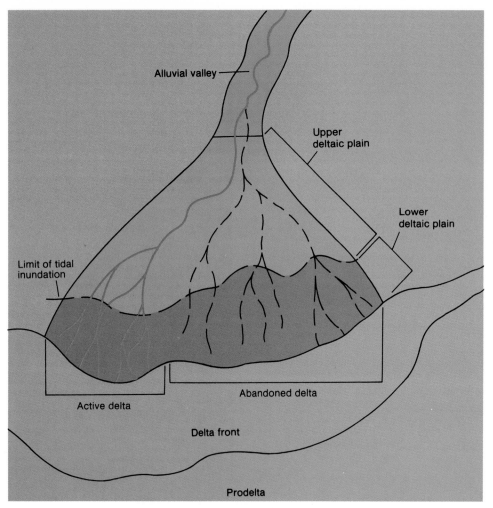

Figure 12.23 Diagrammatic view of a delta showing its principal features. (Source: J.M. Coleman and L.B. Wright, *Analysis of Major River Systems, Their Deltas, Procedures, and Rationale, with Two Examples,* Coastal Studies Institute Technical Report 95, Louisiana State University, 1971.)

seaward precipice leading to deep water that absorbs most of the incoming wave energy. In violent storms, however, the reef crest and the reef flat behind it also feel the impact of wave energy. Typically, water over the crest and flat is well lighted and shallow.

Despite their seeming diversity, reefs can be grouped into three fundamental types: *fringing,* or those linked closely to the land surface with little or no intervening lagoon; *barrier reefs* with large lagoons separating the reef structure from its associated landmass; and *atolls,* or circular reef structures built around a central lagoon with no associated landmass visible at the surface (fig. 12.25). In the mid-nineteenth century, the naturalist Charles Darwin postulated that these three reef types often represent an evolutionary sequence associated with subsidence of the ocean basin floor (see also chapter 10). He reasoned that as the associated landmass (usually an island) was gradually carried downward with the ocean floor, the reef continued to be built upward by corals and other organisms, its position being maintained at or just below sea level. The reef would begin its growth at the shoreline, attached to the land. Vertical growth of the reef combined with subsidence of the island would result in a larger lagoon and smaller island. Eventually, the island would be com-

pletely submerged leaving an atoll surrounding an empty central lagoon.

The entire reef complex is home to many species, some of which are very colorful. Unseen by most human visitors to the reef are the *cryptic,* or hidden, *fauna,* possibly many as yet unnamed species. The reef is a relatively fragile system, and changes in temperature, amount of terrigenous sediment, and pollution can diminish its growth or cause it to die, leaving deposits of sterile carbonate rock and sand.

Ancient reefs, such as the one in figure 12.26, can serve as clues to the marine settings of the geologic past. Other, larger reefs that yield clues to the past lie in Texas and in the Canadian Rockies. Their presence indicates a different, warmer climate than that found in either setting today.

There are many carbonate environments in the sea. Among the most notable are the Great Barrier Reef of Australia and the islands of the Bahamas. The Great Barrier Reef is the largest structure ever made by life (fig. 12.27). It is more than 1,000 kilometers (620 mi) long and consists of numerous individual carbonate reefs in different stages of development. In the warm waters of the Gulf Stream, reefs have developed on a stable platform called the Bahama Banks. The abundant reef material in the

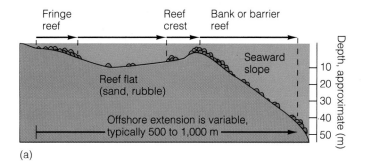

(a)

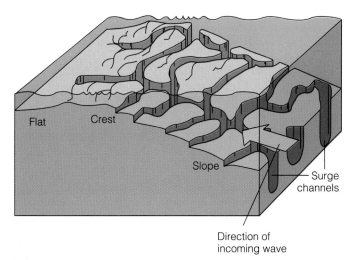

Figure 12.24 (a) Cross section through a typical coral reef. (b) Development of surge channel through the reef.

(b)

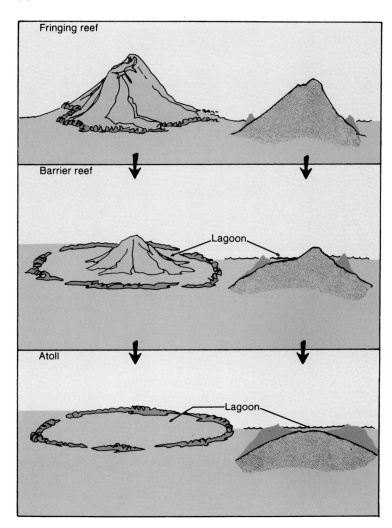

Figure 12.25 The origin of three common types of reefs as first postulated by Charles Darwin. The progression from fringing reef to atoll is evolutionary on a subsiding ocean floor.

Figure 12.26 Dry season aerial view of fossilized coral reef in The Kimberleys, Windjana Gorge National Park, Western Australia.

Figure 12.27 The location of the Great Barrier Reef along the northeastern coast of Australia.

warm, shallow waters has accumulated in both unconsolidated and consolidated forms, making individual islands called *cays*. The cays are essentially collections of carbonate sediments or portions of older reef rock that now lie above sea level. Some have extensive deposits of concentrically developed, carbonate sand grains called *oolites*. Figure 12.28 reveals the rough surfaces of eroded older reef material that make up part of a small island in the Bahamas.

THE CONTINENTAL SLOPES

Continental slopes are the sites of sporadic, but highly effective, geologic work. Essentially areas bypassed by deposition, the slopes house spectacularly large canyons and, at their bases, piles of terrigenous sediment (fig. 12.29).

Turbidity Flows and Submarine Canyons

Turbidity flows develop along continental slopes whenever dense, sediment-filled water migrates under gravity downslope within another, larger body of water. The flow is characterized by a chaotic movement of water and sediment in a dense "plume" with a leading edge called a *head*. Behind this head trails the rest of the flow as an identifiable body with a thinning *tail* of turbulent water and sediment. The speed of the flow, which has been estimated to reach more than 20 kilometers (15.5 mi) per hour, generally depends on the height of the head.

As massive volumes of sediment move occasionally off the continental shelf, they slide over and erode the continental slope, which has an average 6% gradient, dropping 100 meters for every kilometer seaward (325 ft/0.62 mi). Channeled turbidity flows appear to be responsible for cutting some **submarine canyons** into the continental slope and for making large and extensive deposits at the base of the slope.

Figure 12.28 Eroded reef rock on a Bahamian cay.

The discovery of canyons in the continental slope led initially to speculation that the canyons were primarily the product of gouging by migrating river sediment. Further analysis of the distribution of these features, however, revealed that many canyons seem to have no associated river.

Apparently, sediments emplaced by ancient rivers migrate to the shelf break, where, under the stress of earthquake temblors or gravity, they move rapidly downward in turbidity flows toward the deep ocean floor. The arrival of such sediments at the shelf break is hardly a mystery. The Mississippi Delta has prograded to the outer shelf. At times of lower sea level, all rivers wend their way to reach their base level farther seaward across exposed shelves.

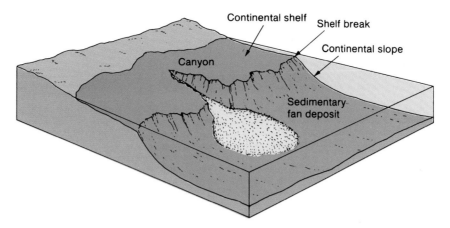

Figure 12.29 Canyon carved in the continental slope with a fanlike accumulation of sediment at its mouth.

Deep-Sea Fans, Turbidites, and Mounds

As turbidity currents reach the bottom of continental slopes, they lose energy, and their sediments settle into a *fan,* or *cone,* lying against the base of the slope and extending onto the deep ocean floor. Ensuing flows of sediment can cut distributary channels and build levees in this accumulation of sediment. Numerous adjacent and connected fans make a sediment accumulation with a gentle gradient called the **continental rise.** The finer sediments of the turbidity currents settle out of the flow to become *distal* (for "distant") *turbidites* tens to hundreds of kilometers from the source of the turbidity current. Coarser grains make up the *proximal* (for "near") *turbidites,* which settle out of the turbidity current closer to the point of flow origin.

In addition to turbidites, moundlike structures dot environments from the continental shelves to the deep ocean floor. Some of these result from the growth of corals that do not require warm water, sunlight, and the symbiotic zooxanthellae. These coral mounds lie as deep as 5 and 6 kilometers (3.1 and 3.7 mi). Other mounds, called *mud mounds,* apparently develop when microbes form bafflelike structures that block the transport of fine sediments. Mud mounds can rise more than 100 meters (328 ft) above the surrounding seafloor.

THE DEEP OCEAN FLOOR

It was only with great difficulty and great expense that humans learned anything about the deep ocean floor. Early attempts to discover just what lies on the ocean bottom included dragging metal cages and boxes over the floor. As technology advanced, crews in submersibles dived to the ocean bottom, where, with the help of artificial lighting and robotic arms, they were able to photograph and take samples of the sediments, rocks, and animals of the deep.

The oceanic deep is the site of the **abyssal plains,** vast areas with relatively little change in elevation (relief), except for volcanic seamounts (chapter 10), islands associated with submarine plateaus, and upwardly warped features of several origins called *diapirs.* In the Pacific, the abyssal floor is dotted by many seamounts and island mountains, such as the Hawaiian Island chain, the Emperor Seamounts, and the Tuamotu Archipelago. In the Atlantic, the relatively flat Hatteras Abyssal Plain along the tectonic trailing edge of North America is the recipient of distal turbidites and hard parts of sea life.

The sediments of the deep, which overlie the predominantly volcanic oceanic crust, include the following:

1. the siliceous and calcareous hard parts (bone and shell) of mammals, fish, and plants (such as diatoms),
2. cosmic particles,
3. fine eolian sediments,
4. terrigenous sediments from turbidity currents and from melting ice rafts calved from glaciers, and
5. **hydrogenous sediments,** which are metal-rich accumulations in the form of encrustations and nodules.

In the cold environment of the abyssal plains are holuthurians (sea cucumbers), clams, worms, and fish, all adapted to the darkness and the great pressure. In contrast with the fisheries of the epipelagic, the abyssal environment apparently has smaller food sources. One of those sources is a detrital rain of organic material and nutrients that falls through the water column. Another source of nutrients ample enough to support a community of varied life forms is the hydrocarbon seeps that have been found on the floor of the Gulf of Mexico. The community that prospers in the vicinity of the seeps is dependent on *chemosynthesizers,* microscopic organisms that can manufacture food without the aid of sunlight. These organisms, initially found associated with the volcanic, hydrothermal vents of the oceanic ridges, require sulfur or methane to start the chemosynthesis. Whereas the volcanic vents supply sulfur dissolved from the volcanic rocks through which the heated water passes, the hydrocarbon seeps appear to furnish the raw materials for food synthesis through methane leaks.

Typically, these chemosynthetic-dependent communities are populated by clams, crabs, and tube worms. Some deep-sea fishes, including grenadiers, or rattails, swim in the vicinity of these ecological havens. Figure 12.30 shows one of these chemosynthetic-based communities photographed by the crew of the *Alvin.*

Figure 12.30 Tube worms and other organisms dependent on chemosynthesis in the deep-sea environment cluster around a thermal vent.

(a)

(b)

Figure 12.31 Microscopic components of deep-sea oozes enlarged many times: (*a*) calcareous foraminifera and (*b*) siliceous radiolaria.

The deep ocean floor is covered by accumulations of fine-grained, organically produced muds derived from various microorganisms. Oceanographers call these muds **oozes** and identify them on the basis of their composition. *Calcareous oozes* are usually white or dirty white muds dominated by accumulations of calcium carbonate in the form of whole and broken hard parts from many organisms (fig. 12.31). Some of the organisms that contribute to oozes are microscopic unicellular foraminifera, small molluscs called pteropods, and algae known as Coccolithophoridae. *Siliceous oozes* are muds dominated by hard parts of opaline silica. Diatoms (chapter 11) and unicellular radiolarians are important sources of these deposits. Extensive diatom oozes occur beneath the cold waters of high latitudes (fig. 12.32).

Beneath the gyres of the Pacific Ocean lie the *red clays.* These fine-rained sediments are terrestrial aluminosilicates that have settled out of the water column after transport in currents. Some clays may also result from the chemical alteration of the basaltic seafloor.

TRENCHES

The deepest points from the surface to the ocean bottom lie at the bottom of trenches, those V-shaped sites of subduction. Along the western coastline of South America, the trench system reach-

es depths in excess of 6 kilometers (3.7 mi). At 11 kilometers (6.8 mi) below sea level, the Mariana Trench in the western Pacific is the world's deepest feature. Even this hostile environment of high pressure and cold water is home to holuthurians and rattails. Since exploration of trenches is very dangerous, additional knowledge about the life in trench waters is scarce.

EPIPELAGIC WATERS

The temperatures and salinities of the epipelagic environment vary widely (chapter 11). Their variations act as controls on some organisms, though others seem to be unaffected by such changes. Those organisms that can tolerate a wide range of salinities are called *euryhaline;* those capable of tolerating wide temperature ranges are *eurythermal.* Animals restricted by narrow salinity and temperature ranges are *stenohaline* and *stenothermal,* respectively.

The penetration by sunlight of the upper 200 meters (656 ft) of open ocean water enhances the proliferation of life wherever nutrients are adequate. The food chain begins with tiny photosynthesizers like the Coccolithophoridae and ends, with the ex-

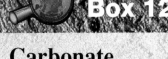
Box 12.1 Investigating the Environment

Carbonate Sediments on the Deep Ocean Floor

Numerous coral reefs, oolitic sands, and sea shells lying in warm, shallow water may give the impression that, with the exception of small coral mounds, carbonate buildups generally do not occur in deep, cold water. On the *Glomar Challenger* during the cruise labeled Leg XVI in the equatorial Pacific Ocean, however, scientists mapped a large bulge of carbonate sediments on the deep-sea floor. The bulge exceeds a thickness of 500 meters (1,640 ft) in its elongate central portion just south of the Clipperton fracture zone (fig. 12.A). The presence of the bulge in cold water enabled scientists to develop a model that combined a number of oceanic and biological processes.

The carbonate bulge in the equatorial and north equatorial Pacific is a knot that ties together the threads of upwelling along the South America, Peru, Equatorial, and Cromwell Currents, the food chain, detrital sedimentation, and ocean chemistry. The bulge reflects a number of interrelationships.

1. Upwelling water enriches surface waters with nutrients.
2. Surface currents spread the nutrient-rich water generally westward.
3. Organisms extract calcium, carbon, and oxygen from the water to build hard parts of calcium carbonate.
4. Floating organisms die and sink, with their journey to the bottom aided when larger organisms ingest then excrete them in heavier fecal pellet form.
5. Lower pH and a higher concentration of dissolved CO_2 in the bottom waters lead to the dissolution of calcium carbonate unless more matter than can be dissolved falls in a detrital rain.
6. The Pacific plate carries sediments from the East Pacific Rise into deeper water away from the ridge crest.

Upwelling water beneath the Peru (Humboldt) Current supplies nutrients to the microscopic organisms of surface waters (chapter 11). The Cromwell Current is also a nutrient-rich current, and a divergence of surface water along its northern boundary allows some of the countercurrent water to upwell. Thus, the epipelagic waters just north of the equator can support an abundance of life. The fertility of these waters and the abundance of life diminishes occasionally when a warming trend called El Niño caps the upwelling with a layer of warm water. For millions of years, however, these nutrient-rich waters have supported organisms that produce hard parts of calcium carbonate.

Because the planktonic organisms are light and have disproportionally large surface areas, they tend to sink slowly unless they are ingested and then excreted by larger organisms. Incorporation in fecal pellets enhances the fall through the water column. As the hard parts and their associated organic matter sink, they undergo attack by bacteria and by more acidic water, which reduces them to their molecular and elemental states. The hard parts collect on the bottom, where they undergo further dissolution or become part of the seafloor sediment.

Bottom water is not conducive to the preservation of calcium carbonate because of a lower pH that is controlled by the presence of dissolved carbon dioxide. At a variable depth in the oceans, the rate of dissolution of calcium carbonate exceeds its depositional rate. This depth of excessive dissolution is called the calcite compensation depth (CCD). In the vicinity of the carbonate bulge, according to oceanographers Tjeerd H. van Andel and G. Ross Heath, the CCD lies between 4,700 and 4,800 meters (15,416 and 15,744 ft). Below this depth, there is no accumulation of calcium carbonate, and the bulge becomes siliceous oozes made primarily of radiolarians. Because the detrital rain of carbonate sediments falls on a moving and subsiding plate, those sediments that accumulate west of the East Pacific Rise are destined to ride into deeper water toward the north and west (chapter 10). With subsidence, the carbonate passes below the CCD.

Apply Your Knowledge

1. How would the CCD of the equatorial and north equatorial Pacific be affected by a decrease in surface productivity?
2. What would a cross section of seafloor sediments look like along a line from the Marquesas Islands past the eastern Hawaiian Islands to the Aleutian Islands?

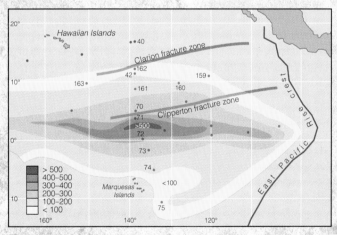

Figure 12.A   Sediment thickness (in meters) in a portion of the eastern Pacific. The numbered dots represent core sample locations.

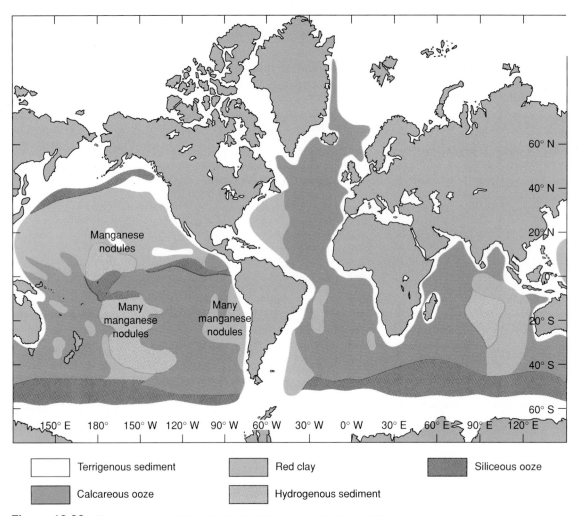

Figure 12.32 The distribution of the principal sediment types in the world's oceans.

clusion of humans, in fast-swimming fishes and mammals. Tunas, billfishes, various sharks, and whales, all euryhaline animals, move with great speed throughout the epipelagic zone, feeding as they go from surface waters into the upper layers of the mesopelagic.

Generally, the large ocean gyres are not as nutrient-rich as coastal waters and areas of divergent currents. Consequently, the epipelagic zone of these gyres is not known as a fishery, and the gyres are called the ocean's deserts. The gyres are not, however, devoid of life, just as continental deserts are not without a variety of life forms.

The epipelagic zone of upwelling waters, however, serves as an abundant fishery. Fast-swimming tunas provide a good example of organisms at the top of the food chain in this fishery. Tunas have fusiform (bullet-shaped) bodies that enable them to cut through the water with hydrodynamic ease. Almost every part of their bodies enhances rapid movement, including tiny finlets that decrease turbulence (fig. 12.33). Tunas also exhibit warm-bloodedness through a network of arteries and veins that act as heat exchangers.

Of the large mammals that also inhabit the epipelagic zone, whales survive by following one of two feeding strategies. Some whales are plankton feeders, using baleen strainers to capture small organisms within the mouth. A large humpback whale that opens its mouth may spread its jaws more than 4 meters (13.1 ft) to take in both food and water. By ejecting the water through its baleen, the whale captures its food. Toothed whales like the sperm whale follow a different strategy for food and seek out prey as deep as the mesopelagic. Other toothed whales, such as the killer whale, frequently attack smaller sea mammals, such as seals, that swim near the surface.

MESOPELAGIC WATERS

The twilight of mesopelagic waters fades to darkness with depth. Some species of fish have adapted to the absence of light in this environment by acquiring organs that generate chemical light or *bioluminescence*. The light apparently serves two purposes: reproduction and food acquisition. Light spots on some fish serve as lures to attract both mates and prey (fig. 12.34).

Generally, the organisms of the mesopelagic are relatively weak swimmers in comparison with those of the epipelagic. Life in the mesopelagic also rises and falls in daily and monthly cycles, making a mass of protein available for deep-diving epipelagics. Sonar waves scatter when they encounter this layer of life that is now known as the *deep scattering layer*.

Figure 12.33 Tuna is an important food source in the world's oceans. The tuna has a body shape designed for constant cruising in search of prey. It reaches a maximum length of approximately 1.25 meters (5.13 ft).

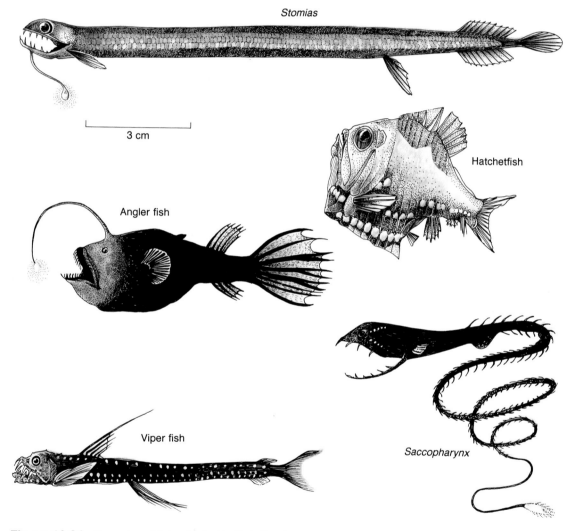

Stomias

3 cm

Hatchetfish

Angler fish

Saccopharynx

Viper fish

Figure 12.34 Examples of fish adapted to life in the deep sea.

Box 12.2 Investigating the Environment

Chatham: Property as Stable as the Shifting Beach Sands

The charm of Cape Cod in Massachusetts is one that few people can miss. The cape was shaped by glaciers and their meltwater, ocean currents, waves, storms, and human activities. The result of natural processes is a stretch of beaches, sandy barriers, sand dunes, and high cliffs of unconsolidated sediments. Human activity has built communities, such as Chatham on the southeastern end of the cape.

Chatham is a model Cape Cad community, with an architecture that has been copied elsewhere in North America. Its Cape Cod–style homes were built on a moderately high, sandy beach in a lagoon behind a narrow barrier island. On January 2, 1987, storm waves from a Nor'easter (a blizzard arriving from the northeast) broke through the offshore bar in the Nauset Beach area, allowing wave energy to strike the shoreline in the residential area. In a short span of a few months, more than 19 meters (65 ft) of the beach area, including homes perched along the low cliff, were washed away.

Figure 12.B shows the southern end of Cape Cod with Chatham on the upper left. Since the breakthrough occurred, sediments have moved to cut off Pleasant Bay from the water north of the break. The loss of their homes has motivated many to seek legal recourse because Massachusetts does not allow artificial structures, such as rock barriers, along the Chatham area. In a desperate attempt to protect homes, some citizens have placed boulders at the base of the sandy cliff. This attempt to save property is, in reality, futile. The boulders will move whenever the sands beneath them are washed away (fig. 12.C).

Cape Cod will always be subject to the movement of its unconsolidated sediments. The Marconi Wireless Station, a set of four tall radio towers that received the SOS call from the *Titanic*, is a historical marker now. The high cliff on which it sat, north of Chatham, has been eroded back beyond two of the tower base pads, a distance of more than 50 meters (164 ft).

Figure 12.B   Aerial view of southern Cape Cod, with Chatham in the upper left.

Figure 12.C   Erosion at Chatham.

Apply Your Knowledge

1. How does a barrier island protect a mainland?

2. Where in North America are communities in danger of repeating Chatham's fate?

3. Are coastal communities built on bedrock subject to the same erosive fate as Chatham?

*F*urther Thoughts: The Human Influence on Marine and Near-shore Environments

Our knowledge about marine and near-shore environments is uneven. We know more about near-shore environments than about marine environments because of accessibility. Many seaside dwellers and transient vacationers are familiar with some features and life in the near-shore environments. That familiarization has meant the intrusion of humans into the natural processes of those environments, often with adverse effects on life that depends on the sea. Our pollution of fragile environments like salt marshes and estuaries has spread into marginal seas and into the epipelagic realm. For several decades prior to the 1990s, for example, Europeans allowed an unchecked pollution of the Elbe River, which annually emptied thousands of tons of pollutants into the North Sea. Even for this relatively shallow environment, however, no one has completely assessed the effects of the Elbe's polluted water. But common sense tells us that pumping heavy metals into near-shore and marginal sea environments is not without adverse influence.

Most of the fish taken for human consumption come from waters that lie no farther than 350 kilometers (217 mi) from coastlines. One particular disaster associated with methyl mercury pollution in the 1950s emphasized the changes humans can force on the near-shore environments. At that time Japanese, who consumed fish and shellfish from Minamata Bay, ingested enough mercury from a local industrial complex to cause birth defects that included both mental retardation and deformities.

In the Caribbean and the Gulf of Mexico, two pollutants attack the environment: petroleum and sewage. The heavy tourist traffic in and around the Caribbean places burdens on sewage systems, and many shellfish ultimately incorporate pathogens dangerous to humans in their tissues. Petroleum enters the Caribbean and the Gulf of Mexico from seeps, spills, and pipeline leaks. In 1978, a blowout beneath an offshore drilling platform owned by a Mexican oil company released approximately 500,000 tons of petroleum into the sea.

There seems to be no way around the conclusion that we are changing both marine and near-shore environments, but the oceans are large and may be capable of absorbing and cycling many of our pollutants. Because they are complex, no one fully understands all the interrelationships within the oceans and between the oceans and other earth systems. Until we gather more information, we can only hope that the capacity of oceans to absorb and cycle will prevent us from detrimentally affecting the planet's ability to support human life.

*S*ignificant Points

1. Life is both affected by and affects marine and near-shore environments.
2. The chemistry of seawater is, in part, controlled by organisms.
3. Marine sedimentation is, in part, the product of life processes that produce hard parts, such as shells.
4. The salt marsh is an ephemeral environment dependent for its survival on sea level. Salt marshes do not become dry, even in the warmest weather.
5. The mean depth of water in a salt marsh is greater during low tide, when the water lies in the tidal channels, than it is at high tide, when water spreads over the entire marsh.
6. Constant bioturbation disrupts the sediments of a salt marsh.
7. Estuaries include bays, lagoons, deltas, and river mouths in which salt water meets fresh water.
8. All modern estuaries are related to the most recent rise in sea level.
9. Both fresh and brackish water flow seaward over denser salt water, which flows landward along the estuary bottom in estuarine circulation.
10. The continental shelves serve as depositional centers.
11. Barrier islands have complex histories related to sea-level rise, modern currents, sediment supply, storms, and urbanization.
12. Deltas are built up by sediment carried in distributaries.
13. Coral reefs are warm-water complexes dominated by the calcium carbonate deposits of coral polyps. Numerous kinds of animals are present in the reef environment.
14. Turbidity currents carry sediment downslope toward the abyssal plains. In the process, they erode the continental slope, cutting canyons.
15. Red clays dominate the sediments beneath the ocean gyres. The ocean floor also has coverings of siliceous and calcareous oozes.
16. Trenches are the lowest points on Earth's surface. The Mariana Trench is Earth's deepest point.

*E*ssential Terms

salt marsh 303	estuarine circulation 306	barrier island 309	abyssal plain 319
tidal channel 303	continental shelf 306	prodelta 315	hydrogenous sediment 319
estuary 305	dip 306	turbidity flow 318	ooze 320
epicontinental sea 305	shelf break 307	submarine canyon 318	
transgression 305	terrigenous 307	continental rise 319	

Review Questions

1. What instruments enable oceanographers to "see" the features of the ocean floor?
2. Does a salt marsh ever dry out?
3. Why does water flow in a tidal channel?
4. What is a tidal range?
5. How does a fiord form?
6. What characterizes estuarine circulation?
7. What is a miogeocline?
8. What is an overwash delta?
9. How did the Sea of Japan form?
10. How did the North Sea form?
11. What is the difference between the shapes of the Nile and Mississippi Deltas?
12. How does an interdistributary bay form?
13. What is the composition of the prodelta? Where does it form?
14. What are zooxanthellae?
15. Do coral structures form in deep water?
16. What is a distal turbidite?
17. How do chemosynthesizers differ from photosynthesizers?
18. What is the composition of an ooze?

Challenges

1. Which North American coastlines favor the development of barrier islands? Why?
2. Why are dunes fragile? How do human activities adversely alter dunes?
3. What road construction problems associated with the nature of the feature do engineers face in deltaic environments?
4. What major canyons lie along the continental slopes of North America?
5. Select a coastal community. What pollution problems must be solved? What construction or use along the coast would disturb the naturally formed features?

PART

6

APPLYING THE EARTH SCIENCES

According to a 1938 speech by Karl Taylor Compton, "Science really creates wealth and opportunity which did not exist before." Knowledge of the composition and processes of our planet has led to the discoveries of many useful materials. These materials now make up an annual worldwide value worth hundreds of billions of dollars.

AT THE END OF PART 6, YOU WILL BE ABLE TO

1. distinguish between renewable and nonrenewable resources;
2. identify the difficulties associated with discovering new sources of raw materials
3. classify soils;
4. identify the types and sources of metallic and nonmetallic resources;
5. identify the types and sources of fossil fuel resources.

Job 28: 1–6. The New Oxford Annotated Bible
(Revised Standard Version)

Surely there is a mine for silver
and a place for gold which they refine.
Iron is taken out of the earth,
and copper is smelted from the ore.
Men put an end to darkness,
and search out to the farthest bound
the ore in gloom and deep darkness.

13

EARTH RESOURCES

Open pit copper mine at Bingham Canyon, Utah.

Chapter Outline

INTRODUCTION

The culmination of earth studies is knowledge about the planet's use and care. In this sense, all of us must read the Earth's package instructions, or the lessons drawn from the work of earth scientists. Earth is our home, and living in any home means using it repeatedly. Generations of humans must reuse the planet, and that use will require the extraction of minerals, some of which are abundant and some of which are scarce.

The planet has undergone many natural changes that have radically altered the environment, constructing, breaking up, and moving continents and emplacing within its crust certain substances that humans find either necessary or desirable. Life, too, has added to the accumulation of valuable substances in Earth's crust, including natural oils, energy-rich coals, and enriched soils. These and other accumulations give humans the materials necessary for the use of our planet. We can continue to live on this world because we have the resources to do so. The word *resource,* appropriately, comes from the Old French for "to arise anew."

When humans began using their home, they found the refrigerator well stocked, ample towels and linen, and more than enough materials for construction, decoration, and energy. Because of the careless and sometimes wanton use of Earth's resources, however, generations to come may find a home with few amenities and diminishing necessities.

RENEWABLE AND NONRENEWABLE NATURAL RESOURCES

A **natural resource** is anything drawn from Earth to support life or a way of life. Earth-derived resources may normally be thought of as being either **renewable** or **nonrenewable resources.** Renewable resources are materials that can be regenerated within a relatively short span of time (crops, timber) or reused (air, water).

Nonrenewable resources (rocks, minerals, and fossil fuels) are found in nature in finite quantities. Their formation within Earth involves such extraordinarily long periods that there is no reasonable expectation that new supplies could form in hundreds or even thousands of years. The discovery and assessment of these resources have traditionally been the special concerns of earth scientists. No matter where or how they may have formed, all nonrenewable resources share three fundamental characteristics, all resulting from an increased human population and its more widespread industrial and technological advancement: (1) increasing rates of consumption, (2) diminishing supplies, and (3) diminishing rates of discovery of new supplies.

Determining the quantity of any resource ultimately available for use is a matter of making educated guesses. The significance of data that relate to making such assessments is often difficult to judge. For example, the number and size of new discoveries of *ore bodies* (economically significant mineral concentrations) over a period of years must be estimated. In addition, technological advances in the processing of ores will allow for use of low-grade ores that have already been discovered but as yet are not economically exploitable. In addition, new uses for any commodity that increase consumption rates will lower the expected life of that resource. Changing economic and technological conditions as well as political, social, and environmental pressures also impact significantly on the useful life of a resource.

Despite the uncertainty inherent in evaluation, there is a portion of each known resource that has already been discovered and measured with some accuracy by earth scientists or engineers. Such measured quantities available for immediate use are designated as **reserves,** the measured, evaluated, and currently exploitable portion of a resource.

SOILS

One convenient approach to the study of resources groups them into metals, industrial minerals, and fossil fuels. To those categories we add one other vital natural resource: **soils.** Without soils, land plants could not grow, and without those plants, the food sources on which we depend would not be available.

Soils may be defined and classified in a number of ways, depending on who defines them and the purpose of the classification. An engineer, for example, considers the soil to be any unconsolidated material overlying bedrock at Earth's surface. The principal concern of the engineer is the ease of removal of such material or its ability to support any structure built upon it, not generally its ability to sustain plant life. This latter consideration is, however, of primary importance to the agronomist and, from the standpoint of its being an "earth resource," to the earth scientist.

In this text, soil is defined as that portion of the loose material at Earth's surface capable of supporting plants with root systems. A number of schemes of varying complexity have been proposed to describe soils in terms of their origin, occurrence, and characteristics. The classification most commonly used today by specialists is called the *7th Approximation* (see appendix H). For the nonspecialist interested only in understanding the general characteristics of soils and their broad patterns of distribution, this scheme is unnecessarily complex and is not used here.

Soils found in most areas of North America typically possess a layered structure, with each layer developed to a greater or lesser degree depending on climate and the length of time the soil-forming processes have been operative. Each of the layers is referred to as a **horizon,** and each has evolved through thousands of years of weathering. The total vertical sequence of horizons is the **soil profile** (fig. 13.1).

The uppermost few centimeters of the soil consists of abundant decaying organic material mixed with some minerals and is a referred to in most soil descriptions as the *O-horizon.* Immediately below this thin surface layer is the first true soil horizon, the *A-horizon.* Because weathering proceeds from the ground surface downward, the A-horizon is the most intensely weathered zone. Water moving downward through this horizon

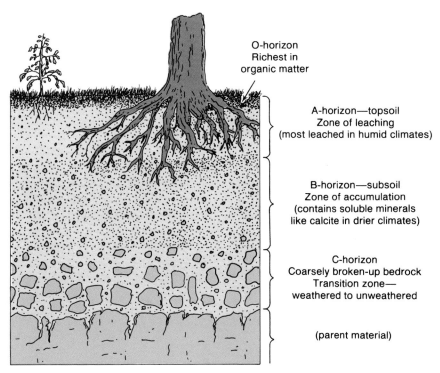

O-horizon
Richest in
organic matter

A-horizon—topsoil
Zone of leaching
(most leached in humid climates)

B-horizon—subsoil
Zone of accumulation
(contains soluble minerals
like calcite in drier climates)

C-horizon
Coarsely broken-up bedrock
Transition zone—
weathered to unweathered

(parent material)

Figure 13.1   A generalized soil profile in a humid climate.

dissolves and carries away certain soil minerals, especially calcite, but also iron oxides and clay particles (which move downward through minute cracks). This downward removal process is **leaching,** and the A-horizon is characterized by being at least partially leached of mineral material. The A-horizon is also the zone of maximum biological activity, housing burrowing animals, plants, and microorganisms. Many generations of plants live, grow, and die in this uppermost soil layer. Dead plant material is broken down by chemical processes and microorganisms in the soil, resulting in a concentration of dark, fine-grained organic material called *humus,* which colors the A-horizon dark brown or even black. This nutrient-rich, dark surface layer is what we also know as the *topsoil,* an incredibly important natural resource but one highly susceptible to loss through erosion.

Below the A-horizon is the *B-horizon,* or the *subsoil.* While the A-horizon is considered the *zone of leaching,* the B-horizon is the *zone of accumulation,* because some of the less-soluble minerals such as the clays are redeposited in this layer. Concentrated zones of these minerals are known as *hardpan* because they are tough and not easily worked. In humid climates, the soluble calcite remains in solution and is completely removed from the soil by the downward moving water. In arid climates, however, the water does not always pass down through the soil but may be drawn back up to the ground surface and evaporated. As the water evaporates, the calcite is precipitated in the soil or, often under true desert conditions, as a white mineral crust on the ground surface. This hard, calcium carbonate crust is called *caliche.* The *C-horizon* represents a transition zone between the weathered horizons above and the unweathered *parent material* (usually bedrock) below, from which the soil has developed.

Five factors seem most important in the formation of any soil and in the development of different soil characteristics: time, parent material, topography (slope and drainage), climate, and organisms. Of these, the most fundamental is climate. Regardless of which soil classification scheme is employed, the broad regional differences in soil types are determined by the climatic factors of temperature and moisture. Figure 13.2 shows a very simple but useful subdivision of soils in the United States based on precipitation. The north-south line that divides the United States into two nearly equal parts is the 64-cm annual rainfall line (approximately 100°W). The area to the east of this line receives on the average more than 64 centimeters (25 in) of rainfall each year, and the area to the west of the line (with the exception of some of the higher mountains and coastal regions) receives less than 64 centimeters annually.

Soils in the east are loosely grouped as **pedalfers** (*ped,* meaning soil; *Al,* for aluminum, *Fe,* for iron). The pedalfers are characterized by humus-rich A-horizons and clay-rich B-horizons that have been completely leached of calcite. Because the calcite has been removed, these soils are highly *acidic* and, for agricultural purposes, have to be regularly treated with anhydrous ammonia and lime to counteract the acidity. Soil horizons in pe-dalfers are well defined, and under natural conditions these soils support both broadleaf and coniferous forests.

West of the 64-centimeter annual rainfall line lie the **pedocals** (*ped,* meaning soil; *cal,* from calcite): Generally, the farther west, the drier the climate and the less complete the covering of vegetation. Because of the drier climate, leaching has not been totally effective and calcite remains in or on the soils. As a result, the soils are naturally *alkaline* rather than acid. With little

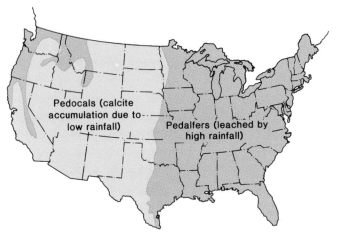

Figure 13.2   Simple subdivision of soils in the United States.

vegetation, the accumulation of humus becomes less significant, and the horizons become more difficult to distinguish from one another. Under natural conditions, the pedocals support grass or scrubby desert vegetation.

Laterites are the soils of the humid tropics. Because they have been subjected to intense chemical weathering for a very long time, they may be tens of meters thick. This same intense weathering has leached these soils of most minerals and has left them as little more than reddish concentrations of iron and aluminum oxides. Abundant soil microorganisms break down and destroy the vegetation immediately; thus, humus does not accumulate, and individual horizons are virtually impossible to distinguish. While the laterites seem to support lush vegetation in the form of tropical rain forests, they are, in fact, very infertile soils. Much of the natural vegetation grows in a litter of decaying vegetation on the forest floor and not within the true soil. When such soils are cleared for intensive agriculture, grazing, or timber, as they have been in the Amazon Basin of South America, they very quickly dry up and become bricklike (and, indeed, have historically been used as building stone in parts of southeast Asia).

METALS

Metals are those elements characterized physically by the properties of malleability, ductility, and conductivity (to be discussed later in this section) and chemically by having the tendency to lose outer shell electrons and form positively charged ions. The metals used in industry may be classified for discussion in various ways. A useful first subdivision is by their relative abundance in nature. Such abundances have been determined by the chemical analysis of thousands of rock and mineral samples. On this basis, metals are arbitrarily separated into those that are **abundant** (average abundances greater than 0.01% of Earth's crust) and those that are **rare** (abundances less than 0.01%). This difference in abundance leads to a fundamental distinction between the groups. When natural processes concentrate the abundant metals just a few times above their average abundances, an

economically exploitable **ore mineral** results. Indeed, the definition of an ore mineral is one that can be mined and refined at a profit. A rare metal must generally be concentrated hundreds or even thousands of times above its average abundance to be profitably worked.

Abundant Metals

The two best known of the **abundant metals** are very different in terms of their histories of exploitation and can be used to illustrate most of the characteristics of this group, as well as some significant differences between group members. Aluminum, the most abundant metal and the third most abundant element (after oxygen and silicon) in Earth's crust, was until this century merely a scientific curiosity. Its use in the last century was so rare that Napoleon III is reported to have given his nephew an aluminum rattle worth thousands of dollars. Iron, the second most abundant metal and fourth most abundant element, on the other hand, has been used by humans for thousands of years.

Aluminum   The chemical element aluminum is found in nature in many minerals but especially in the feldspars and clays. Both of these mineral families are, in turn, extremely abundant, with the feldspars forming the bulk of most igneous rocks and the clays the most common minerals in sedimentary rocks. This would seem to mean that there is a ready supply of aluminum ore present virtually everywhere at Earth's surface. As it turns out, however, this is not the case. Abundance alone does not make an ore. Minerals must be mined and refined at a profit to be considered as ores. If the desired metal cannot be extracted easily and cheaply from the enclosing mineral, then that mineral is not an ore. Such is the case for aluminum in both feldspars and clays. These are silicate minerals in which the aluminum ions are very tightly locked in the mineral structure and from which they cannot be easily removed. Therefore, the abundance of these minerals is of little consequence from the standpoint of aluminum production.

Although other aluminum-bearing minerals are much less abundant than either feldspar or clay, some can be more easily processed. One of these is the important ore *bauxite,* which is a complex mixture of aluminum oxide and hydroxide minerals. Bauxite formed as a tropical weathering product and may be considered as "fossil" soil (fig. 13.3). Because they formed in Earth's surface environment, bauxite deposits are shallow and can normally be mined by large-scale, low-cost open pit operations.

The separation of aluminum ions from oxygen requires extraordinary amounts of electrical energy. Thus the availability and cost of electrical power is a controlling factor in the profitability of aluminum production. For this reason, aluminum ore has historically been transported from the tropical areas where it is mined (Jamaica, South America, West Africa) to areas of abundant, cheap hydroelectric power (western Canada, Scandinavia, the United States) for processing. This procedure is very different from that employed for most metals, for which

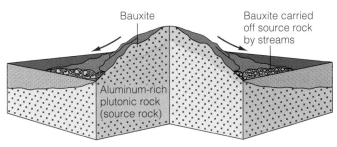

Figure 13.3 The formation of the ore mineral bauxite by the weathering of aluminum-rich source rocks under humid tropical conditions.

considerable processing is done in the immediate vicinity of the mine before the upgraded ore is sent to the final processing location (smelter, refinery, blast furnace). Because of the high cost of electricity, it has been cheaper to ship the raw aluminum ore, still mixed with large amounts of waste material, thousands of kilometers for processing.

The invention of the airplane, more than any other event, led to the widespread use of aluminum. This metal is strong yet lightweight, a requirement for anything that is going to get off the ground. In addition to its use in the transportation industry in general (decreased weight means decreased fuel requirements), which accounts for 24% of total usage in the United States, aluminum is also used extensively in packaging (31%), in the building industry (20%), and in the electrical industry (10%) as a cheaper, but generally less-effective, substitute for copper. Numerous other uses account for the remaining production (1991 data). Recycled scrap metal has become an increasingly important source of aluminum, accounting for approximately 20% of annual metal production. It is estimated that on average 50% of the aluminum used in beverage cans is now recycled, and in those states with beverage container deposit laws, the percentage recycled is much higher. Recycling uses far less electrical energy than the electrolytic processing of raw ore.

Iron   Iron has been the mainstay of industry since humans first (probably accidentally) refined iron ore by placing it in a fire. Iron ores, most commonly as the oxide minerals hematite, limonite, and magnetite, are abundant and widespread. Hematite is found concentrated in certain ancient sedimentary rock layers, limonite in tropical laterites and other soils, and magnetite in many, but especially dark, igneous rocks. By far, the most important type of iron ore, exceeding all others combined, is a sedimentary ore known as **banded iron formation (BIF),** or **taconite** (fig. 13.4*a*). This is a low-grade ore in which iron (in the form of hematite and magnetite) is concentrated in dark layers that contrast with light, iron-poor, quartz-rich layers, giving the rock an overall banded appearance (fig. 13.4*b*). All BIF deposits, including those of the Lake Superior region, are geologically very old. They are thought to represent chemical precipitates formed in ancient seas on or at the margins of the original continents. The iron was transported to the oceans dissolved in streams at a time when Earth's atmosphere contained much less

oxygen than it does today. The soluble iron, therefore, was not converted (oxidized) during this period of transport to the insoluble oxidized variety that dominates Earth's surface environment today.

Because BIF is a low-grade ore, it is mined almost exclusively in large, open pits within the iron ranges of the western Lake Superior region. Because the ore is transported some distance to the iron and steel mills, located closer to the coal and limestone deposits necessary for its refining, it must be upgraded. Before being transported, most BIF ores are crushed and the iron-rich minerals are magnetically separated from the waste rock. The iron-rich material is then mixed with special clays and crushed limestone and rolled into pellets, which are transported to steel mills and fed directly into blast furnaces.

The processing of iron ore is simple and cheap compared to the processing of aluminum ore. It requires only that the oxygen be burned, thereby freeing the iron. This is done in a blast furnace by burning *coke,* a form of coal from which the gases have been driven by heating in special ovens, (fig. 13.5). To the iron ore–coke mixture are added limestone and a small amount of manganese. These materials act as scavengers, picking up oxygen, sulfur, and other impurities from the iron ore and coke. The limestone forms waste material, or *slag,* which floats on top of the liquid iron and is removed. Disposal of large volumes of slag has always been a major environmental problem. One solution in recent years has been to build on top of these layered and compacted slag piles (fig. 13.6).

The molten iron produced from the blast furnace is *pig iron,* so-called because early iron workers were reminded of a sow and her piglets when the molten iron from the furnace ran into molds. Most pig iron, or *cast iron,* is moved in liquid form to nearby oxygen, open hearth, or electric furnaces, where it is converted to *steel.* Steel is formed by the addition of carbon and one or more metals. *Plain carbon steel,* which accounts for more than 90% of steel production and is the backbone of the construction and automobile industries, contains about 1.5% carbon and small amounts of manganese and silicon. *Stainless steel* contains between 12% and 60% chromium.

Manganese, chromium, and several other metals are routinely mixed, or *alloyed,* with iron to give steel certain desirable qualities. Chromium, for example, in addition to providing corrosion resistance ("stainless"), also gives strength. Nickel makes steel more wear-resistant. Manganese, the most important of the alloying metals, lengthens crystal size and hardens the steel, as do molybdenum and tungsten. Because these metals are produced principally for use in the steel industry, they are grouped together under the name of **ferroalloys** (table 13.1).

Most nations possess some iron ore deposits. What many lacked was the infrastructure (roads, railroads, ports, etc.) necessary to develop and exploit those deposits. This situation has now changed, and steel industries have developed in many countries, much to the detriment of that in the United States. Because of higher labor costs, aging facilities, lack of government subsidies, and environmental constraints, the United States' steel

Box 13.1 Investigating the Environment

Agricultural Capability Units: How Best to Use the Land

The rationale for including soils among Earth's natural resources is that they have been used for agricultural purposes for thousands of years. Not all soils are equally suited for such use, however. While thousands of soil types have (or can be) described, it is often beneficial to have a method whereby the agricultural usefulness of any soil can be quickly assessed and communicated. Such a system is that of *agricultural capability units,* as defined by the U.S. Department of Agriculture (table 13.A). The eight classes are characterized in terms of such factors as slope, erodibility, and drainage, with Class I being the best suited for agriculture and Class VIII the least suitable (fig. 13.A). Subdivisions of these classes may be formed by indicating with a lowercase letter the specific limitation: *e* for erosion, *w* for wetness, *s* for shallow, and *c* for climate. A class IV soil for example, with an extreme risk of erosion would be designated *IVe*.

Apply Your Knowledge

1. Which of the eight classes in this scheme best describes the land in your immediate vicinity?

2. Can you cite an example from your own knowledge of a location that would be considered as Class II? Can you cite an example of Class V?

3. Cite examples from anywhere in the world where humans have modified the natural environment or adapted crops to areas where you would not normally expect agriculture to occur.

Table 13.A

Agricultural Capability Units

	Class	Land Quality (soils)	Slope	Drainage	Productivity (cultivation)	Precautions	Primary Uses	Secondary Uses
Farmland	I	Excellent	Flat	Well drained	Easily cultivated, very productive	None	Agriculture	Pasture, recreation, wildlife
Farmland	II	Good (some sandy)	Gently sloping	Less well drained	Productive but smaller range of crop suitability	Some (contour farming)	Agriculture, pasture	Recreation, wildlife
Grazing Land	III	Moderately good	Sloping	Less well drained	Regularly cultivated, much less crop diversity (rotation)	Special erosion prevention	Agriculture, pasture, watershed	Recreation, wildlife
Grazing Land	IV	Fair (thin or sandy)	Steeper slopes	Poorer drainage	Cultivated only occasionally, very little diversity	Requires careful management	Pasture, tree crops	Recreation, wildlife, watershed
Limited-use Land	V	Low fertility	—	Wet	Unsuited for agriculture	Requires good management	Forest, range, watershed	Recreation, wildlife
Limited-use Land	VI	Very low (shallow soil)	Steep and easily eroded	—	Completely unsuited for agriculture	Requires careful management	Forest, range, watershed	Recreation, wildlife
Limited-use Land	VII	Very low (shallow and/or sandy)	Steep	Low rainfall	Pasture and forest	Very careful management	Watershed, recreation, wildlife	Forest, range
Limited-use Land	VIII	Absent or thin	Steep	Extremely wet or dry	Scenery or recreation	Little return for management input	Recreation, watershed, wildlife	—

(Source: After U.S. Department of Agriculture.)

industry is in many cases no longer economically competitive in the world market.

Iron and steel account for approximately 95% of all the metal products consumed worldwide. The construction (31% of 1991 usage in the United States), transportation (27%), machinery (18%), and appliance (7%) industries have their foundations in iron and steel. Other metals could substitute for iron and steel, but all are more expensive. As abundant and widespread as iron ore deposits are, there is no realistic possibility that we will exhaust them in the foreseeable future. A far more immediate constraint, at least from the standpoint of the United States and Canada, is the rather limited and unevenly distributed supplies of the important ferroalloys. The United States is self-sufficient only in molybdenum, and Canada possesses significant quanti-

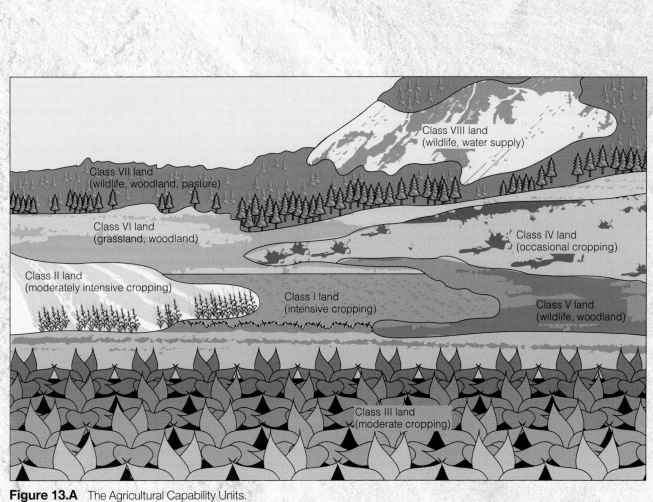

Figure 13.A The Agricultural Capability Units.

ties only of nickel and cobalt. The other ferroalloys must be imported, sometimes from politically unstable or otherwise unreliable sources.

Russia is the world's leading producer of steel, followed by Brazil. In 1991 the United States produced approximately 6% of the world's total. While the United States possesses enough reserves to be self-sufficient in steel production, it nevertheless imports both raw ore (from Canada) and approximately 15% of the finished steel it uses. The steel industry in the United States is intensely localized. Because of the ease of transportation afforded by the Great Lakes for ore from the great iron deposits in northern Minnesota and the proximity of major coal and limestone deposits, 47% of the steel-making capacity has been concentrated in the

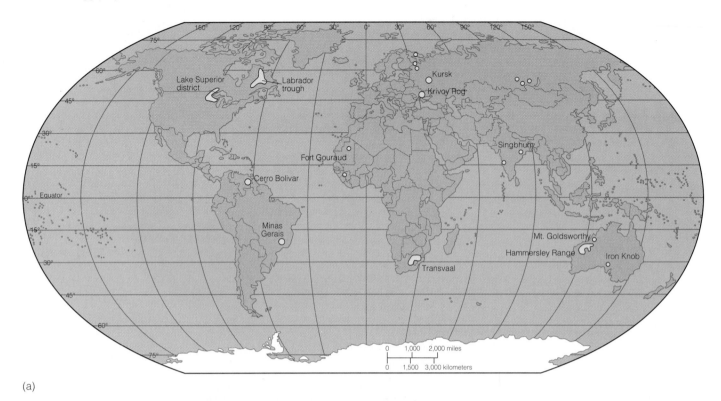

(a)

(b)

Figure 13.4 (*a*) Locations of important banded iron formation deposits. (*b*) Mass of BIF on display outside of the Smithsonian Institution, Washington, D.C.

Illinois-Indiana-Michigan area. An additional 31% is located in Ohio and Pennsylvania.

Rare Metals

The rare metals, which include copper, tin, gold, silver, lead, zinc, and a number of others, must be concentrated to many times their average abundance by natural processes to be economically exploited. As with the abundant metals, two examples, copper and gold, are considered here to illustrate the important characteristics of this group.

Copper   Copper plays an important role in our everyday lives because of its reasonable cost and peculiar characteristics. Copper is *malleable* and *ductile,* that is, it can be shaped and drawn out into fine wire. While not as good an electrical conductor as gold or silver, it is still effective enough for most purposes and its much cheaper cost makes it available for use in large quantities. Its widespread use in the electrical industry (about 60% of the total usage of this metal), in a sense, makes copper responsible for our high standard of living with its many labor-saving devices. Copper has also found applications in the plumbing and transportation industries, in coins, and in munitions. Copper

Table 13.1

The Principal Ferroalloys and Their Uses

Ferroalloy	Steel Type	Characteristic	Ore Mineral
Chromium	Stainless steels, high-speed tool steels	Strength, hardness, corrosion resistance	Chromite
Cobalt	Magnet steels	Hardness at high temperatures, magnetism	By-product of copper mining
Manganese	Carbon steels	Removes O and S, resistance to wear	Pyrolusite
Molybdenum	Molybdenum steels	Hardness, strength, corrosion resistance	Molybdenite
Nickel	Nickel stainless steels	Corrosion and stain resistance	Pentlandite
Tungsten	Tungsten steel	Hardness, high-speed cutting tools	Scheelite, wolframite
Vanadium	High-temperature steels	Tarnish strain and fatigue resistance, grain size control	Carnotite

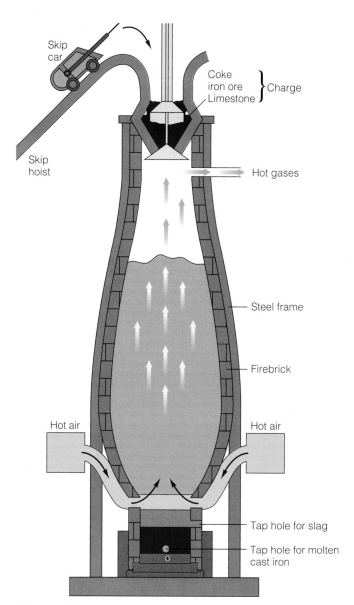

Figure 13.5 Diagrammatic representation of a blast furnace.

mixes easily with other metals and has been alloyed with zinc to form *brass* and with tin and zinc to form *bronze*.

The element copper occurs in well over 100 minerals, but the most important ore mineral is *chalcopyrite,* a copper-iron sulfide. Most importantly, chalcopyrite is found disseminated as small grains and concentrations in large masses of granitelike rock known as *porphyry copper* deposits. These rock bodies are geologically young (less than 225 million years old) and invariably parallel to subduction zones. They occur throughout the western United States and southward into the Andes Mountains of Peru and Chile. They are especially common in Montana, Utah, and southern Arizona, the latter state accounting for more than half of the total production of copper in the United States.

Porphyry copper deposits formed from hydrothermal solutions that worked their way through cracks and crevices in the now solidified granite and the surrounding country rock (fig. 13.7). As these fluids cooled, metal-bearing minerals usually in a matrix of quartz were precipitated. These small, irregularly shaped bodies, called **veins,** contain not only copper-bearing minerals but also minerals bearing gold, silver, tin, mercury, molybdenum, beryllium, and many other metals. The metallic ions could not be accommodated in the structures of the abundant rock-forming minerals that crystallized earlier. Thus, they were concentrated in many rare minerals and far above their average abundances.

The porphyry coppers are low-grade deposits, meaning that the percentage of copper present in the ore body is very low, often less than 0.5%. Consequently, it is necessary to mine very large tonnages of material to derive relatively small amounts of copper. Even so, staggering wealth has resulted over the years from some of these mines. The city of Butte, Montana, lies on top of what has been called "the richest hill on Earth." More than $3 billion worth of metals have been extracted from the mines at Butte. But even this pales when compared to the open pit mine at Bingham Canyon, Utah, the largest mine in the world (fig. 13.8). At Bingham Canyon, where the average copper content is only 0.65%, more than $15 billion worth of metals (mostly copper, but also gold and silver) have been removed from a mine 1.7 by 2.5 kilometers (1.0 by 1.5 mi) in area and a little over 1 kilometer (0.6 mi) deep.

Because reserves of copper ore in the United States are relatively low, there has been for a number of years a strong

Box 13.2 Further Consideration

By-Products: A Use for Everything

While large mining operations are often developed for the production of a single metal resource, such as iron or copper, the much lower natural concentrations of many other metals preclude the expenditure of large capital to develop specific sources for them. They are retrieved from Earth entirely or in large measure incidentally during the mining and/or processing of ores of other metals, that is, as *by-products*. In some instances, the by-product metal is present in a mineral that is intimately associated with, the principal ore mineral and must be separated from it before processing begins. In other cases, the by-product metal is present within the principal ore mineral and must be separated during the refining process or from the refinery wastes.

Of the ferroalloys, only manganese and nickel are produced as primary products. Cobalt is mostly a by-product from the waste (gangue) minerals associated with copper ores and from nickel-bearing laterites. Vanadium is primarily retrieved from uranium ores and titaniferous magnetite deposits. There are large deposits of molybdenum ore at Climax, Colorado, but much of the production of this metal comes as a by-product of copper ores because of the large quantities of copper that are processed.

Silver, a precious metal, is rarely produced as the primary metal but is separated from lead, copper, and zinc ores after smelting. Consequently, the production rate of silver is dependent on the production of these other metals. The amount of silver held within the ore of lead, galena, is dependent on the presence of bismuth, thus bismuth becomes a by-product of lead refining as well. While there is some primary production of the platinum group metals in Russia and South Africa, most comes as a by-product of nickel refining.

For many rare elements important to today's technology there are no sources other than as by-products from the wastes from ore processing. Rhenium is retrieved from the flue dust of molybdenite smelters and along with gallium from coal ash. Germanium, thallium, and indium come from zinc refinery residues and flue dust. Cadmium is salvaged from lead smelter dust and selenium and tellurium from the slimes (residues) of copper refining. Hafnium is recovered during the production of zirconium metal and scandium from fluorspar wastes.

If two or more metals of approximately equal value are produced by working the same deposit, they are referred to as *co-products*. The joint production of zinc and lead from deposits at many localities and copper from many nickel and tin deposits are examples.

Apply Your Knowledge

1. Why is it not feasible to produce many of the rare elements as primary ores?

2. Discuss the economic and technological problems associated with the production of materials as by-products.

3. What is an environmental benefit of the processing of refinery and manufacturing wastes for their by-product content?

Figure 13.6 Commercial development (shopping center) on a slag dump south of Pittsburgh.

Figure 13.7 Formation of hydrothermal mineral deposits.

Ore and silicate minerals deposited by hot magmatic fluids in cracks in surrounding rocks

Veins

Cooling magma (granite)

impetus to conserve and recycle copper. Today, approximately 20% to 25% of the annual copper production of the United States is recycled metal. Ironically, world reserves of copper are actually increasing, not just because of the discovery of new deposits but, in part at least, because of the replacement of copper by glass fiber optics in transmission lines.

Gold   Gold is not only a rare metal, but, because of its high economic value (measured in dollars per ounce), it is also, along with silver and the members of the platinum group, considered to be a *precious metal*. Gold derives its long history of use from its beauty, its rarity, and from the fact that it is virtually indestructible. With the exception of that in sunken Spanish treasure

Figure 13.8   Open pit copper mine at Bingham Canyon, Utah. The largest mine in the world.

ships on the seafloor, much of the gold that has been mined throughout history, an estimated 95,000 metric tons (105,000 tons) is still in circulation. Its principal use today is as a backing for the world's monetary systems and consequently an estimated 32,600 metric tons (36,000 tons) are stored in vaults, such as that at Fort Knox, Kentucky. Other important uses are in jewelry, dentistry, plating, coinage, electronics, and the aerospace industry.

Like chalcopyrite, pure gold and rare gold-bearing minerals occur primarily in hydrothermal vein deposits. A secondary source is in small sedimentary units known as **placer deposits,** which are distributed along stream channels and shorelines. Placer deposits represent the accumulation of heavy mineral grains that were eroded from the primary vein deposits, transported some distance downstream, and then dropped when the stream currents lost energy and could no longer carry them (fig. 13.9). The largest gold-bearing deposits, those in the Witwatersrand District of the Union of South Africa (known as "the Rand"), are found in ancient bodies of metamorphosed conglomerates. These have the appearance of placer deposits, but because their size is so much larger than normal placers, their exact origin remains a mystery. South Africa possesses approximate-

ly one-half of the world's gold reserves and has led the world in the production of gold since the late 1800s. Other important gold-producing countries include Russia, Australia, Canada, and the United States.

The United States currently produces approximately 35% of its own needs. About 10% of this gold is a by-product of the refining of other metals (principally copper), and the remainder represents primary mine production and recycled metal. Most gold mines today are open pit operations that produce large tonnages of very low-grade ore. Indeed, the great increase in the number of such mines in recent years has led to a major controversy within the metal ("hard-rock") mining industry and within the halls of Congress (box 13.3).

INDUSTRIAL MINERALS

Because all industrial minerals are abundant in nature, they are classified by use rather than by abundance. In this text, we recognize three categories of industrial minerals with some subdivisions: raw chemicals, fertilizers, and construction materials.

Box 13.3 Further Consideration

The General Mining Law of 1872: Too Much of a Bad Thing?

A major controversy boils today in the field of mineral resource development over a movement to repeal or modify in a significant way the General Mining Law of 1872. This law was one of a series of acts designed to provide an economic stimulus following the Civil War by opening the 260 million square kilometers (100 million sq. mi) of federally owned land in the western United States to development. That it has been a success is a matter of historical record. That it has been (and still is) abused is equally certain.

The basic provisions of the law are such that any United States citizen can file claim to a 8-hectare (20-acre) (placer claim) or 457-meter-by-183-meter (1,500-by-600-ft) (lode claim) area of federally owned land for the expressed purpose of developing a mineral deposit. The only costs to the claimant are a $100 annual maintainance assessment and a $2.50 per acre (placer claim) or $5.00 per acre (lode claim) fee paid at the time of applica-

tion for patenting. Receiving the patent to the land provides the claimant with full title to the land, but even if the patent is never applied for or is refused, the claimant still retains full rights to the mineral resource as well as unrestricted use of the land surface provided that such use is even remotely related to mining activity. Thus, for a $100 annual fee, a significant tract of land could be easily diverted to recreation or some other activity. Further, any minerals that are taken from the claim are not subject to royalty payments to the federal government as is the case for coal, oil, natural gas, and other nonmetallic resources.

Proponents of reform have suggested three fundamental changes. First, they would impose a significant annual "rent" on a standardized claim size of 16 hectares (40 acres) that would rise from $800 in each of the first 5 years to $12,800 after 25 years. Second, they would impose a leasing-royalty payment system of perhaps 8%. Third, any new law or major revision would include stringent provisions to protect the environment, something that was not a concern when the law was originally formulated.

Those who argue against changing the provisions of the law as they apply to the "hard-rock" (metallic) mineral industry, primarily those who work in that industry, offer the following counterarguments. The hard-rock mining industry needs the economic incentive provided by this law to compete in an

international minerals market where foreign mining companies may operate with significant government subsidies. Although no royalties are paid on minerals produced, the economic benefits in the form of jobs and taxes more than outweigh such potential royalties. The General Mining Law has no provisions for maintaining the environment, but all mining on federal land falls within the purview of other environmental legislation, including the National Environmental Policy Act (NEPA), the Clean Air Act, and the Clean Water Act.

Regardless of the outcome of this debate, it provides us with an opportunity to examine an important and controversial subject in the earth sciences from opposite viewpoints and to evaluate the merits of the arguments raised by both sides.

Apply Your Knowledge

1. Take a stand on this issue and defend your position.
2. Regardless of your current position on this issue, do you think that the law when it was passed was justifiable on economic and social grounds? Why?
3. Should the mining of "hard rock" metals be treated differently from other mineral resources? Why?

Raw Chemicals

The raw chemical products are dominated by two mineral substances, halite (common salt) and sulfur. Halite (NaCl) is one of a group of economically important mineral products, known collectively as *evaporites,* that formed when large portions of ancient seas underwent periodic isolation and evaporation. Seaways or small basinal areas may have been detached for an extended period of time from the main ocean basin by uplift of the seafloor, the building of reef masses, or some other natural process. Intense evaporation left the isolated waters greatly enriched in dissolved mineral material to the point where these minerals could no longer remain in solution. They precipitated out to form layers on the seafloor. The seaway need not have dried up completely (although this perhaps occurred on occasion). The concentration of dissolved mineral matter only had to exceed some critical level at which the water was saturated with salts. As such a level was reached and exceeded for each member of the mineral group, that mineral came out of solution (chapter 11). In North America, significant deposits of halite and related evaporites occur in lower Michigan (the Michigan, or

Salina, Basin) and stretch east and southeast to underlie parts of Ohio, Pennsylvania, and New York; in West Texas, eastern New Mexico, and western Kansas (the Permian Basin); in western North Dakota, eastern Montana, and southern Canada (the Williston Basin); and in belts paralleling the Gulf Coast (fig. 13.10).

Along the Gulf Coast, the salt deposits show a unique characteristic. The salt layer, buried at great depth, has been subjected to the enormous pressure exerted by the weight of thousands of meters of overlying sedimentary rocks. As a result, the salt has slowly flowed and has been squeezed upward to form hundreds of vertical pillars, many of which reach the surface. These are referred to as **salt domes,** and not only do they form important sources of halite, but they also are the location of other important resources such as oil, gas, and sulfur (fig. 13.11).

Salt has been an important commodity for thousands of years, as evidenced by the English word "salary," which derives from the Latin for salt. It has served as a basis for economic wealth, trade, and taxation. Its value has made it subject to government monopolies (as in India and China today). Approximately 50% of the salt used in the United States is con-

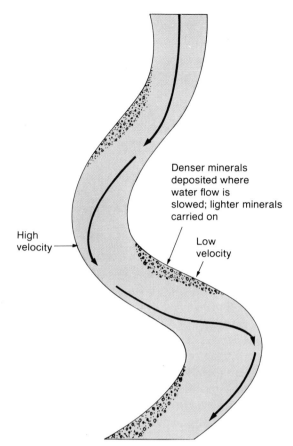

Figure 13.9 The formation of placer deposits along a stream channel.

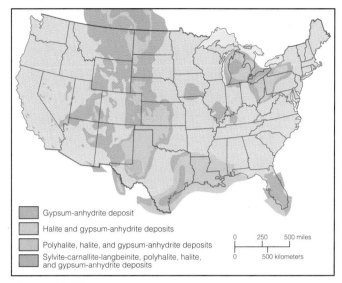

Figure 13.10 Location of various types of marine evaporite deposits in the United States.

Gypsum-anhydrite deposit

Halite and gypsum-anhydrite deposits

Polyhalite, halite, and gypsum-anhydrite deposits

Sylvite-carnallite-langbeinite, polyhalite, halite, and gypsum-anhydrite deposits

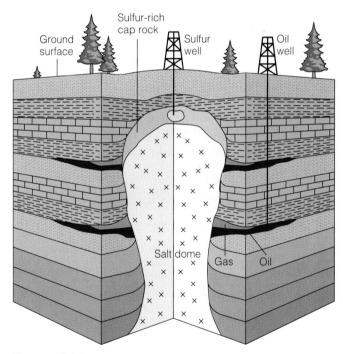

Figure 13.11 Typical Gulf Coast salt dome with associated sulfur and petroleum deposits.

sumed by the chemical industry, where 96% of this amount is manufactured into chlorine-based bleaches, polyvinyl chloride, and caustic soda (NaOH). De-icing of highways accounts for 26% of the salt consumed, with most of the remainder used in food seasoning and preservation, water treatment, ceramic glazes, fertilizers, medicines, and insecticides.

In the United States, the world's leading salt producer, 55% of production is from *brines,* highly saline groundwater (natural or formed by injecting water) found in the pore spaces of sedimentary rocks; 29% is from the mining of rock salt deposits; and 15% is produced by solar evaporation from shallow ponds as at Great Salt Lake in Utah and along San Francisco Bay. There are no economic substitutes for salt. Fortunately, world resources of salt, including the waters of the oceans, are limitless.

In its pure form, sulfur is often found in veins and filling fractures in rocks at the tops of salt domes. It forms from the alteration of the mineral *anhydrite* ($CaSO_4$) by anaerobic, sulfate-reducing bacteria. Anhydrite is another evaporite mineral usually associated with halite. Texas and Louisiana, together accounting for 55% of U.S. sulfur production, lead the country and the world in the production of sulfur. For much of this century, most sulfur was produced by the *Frasch process.* In this method, air and hot water are pumped down separate pipes in a well bore and into the sulfur-bearing formation. Under pressure, they dissolve the sulfur to form a foamy mixture that is pumped to the surface

through the same well bore (fig. 13.12). Sulfur produced by this technique is extremely pure (99.8% sulfur). In recent years, the percentage of sulfur produced by the Frasch process has declined to approximately 35%. This reduction is being made up by a dramatic increase in the amount of sulfur recovered from oil and natural gas, currently 53% of United States' production. While the supply of sulfur that can be obtained by the Frasch process and from natural gas is limited, that which could be produced from the sulfate minerals gypsum and anhydrite is virtually limitless.

Most of the sulfur produced in this country goes into the making of sulfuric acid (H_2SO_4), and, of that, 60% goes to the

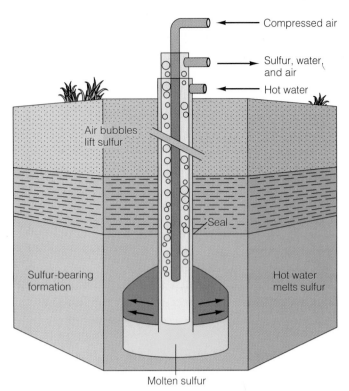

Figure 13.12 Recovery of sulfur by means of the Frasch process.

production of fertilizer. Sulfuric acid is reacted with phosphate rock to yield phosphatic acid, H_2PO_4, the basis for the phosphate-based fertilizers discussed in the next section. Other uses of sulfur include plastics, paper, synthetic fabrics, explosives, medicines, and insecticides.

Fertilizers

Modern intensive agriculture places great stress on soils. As crops are harvested, nutrients that plants extract from the soil and incorporate in plant tissue are lost to the soil. If a soil is to remain a useful resource capable of sustaining future generations of plant growth, these nutrients must be replaced. This is done in large measure by the use of inorganic **fertilizers,** chemical resources taken from Earth in one form and returned to it in a form usable by plants. Three nutrients essential for plant growth are the elements potassium, nitrogen, and phosphorous.

Potassium (for marketing purposes referred to as *potash*) is one of the most common elements found in the rocks and minerals of Earth's crust. Although it does not occur as elemental potassium in nature because it reacts readily with water, potassium is an important constituent of micas, feldspars, and clays. As is the case in the processing of aluminum, these abundant silicate minerals are *refractory,* or difficult to break down, and the potassium in them is not easily released. Consequently, a better but less abundant source of potassium for fertilizer is the mineral *sylvite*. Sylvite (KCl) forms in the same manner as halite, and, thus, it too is classified as an evaporite mineral. Sylvite, however, is even more soluble than halite, and an even greater

amount of evaporation is required before it precipitates onto the seafloor.

In the Permian Basin, a blanket of subsurface evaporite deposits is found covering more than 260,000 square kilometers (100,000 sq. mi). Conventional underground mining of these deposits accounts for almost 90% of the potash production in the United States. Even larger deposits of potash are known from central Europe, Russia, and western Canada, with these last two countries being the world's leading producers. Approximately 95% of the potash marketed in the United States goes directly into the manufacturing of fertilizer. The remaining 5% finds various applications in the chemical industry. World reserves of potassium salts are enormous and are certainly sufficient to last for several centuries.

For many years, until the time of World War I (1914–18), the world's demand for nitrogen salts was supplied almost exclusively from vast deposits in the Atacama Desert of Chile. These Chilean nitrate deposits, consisting primarily of sodium nitrate ($NaNO_3$), were formed by the evaporation of groundwater and the consequent precipitation of soluble salts under arid conditions, forming a type of caliche. During the war, these deposits were more important for the manufacturing of explosives than they were for fertilizers. When the British naval blockades prevented German ships from reaching these nitrates, German scientists developed methods whereby nitrogen could be *fixed*, or extracted from the air by combining chemically with other elements, and used in the munitions industry. These same techniques were subsequently employed for the manufacturing of nitrogen-based fertilizers, and the Chilean nitrate deposits eventually lost their significance. The major difficulty with using what is an essentially inexhaustible supply of atmospheric nitrogen is the large energy requirement of processing.

Most of the fixed nitrogen produced in the United States is in the form of ammonia (NH_3), with the hydrogen derived from natural gas. The United States produces most of the fixed nitrogen products that it consumes, with 60% of that production coming from the states of Texas, Oklahoma, and Louisiana, the sites of large natural gas fields. Approximately 80% of the nitrogen goes to the fertilizer industry for manufacturing into the anhydrous ammonia that is applied to fields in liquid form. The remaining ammonia is used in synthetic fabrics, in plastics, and in explosives.

Like nitrogen and potassium, the potential reserves of phosphorous for use in fertilizers is enormous. Most of these reserves are in marine sedimentary rocks. Florida and North Carolina produce almost 90% of the phosphate rock in the United States, with Idaho, Montana, Tennessee, and Utah producing much of the remainder. Vast amounts of phosphate-rich sediment have been identified on the continental shelf off the eastern coast of the United States and almost certainly will be mined some day. The United States is the world leader in the production of phosphate rock followed by Russia, Morocco, and Israel. More than 90% of the world's production of phosphate rock goes to the fertilizer industry. The remainder is used in the manufacturing of products as varied as scouring powders, insecticides, and ceramics.

Construction Materials

Construction materials represent the largest tonnage of all of the natural resources produced, with an aggregate value in 1991 of more than $11 billion in the United States alone. The simple reason for this is that virtually every rock type can be used for building something. Such usage is governed largely by local supply, because high transportation costs preclude the long-distance transport of these large tonnage materials, except when needed for specialized applications. A convenient subdivision of building materials separates them into two categories: natural products and prepared products.

Natural Products The *natural products* are those rocks and minerals that require a minimal amount of preparation for use. Such preparation normally consists of cutting, polishing, crushing, sorting, and/or washing. Humans have always used locally available materials, fashioned to the desired size and shape, for the construction of dwellings and monuments. Architecturally, all such building stone is referred to as *dimension stone*. The most commonly used dimension stones are various coarse-grained igneous rocks generically termed "granite" (52% of usage in the United States) and limestone (28%). Because brick and concrete are far cheaper, use of dimension stone for most construction purposes has diminished. Exceptions to this trend are as decorative facing in building lobbies and other interior areas, where quality dimension stone is locally available, or where considerations of local or national pride outweigh the considerations of cost.

At one time, the metamorphic rock slate, because it splits easily along parallel surfaces, was widely used for roofing. Asbestos shingles have now almost totally replaced it for that purpose. Slate is still widely used, however, for billiard tables, blackboards, and countertops. Indiana (limestone) and Georgia and Vermont (granite) together account for 40% of the dimension stone production in the United States. Resources of dimension stone are inexhaustible, but specific varieties are limited on a local or regional basis.

Crushed stone, used primarily as aggregate in highway construction and maintainance (52% of usage) is generally whatever rock material is available locally that can be crushed easily to the desired size. In the coal-producing areas of Pennsylvania and West Virginia, driveways, country roads, and parking lots are often covered with a material known locally as "red dog," shale rock that has been hardened and given a reddish color by the combustion of coal mine waste deposits, or "gob" piles. It is available and cheap, and it serves the purposes of communities and individuals who do not wish to bear the costs of using more expensive materials. In addition, using the red dog helps to eliminate the unsightly debris piles. Crushed rock is mined from quarries in 48 of the 50 states, with the most commonly used rock types being limestone and dolostone (71% of usage), granite (14%), and basalt (8%). Crushed rock also presents a relatively new and interesting example of recycling. Increasingly, when a highway is resurfaced, the old aggregate is saved and reconstituted into new aggregate material.

Sand and gravel, which generally are washed free of their fine-grained impurities and sorted by size, are available virtually everywhere. They particularly occur as glacial, stream channel, and beach deposits. Sand and gravel have many of the same uses as crushed rock, and like crushed rock they have a low cost per ton value. This means that they must be mined close to their points of consumption, usually in or near urban areas. This proximity often leads to environmental problems associated with noisy and dirty quarrying operations and heavy truck traffic. There are, however, notable examples of sand and gravel pits that, when they were exhausted, were converted into community assets by wise planning and the expenditure of some money. Indeed, one location in the city of Denver, Colorado, was used sequentially for gravel pits, for sanitary landfills, and finally for a stadium parking lot complex.

Prepared Products Prepared products are manufactured, usually by mixing, heating, and/or chemical conversion. Such products would include (among many others) cement, clay products, and glass. *Cement* is the binding material in concrete. It is formed by roasting a crushed mixture of limestone (or dolostone) and clay (or shale) in a kiln at a temperature of approximately 1500° C (2,732° F). This drives off the water present in the clays as well as the carbon dioxide generated in the roasting process. The end product is *clinker,* which is ground to the desired size and mixed with a small amount of gypsum to form the final cement. Unfortunately, a considerable quantity of fine cement dust is also given off in the process. This dust is vented high into the atmosphere through a very tall stack, or chimney, so that its dispersal occurs across a broad area in the downwind direction rather than accumulating in great quantity near its source.

Clay is the group name for a number of related minerals. All are hydrous aluminum silicates, and most have been formed by the weathering of igneous rocks. Because there are compositional differences between clays, their chemical and physical properties vary, and each is best suited for certain applications. Clays are widely used not only in the making of cement but also in the manufacturing of bricks, ceramics, chinaware, and a host of other products (table 13.2). Reserves of most types of clay are enormous because the soils that cover the ground surface consist in large part of clay.

The principal component of *glass* is silica, in the form of quartz sand (fig. 13.13). The basic process of producing glass involves melting the silica at high temperatures and then cooling ("quenching") it quickly so as to inhibit the growth of crystals. Although seemingly solid, glass is, in fact, a supercooled liquid, a material in which the individual ions have not arranged themselves into the orderly crystalline pattern that characterizes true solids. Various chemical substances are added to the silica to speed its hardening and to impart to the glass the desired qualities of color, weight, and resistance to heat and breakage. The United States' reserves of the necessary raw materials for glass manufacturing are sufficient for the foreseeable future.

Table 13.2

The Various Categories of Clay and Their Uses

Clay Type	Use
Common clay (illite-rich)	Bricks, tiles, pipes, cement
Kaolin "china clay" (includes kaolinite and halloysite)	Chinaware, refractories (furnace lining)
Montmorillonite . Bentonite Fuller's earth (includes attapulgite)	Drilling mud (oil industry) Adsorbent (grease, etc.)
Ball clay	Tiles and refractories
Fire clay	Refractories

(Source: U.S. Department of Agriculture.)

Figure 13.13 Glass sand quarry near Berkeley Springs, West Virginia.

ABRASIVES AND GEMS

Abrasives and gems are considered together because they generally share the same characteristic of hardness and in some cases even represent the same mineral used for different purposes. Hardness is essential in abrasives because of the nature of their use. In gemstones, hardness is an asset because the cost of most gems makes it important that their useful lives be extremely long.

Abrasives

Abrasives include a variety of materials, natural and manufactured, that perform one of several functions either in industry or in the home: cleaning, polishing, grinding, cutting, or drilling. Although all lie toward the high end of the Mohs scale, with respect to one another they can be further categorized as hard, siliceous, or soft. This subdivision is based in part on their natural hardnesses but also in part on their compositions and on the form in which they are used. The principal abrasives included in the "hard" category are diamond, corundum, emery, and garnet.

Diamond is the hardest naturally occurring substance (10 on the Mohs scale) and is used in one of three size categories, from largest to smallest: stone, bort, or dust. Because the United States has no diamond deposits that can be exploited under current economic conditions, all natural diamond stone (60% of that consumed in the United States) is imported. The remainder is synthetic diamond. Natural industrial diamonds, those unsuited for gem use because of color or flaws, occur, like the gem variety, in the igneous rock known as kimberlite. Because of the tremendous pressure required to form diamond, it is assumed that this rock originates at a depth of several hundred kilometers in Earth and is somehow intruded, perhaps explosively, through Earth's crust in cylindrical masses known as *diamond pipes* (fig. 13.14). Diamond stones are also typically found in placer deposits, either along stream courses or in beaches.

Bort consists of small fragments and, like stone, is used mostly in rock drill bits (approximately 75% of total usage). Diamond dust, used mostly for cutting and polishing gemstones, is formed from crushed synthetic diamonds. The United States manufactures more than enough diamond dust to satisfy its needs.

Corundum, the second hardest naturally occurring substance, and *emery,* a natural mixture of corundum and magnetite, are used mostly on paper or cloth for polishing and finishing or on grinding wheels. In addition, emery is widely used for nonskid surfaces on stairs and pavements. Corundum occurs in metamorphic rocks, and virtually all is imported into North America, chiefly from South Africa. The largest deposits of emery are found in Turkey, but minor deposits also occur around Peekskill, New York. Both corundum and emery are being replaced in many uses by synthetic alumina, Al_2O_3.

Garnet, primarily the variety almandine, is found in both igneous and metamorphic rocks. It is used mostly for sandblasting (45%); for the finishing of hardwoods, in which its cutting power is several times greater than that of sandpaper; and for cutting marble and slate. The United States possesses the world's greatest reserves of garnet and produces almost 50% of the world's annual supply. The principal deposits of garnet currently being exploited are in the Adirondack Mountains near the town of North Creek, New York (fig. 13.15).

Sand and sandstone are examples of siliceous abrasives. Quartz sand is widely used as a sandblasting medium, while sandstone is shaped into grindstones, whetstones, and millstones. In fact, a sandstone unit in England that has been so used for hundreds of years is known as the "Millstone Grits." The use of siliceous rocks and sand for polishing and pulverizing has been replaced to a large extent by steel balls.

Among the soft abrasives are feldspar, fuller's earth, and diatomite. Feldspar, from granite pegmatites, is crushed and used in soaps and as scouring powders. *Fuller's earth,* a variety of clay, is used as a soft polishing medium for metals. *Diatomite,* or *diatomaceous earth,* is an accumulation of the siliceous remains of shell-secreting freshwater and marine algae (diatoms). It is used not only as an abrasive in polishes and toothpaste but as a filtering medium in water and waste treatment plants and as a filler in paints and plastics. The United States possesses the world's most important reserves of diatomite.

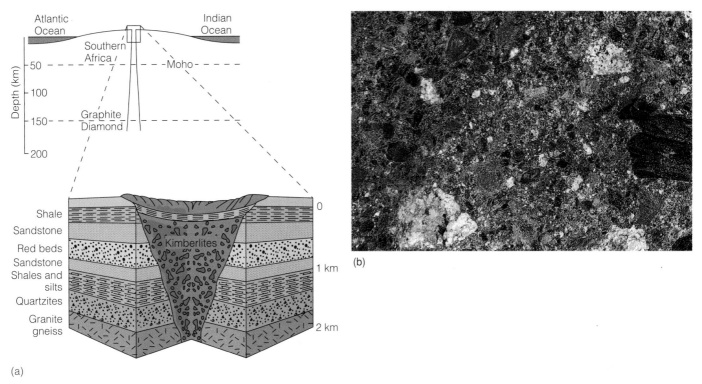

Figure 13.14 (*a*) Cross-sectional view of a South African diamond pipe. (*b*) The diamond-bearing rock kimberlite (microscopic enlargement of mineral components).

Figure 13.15 Red patches of garnet in dark igneous rock at Barton Mines, North Creek, New York.

Gems

Gemstones, or **gem minerals,** are those natural substances that can be cut and polished to form *gems,* which, in turn, can be set in *jewelry.* With few exceptions, they possess the special characteristics of hardness (or durability), beauty (color, clarity, etc.), and rarity. Many have been used for adornment for thousands of years. Others have found widespread usage only recently. The great majority of gem minerals have their geologic origin either in pegmatites (emerald) or in metamorphic rocks (ruby).

Diamond is a significant exception to this, having formed, as noted earlier, in kimberlites. Because of the inherent hardness of the minerals in this group, many of them are also found as placer accumulations. A few materials such as *amber* (hardened tree sap) and *pearl* (formed by secretions of oysters within their shells) do not fit the strict definition of a mineral, but because of their widespread use in jewelry they are included here.

Some gem minerals can be manufactured and are thus considered to be *synthetic gems.* Synthetic rubies are especially important as lasers, and synthetic diamonds and diamond dust are widely used as abrasives. *Simulated gems* are materials (cheaper stones or even glass) that have the appearance of one of the precious gems. Because these are readily available on the market, buyers must exercise caution to avoid confusing them with their more valuable namesakes. Certainly, gems marketed under a trade name such as "Cape May Diamonds" are not real diamonds but inexpensive imitations (in this case, crystals of quartz).

For convenience, gem minerals have long been divided into two basic groups: *precious gems* (diamond, ruby, sapphire, and emerald) and *semiprecious gems* (all others). Such a subdivision, seemingly based on economic value, is not necessarily foolproof. There are exceptions. For example, some specimens of alexandrite (the mineral chrysoberyl) from the Ural Mountains of Russia surpass emeralds in value.

Diamonds are, by far, the most important of the precious stones, accounting for approximately three-quarters of the world's gem trade. Until the eighteenth century, diamonds came exclusively from alluvial deposits in Brazil and India. The latter went to the collections of local princes, and in some instances, such as the Hope diamond now in the Smithsonian Institution, became spoils of war. After the discovery of the diamond fields around Kimberly, South Africa, in 1871, that country became the world's leading diamond producer and held that position for many years. Recently, South Africa has been surpassed in output by Australia, Botswana, and Russia.

Ruby and *sapphire* are, respectively, the red and blue clear gem varieties of the mineral corundum. They typically occur in marbles and alluvial deposits formed from them. Large, blood-red rubies are extremely rare and command extraordinarily high prices. Both rubies and sapphires may contain tiny needles of the mineral rutile, which reflect light to give a six-rayed star pattern. These are the highly prized "star rubies" and "star sapphires." The most important production of both of these stones has been from placer deposits in Burma and Sri Lanka.

Emerald is the deep green variety of the mineral beryl. One of the most ancient of stones in terms of its usage, emerald mines were operated in Egypt as early as 1600 B.C. Emeralds occur most commonly in granite pegmatites and in the metamorphosed limestones into which they have been intruded. The most important deposits worked in recent years have been those around Muzo, Colombia. Beryl may also be found in a lighter blue-green gem variety known as *aquamarine,* one of the semiprecious gems. The significant characteristics of aquamarine, as well as those of a number of other semiprecious gems, are listed in table 13.3.

FOSSIL FUELS

Coal, oil, and natural gas constitute the **fossil fuels.** These fuels represent solar energy used, stored, and preserved for millions of years in plant and animal tissues. While not strictly speaking minerals, because they are of organic origin, they are usually considered along with other earth materials as mineral resources or as mineral fuels. In small ways, humans have used fossil fuels for thousands of years. Such usage has increased enormously beginning with the Industrial Revolution of the mid-eighteenth century. Fossil fuels today represent the backbone of industry and the lifeblood of our technological society.

Coal

Coal formed from the accumulated remains of plants that once grew in vast coastal swamps (fig. 13.16). Generations of such plants lived, died, and were preserved in a thickening pile of debris beneath the shallow waters of the swamps. These waters must have lacked the dissolved oxygen necessary to sustain the chemical and biochemical processes that normally cause dead plant tissue to decay. Accumulations of hydrogen sulfide gas prevented the presence of organisms that consume dead vegetation. A modern analog of such an environment is the Great Dismal Swamp of Virginia and North Carolina.

Over hundreds of thousands of years, the geologic environment changed, with shallow coastal seas flooding the land, de-

(a)

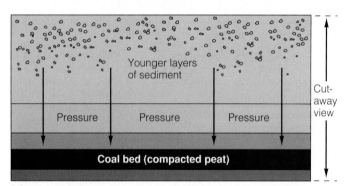

(b)

(c)

Figure 13.16 The formation of coal. (*a*) Plant debris accumulates in the reducing environment of a coastal swamp. (*b*) Rising sea level floods the area and brings normal marine deposition.(*c*) The pressure (weight) of overlying sedimentary layers converts the vegetation to coal.

stroying the coastal swamps, and allowing for normal marine sedimentation processes. The weight of accumulating thicknesses of sandstones, shales, and limestones compressed and heated the underlying vegetation. The compression of the vegetation squeezed out most of the water and gases, which make up a large proportion of plant tissue, and left the remaining material enriched in carbon and a number of minor solid impurities.

This is a progressive process. As heat and pressure increase, vegetation, with a carbon content of approximately 50%, is converted to *peat* (carbon content 60%), a substance in which the original vegetable material is still recognizable. Peat is widely used as a soil conditioner, and, where better grades of coal are unavailable, it is burned as a fuel. Deeper burial of peat to still higher temperatures and pressures forms first *lignite* (carbon content

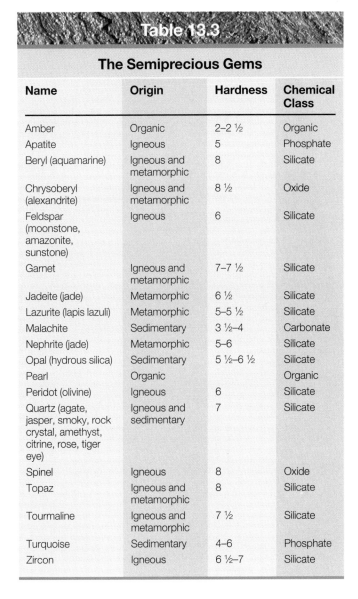

The Semiprecious Gems

Name	Origin	Hardness	Chemical Class
Amber	Organic	2–2 ½	Organic
Apatite	Igneous	5	Phosphate
Beryl (aquamarine)	Igneous and metamorphic	8	Silicate
Chrysoberyl (alexandrite)	Igneous and metamorphic	8 ½	Oxide
Feldspar (moonstone, amazonite, sunstone)	Igneous	6	Silicate
Garnet	Igneous and metamorphic	7–7 ½	Silicate
Jadeite (jade)	Metamorphic	6 ½	Silicate
Lazurite (lapis lazuli)	Metamorphic	5–5 ½	Silicate
Malachite	Sedimentary	3 ½–4	Carbonate
Nephrite (jade)	Metamorphic	5–6	Silicate
Opal (hydrous silica)	Sedimentary	5 ½–6 ½	Silicate
Pearl	Organic		Organic
Peridot (olivine)	Igneous	6	Silicate
Quartz (agate, jasper, smoky, rock crystal, amethyst, citrine, rose, tiger eye)	Igneous and sedimentary	7	Silicate
Spinel	Igneous	8	Oxide
Topaz	Igneous and metamorphic	8	Silicate
Tourmaline	Igneous and metamorphic	7 ½	Silicate
Turquoise	Sedimentary	4–6	Phosphate
Zircon	Igneous	6 ½–7	Silicate

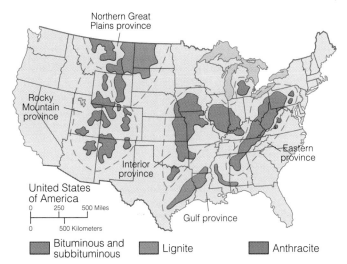

Figure 13.17 The distribution of coal fields and coal types in the United States.

70%), then *subbituminous coal,* and finally *bituminous coal* (carbon content 80% to 85%). Bituminous, or *soft,* coal, is the common variety in the extensive coalfields of western Pennsylvania, Ohio, West Virginia, Kentucky, and Illinois. Lignite and subbituminous coal are common in the western states of Wyoming and Montana (fig. 13.17).

Increasing carbon content yields higher heat values as measured in British Thermal Units (BTUs). A coal with 28,600 BTUs per kilogram (13,000 BTUs per lb) is theoretically more valuable than one with a lower rating. The burning of coal, however, has an environmental impact that must be taken into consideration. Unfortunately, many of the bituminous coals of the East and Midwest contain excessive amounts of sulfur and when burned yield sulfuric acid, a prime contributor to acid rainfall and other types of pollution. In general, the western coals, although lower in grade, contain less sulfur and occur at shallower depths and in thicker layers. These last two qualities make them exploitable on a large scale and at relatively low cost.

In northeastern Pennsylvania and in a few locations in Virginia, where the coal beds and other sedimentary rock layers have been subjected by continental collision to the additional pressures of intense folding, the bituminous coal has been metamorphosed to *anthracite (hard)* coal with a carbon content of approximately 95% (fig. 13.18). Additional stress caused by very deep burial and/or intense folding would, presumably, lead to the formation of the mineral graphite (pure carbon).

Even though coal formation is directly related to the growth of plants (a renewable resource), the quantity of plant material and the time involved make it a nonrenewable resource. It has been estimated that it requires 330 years of plant growth to form a 1-meter-(3.3-ft-) thick layer of peat and that it requires the compression of 30 meters (100 ft) of peat to form a 1-meter (3.3-ft) layer of soft coal. Thus, a meter of soft coal requires the accumulated plant growth of almost 10,000 years. To form a 3-meter (10-ft) coal seam such as the Pittsburgh coal, a major economic resource in Pennsylvania, West Virginia, and Ohio, would require the accumulated plant growth of 30,000 years.

Although nonrenewable, coal still represents a resource that is present in great abundance in the United States as well as in many other parts of the world. Its distribution, however, is not uniform. Some 95% of the world's coal reserves lie in Asia, North America, and Europe. Little is found in the Southern Hemisphere, but two such occurrences are rather interesting. South Africa, a country poor in oil but rich in coal, has for many years chemically converted enough coal to petroleum to satisfy most of its oil requirements. Antarctica, a continent almost totally buried beneath an ice sheet thousands of meters thick, possesses coal deposits, indicating a far warmer climate at various times in the geologic past and providing one line of evidence to substantiate the idea that continental positions are not fixed but have changed.

The United States could be self-sufficient in its production and usage of coal. Most coal produced today is burned in power plants to generate electricity (more than 50%) or, in the form of

Box 13.4 Further Consideration

Karats and Carats

Precious metals and gemstones are dealt with in such small quantities that special units of measurement are necessarily employed. The precious metals are traded in dollars per troy ounce,* with gold being approximately 100 times more valuable than silver and platinum somewhat more valuable than gold. Current values for the precious metals may be obtained from tables in the financial sections of most daily newspapers. The *purity* of precious metals, or objects made from them, is indicated in parts per thousand (*fineness*) or, in the case of gold, in number of parts per 24 (*karats*). Sterling silver, for example, has a fineness of 925, being alloyed with 75 parts of copper. Silver coinage, used for many years in the United States, was 900 fine, the other 100 parts being copper.

Gold, because of its inherent softness, is almost always alloyed with silver, platinum, copper, and/or nickel. Most commonly gold alloys are 12K (12 karat), 14K, 18K, or 21.6K, meaning that they contain 12, 14, 18, and 21.6 parts of gold out of a total of 24, respectively. These alloys equate to fineness values of 500, 585, 750, and 900. Gold coins, when they were still minted by the United States government, were 21.6K, or 900 fine.

The *weight* (not parts or percentages) of gemstones is universally measured by the other *carat*. This "metric carat" is equal to 200 milligrams (0.2 gr) and, in turn, is divisible into 100 *points*. The term "carat" is thought to derive from the "carob" tree (a member of the locust family), which is native to the Mediterranean region. Carob seeds were used from ancient times as a standard measure for gemstones, because the seeds are remarkably uniform in weight. Even so, there was always some slight variation in the value of the carat, because of the imprecise standard on which it

was based. To end this confusion, the metric carat was adopted by France, Italy, and Germany in 1907, by the United States in 1913, and by most other countries by the 1920s.

*1 troy ounce = 31.1 gram = 1.09714 avoirdupois ounce (1/16 of a pound).

Apply Your Knowledge

1. Is there any relationship between the carat and the karat?

2. Does the value of a gemstone increase arithmetically or geometrically as the carat size increases? Why?

3. Examine any gold or silver object to which you have access. How is its purity marked? If it is in karats, convert this to fineness or vice versa.

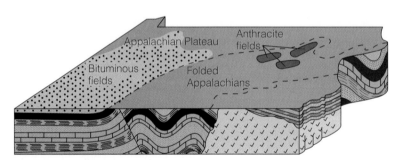

Figure 13.18   Location and geological setting of the Pennsylvania coal fields. High pressures associated with folding have converted bituminous coal to anthracite in the folded Appalachians.

coke with its upgraded carbon content, burned in blast furnaces to produce iron and steel. Minor amounts of coal are still used for the direct heating of homes and for a variety of industrial purposes.

A number of environmental problems, besides that of acid rainfall, are traceable to the mining and burning of coal. **Strip mining,** the common method by which shallow coals are removed from the ground, requires excavation of the soil and overlying rock layers, or *overburden,* to expose the coal. This procedure destroys the vegetation cover, retards soil development, and disrupts groundwater flow. Soil erosion and stream siltation become significant problems unless great care is taken both during the mining activity and with reclamation of the land. While the coal industry has made great strides in reclamation in recent years under the impetus of new federal and state laws, many parts of the old coal-mining regions of the East and

Midwest still bear the scars of careless strip mining as well as piles of waste rock from both surface and underground mines (figs. 13.19 and 13.20). New technologies for separating coal included in these waste dumps have made reclamation feasible. Screening and washing the waste rock in a water to which a special mud has been added allows the coal to float and separate from the waste in response to density differences.

The mineral pyrite (FeS_2), found both in the coals and in shale beds associated with the coals, is converted to sulfuric acid upon exposure to oxygen and water in the atmosphere and carried into the local stream drainage network. This condition is known as *acid mine drainage* and is a problem associated with both surface and deep coal mining. In addition to the acid, the iron present in the pyrite is oxidized and also becomes a stream pollutant, coloring the water red, giving the water a disagreeable taste, and inhibiting plant and animal life.

Figure 13.19 Regraded and reclaimed area in eastern Ohio that was once stripped for coal.

Figure 13.20 Mine refuse, or "gob," pile in eastern Ohio.

Subsidence, an environmental problem specifically related to underground mining, was discussed as a form of mass wasting in chapter 8. An additional environmental problem encountered, especially in the older coal-mining regions, is fire. Both the underground mines and the surface piles of waste rock that came from them are susceptible to catching fire either by lightning strikes or human carelessness. Mine fires can be most hazardous, and in extreme cases they have led to the abandonment of buildings or even towns situated above the burning mine or adjacent to the burning spoil bank.

Petroleum

The other major fossil fuel is **petroleum.** While not a sedimentary rock in the sense that coal is, petroleum is found almost exclusively *in* sedimentary rocks. Petroleum is a complex mixture or hundreds of chemical compounds representing all three states of matter: solid (*asphalt* or *bitumen*), liquid (**oil**), and gas (**natural gas**). With the exception of some contaminants, these materials are all **hydrocarbon compounds,** that is, compounds composed predominantly of hydrogen and carbon.

It is thought that the hydrocarbons represent the altered remains of microscopic organisms that lived in the seas. When the organisms died, their remains settled to the sea bottom, where they escaped complete decay and became incorporated into the accumulating sediment. With time, heat, and pressure these remains were converted to hydrocarbons, which were squeezed out of the rock layer, or **source rock** (usually shale), in which they originated to begin a slow migration upward through the surrounding rocks. As long as these rocks were permeable, the hydrocarbons continued to move. Eventually, however, they would come up against impermeable rock that would prevent their further migration. The hydrocarbons would come to rest and accumulate in environments that petroleum geologists refer to as **traps,** awaiting only the enterprising geologist (and driller) to discover their presence. Figure 13.21 shows some of the common petroleum traps, the geological origins of which have been considered in earlier chapters. The rock layer within which the petroleum eventually collects is called the **reservoir rock** and is most often a sandstone or a limestone.

Oil and natural gas (and the two almost always occur together) within the pore spaces of a rock layer thousands of meters deep are under extremely high pressure. When the reservoir rock is penetrated by a drill, it is very much like sticking a pin in an inflated balloon. There is a sudden drop in pressure and oil and gas rush up the borehole toward Earth's surface in response to the pressure difference between reservoir rock and ground surface. If uncontrolled, the uprushing oil and gas may reach and destroy the drilling platform at the surface. This condition is known as a *blowout,* and modern drilling methods anticipate and prepare to counteract this. Under controlled conditions, then, oil and gas may flow all the way to the surface, where they are led through a series of valves into collecting lines and holding tanks. If oil in the reservoir rock is under less pressure, it may need to be brought to the surface by pumping.

Normally 80% to 85% of the oil in the reservoir rock remains in place and does not flow or cannot be pumped to the surface. A number of ingenious methods, collectively called *enhanced recovery techniques,* have been devised to recover additional amounts of this oil. In one widely used and highly successful technique, the *waterflood,* water is pumped under pressure down a well bore and into the reservoir rock (fig. 13.22a). The hope is that this water will push some of the oil ahead of it to another well, from which the oil can be pumped. *Gas injection* techniques employing natural gas, carbon dioxide, or nitrogen can achieve the same result. In fact, natural gas that is normally produced with the oil can be separated and pumped back into the reservoir rock to force out additional oil. Other procedures attempt to make the oil more fluid, and therefore more mobile, by heating it. This is accomplished by injecting steam into the reservoir (fig. 13.22b) or igniting some of the oil in the reservoir, causing it to burn and heat the remaining oil. Newer methods, mostly still in the experimental or pilot stage,

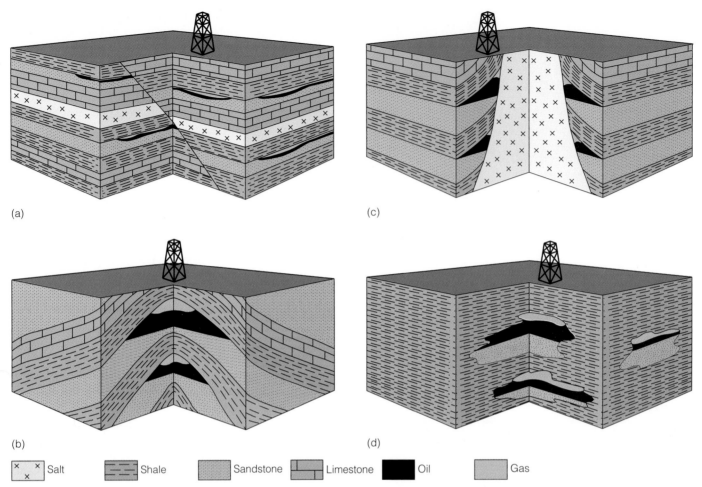

(a)

(b)

(c)

(d)

| × × Salt | Shale | Sandstone | Limestone | Oil | Gas |

Figure 13.21 Principal types of petroleum traps: (a) fault, (b) anticline, (c) salt dome, and (d) stratigraphic, or "pinchout."

involve the use of chemical solvents to dissolve and remove the oil from the reservoir (fig. 13.22c).

We are quite familiar with the uses of petroleum and natural gas for transportation, heating, and electric power generation, but their uses go far beyond that. These fossil fuels also form the basis of the petrochemical industry and as such are the raw materials from which thousands of widely used products are manufactured. These include plastics, medicines, fertilizers, explosives, synthetic fabrics (rayon, nylon, etc.), rubber, teflon, alcohol, antifreeze, and a host of others.

Most of the world's known reserves of petroleum lie in the Middle East (Saudi Arabia, Kuwait, Iran, and Iraq). The United States' position with respect to these liquid fuels is less than desirable, but enormous deposits of hydrocarbon-rich rocks do lie beneath southwestern Wyoming, northeastern Utah, and northwestern Colorado. These are the so-called **oil shales** (fig. 13.23).

Although the potential for oil shale usage is enormous, the developers of these deposits will not only have to overcome problems of technology and economics but those of a social and environmental nature as well. Several methods of exploiting the oil shales have been considered. Most likely they would be

mined by conventional strip (where overburden is not too thick) or deep mining techniques and the organic compounds converted to hydrocarbons by some type of retorting (heating) process. A difficulty with this approach is that the removal of the hydrocarbons leaves a pulverized waste rock that actually occupies a greater volume than the original mined rock. Thus, a significant waste disposal problem would exist. Processing of the oil shales uses a large quantity of water, which is not readily available in a semiarid region. Such water usage would have to compete with ranching and irrigated farming. Lastly, the infrastructure of this region would have to be developed and social problems related to housing, schools, and services for growing communities in an area of very low population density would have to be solved. Current economic conditions do not allow for the development of oil shale resources beyond the pilot feasibility stage. Whenever economic and social pressures outweigh the obvious environmental concerns about mining oil shale, the United States may resort to the exploitation of this valuable oil reserve.

A similar nonliquid petroleum resource is found in abundance along the Athabasca River in the Canadian province of Alberta. These are *tar sands,* sandstones impregnated with solid

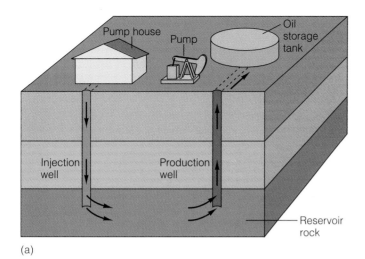

(a)

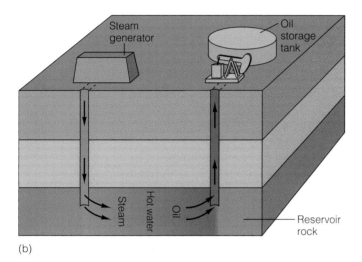

(b)

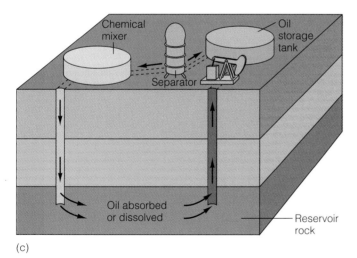

(c)

Figure 13.22 Enhanced oil recovery techniques: (*a*) waterflood, (*b*) steam injection, and (*c*) chemical solvents.

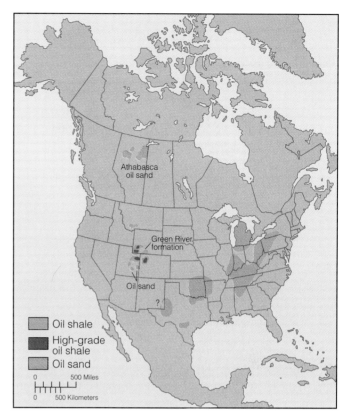

Figure 13.23 Oil shale and tar sand deposits in North America.

asphaltic hydrocarbons. These deposits have already been exploited for a number of years by strip mining. The mined rock is transported by an extensive system of conveyor belts to a processing plant, where the petroleum is separated from the rock, liquified, and sent by pipeline to conventional oil refineries around the city of Edmonton.

Further Thoughts: Conserve, So That All Can Share

Earth, during its formation and throughout its history, was blessed with abundant resources. Largely untapped until recent years, these resources now are being consumed at varying, but generally very high, rates. In but the rarest of cases nature cannot replace them. As world population increases and as developing countries attempt to reach the level of development enjoyed by the United States, Canada, and western Europe, these resources are being stretched perilously thin. Whether such development can ultimately extend to all of the peoples of the world is doubtful. What is certain is that we must be far more careful stewards of nature's gifts than we have been in the past. We must be less wasteful of what we use, and leave uncontaminated what we do not need. Care of the planet's resources can ensure an adequate supply of resources for future generations.

Significant Points

1. A natural resource is anything taken from Earth to support life or a way of life. Nonrenewable resources form over long periods of time and are present in finite quantities.
2. A soil is that portion of the loose surface material that is capable of supporting plants with root systems. It evolves by weathering over thousands of years and typically consists of several distinct layers known as horizons.
3. Of the factors that lead to soil development (time, parent material, topography, climate, and organisms), climate (temperature and rainfall) is the most important.
4. The principal ore mineral of aluminum is bauxite, a tropical weathering product, that is mined from large, shallow, open pit mines.
5. The most important iron ores are banded iron formations (BIF), or taconite, and are typified by those mined in the Lake Superior region. They consist of the oxides hematite and magnetite mixed with quartz-rich rock.
6. Gold, diamonds, and other heavy, resistant minerals are found along stream channels and beaches in secondary sedimentary deposits known as placers.
7. The industrial minerals, because they are all abundant, are classified by use rather than by abundance.
8. Halite and sylvite, examples of industrial minerals, are classified as evaporites because they were precipitated from marine waters as those waters underwent evaporation.
9. Potassium (potash), phosphorous, and nitrogen are three elements necessary for plant growth that must be returned to the soil in the form of inorganic fertilizers.
10. Because a use can be found for virtually every type of rock and mineral, building products represent the largest volume production of all the natural resources.
11. Abrasives are characteristically hard minerals and are used for cleaning, cutting, grinding, polishing, and drilling.
12. Coal formed from the compression of plant material that once grew in coastal swamps.
13. Petroleum is a complex mixture of hydrocarbon compounds formed from the organic remains of microorganisms that once lived in the seas. After being squeezed out of its source rock, it migrates through permeable rock strata until it is trapped in a reservoir rock by an impermeable rock barrier.
14. The oil shale of the western United States represents an important future petroleum resource if economic, technological, and environmental problems can be overcome.
15. The tar sands of northwestern Canada have been mined and processed for their petroleum content for a number of years.

Essential Terms

natural resource 330	pedocal 331	salt dome 340	oil 349
renewable resource 330	laterite 332	fertilizer 342	natural gas 349
nonrenewable resource 330	metal 332	abrasive 344	hydrocarbon compound 349
reserve 330	abundant metal 332	gemstone (gem mineral) 345	source rock 349
soil 330	rare metal 332	fossil fuel 346	trap 349
horizon 330	ore mineral 332	coal 346	reservoir rock 349
soil profile 330	ferroalloy 333	strip mining 348	oil shale 350
leaching 331	vein 337	petroleum 349	
pedalfer 331	placer deposit 339		

Review Questions

1. What features characterize all of the nonrenewable resources?
2. What are the two fundamental soil types in the United States? What are their essential characteristics? How do they differ from one another? How do they relate to climate?
3. What are the soils of the humid tropics? How do they differ from those of the United States and Canada?
4. What is the definition for an ore mineral?
5. What scrap metals are commonly recycled?
6. Name the important ferroalloys. What is the principal use of each?
7. What is the principal ore mineral of copper? What is its geologic origin? Where are some important geographic occurrences?
8. Describe the two common occurrences of the evaporite mineral halite.
9. What are the most commonly used dimension stones?
10. What are the most common uses of crushed rock and sand and gravel?
11. How do prepared rock products differ from natural rock products?
12. What are gemstones? What are the principal categories of gemstones? Give some examples.
13. What are some of the environmental problems associated with the mining and burning of coal?
14. What are some of the enhanced recovery techniques for petroleum?

Challenges

1. Make a survey of your town and its environs to ascertain what metal processing plants, if any, exist there (hint: you might start by consulting the telephone directory). You could pursue this further by talking with plant officials to determine where the raw materials (ore, fuel, etc.) come from.
2. Are there any mining operations near where you live? If so, you might investigate how the mining is done and what measures are taken to protect the environment. Do these measures meet the standards specified by federal, state (or provincial), and local laws?
3. Investigate any abandoned mining operations to determine whether they still pose a hazard to the environment and in what ways. If they still do, what are the reasons that nothing has been done about them?
4. What are some observable characteristics of the soil in your area? Can you identify its type from those characteristics? If you live in an agricultural region, ask the farmers what treatments are necessary to maintain the soil's fertility.
5. Is oil or natural gas produced near where you live? If so, consult with local geologists, petroleum engineers, or oil field workers to compile a report that includes information about the reservoir rock, type of trap, amount of production, and any enhanced recovery techniques employed.
6. Pick a metal and compile a report that deals with its geologic origin, mining, processing, and final manufactured products. Pay particular attention to the geography of where all of this occurs and what role the metal plays in the world economy.

THE HISTORY OF NORTH AMERICA

In the late nineteenth century, in a letter to Bishop Mandell of the Church of England, Lord Acton wrote, "Advice to Persons About to Write History—Don't." The thousands of years of human history that motivated Lord Acton to make that comment are just a small fraction of Earth's history. If a history of human existence, for which a considerable written record has been kept, is difficult, then how difficult is the task to summarize billions of years with no written record? To avoid the obvious problems involved in writing a comprehensive and complete history of our planet, we have chosen to highlight some of the significant events that have shaped our continent.

AT THE END OF PART 7, YOU WILL BE ABLE TO

1. identify the basic questions that led to the development of earth history;
2. distinguish between absolute and relative age dating;
3. classify geologic time;
4. relate Earth's early atmosphere to the rise of life;
5. list the major geologic events in the history of North America.

from "A Song of the Rolling Earth"
from Walt Whitman's Leaves of Grass

The earth does not withhold, it is generous enough,
The truths of the earth continually wait, they are not so
conceal'd
either,
They are calm, subtle, untransmissible by print,
They are imbued through all things conveying themselves
willingly.

EARTH HISTORY: HOW THINGS GOT TO BE THIS WAY

Dinosaur footprint preserved in rock at Dinosaur State Park, Rocky Hill, Connecticut.

Chapter Outline

INTRODUCTION

When Alex Haley's book *Roots* became a popular American television movie in the 1980s, many who saw the production became newly interested in their origins. Of course, it might seem natural for every generation to find some meaning in the past, and museums, antiques, and historical records survive because a continuous line of preservationists maintain the artifacts of life for posterity.

The planet, too, has its roots, and interest in the origins and history of Earth precede modern science. In fact, explanations for Earth's origins are among the earliest of writings and oral traditions, many of which serve as the bases of religions and mythologies. Scientists, however, must rely on quantifiable information and testable hypotheses when they propose a model of the physical world. Unfortunately, much of the planet's history appears to have preceded the advent of intelligent life, so no one has gathered firsthand observations on the processes of planetary formation and early development.

For at least the last 300 years scientists, philosophers, mathematicians, and others have not only struggled to understand the natural laws that govern Earth, but they have also tried to answer the more fundamental questions of how Earth formed and how it attained its present form. From your reading of chapter 2, you now know that Earth had its origins in the Nebular Cloud and the accretion of planetesimals. You also know that the young planet underwent an intensive outgassing of its interior (chapter 4). The volatile gases that escaped during volcanic activity on the early Earth led to the formation of both the atmosphere and the oceans, initiating the hydrologic cycle and establishing the conditions under which life first evolved.

An encompassing earth history must account for more than the planet's origins. Shielded in the darkness of an unobserved past, the story of this planet also must tell of life and its relationship to the home planet. This chapter, therefore, recounts significant physical and biological events that mark the path to the present.

History is an orderly account of both events and relationships connected by either time or cause. In human history oral traditions, documents, artifacts, and architecture are the clues used in organizing events. The clues for earth history, however, often provide only partial information about events and relationships. Thus, to relate causes and their effects and to place earth events chronologically requires scientists to adopt the principle of **uniformitarianism,** a doctrine that today's earth processes are not different in kind from ancient processes. If streams erode the landscape today and deposit their sediments in basins, for example, they did so in the past, leaving evidence of their action as valleys and sedimentary deposits.

Because the historical clues are frequently unclear—and sometimes missing entirely—earth historians are still writing the story of our planet. In their efforts they are guided by three basic questions.

1. How can we date Earth's origin and the events that shaped our planet?

2. What sequence of events produced the features we observe on our planet?

3. How did life evolve and persevere?

To answer these larger questions, earth historians seek the answers to many smaller ones. When did life first invade the land? Are all rocks the same age? What is the significance of fossils, that is, those relicts and traces of ancient life found in many sedimentary rocks? Are these mysterious ancient life forms related to the life of our time? Ultimately, historical inquiries must be framed on some type of *calendar* that organizes events. The first question, therefore, is about time. What is the temporal measuring stick that enables an unwritten history to unfold?

RELATIVE AND ABSOLUTE DATING

No history is possible without a point of reference. Our count of years, for example, begins approximately with the birth of Christ. Any reference, however, would suffice, and other cultures and civilizations outside the influence of Christianity begin their account of time on different bases. Like social historians, those who trace the planet's history need an acceptable reference.

Relative Dating Methods

Efforts to understand the history of Earth were advanced by two principles developed in the seventeenth century. The first is the **principle of superposition,** which explains that in a sequence of sedimentary rock layers the oldest layers, or beds, are overlain by younger sediments. For example, if you travel downward to the bottom of the Grand Canyon, you will encounter increasingly older rocks (fig. 14.1). The second, called the **principle of lateral continuity,** is an acknowledgment that sedimentary strata extend in all directions until they encounter an obstacle like a basin edge or thin to zero thickness. You can experience something similar to this in the microbasin of a lagoon behind a barrier island. The muds thin toward the mainland and seaward toward the sands of the island. These two principles enable scientists to trace sedimentary layers over large regions and identify their time relationships. Whenever geologists use both principles for identification, they *correlate,* or match, sedimentary rocks.

Correlation, therefore, is a method for the **relative age dating** of sedimentary layers and the fossils found within them through a comparison with similar rocks and fossils. Such relative age dating serves a very useful purpose. You can say, for example, that one person is older or younger than another without knowing an exact age for either. If, according to the principle of superposition, you know that sedimentary bed A lies below bed B, then you know that the fossils within A preceded those in B (fig. 14.2). Similarly, the fossils themselves may be useful correlators. You can recognize that two widely separated beds are from the same time period by identifying similar fossils in both.

Relative dating can also be used for crystalline rocks. A dike that cuts across sedimentary layers must be younger than the rock it intrudes according to the **principle of crosscutting** (fig. 14.3). Crosscutting cannot occur unless the older rocks already exist. Although relative age dating is a valuable process for historical geologists, it does not reveal the exact absolute age in numbers of years for earth materials.

Absolute Dating Methods

Absolute age dating is essential to historical earth studies because it gives us the number of years ago that something happened. When no correlation is available, such absolute dates reveal the chronology of Earth's development. The task of dating something definitively is not an easy one. Finding precise ages for earth materials requires a system that does not vary over time, so that scientists can use it as a measuring device. If you want to measure a desk or a room, you know you can rely on the dimensions of a meterstick or a yardstick. Neither stick changes length, and both can be used year after year to obtain accurate measurements. Only in the twentieth century has such a measuring "stick" become available for geologic time studies. That measure is **radioactive decay,** the process through which **parent atoms** ultimately change to **daughter atoms** by losing or gaining subatomic particles.

A key element in radioactive age dating or radiometric dating is uranium. One isotope of this element, *uranium 238,* is very unstable, meaning that it splits apart or decays to become an isotope of another element. The process of decay occurs in a series of changes, each of which involves either the loss or the addition of subatomic particles to the nucleus of the atom.

Two important kinds of decay occur. One is **alpha decay,** which is the loss of a helium nucleus (two protons and two neutrons) from the uranium nucleus. With the loss of two protons, element 92 becomes element 90. In fact, every time an alpha particle leaves the parent atom, the atomic number decreases by two. The other type of decay is **beta decay,** which is essentially the loss of an electron from the nucleus as a neutron becomes a proton. Beta decay results in an increase by one in the atomic number. The ultimate product of radioactive decay is stabilization in a daughter isotope. In the case of uranium 238, that daughter isotope is *lead 206* (fig. 14.4). The significance of this decay series is that the *rate* at which it occurs is known. Although one cannot predict just which U 238 atom will decay at any given time, scientists do know that one-half of the atoms decay within a particular period. This period of time is the **half-life** (fig. 14.5). Note that the line in figure 14.5 is curved rather than straight. This indicates that the radioactive decay process is not a constant time process but one in which less material decays during each half-life time interval. In other words, during the first half-life interval, one-half of the original material decays. During the second interval, one-half of the remaining material decays (one-half of one-half, or one-fourth of the original material).

Only minerals that contain radioactive isotopes (usually in very small amounts), such as U 238, can be used for radioactive dating. U 238 has a known half-life of 4.5 billion years. Dating of rocks and minerals older than approximately 100 million years

is possible because of this long half-life. Dating events younger than 100 million years is difficult because very little of the parent isotope will have decayed in this "brief" interval. Other elements, such as potassium 40, thorium 232, and carbon 14, are also used in obtaining absolute dates. Carbon 14 is especially useful for dating materials younger than 50,000 years because it has a short half-life, only about 5,800 years. Dating events older than this with C_{14} is impossible because little of the original parent isotope remains.

With the tools of relative and absolute dating, scientists have worked out a **geologic time scale,** that is, a time line that runs back to Earth's origin 4.55 billion years ago. This time scale was initially a relative one, but radiometric dating has fixed reasonably accurate dates onto the great calendar of events in Earth's history.

THE GEOLOGIC TIME SCALE

The geologic time scale is the product of many minds, and it is still being refined by earth scientists. The scale evolved from studies of rocks and fossils in numerous locations, and it reads, in part, like a geography lesson. For example, the time between 360 and 250 million years ago is divided into the Mississippian and Pennsylvanian periods based on significant occurrences of rocks of those ages in the upper Mississippi River valley and the state of Pennsylvania, respectively. The ensuing 45-million-year span is called the Permian period, from rocks of that age in the Perm Basin of central Russia. Other geographical names include the Cambrian period, after a Roman name for Wales and the Jurassic period, for the Jura Mountains in France. The naming system is not, however, solely geographical. Some names reflect the character of the rocks themselves. For example, the name Cretaceous is derived from the Latin word *creta,* meaning "chalk," in reference to the chalk cliffs along the English channel. Other names on the geologic calendar refer to the character or abundance of the life of that time as expressed in the fossil record (e.g., *Mesozoic* means "middle life"). The geologic time scale now categorizes time, in descending order of magnitude, into eons, eras, periods, epochs, and ages.

Geologists recognize two eons. The first is called Prephanerozoic (*pre,* meaning "before"; *phaneros,* meaning "visible"; *zoio,* meaning "animal") because it covers the time for which there is no sign of *metazoan,* or multicellular, life. The other eon is the Phanerozoic, an eon of metazoan life forms. The Phanerozoic is further subdivided into three eras: the Paleozoic ("ancient life"), the Mesozoic, and the Cenozoic ("recent life"). The Prephanerozoic is also known as Precambrian (table 14.1). The terms have evolved over the last 300 years and have provided scientists with those temporal reference points so necessary for the study of Earth's history.

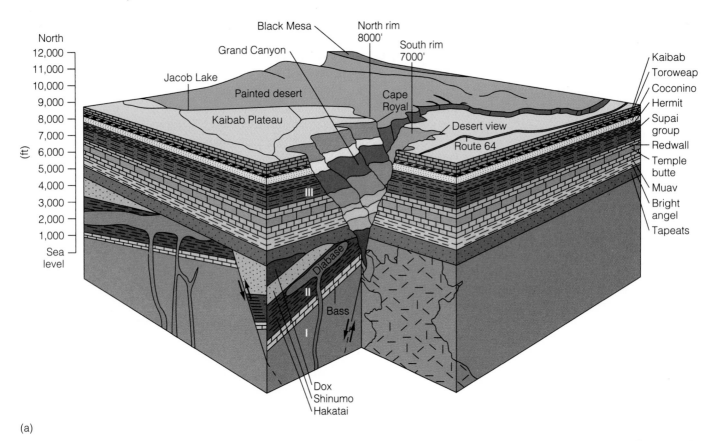

North
12,000
11,000
10,000
9,000
8,000
7,000
(ft) 6,000
5,000
4,000
3,000
2,000
1,000
Sea
level

Black Mesa
Grand Canyon
Jacob Lake
Painted desert
Kaibab Plateau

North rim
8000'
South rim
7000'
Cape
Royal
Desert view
Route 64

Kaibab
Toroweap
Coconino
Hermit
Supai
group
Redwall
Temple
butte
Muav
Bright
angel
Tapeats

Diabase
Bass
Dox
Shinumo
Hakatai

(a)

Figure 14.1 (a) The geology and regional setting of the Grand Canyon in Arizona. (b) Cross section showing the sequence of rocks and their geologic ages.

The Early Atmosphere and the Rise of Life

Just as the atmosphere supports life today, its initial constituents provided the circumstances necessary for the rise of life more than 3 billion years ago. Studies of volcanic outgassing and atmospheric chemistry have provided clues about the composition of Earth's early atmosphere and the effect it most likely had on the rise of life.

Early Reducing Atmosphere and Organic Compounds

Although the exact composition of Earth's early atmosphere is difficult to ascertain, most scientists agree that it differed from our present gaseous envelope. In our present atmosphere, oxygen plays an important role because it is highly reactive. The abundance of oxygen (21%) gives us the *oxidizing atmosphere* that allows wood to burn (rapid oxidation) and apples to rot (slow oxidation). Because molecular oxygen (O_2) in our atmosphere is predominantly a by-product of photosynthesis, increasing amounts of this gas were produced as plants converted sunlight, carbon dioxide, and water into food.

Before the evolution of plants, however, no known source of molecular oxygen could have established an oxidizing con-

dition similar to what we experience today. Even if such a source were available, it would have inhibited the rise of life because of oxygen's affinity for organic compounds. Instead, a *reducing atmosphere,* that is, one that contained carbon dioxide, methane, and other gases but little or no free oxygen, most likely contributed to the abiotic production of life. Although no one knows the precise mechanism that triggered the rise of life, many scientists believe that the stuff of life was readily available on the early Earth. The six elements deemed essential to life—hydrogen, oxygen, carbon, nitrogen, sulfur, and phosphorus—were present in both the early atmosphere and oceans because of outgassing and chemical reactions.

The earliest life forms were *anaerobes,* that is, organisms such as bacteria that can live in the absence of free oxygen. Although no one can account for the initial rise of these life forms, Stanley L. Miller, as a graduate student working under Professor Harold C. Urey at the University of Chicago, combined the elements of the suspected early atmosphere to form organic compounds. Miller's 1953 experiment did not produce life, but it did show that organic compounds could form under abiotic (or prebiotic) conditions. To fuel the reactions that produced these compounds, Miller used an electric sparking device. On the newly formed Earth, lightning and ultraviolet radiation could have served as the source of energy to drive the reactions. Because some organic molecules are relatively short-lived and all organic molecules dissociate under high temperatures, some

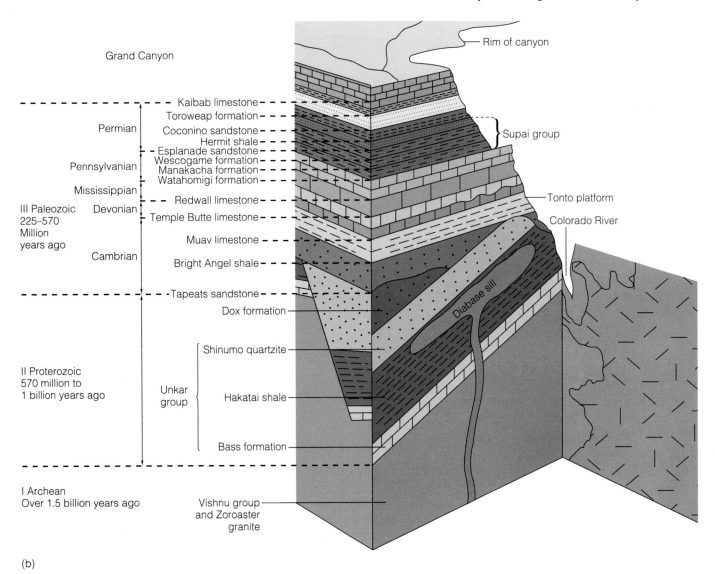

Grand Canyon

Rim of canyon

Kaibab limestone
Toroweap formation
Permian — Coconino sandstone
Hermit shale
Esplanade sandstone
Wescogame formation
Pennsylvanian — Manakacha formation
Watahomigi formation

Mississippian

III Paleozoic   Devonian — Redwall limestone
225–570
Million   — Temple Butte limestone
years ago

Cambrian — Muav limestone

Bright Angel shale

Tapeats sandstone

Dox formation

Shinumo quartzite

II Proterozoic
570 million to
1 billion years ago   Unkar
group

Hakatai shale

Bass formation

I Archean
Over 1.5 billion years ago

Vishnu group
and Zoroaster
granite

Supai group

Tonto platform

Colorado River

Diabase sill

(b)

Figure 14.1   continued.

scientists, including Stanley Miller, suggest that life formed over a relatively brief period, possibly within tens of thousands of years. The initial life, therefore, appeared after Earth's surface cooled sufficiently for organic molecules to survive.

Once life obtained a niche on the planet, it altered its forms and processes, eventually becoming capable of turning sunlight, water, and carbon dioxide into sugars. Thus, the carbon dioxide produced by outgassing and dissolved in ocean water made life possible for photosynthesizing plants and enhanced the evolution and diversification of the plant kingdom.

The Oxidizing Atmosphere and the Proliferation of Life

As photosynthesizers produced oxygen, the reactive gas would have first combined with mineral matter in seawater and in surface rocks rather than accumulating in the atmosphere. Only after most surface mineral material had been oxidized was oxygen free to accumulate.

With the rise in atmospheric oxygen we might expect the evolution of air-breathing animals, the *respirers,* but Earth had to pave the way for its land dwellers. Instead, respiration occurred first in the sea partly because the land was an inhospitable place. Because ultraviolet radiation is toxic to both plants and animals (see chapter 1), photochemical reactions that produced a stratospheric ozone layer had to occur before Earth's *subaerial* (under air) environments became safe for life. All of these chemical reactions took considerable time, and Earth did not house land-dwelling respirers until amphibians evolved and proliferated about 370 million years ago (fig. 14.6).

Living on the land meant adapting to an environment radically different from that in which life first evolved. In addition to the changes that took place in the composition of the atmosphere, life itself had to adapt to an environment in which water was not always readily available and dessication was a constant threat. In addition, life on land means adapting to the pull of gravity, resulting in strong skeletons for large vertebrates. Life's invasion of the land, therefore, depended on

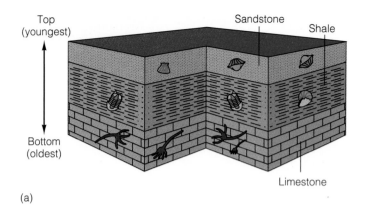

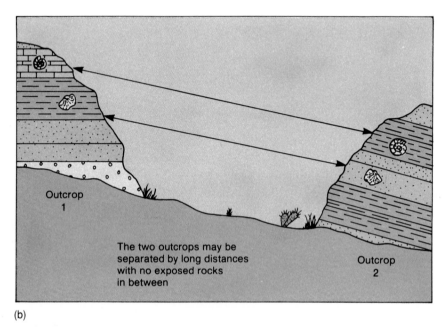

Figure 14.2 (a) The law of superposition states that in an undisturbed sequence of sedimentary rocks the oldest layer is on the bottom and the youngest layer on top. A corollary is that the fossils contained in the lowest layer (limestone) are older than those in the highest layer (sandstone). (b) The rock layers indicated by arrows are correlative because they contain the same fossils, even though the lithologies (rock types) are different.

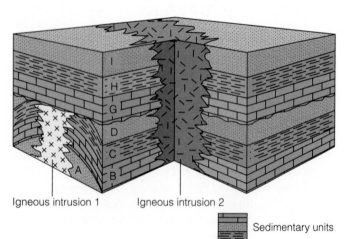

Figure 14.3 Principle of crosscutting relationships. Igneous intrusion 1 must be younger than layers A, B, and C, but not G, H, and I. Igneous intrusion 2 must be younger than all the rock layers since it cuts them all.

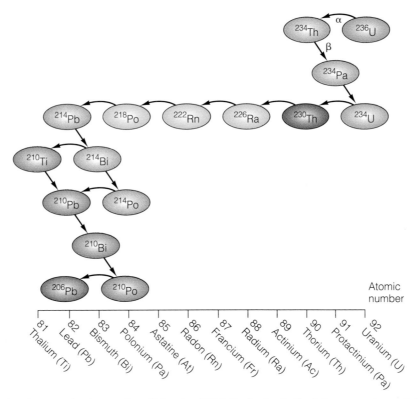

Figure 14.4 The radioactive decay series for uranium 238 to lead 206. Each step to the left represents an alpha decay. Each step down and to the right is a beta decay.

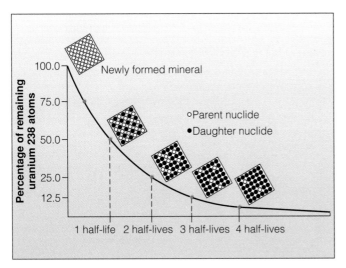

Figure 14.5 The results of radioactive decay across four half-lives (18.0 billion years) in terms of the amount of parent nuclide (U 238) remaining and daughter nuclide (Pb 206) formed.

Figure 14.6 Model of *Diplovertebron punctatus,* a Pennsylvanian age amphibian representative of the ichthyostegid body style.

Table 14.1

The Geologic Time Scale

Eon	Era	Period	Millions of Years Before Present
PHANEROZOIC	Cenozoic	Quaternary	2
		Tertiary	65
	Mesozoic	Cretaceous	144
		Jurassic	208
		Triassic	245
	Paleozoic	Permian	286
		Pennsylvanian	320
		Mississippian	360
		Devonian	408
		Silurian	438
		Ordovician	505
		Cambrian	570
		Ediacaran	700
PREPHANEROZOIC	Late Proterozoic		900
	Middle Proterozoic		1600
	Early Proterozoic		2500
	Late Archean		3000
	Middle Archean		3400
	Early Archean		3800
	Earth forms		4550

more than the mere accumulation of molecular oxygen in the atmosphere.

THE PRECAMBRIAN OR PREPHANEROZOIC EON

The Precambrian eon accounts for more than 85% of all earth history. In some localities, Precambrian rocks are quite ordinary looking shales, sandstones, and limestones distinguishable from their younger counterparts only by a general lack of fossils. Indeed, such sequences might stretch without noticeable break across the time boundary and into the Phanerozoic. In most places, however, Precambrian rocks are great masses of igneous and metamorphic rocks, characterized not only by a lack of fossils but by showing signs of exposure to long periods of erosion. Precambrian rocks are found in the lower parts of deep canyons, in the cores of mountains, and, most especially, as broad, low-lying expanses called **shields,** which form the cores of all the continents (fig. 14.7).

Because the fossils used to subdivide younger rock sequences are missing in Precambrian rocks, geologists have subdivided the shields on the basis of rock types, rock structures, and especially *rock ages* into distinct areas known as *mineral age provinces*. The central core region of the North American continent is the expansive Canadian Shield on which geologists have identified various provinces (fig. 14.8).

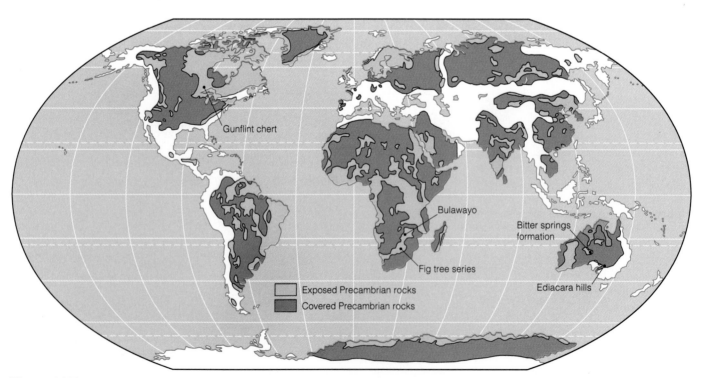

Figure 14.7   Precambrian shield areas of the world.

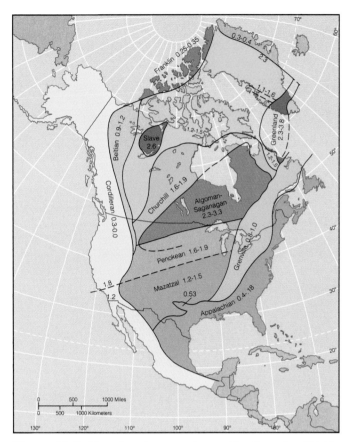

Figure 14.8 The mineral age provinces of the Canadian Shield.

The Canadian Shield

The Canadian Shield is a low-lying, glaciated area of 4.66 million square kilometers (1.8 million sq. mi) covered at various places by coniferous forests, muskeg, or tundra vegetation. The rocks are largely granites and gneisses but with infolded and preserved pockets of sedimentary and volcanic rocks. Each mineral age province probably represents a rock cycle consisting of the following:

1. deposition of sedimentary rocks and volcanism;
2. mountain building, regional metamorphism, and batholithic intrusion;
3. erosion to a low-lying surface;
4. stabilization and attachment of the region to the continent.

Such a cycle probably encompassed several hundred million years.

Rocks (and the history) of the Canadian Shield fall into two broad time frames: the Archean and the Proterozoic (table 14.1). The Archean sequences contain the oldest rocks known from North America: gneiss from western Greenland dated at 3.98 billion years. Other rocks of interest include *greenstone belts* composed of metamorphosed basaltic lava flows and ultrabasic sills and dikes.

This early phase of North American Precambrian history was closed by a period of extraordinarily great mountain build-

ing referred to by geologists as the *Kenoran* or *Algoman orogeny*. The latter portion of the Precambrian eon, the Proterozoic, is known in somewhat more detail, if only because the rocks are more widely exposed and less eroded.

The Proterozoic

Significant rock types formed during the early Proterozoic include the banded iron formations (BIF) of the Mesabi and Cuyuna ranges of Minnesota, chert (microcrystalline quartz) beds containing what are thought to be the remains of one of Earth's first life forms, the primitive unnucleated cells of *cyanobacteria,* and hardened glacial deposits called *tillite,* which indicate an early, widespread period of glaciation.

During the middle portion of the Proterozoic, extensive basaltic lava flows were extruded in the area that is today the Keweenaw Peninsula of Michigan. The *vesicles,* or openings (caused by gas bubbles), of these basalts were subsequently filled with native copper. In addition to the lava flows, plutonic activity also occurred as intrusions of dark igneous rock. Rocks of the late Proterozoic look increasing like the sedimentary rocks of the younger geological eras and in several areas they pass without break into those younger rock sequences. Near the end of the Proterozoic era there is evidence once more of a worldwide episode of glaciation, with tillites of this age being found on all of the continents except Antarctica (fig. 14.9).

Life in the Precambrian

At some point, probably very early in the Precambrian, single-celled life evolved. Because all Precambrian fossils are of marine origin, it follows that life probably originated in the oceans, in the atmosphere, or, if on land, then along the margin of the ocean.

Throughout most of Precambrian time, life was relatively primitive, consisting largely of algae and bacteria. Probably the most significant change in life during this vast eon was the evolution of *nucleated cells, or eukaryotes.* Toward the close of the Precambrian, however, a variety of multicellular organisms evolved in what must have been a period of rapid, or "explosive," evolution. Some forms of the new multicellular life probably did not survive as fossils, and many life forms were failed natural experiments with little chance of prolonged survival. Other multicellular forms may have endured through long periods only to be lost as fossils because of the chancy nature of preservation. The fossil record is obviously lopsided, and though many organisms are well preserved, others are totally or partially lost.

As the geologic clock ticked from the Precambrian into the Phanerozoic, the progressively more abundant and relatively complex new life forms began to leave a fossil record in the rocks. This sudden increase in the fossil record at the Prephanerozoic-Phanerozoic boundary has long been the evidence used by geologists to delineate these eons.

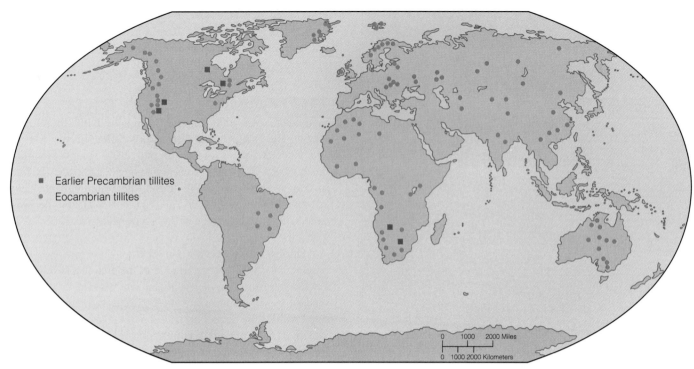

Figure 14.9 Precambrian tillite exposures.

Precambrian Resources

Interest in Precambrian rocks stems from more than a recognition of their relationship to early life. The *economic resources of the Precambrian are extensive* but confined largely to the metals, such as the iron of the BIF (see chapter 13). Of special importance are the precious metals, deposits of which include the gold-bearing "reefs" of South Africa; gold deposits around Porcupine, Ontario, and Lead, South Dakota; silver (plus cobalt) from Cobalt, Ontario; and platinum in the Bushveldt Complex of South Africa. The Bushveldt Complex also produces chromium. Uranium is derived from Precambrian rocks in Canada and South Africa, copper from Michigan and Canada, and nickel from the world's most important deposits around Sudbury, Ontario. In fact, most of Canada's great mineral resources are found in Prephanerozoic rocks.

THE FIRST PHANEROZOIC ERA: THE PALEOZOIC

Earth historians divide the time after the Precambrian primarily on the basis of life, and they recognize three geological eras: the Paleozoic, the Mesozoic, and the Cenozoic. In addition to differences in life forms, each is characterized by certain physical events. These events, while distinctive in time and recognizable by geologists, nonetheless are of several common types:

1. tectonic events, such as mountain building episodes and the breakup of continents;
2. geomorphologic events, such as the erosion of landforms by ice and water;
3. depositional events, such as the infilling of basins by sediments
4. eustatic or sea-level events, such as incursions of seas onto lands.

Geologists interpret such ancient events from the record preserved in the rocks themselves: their *types, compositions, locations, and orientations*. While geologic histories have now been worked out for most continental areas, for reasons of brevity, familiarity, and availability of information, North American geologic history is the focus of this chapter.

As Earth passed the time boundary from the Precambrian into the Paleozoic, North America was rather smaller than it is today and was dominated by a vast, relatively featureless plain, the product of millions of years of weathering and erosion. The continent also lay farther to the south than it lies today. Large, sluggish streams meandered across surfaces barren of life, since neither land plants nor land-dwelling animals had yet evolved. Oceans lapped at the shores, including the *Iapetus Ocean,* which separated North America from Europe and Africa.

The Development of Eastern North America

At some point, early in the Paleozoic era, opening of the Iapetus Ocean ceased as the direction of plate motion reversed. The Iapetus began to close as the ancestral African continent started to "drift" inexorably westward, and a subduction zone formed beneath the eastern edge of the North American continent (fig. 14.10). At several times during the long Paleozoic era, portions of the westward drifting African plate collided with North America before being partially subducted. Each collision pro-

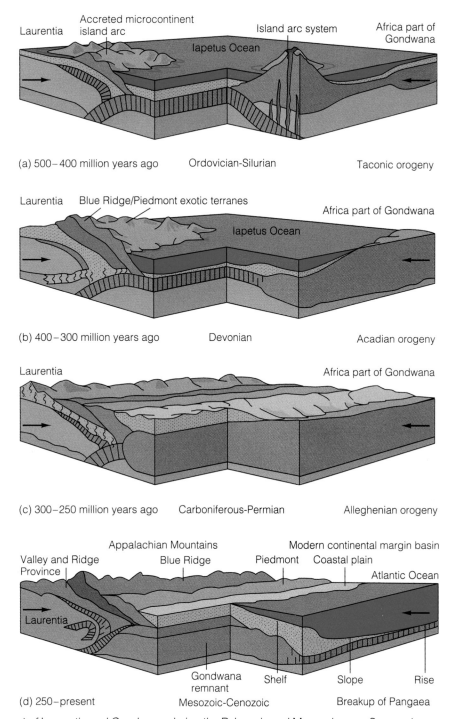

Laurentia   Accreted microcontinent island arc   Island arc system   Africa part of Gondwana

Iapetus Ocean

(a) 500–400 million years ago   Ordovician-Silurian   Taconic orogeny

Laurentia   Blue Ridge/Piedmont exotic terranes   Africa part of Gondwana

Iapetus Ocean

(b) 400–300 million years ago   Devonian   Acadian orogeny

Laurentia   Africa part of Gondwana

(c) 300–250 million years ago   Carboniferous-Permian   Alleghenian orogeny

Appalachian Mountains   Modern continental margin basin
Valley and Ridge Province   Blue Ridge   Piedmont   Coastal plain
Atlantic Ocean
Laurentia
Gondwana remnant   Shelf   Slope   Rise

(d) 250–present   Mesozoic-Cenozoic   Breakup of Pangaea

Figure 14.10   Movement of Laurentia and Gondwana during the Paleozoic and Mesozoic eras. Successive movements and collisions caused the Taconic, Acadian, and Alleghenian orogenies.

duced a major mountain system along the eastern margin of North America.

During and after each orogenic (mountain-building) event, the mountains were reduced by weathering and erosion over millions of years to low-lying, relatively featureless plains. The rock and mineral products of these long periods of erosion were spread as thick wedges of clastic sediment both eastward and westward from the gradually lowered mountains. Today, the **clastic wedges** that spread to the west are recognized as sedimentary rocks of Ordovician, Silurian, Devonian,

and Permian age, representing, respectively, the eroded mountains of the *Taconic, Caledonian, Acadian,* and *Alleghenian* orogenies. Sediments shed to the east were carried into the ocean basin and all record of them has been lost. Figure 14.11 illustrates the Acadian mountain system and clastic wedge of late Devonian time. Evidence of these events (particularly the Alleghenian) is preserved as the rocks, faults, and spectacular folds that we see in the Appalachian Mountains today (see box 9.20). While the original Appalachian Mountains have long since been eroded to nothing, the rocks that formed their

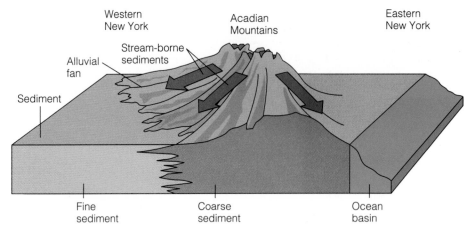

Figure 14.11   The Acadian Mountains formed in late Devonian time. Sediments are carried off the highlands in all directions to accumulate as thick wedges.

foundation have been elevated from time to time. Streams then carve out areas of weak rock to form valleys that separate ridges underlain by more resistant rock. This is what we now see in central Pennsylvania (fig. 14.12).

Early Development of Western North America

Along the western margin of North America, similar but less-frequent and less-dramatic events were occurring during the Paleozoic. Evidence for mid-Paleozoic *Antler* and late Paleozoic *Sonoma* orogenies can be found in the rocks of the West. The truly great western orogenies, however, came about during the following eras, the Mesozoic and the Cenozoic.

Central North America

Throughout the Paleozoic, the central portion of the North American continent remained for the most part quiet, escaping the ravages of continental collision and mountain building. Nevertheless, events of geologic significance occurred even here. Much of the continental interior was, from time to time, inundated beneath a gradually encroaching sea, so that areas today that are more than a thousand kilometers from the nearest sea were under shallow, *epicontinental seas.* Such submergences were especially prominent in late Cambrian, late Ordovician, and middle Mississippian time and are known as *transgressions.* Geologic evidence for these events can be found in beach sandstones in central Wisconsin and marine limestones blanketing much of the Midwest and West (fig. 14.13).

Other areas, such as lower Michigan, southern Illinois, western North Dakota, and west Texas showed some crustal instability by undergoing gradual subsidence (fig. 14.14). Within these subsiding "basins," thick sequences of sediments accumulated, incorporating large quantities of organic material that would be converted through time to petroleum. Historically, these regions of the continent have been major oil- and gas-producing provinces, and they remain so today.

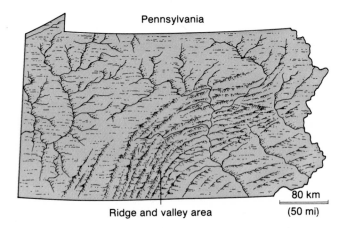

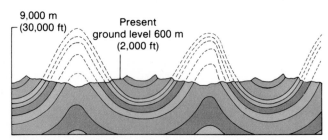

Figure 14.12   The geographic setting and underlying rock structure of the ridges and valleys of central Pennsylvania.

As some portions of the continental crust subsided, other portions were elevated as isolated, relatively short-lived mountain systems. The formation of such features was especially common during the Pennsylvanian period in Texas, Oklahoma, and Colorado. The Pennsylvanian period was also characterized by vast swamps that stretched from Oklahoma to Pennsylvania (fig. 14.15). Within these swamps lived the plants whose remains, after death, would accumulate to form the coal deposits upon which the Industrial Revolution was founded and upon which we still greatly depend.

Figure 14.13 Beach sandstone of late Cambrian age exposed in central Wisconsin.

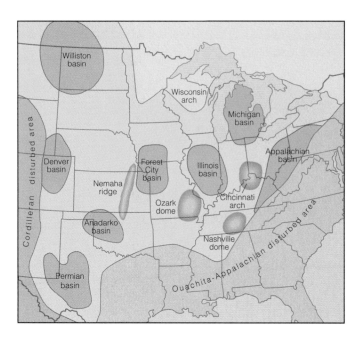

Figure 14.14 The locations of major sedimentary basins and uplifts that were active in the stable central portion of North America during middle and late Paleozoic time.

Life in the Paleozoic

Life during the Paleozoic was dominated by a wide variety of multicellular animals that evolved in and inhabited the world's oceans. These ranged from simple sponges to relatively complex clams, worms, and snails. Because none of these have internal backbones, they are referred to collectively as *marine invertebrates*. Many of the marine invertebrates have living de-scendants that are quite familiar to us. Others that were prominent during this time, such as the *trilobites* (relatives of the crabs and lobsters), became extinct at the close of the Paleozoic era (fig. 14.16). Groups that have minor representations in today's oceans, such as *crinoids* ("sea lillies") and *brachiopods* ("lamp shells"), dominated the seafloor for much of Paleozoic time. Figure 14.17 shows a typical seafloor environment during the Devonian period.

Figure 14.15 Diorama of Pennsylvanian age coal-forming swamp, showing typical flora.

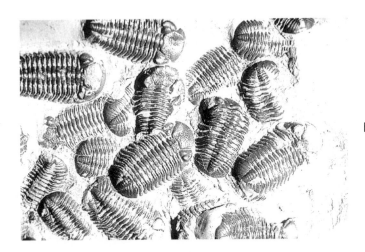

Figure 14.16 Trilobites.

During the Ordovician period, the first primitive, armor-plated fish evolved, and by Devonian time fish had become such a prominent part of the marine fauna that paleontologists call this period the Age of Fishes (fig. 14.18). At some time during the Silurian period, oxygen had accumulated in sufficient amounts in Earth's atmosphere so that primitive relatives of today's frogs and toads evolved to the extent that they were able to occupy the land surfaces during at least part of their life cycles. These first land animals had to return to the water to protect their eggs from drying. Once opened to colonization, the continents were exploited, in turn, by the evolving groups of reptiles and, early in Mesozoic time, by the mammals.

Late in the Silurian period, representatives of the plant kingdom also developed the capability of surviving on the land. Once having attained a "roothold," they evolved explosively to inhabit their new environments. The results were vast forests in Devonian time and the extensive coal-forming swamps of the Pennsylvanian period. These were largely giant ferns and "scale trees," (fig. 14.19) but in late Paleozoic time the first *gymnosperms,* nonflowering, cone-bearing trees, evolved.

MIDDLE PHANEROZOIC TIME: THE MESOZOIC

Perhaps the most interesting aspect of the physical history of the Mesozoic era in North America is the dramatic shift in tectonic activity from the eastern to the western part of the continent. The

Figure 14.17 Restoration of the floor of a shallow epicontinental sea during Devonian time. A spiny trilobite is shown in the foreground along with straight and coiled cephalopods and corals with tentacles.

(a)

(b)

Figure 14.18 (*a*) Fossil Devonian fish. (*b*) Artist's representation of a typical Devonian fish.

shift was the product of seafloor spreading in the newly forming Atlantic as Pangaea broke apart. Some block faulting and small-scale igneous activity occurred during the Triassic period in the Appalachian region, but this diminished in intensity during the Jurassic period and finally ceased before the close of the era. Major orogenic events were occurring in the west, however, where episodic mountain building and batholithic intrusion began in late Jurassic time and continued through the remainder of the Mesozoic era and into the Cenozoic era.

Two major orogenic pulses have been designated by geologists. The *Nevadan Orogeny* of the late Jurassic and early Cretaceous is recognized by the intrusion of many large granite batholiths, including the Sierra Nevada batholith of eastern California, which is spectacularly exposed at Yosemite National Park. The late Cretaceous to early Cenozoic *Laramide Orogeny* saw the creation of the basic structures of the Rocky Mountains, although, like the Appalachians in the East, these mountains were to be worn down and reelevated at least two times since their initial formation.

During the Jurassic period, much of the western portion of the United States was the site of a vast desert, the evidence for which exists in the form of sandstones that represent "fossilized" sand dunes. At other times during the Jurassic and the Cretaceous periods, much of this part of the continent was covered by the vast, shallow **Sundance Sea** (fig. 14.20).

Life in the Mesozoic

The transition from the Paleozoic to the Mesozoic era is marked by the extinction of as many as 50% of the known invertebrate animal groups, including the trilobites. Indeed,

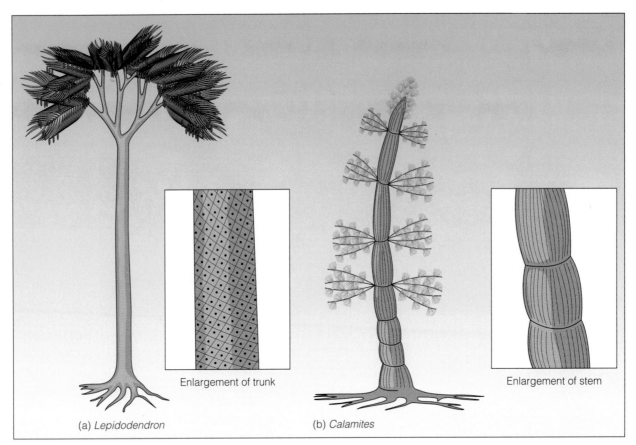

Figure 14.19 Common plants of the Pennsylvanian coal-forming swamps: (*a*) *Lepidodendron,* a scale tree, and (*b*) *Calamites,* a giant rush.

it is this dramatic change (of unknown cause) in the nature of the fossil record that led geologists many years ago to recognize this as a significant point in time. Along with the change in the marine invertebrate fauna, there was also a change in dominance among the land-dwelling fauna. Reptiles became increasingly important, and that branch known as the **dinosaurs** ("terrible lizards") began a 100-million-year reign that was to end at the close of the Mesozoic era. The dinosaurs evolved into many diverse forms that would dominate all environments: land, sea, and even the air (fig. 14.21). At some point during the Jurassic period, the first bird (known as *Archaeopteryx*) evolved, most likely from small, light dinosaurs that had evolved the ability to glide from cliffs or trees just as a flying squirrel does today.

Mammals likewise evolved from small reptiles, probably in late Triassic time. Throughout the Mesozoic era, however, as long as the dinosaurs walked the land, mammals remained very small, seemingly insignificant creatures living in the shadow of their larger, more numerous contemporaries. Only with the dramatic extinction of the dinosaurs at the close of the Mesozoic did mammals undergo rapid evolution and diversification to fill the ecological niches left by these reptiles. Mammals are discussed further in the section on the Cenozoic era.

The Extinction of the Dinosaurs

Geologists have yet to arrive at an acceptable theory to explain the sudden demise of all dinosaur lineages at (or very near) the Mesozoic-Cenozoic boundary. Most hypotheses on dinosaur extinction have either been discarded or modified to fit into one of two basic causes: meteorite impact or intense volcanism. In both cases, scenarios are envisioned in which large quantities of dust and ash (from fires or volcanic activity) were present in the atmosphere, blocking sunlight, causing a sharp drop in surface temperatures, and destroying much of the vegetation on which the food chain depends.

The heart of the *meteorite impact hypothesis* is based on the presence at many places around Earth of a thin layer of clay at the Mesozoic-Cenozoic boundary that contains an abnormally high concentration of the element *iridium.* Iridium is rare in surface rocks on Earth but is known to occur in concentrations many times greater in meteorites. The weakness of this theory is that extinction of the dinosaur families was probably not as "instantaneous" (a few months to a few years), as would be required under this scenario, but probably stretched over thousands or as much as several hundred thousand years (still a very short time geologically speaking). The discovery of iridium in rocks not associated with extinctions also casts a shadow of doubt on the impact theory. A gradual increase in the level of worldwide vol-

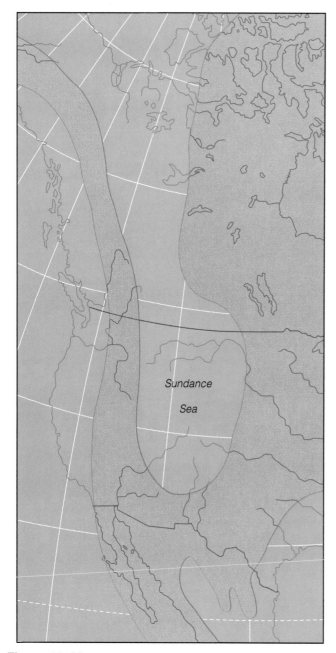

Figure 14. 20   Map shows the extent of a shallow epicontinental sea, the Sundance Sea, that inundated most of west central North America during mid-Jurassic time.

canism would better explain this extended time interval. The other major shortcoming of the meteorite theory is the lack of an identified crater that would mark the site of impact of the meteorite, though some argue that it landed in an ocean basin. In support of an ocean impact, researchers have discovered coarse sediments at some sites around the Gulf of Mexico. These may have resulted from tsunami-sized waves generated by the impact of a meteorite estimated to have been 6 kilometers (3.7 mi) in diameter.

Flowering Plants and Grasses

One other event of major significance in the history of life punctuated the Mesozoic era. This was the evolution of the *angiosperms,* plants that produce flowers and contain seeds enclosed in "fruits." Included in this category, in addition to flowers, are grasses, trees, and most of the other plants with which we are familiar today. The development of extensive "grasslands," in fact, set the stage for the evolution of the many types of grazing mammals that came to dominate the succeeding Cenozoic era.

Economic Resources of Mesozoic Rocks

Like each of the preceding eras, Mesozoic rocks possess their share of important economic resources. Much of the oil and gas produced in the Middle East comes from rocks of Jurassic and Cretaceous age. Many of the widespread coal deposits of the western United States are also of Cretaceous age. The California gold deposits are associated with the late Jurassic intrusion of the Sierra Nevada batholith. The Jurassic-age Morrison formation of the western United States, which is noted for its dinosaur fossils, is also well known for its uranium-bearing ores.

THE LAST 65 MILLION YEARS: THE CENOZOIC

The Cenozoic era in North America was characterized by a number of dramatic physical events, most of which took place in the western portion of the continent. Figure 14.22 shows some significant Cenozoic geologic features of western North America. The Rocky Mountains, which had been raised at the end of the Mesozoic era, underwent erosion and subsequent reelevation, probably at least two times during the Cenozoic. The rock and mineral debris of these erosional levelings spread eastward across what is today the Great Plains, forming thick sheets of clastic sediment. Among these sedimentary layers is the *Ogallala Sandstone,* an important aquifer in the Great Plains region today. Between the numerous small ranges of Colorado, Wyoming, and Utah, which together comprise parts of the Rocky Mountains in the United States—they also extend into Canada—several large freshwater lakes existed during the Eocene epoch. The sediments that accumulated to thicknesses of thousands of meters in these lakes are famous for beautifully preserved fossil specimens of the fish that once swam in the lakes. More importantly, these same sediments, the oil shales, contain unusually high quantities of organic material.

In the Great Basin region, centered on the state of Nevada, numerous fault-block mountain ranges were created when tensional forces, presumably associated with a change in plate motion along the western margin of North America, stretched the continent and caused it to break up into a series of blocks. The largest and most majestic of these mountains is the Sierra Nevada of eastern California, which rises over 3,048 meters (10,000 ft) above the adjacent desert floor. Other mountains, formed by folding and crumpling of rocks, developed along the continent's edge

Box 14.1 Further Consideration

The Evolution of the Horse

The horse family (Equidae) illustrates many important principles of evolution. Its history is one of changing physical characteristics over some 50 million years to adapt to everchanging environmental conditions. This evolutionary progression is extremely well documented by fossil evidence, especially in North America because terrestrial sedimentary layers in which the fossil remains of horses would be preserved are especially abundant in the intermontane basins and Great Plains of this continent. North America was the center of evolution for the horse family, but from time to time representatives would migrate to Eurasia, where they might flourish for a time before becoming extinct. In time a new wave of immigrants, of yet another species of horses, would arrive from North America.

The horse is a member of the mammal group known as ungulates, or hooved grazers, which evolved in late Cretaceous or early Paleocene time. Ungulates, in turn, are subdivided into two orders: the artiodactyls, or even-toed grazers (cows, deer, pigs, etc.), and the perissodactyls, or odd-toed grazers (horses, rhinos, and tapirs). The tapirs and rhinos are characterized by three toes on each foot. Horses possess a single toe on each foot that has evolved into a hard hoof.

The earliest perissodactyl (and also the earliest member of the horse lineage), *Hyracotherium,* or *Eohippus,* evolved in late Paleocene or early Eocene time. It was a small animal, about the size of a fox or small dog, with a curved back and a small head (brain). It possessed four toes on its front feet and three on each back foot. *Hyracotherium* lived on the forest floor and subsisted on a diet of soft leaves from shrubs and other forest plants.

During Eocene time, much of North America was covered by verdant forests, natural homes for the primitive perissodactyls (and other mammals) as they were for the dinosaurs before them. Through Oligocene and Miocene time, however, the climate became less hospitable—somewhat drier and colder. The abundant forests began to shrink, replaced by grasslands and prairies. (The various types of grasses also underwent a period of explosive evolution in response to this climatic change.) By the middle of the Miocene epoch vast new areas had become inhabited by those organisms able to adapt, through evolution. The evolving species of horses were in the forefront of this wave of colonization. Very slowly over many generations individuals exploited minor physical advantages that they held over their competitors in order to adapt to and survive in this changing world.

What physical changes evolved to allow horses to do this? There were several, but each served one of two possible functions: to nourish their bodies and to survive their natural enemies. The first type of change involved adapting to a new diet, from the soft leaves of forest shrubs to the hard, silica-rich grasses of the plains: from a *browser* to a *grazer.* Grasses are abrasive to the teeth of grazers and wear them down quickly. The slow evolutionary response to this was to develop broad, high-crowned, enamel-capped molars and premolars that would continue to grow throughout the lifetime of the animal. As these large grinding surfaces wore down, they would be continuously replaced.

Life on the plains not only exposed horses to new diets but to new dangers as well. No longer could they hide from predators in the protective undergrowth of the forest. They were now in the open and came to depend on methods of escape rather than concealment. This meant that they must be fast, agile, and alert. Evolutionary responses to these pressures were several. The body of the horse became larger with much longer legs. Longer legs meant longer strides and more speed. To increase the stride still further, the foot evolved into an elongated extension of the leg. Your own experience will tell you that the faster you run, the higher up on your toes you go. The horse actually runs on the tip of its middle toe, which evolved into a single hard hoof capable of withstanding the abuse of running on the hard prairie surfaces. The other toes became of no importance and were gradually lost. Along with increased body size there was also a pronounced increase in the size of the head. This served two purposes: to increase cranial capacity allowing for a larger brain and to accommodate the larger grinding teeth (fig. 14.A).

in Oregon and California. These are the Coast Ranges (fig. 14.23).

Volcanism also played a significant part in the physical evolution of North America during the Cenozoic era. In the Pacific Northwest, a series of volcanic cones, the Cascade Range, formed by the outpouring of lava and volcanic ash over millions of years. Indeed, the 1980 eruption of Mount Saint Helens indicates that this activity still continues. Just to the east of the Cascades, and certainly related to it, are extensive plateaus, the Columbia River plateau and the Snake River plain. These features were formed by the piling up of layer upon layer of basaltic lava flows to thicknesses of more than 3 kilometers (1.86 mi). Geologists have estimated that these flows represent an aggregate of 150,000 cubic kilometers (36,000 mi³) of lava. In the southwestern United States, other instances of volcanism, as represented by basaltic flows and isolated volcanoes, were common events during this era. In this same region, a late Cenozoic uplift of some 1,000 meters (3,280 ft) began the episode of downcutting by streams that produced the Grand Canyon of the Colorado River and other spectacular canyons. This downcutting has not ceased, and the Colorado continues to erode its famous canyon and the surrounding plateau.

The eastern and southern portions of the continent were tectonically quiet during this interval. Seas that had covered much of the continental interior in late Mesozoic time began a slow but steady withdrawal, so that *in the last few million years more of the North American continent has been exposed than at any other time.*

A sequence of shallow water sediments was deposited along the Atlantic seaboard and a much thicker sequence, in places more than 18 kilometers (11 mi) thick, was deposited along the Gulf Coast, burying the Jurassic and Cretaceous deposits, including great thicknesses of salt. Under the great weight of these sediments the salt began to flow, eventually forming the vertical structures called *salt domes* (chapter 13).

In addition to the resources already cited, others should be mentioned. More than 40% of the world's petroleum comes from

The evolution of the horse is not a story of success without failures. There were many false starts, evolutionary lineages that went only to extinction (such as the emigrant groups that went from North America to Eurasia). Some ungulate ancestors never left the forest but adapted to a premanent existence in that realm. The great irony of this story is that as successful as the horse family was in North America for 50 million years, the modern horse genus *Equus* became extinct on this continent only about 10,000 years ago, while managing to live on in Eurasia. The modern horse was reintroduced into North America in the 1600s as the progeny of horses that escaped from the early Spanish explorers and conquerors. What caused the extinction of the horse in North America is one of the mysteries of paleontology. One widely held view is that since their demise coincided roughly with the immigration of humans into North America across the Bering land bridge that connected Siberia to Alaska at this time, they may have been hunted to extinction by our ancestors.

Apply Your Knowlege

1. If human beings killed off horses in North America, why didn't this happen in Eurasia as well?

2. What other cause(s) might have led to the disappearance of the modern horse in North America?

3. What other groups of large mammals became extinct in North America at approximately the same time as the horse?

4. Can you postulate how environmental pressures such as those that operated on the horse family might have affected the evolution of human beings?

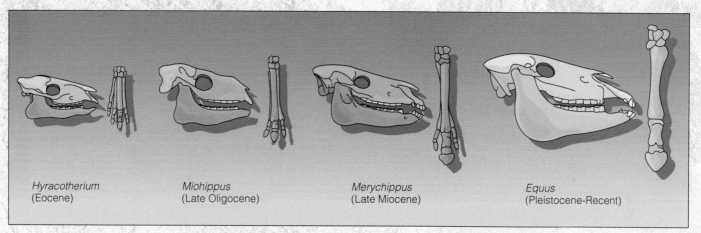

Hyracotherium
(Eocene)

Miohippus
(Late Oligocene)

Merychippus
(Late Miocene)

Equus
(Pleistocene-Recent)

Figure 14. A Examples of 4 genera in the horse lineage from Eocene to recent showing progressive changes.

rocks of Cenozoic age. Many of the extensive and thick coal seams of the western United States are Cenozoic in age as are many of the gold, silver, and copper deposits distributed throughout the western portions of both North America and South America. In the United States and Canada, the sands and gravels sorted and dumped by glaciers and glacially derived streams have been a valuable asset to the construction industry for many years.

The Ice Age

Throughout the Cenozoic era, climates, in general, grew progressively cooler. This gradual deterioration led to the culminating physical event of the era. Within approximately the last 1 or 2 million years, great quantities of snow and ice accumulated over eastern Canada in four major cycles. When this mass became sufficiently thick, it began to flow outward under its own weight as an extensive continental glacier. As the climate periodically warmed, the ice sheet melted away, exposing the greatly modified land surface once again. To the south these ice sheets were to cover at various times much of the upper midwestern United States, as far as the southern tip of Illinois, and all of New England. Mountainous areas such as the Canadian Rockies, Sierra Nevada, Rocky Mountains, Adirondacks, and parts of the Appalachians were to experience a great increase in the number and size of ice streams or alpine glaciers. Because climates were altered and more moisture was available, areas that are deserts today were, then, the locations of large lakes. Most of the topography and some of the economic resources (such as sand and gravel deposits) are the direct consequence of this glaciation.

Major rivers such as the Ohio and Missouri had their present courses largely determined by the position of the ice sheet. Water that was withdrawn from the oceans to form the ice sheets left sea level as much as 150 meters (490 ft) lower than it is today.

Figure 14. 21   The evolutionary lineages of some important dinosaur types.

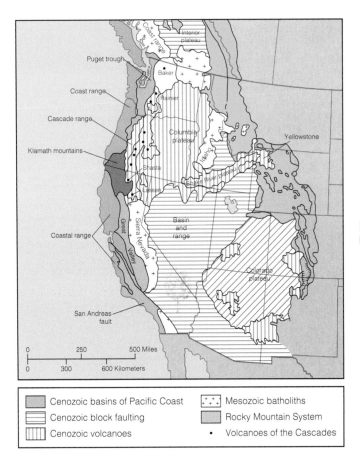

Figure 14.22 Major geologic provinces and Cenozoic features in western North America.

▨ Cenozoic basins of Pacific Coast	⊞ Mesozoic batholiths	
▤ Cenozoic block faulting	▨ Rocky Mountain System	
▥ Cenozoic volcanoes	• Volcanoes of the Cascades	

This lowering of sea level exposed areas of continental shelves that are today covered by shallow seas, such as the Bering Strait between Asia and North America, thereby enabling large land mammals, including humans, to migrate freely between continents (fig. 14.24).

Life in the Cenozoic

Without question, the most significant biological event of this time was the extinction of the dinosaurs at the Mesozoic-Cenozoic boundary and their replacement as the dominant group on the lands by the mammals. It was as if the mammals were hiding in the wings waiting to upstage the reptiles. A second factor that favored the explosive evolutionary diversification of the mammals during the Cenozoic era was the gradual climatic cooling that affected the middle latitudes, reducing the size of tropical forests and providing weather conducive to the spread of grasslands.

The evolution of mammals most likely derived from a group of reptiles called *therapsids*. Various therapsids had existed in the late Paleozoic, but many became extinct at the end of the Permian period. Among those that survived, certain species apparently evolved skins with hair and warm-blooded bodies sometime in the Triassic period. True mammals probably existed by the Jurassic period, a time when many large dinosaurs evolved.

Since their first evolutionary appearance in late Mesozoic time, the flowering plants, or angiosperms, had diversified and spread widely to include vast forests. The replacement, in part, of the forests by widespread, grass-covered plains about 25 million years ago favored the evolution of grazing mammals (see box 14.1). At about the same time horses evolved from smaller, browsing animals into larger grazers. A number of large carnivores, such as the saber-toothed cat shown in figure 14.25, appeared during the last few million years, preying on the numerous grassland dwellers.

The question of human origins is a difficult one even for the most experienced researchers. The reason for the difficulty lies in the contemporaneous rise of species that have many features seemingly common to human beings. Under the closest inspection, however, human beings do differ from the other primates. Our *bipedalism,* or ability to walk upright on two feet, is a distinguishing characteristic, even though other nonprimate species, such as bears, can walk for short distances on their hind legs. As primates, chimpanzees also show an ability to walk upright, but their stride is noticeably different. Their forelimbs (arms) are noticeably longer than their legs, whereas human legs are more or less equal to human arms. An early example of a

Box 14.2 Further Consideration

Cladistics: A Special Kind of Family Tree

One way to show relationships among life forms is a plot of similarities known as a *cladogram*. For example, humans have a body style that has four limbs. All other body styles that have four limbs would be related by the similarity in number of limbs. Therefore, a crocodile, a dinosaur, and a gorilla are all related to humans on a cladogram because of their four limbs. These organisms, including you, make up a group of life forms called tetrapods (from *tetra*, meaning "four," and *pod*, meaning "foot"). Obviously, your four limbs are not the same as those on a crocodile, and the difference between the limbs you call arms and the forelegs of a crocodile is one of the dissimilarities that separates you from the crocodiles.

Grouping tetrapods together because of their *shared characteristic* of four limbs is one example of a classification scheme known as *cladistics*. All the tetrapods that ever lived make up a *clade*, or group of species. Each clade can be further subdivided into subclades. With regard to tetrapods, for example, reptiles make up a subclade, as do the mammals, and further subdivisions can be determined.

Each subclade is like a branch on a tree. The larger branches from which smaller branches derive represent the common ancestors. A trace of the branching backward in time eventually reveals a *stem group* that includes the first, or *primitive*, members of a clade. The characteristics of the primitive members extend through time to the many species that have branched off the main stem. These primitive characteristics (those shared by all members of a clade) tie all the species of a clade together, but they do not show any one species to be the direct evolutionary ancestor of any other species.

To diagram a clade, cladists plot all members along a horizontal line at the top of the cladogram (fig. 14.B). A newly derived characteristic marks a change among the species of the vertebrate (species with backbones) subclade. The vertebrates have shared characteristics, but some are jawless, such as the hagfish and lampreys. The introduction of the new characteristic, jaws, is marked on the cladogram.

Apply Your Knowledge

1. Draw a cladogram for the horses (see box 14.1). Remember that the horses are perissodactyls, as are the rhinos. These mammals have odd numbers of toes, one of the distinguishing characteristics that separates them from the artiodactyls (even-toed mammals like pigs).

2. Place the following groups of species on a cladogram: primates, anthropoids (a group that includes human beings), and lemurs (a prosimian group). What are the shared characteristics of the primate subclade?

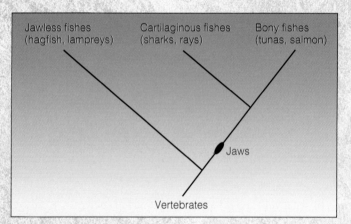

Figure 14. B   A cladogram of the subclade vertebrates.

Figure 14.23   A view of the Coast Range north of San Francisco.

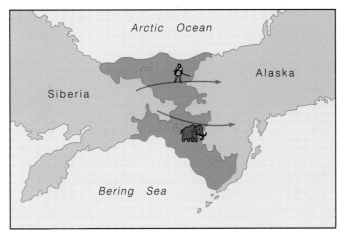

Figure 14.24   The Bering land bridge. Shaded area, now below sea level, was exposed as dry land during the times of glacial maxima. Arrows indicate migration routes.

Figure 14.25   The saber-toothed tiger (*Smilodon*).

bipedal, humanlike species was found in Ethiopia. This 3.5-million-year-old fossil would place our earliest ancestors in Africa before the Ice Age. Evidence that points toward Africa as the point of origin for human ancestors comes from several types of research. The oldest humanlike skeletons and skulls have been found in eastern Africa. Also, some genetic researchers believe that evidence from mitochondrial deoxyribonucleic acid (DNA, the replication material in living cells) points to an "Eve," or first mother, living in Africa. Just prior to the appearance of the first bipedal hominids, the Mediterranean dried up (see chapter 11), and Africa entered a period of extensive drought. A number of species appeared on the continent for the first time, some as a result of the land bridge across the partially dried Mediterranean Sea to the European landmass.

Further Thoughts: The Influence of Our Species

No species has ever affected Earth as much as *Homo sapiens* in as short a time. Through photosynthesis, plants changed the composition of the atmosphere, but that change took billions of years. Corals along the eastern coast of Australia built a series of islands and reefs over tens of millions of years. Our influence, however, has been drastic just within this century. We have altered the ozone layer, dammed giant rivers, such as the Nile and the Colorado, overturned rock to obtain mineral resources, and polluted the seas. Both South and North America have been extensively deforested, leading to increased erosion, and, in Brazil's rain forest, to the extinction of many species.

Extinction is not unusual in earth history. At the end of the Permian period, for example, the majority of faunal species became extinct. At the end of the Cretaceous more than just the dinosaurs became extinct. But during the past few hundred years we have been responsible for the extinction of numerous species and the endangerment of many more.

Human beings have attempted to control their world and bypass their connectedness to the planet in many ways. Inevitably, however, the planet, which is more encompassing than any of its systems or organisms, may produce effects that humans could not or did not care to foresee. The damming of the Nile River at Aswan was a spectacular engineering accomplishment, but the dam blocked the flow of sediment downstream, and the Nile Delta has eroded without a continuing replenishment of that sediment. The damming of the Colorado and the subsequent use of its water for irrigation has changed the salinity of the northern end of the Gulf of California. The building of recreational facilities on barrier islands has changed the flow of sands in the longshore current and altered the shapes of those islands. We have lowered water tables, caused subsidence through drilling and mining practices, and even initiated earthquakes. The effects of human activities extend even beyond Earth. They pervade the planet and the local regions of outerspace, where the remnants of spacecraft move in decaying orbits.

In 1992, almost all nations sent representatives to Brazil for a conference on the health of our planet. Although they found no permanent solutions to the massive environmental problems caused by the proliferation of human life, they did make many millions aware of the problems we face, problems of our own making.

We know that our changing world exhibits an interconnectedness of processes, features, and life. If we alter a process, we change a feature or influence life. Almost every recognizable change that has occurred throughout earth history has precipitated another related change.

Significant Points

1. The nebular hypothesis is widely accepted as the scientific explanation of Earth's origin.
2. Many earth scientists believe Earth's atmosphere and oceans derived from volatile gases that escaped from the mantle through volcanic activity in a process called degassing, or outgassing.
3. Carbon dioxide, methane, ammonia, and water vapor may have dominated the early atmosphere. The accumulation of oxygen in the atmosphere had to wait for the evolution of photosynthesizing life.
4. The principle of superposition states that for sedimentary rocks that have undergone no faulting or folding, the oldest beds are on the bottom of a sequence.
5. Relative age dating can be accomplished by comparing one rock unit with another. The use of fossils in correlations enhances the relative age dating of many sedimentary rocks.
6. Radioactive isotopes of uranium, potassium, and carbon have become useful indicators of absolute ages.
7. Through the radiometric dating of meteorites, moon rocks, and some very ancient (3.9 billion years) earth rocks, we know that the planet formed approximately 4.55 billion years ago.
8. The geologic time scale categorizes Earth's history in eons, eras, periods, epochs, and ages.
9. The Precambrian accounts for almost 85% of Earth's history.
10. Rocks of the shields are divided into distinct mineral age provinces. Rocks of the Canadian Shield are subdivided into those formed during the Archean and the Proterozoic.
11. Life evolved in the Precambrian under a reducing atmosphere. During the Cambrian period, multicellular life became abundant and left a significant fossil record.

12. The Paleozoic era closed with a great extinction, opening the way for the rise of the dinosaurs in the Mesozoic era.
13. During the Mesozoic, the great supercontinent of Pangaea began to separate under the influence of tectonic activity.
14. Angiosperms evolved at the end of the Mesozoic, and dinosaurs became extinct about 65 million years ago.

15. With the extinction of the dinosaurs the mammals, which had played a minor role in the Mesozoic, began a rapid evolutionary expansion from their therapsid origins.
16. The last 2 million years have been marked by major glacial advances and interglacial ice retreats.

*E*ssential Terms

uniformitarianism 358	correlation 358	radioactive decay 359
principle of superposition 358	relative age dating 358	geologic time scale 359
principle of lateral continuity 358	principle of crosscutting 359	shield 364
	absolute age dating 359	clastic wedge 367
		dinosaur 372

*R*eview Questions

1. How does relative age dating differ from absolute age dating?
2. How does the composition of the early atmosphere differ from Earth's present atmosphere? How is the early atmosphere characterized?
3. What process was responsible for the buildup of oxygen in the atmosphere? Why was a reducing atmosphere more conducive than an oxidizing atmosphere to the origin of life?
4. What is the principle that allows earth historians to determine the relative ages of sedimentary beds?
5. What is uniformitarianism?
6. What is the principle of crosscutting?
7. What is a fossil?
8. How do scientists use radioactivity to determine the absolute ages of rocks and fossils? What parent elements are used in radiometric age dating?
9. What is the estimated age of Earth?
10. What is the meaning of Phanerozoic?
11. What is an anaerobe?
12. What part of North America is the core of the continent?
13. When did organisms begin to leave abundant fossil hard parts?
14. When did vertebrates first inhabit land?
15. What temporal boundary marks the extinction of the dinosaurs?
16. When did the first flowering plants appear?
17. Did any mammals inhabit Earth prior to the extinction of the dinosaurs?
18. How did the development of grasses affect the evolution of mammals?
19. When did Earth enter into its most recent glacial age?
20. What continent probably served as the birthplace of hominids?

*C*hallenges

1. What processes might reduce the chances of preservation of a fossil?
2. Using a historical geology text from the library, determine the geologic history of your region. Was your area ever covered by an epicontinental sea?
3. Would the rocks exposed in road cuts or in rock outcrops in your region lend themselves to relative age dating?
4. Earth's processes do not always occur at the same rate. What might alter the rate of erosion in a stream valley and the deposition of sediments in a basin?
5. How could tectonic activity, such as seafloor spreading, affect life?

APPENDIX A

WORLD MAP

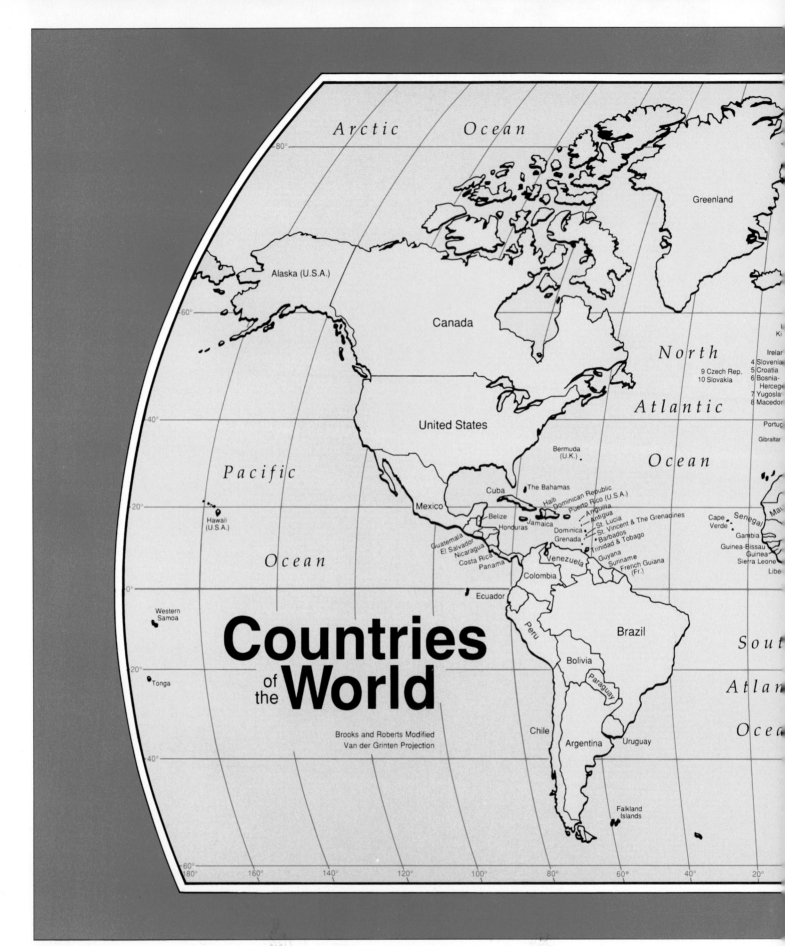

Figure A.1 (Source: From Arthur Getis, Judith Getis, and Jerome Fellmann, *Introduction to Geography,* 3d ed. Copyright © 1991 Wm. C. Brown Communications, Inc., Dubuque, Iowa. All Rights Reserved. Reprinted by permission.)

APPENDIX B

TABLE OF THE PLANETS

Property	Mercury	Venus	Earth	Mars	Jupiter	Saturn	Uranus	Neptune	Pluto
Mean distance from sun (10^6 km)	57.9	108.2	149.6	227.9	778.3	1,427	2,873	4,498	5,914
Mean distance from sun (10^6 mi)	36	67.2	93	141.5	483.3	886.1	1,783	2,793	3,666
Mean distance from sun (AU)	0.387	0.723	1.0	1.524	5.203	9.539	19.18	30.06	39.44
Equatorial diameter (km)	4,880	12,104	12,756	6,787	142,800	120,600	51,300	49,100	2,300(?)
Equatorial diameter (mi)	3,033	7,523	7,928	4,222	89,424	74,914	31,770	31,410	1,519
Mass (g)	3.3×10^{26}	4.9×10^{27}	6.0×10^{27}	6.4×10^{26}	1.9×10^{30}	5.7×10^{29}	8.7×10^{28}	1.0×10^{29}	1.5×10^{25}
Mass (Earth = 1)	0.055	0.815	1.0	0.1	317.9	95.2	14.4	17.2	0.1(?)
Volume (Earth = 1)	0.06	0.88	1.0	0.15	1,316	755	67	57	0.1(?)
Specific gravity	5.4	5.2	5.5	3.9	1.3	0.7	1.3	1.6	2.0
Period of revolution	88 days	224.7 days	365.26 days	687 days	11.86 year	29.46 year	84.01 year	164.8 year	247.7 year
Period of rotation	59 days	243 days*	23 hr, 56+ min	24 hr, 37+ min	9 hr, 50+ min	10 hr, 14 min	17 hr, 12 min*	22 hr	6 days, 9 hr
Inclination of axis	0°	178°	23°27'	23°59'	3°5'	26°44'	82°5'	28°48'	?
Inclination of orbit to Earth's orbital plane	7°	3°24'	0°	1°51'	1°18'	2°29'	0°46'	1°45'	17°12'
Orbital eccentricity	0.206	0.007	0.017	0.093	0.048	0.056	0.047	0.009	0.25
Orbital speed (km/sec)	47.9	35.0	29.8	24.1	13.1	9.6	6.8	5.4	4.7
Number of satellites	0	0	1	2	16	23	15	8	1
Surface temp (° C)	430 days,-180 night	457	25	−50	−130	−185	−200	−200	−230
Ring structure	no	no	no	no	yes	yes	yes	yes	?
Gravity (Earth = 1)	0.38	0.90	1.00	0.38	2.54	1.07	1.17	1.2	low
Escape velocity (km/sec)	4.3	10.4	11.2	5.0	60.2	37.0	22.5	23.9	low
Atmosphere (%)	He(98),H_2	CO_2(96),N_2	N_2(77),O_2(21)	CO_2(95),N_2	H_2(89),He	H_2(87),He	H_2,He,CH_4,NH_3	H_2,He,CH_4	CH_4
Surface atmospheric pressure (bars)	10^{-6}	90	1	6×10^{-3}	>>100	>>100	>>100	>>100	?
Magnetic field	weak	yes	strong	weak	strong	strong	strong	?	?
Internal composition (estimated)	Fe core silicates	Fe-Ni core silicates	Fe-Ni core silicate mantle and crust	solid Fe-FeS core, silicates	ice-silicate core, liquid H_2 mantle	ice-silicate core, liquid H_2 mantle	rocky core, liquid H_2O mantle	rocky core, ice mantle, H + He crust	CH_4

*Indicates retrograde rotation.

APPENDIX C

WEATHER MAP SYMBOLS

A weather scientist or forecaster depends upon a number of important charts and maps to analyze the development and movement of weather systems. Such maps and charts display conditions at various levels of the atmosphere and cover areas as small as the local region to as large as the entire hemisphere. Included among these tools are analysis charts, prognosis charts, satellite imagery, upper-air charts, and extended forecast charts. The information for the maps and charts is collected by observers at approximately 500 Weather Service offices located throughout the United States. Many of the offices use the Automated Surface Observation System (ASOS) to collect the data. The data are then converted to maps, charts, and text and distributed to the users, by wire, in a timely fashion by the National Weather Service.

One product generated by the National Weather Service that is most familiar to the general public is the Surface Analysis Chart, or "weather map." There are many printed forms of this map that range from the simplified version found in the daily newspaper to the Official Weather Map. Perhaps its greatest dissemination occurs through local and national television weather programs. Often the focus of attention is the surface weather map that is used by the presenter to explain current weather patterns or to predict future weather conditions.

CODES AND MAP PLOTTING

In order that the necessary speed and economy of space and transmission time may be realized, codes have been devised for sending the information and for plotting it on the maps.

A great deal of information is contained in a brief coded weather message. If each item were named and described in plain language, a very lengthy message would be required; and it would be confusing to read and difficult to transfer to a map. Use of a code permits the message to be condensed to a few five-figure numeral groups, each figure of which has a meaning depending on its position in the message. Persons trained in the use of the code can read the message as easily as plain language.

The location of the reporting station is printed on the map as a small circle (the station circle). A definite arrangement of the data around the station circle, called the station model, is used (fig. C.1). When the report is plotted in these fixed positions around the station circle on the weather map, many of the code figures are transcribed exactly as sent. Entries in the station model that are not made in code figures or actual values found in the message are usually in the form of symbols that graphically represent the element concerned. In some cases, certain of the data may or may not be reported by the observer, depending on local weather conditions. Precipitation and clouds are examples. In such cases the absence of an entry on the map is interpreted as nonoccurrence or nonobservance of the phenomena.

The station model is based on international agreements. Through such standardized use of numerals and symbols, a meteorologist of one country can use the weather reports and weather maps of another country even though he or she does not understand the language. Weather codes are, in effect, an international language making possible complete interchange and

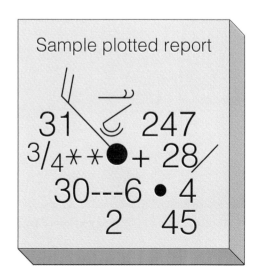

Figure C.1

Cloud abbreviation		C_L	Description (abridged from W.M.O. Code)
St or Fs-Stratus or Fractostratus	1		Cu of fair weather, little vertical development and seemingly flattened
Ci-Cirrus	2		Cu of considerable development, generally towering, with or without other Cu or Sc bases all at same level
Cs-Cirrostratus	3		Cb with tops lacking clear-cut outlines, but distinctly not cirriform or anvil-shaped; with or without Cu, Sc, or St
Cc-Cirrocumulus	4		Sc formed by spreading out of Cu; Cu often present also
Ac-Altocumulus	5		Sc not formed by spreading out of Cu
As-Altostratus Sc-Stratocumulus	6		St or Fs or both, but no Fs of bad weather
Ns-Nimbostratus	7		Fs and/or Fc of bad weather (scud)
Cu or Fc-Cumulus or Fractocumulus	8		Cu and Sc (not formed by spreading out of Cu) with bases at different levels
Cb-Cumulonimbus	9		Cb having a clearly fibrous (cirriform) top, often anvil-shaped, with or without Cu, Sc, St, or scud

	C_M	Description (abridged from W.M.O. Code)		C_H	Description (abridged from W.M.O. Code)
1		Thin As (most of cloud layer semi-transparent)	1		Filaments of Ci, or "mares tails," scattered and not increasing
2		Thick As, greater part sufficiently dense to hide sun (or moon), or Ns	2		Dense Ci in patches or twisted sheaves, usually not increasing, sometimes like remains of Cb; or towers or tufts
3		Thin Ac, mostly semi-transparent; cloud elements not changing much and at a single level	3		Dense Ci, often anvil-shaped, derived from or associated with Cb
4		Thin Ac in patches; cloud elements continually changing and/or occurring at more than one level	4		Ci, often hook-shaped, gradually spreading over the sky and usually thickening as a whole
5		Thin Ac in bands or in a layer gradually spreading over sky and usually thickening as a whole	5		Ci and Cs, often in converging bands, or Cs alone; generally overspreading and growing denser; the continuous layer not reaching 45° altitude
6		Ac formed by the spreading out of Cu	6		Same as 5, but the continuous layer not exceeding 45° altitude
7		Double-layered Ac, or a thick layer of Ac, not increasing; or Ac with As and/or Ns	7		Veil of Cs covering the entire sky
8		Ac in the form of Cu-shaped tufts or Ac with turrets	8		Cs not increasing and not covering entire sky
9		Ac of a chaotic sky, usually at different levels; patches of dense Ci are usually present also	9		Cc alone or Cc with some Ci or Cs, but the Cc being the main cirriform cloud

Figure C.2 (Source for figures C.2, C.3, C.4, C.5, C.6, C.7, and C.8: Adapted from *World Meteorological Organization Code.*)

use of worldwide weather reports so essential in present-day activities.

An explanation of the symbols with remarks on map entries appears in figure C.2.

Many of the elements in the plotting model are entered in values that can be interpreted directly; however, some require reference to code tables. These tables are shown in figures C.3 to C.8.*

h	Height in feet (rounded off)	Height in meters (approximate)
0	0 - 149	0 - 49
1	150 - 299	50 - 99
2	300 - 599	100 - 199
3	600 - 999	200 - 299
4	1,000 - 1,999	300 - 599
5	2,000 - 3,499	600 - 999
6	3,500 - 4,999	1,000 - 1,499
7	5,000 - 6,499	1,500 - 1,999
8	6,500 - 7,999	2,000 - 2,499
9	At or above 8,000, or no clouds	At or above 2,500, or no clouds

Figure C.3

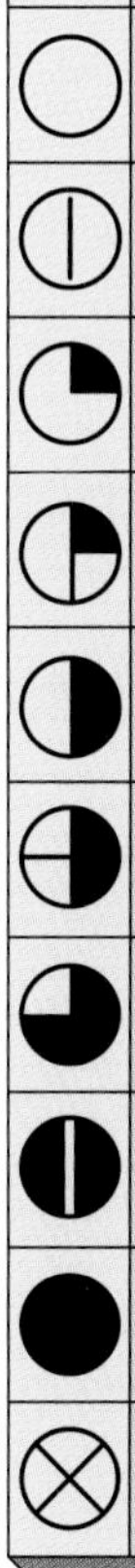

N	Sky coverage (total amount)
	No clouds
	Less than one-tenth or one-tenth
	Two-tenths or three-tenths
	Four-tenths
	Five-tenths
	Six-tenths
	Seven-tenths or eight-tenths
	Nine-tenths or overcast with openings
	Completely overcast
	Sky obscured

Figure C.4

N_h	Sky coverage (low and/or middle clouds)
0	No clouds
1	Less than one-tenth or one-tenth
2	Two-tenths or three-tenths
3	Four-tenths
4	Five-tenths
5	Six-tenths
6	Seven-tenths or eight-tenths
7	Nine-tenths or overcast with openings
8	Completely overcast
9	Sky obscured

Figure C.5

* Text exerpted from L. Moses and John Tomikel, *Basic Meteorology, An Introduction to the Science*. Copyright © 1981 Allegheny Press. Used by permission.

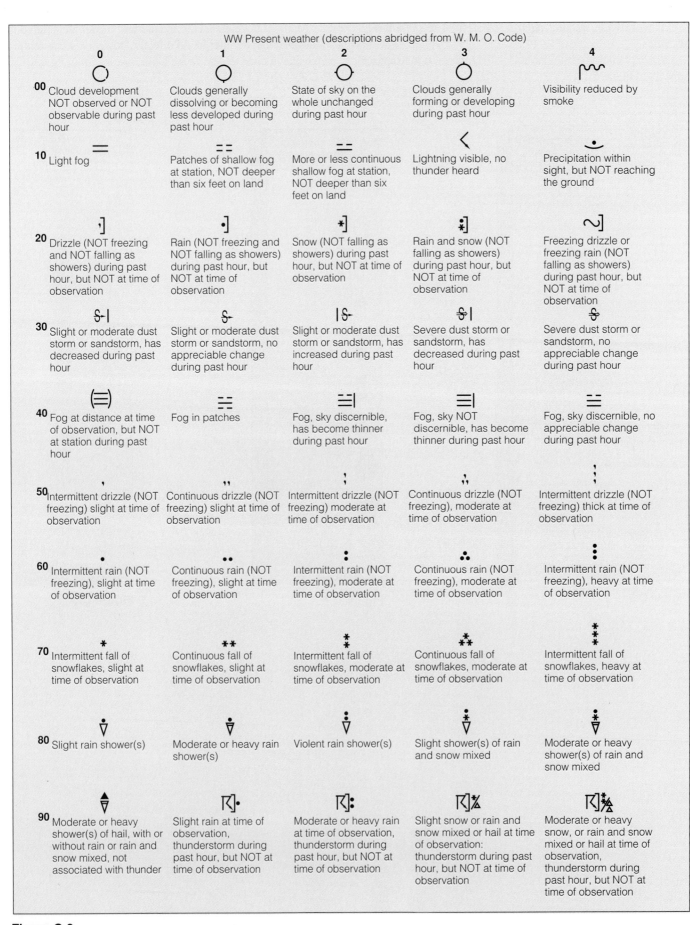

WW Present weather (descriptions abridged from W. M. O. Code)

	0	1	2	3	4
00	Cloud development NOT observed or NOT observable during past hour	Clouds generally dissolving or becoming less developed during past hour	State of sky on the whole unchanged during past hour	Clouds generally forming or developing during past hour	Visibility reduced by smoke
10	Light fog	Patches of shallow fog at station, NOT deeper than six feet on land	More or less continuous shallow fog at station, NOT deeper than six feet on land	Lightning visible, no thunder heard	Precipitation within sight, but NOT reaching the ground
20	Drizzle (NOT freezing and NOT falling as showers) during past hour, but NOT at time of observation	Rain (NOT freezing and NOT falling as showers) during past hour, but NOT at time of observation	Snow (NOT falling as showers) during past hour, but NOT at time of observation	Rain and snow (NOT falling as showers) during past hour, but NOT at time of observation	Freezing drizzle or freezing rain (NOT falling as showers) during past hour, but NOT at time of observation
30	Slight or moderate dust storm or sandstorm, has decreased during past hour	Slight or moderate dust storm or sandstorm, no appreciable change during past hour	Slight or moderate dust storm or sandstorm, has increased during past hour	Severe dust storm or sandstorm, has decreased during past hour	Severe dust storm or sandstorm, no appreciable change during past hour
40	Fog at distance at time of observation, but NOT at station during past hour	Fog in patches	Fog, sky discernible, has become thinner during past hour	Fog, sky NOT discernible, has become thinner during past hour	Fog, sky discernible, no appreciable change during past hour
50	Intermittent drizzle (NOT freezing) slight at time of observation	Continuous drizzle (NOT freezing) slight at time of observation	Intermittent drizzle (NOT freezing) moderate at time of observation	Continuous drizzle (NOT freezing), moderate at time of observation	Intermittent drizzle (NOT freezing) thick at time of observation
60	Intermittent rain (NOT freezing), slight at time of observation	Continuous rain (NOT freezing), slight at time of observation	Intermittent rain (NOT freezing), moderate at time of observation	Continuous rain (NOT freezing), moderate at time of observation	Intermittent rain (NOT freezing), heavy at time of observation
70	Intermittent fall of snowflakes, slight at time of observation	Continuous fall of snowflakes, slight at time of observation	Intermittent fall of snowflakes, moderate at time of observation	Continuous fall of snowflakes, moderate at time of observation	Intermittent fall of snowflakes, heavy at time of observation
80	Slight rain shower(s)	Moderate or heavy rain shower(s)	Violent rain shower(s)	Slight shower(s) of rain and snow mixed	Moderate or heavy shower(s) of rain and snow mixed
90	Moderate or heavy shower(s) of hail, with or without rain or rain and snow mixed, not associated with thunder	Slight rain at time of observation, thunderstorm during past hour, but NOT at time of observation	Moderate or heavy rain at time of observation, thunderstorm during past hour, but NOT at time of observation	Slight snow or rain and snow mixed or hail at time of observation: thunderstorm during past hour, but NOT at time of observation	Moderate or heavy snow, or rain and snow mixed or hail at time of observation, thunderstorm during past hour, but NOT at time of observation

Figure C.6

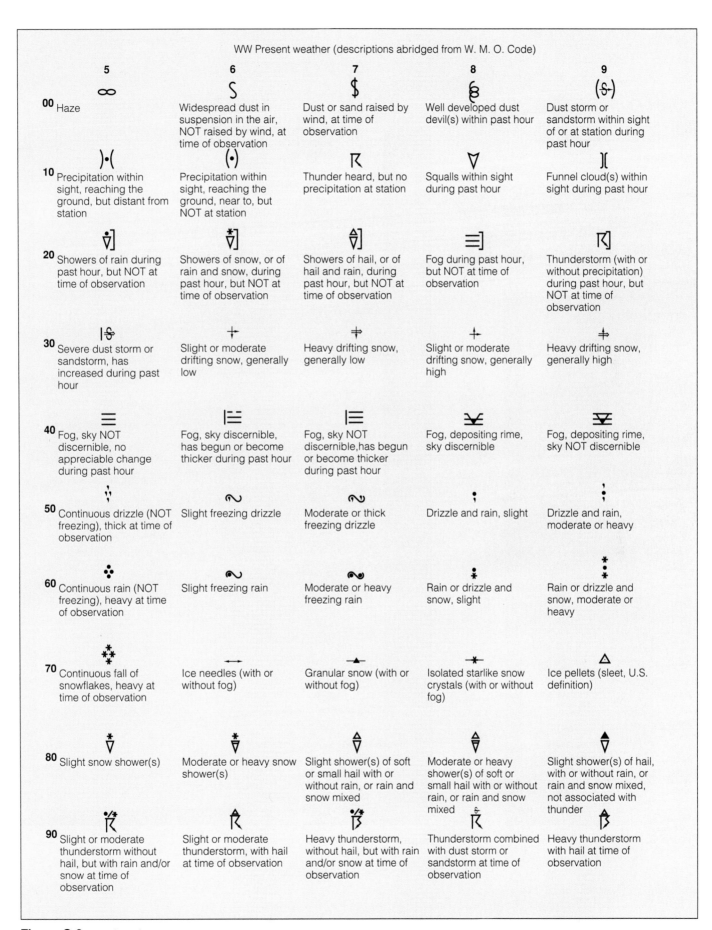

WW Present weather (descriptions abridged from W. M. O. Code)

	5	6	7	8	9
00	Haze	Widespread dust in suspension in the air, NOT raised by wind, at time of observation	Dust or sand raised by wind, at time of observation	Well developed dust devil(s) within past hour	Dust storm or sandstorm within sight of or at station during past hour
10	Precipitation within sight, reaching the ground, but distant from station	Precipitation within sight, reaching the ground, near to, but NOT at station	Thunder heard, but no precipitation at station	Squalls within sight during past hour	Funnel cloud(s) within sight during past hour
20	Showers of rain during past hour, but NOT at time of observation	Showers of snow, or of rain and snow, during past hour, but NOT at time of observation	Showers of hail, or of hail and rain, during past hour, but NOT at time of observation	Fog during past hour, but NOT at time of observation	Thunderstorm (with or without precipitation) during past hour, but NOT at time of observation
30	Severe dust storm or sandstorm, has increased during past hour	Slight or moderate drifting snow, generally low	Heavy drifting snow, generally low	Slight or moderate drifting snow, generally high	Heavy drifting snow, generally high
40	Fog, sky NOT discernible, no appreciable change during past hour	Fog, sky discernible, has begun or become thicker during past hour	Fog, sky NOT discernible, has begun or become thicker during past hour	Fog, depositing rime, sky discernible	Fog, depositing rime, sky NOT discernible
50	Continuous drizzle (NOT freezing), thick at time of observation	Slight freezing drizzle	Moderate or thick freezing drizzle	Drizzle and rain, slight	Drizzle and rain, moderate or heavy
60	Continuous rain (NOT freezing), heavy at time of observation	Slight freezing rain	Moderate or heavy freezing rain	Rain or drizzle and snow, slight	Rain or drizzle and snow, moderate or heavy
70	Continuous fall of snowflakes, heavy at time of observation	Ice needles (with or without fog)	Granular snow (with or without fog)	Isolated starlike snow crystals (with or without fog)	Ice pellets (sleet, U.S. definition)
80	Slight snow shower(s)	Moderate or heavy snow shower(s)	Slight shower(s) of soft or small hail with or without rain, or rain and snow mixed	Moderate or heavy shower(s) of soft or small hail with or without rain, or rain and snow mixed	Slight shower(s) of hail, with or without rain, or rain and snow mixed, not associated with thunder
90	Slight or moderate thunderstorm without hail, but with rain and/or snow at time of observation	Slight or moderate thunderstorm, with hail at time of observation	Heavy thunderstorm, without hail, but with rain and/or snow at time of observation	Thunderstorm combined with dust storm or sandstorm at time of observation	Heavy thunderstorm with hail at time of observation

Figure C.6 continued

ff	Miles (statute) per hour	Knots
◎	Calm	Calm
—	1 - 2	1 - 2
⊸	3 - 8	3 - 7
⟍	9 - 14	8 - 12
⟍	15 - 20	13 - 17
⟍	21 - 25	18 - 22
⟍	26 - 31	23 - 27
⟍	32 - 37	28 - 32
⟍	38 - 43	33 - 37
⟍	44 - 49	38 - 42
⟍	50 - 54	43 - 47
⟍	55 - 60	48 - 52
⟍	61 - 66	53 - 57
⟍	67 - 71	58 - 62
⟍	72 - 77	63 - 67
⟍	78 - 83	68 - 72
⟍	84 - 89	73 - 77
⟍	119 - 123	103 - 107

Figure C.7

Code number	a	Barometric tendency	
0	⟋‾	Rising, then falling	
1	⟋	Rising, then steady; or rising, then rising more slowly	
2	╱	Rising steadily, or unsteadily	Barometer now higher than three hours ago
3	⟍	Falling or steady, then rising; or rising, than rising more quickly	
4	—	Steady, same as three hours ago	
5	⟍	Falling, then rising, same or lower than three hours ago	
6	⟍_	Falling, then steady; or falling, then falling more slowly	
7	⟍	Falling steadily, or unsteadily	Barometer now lower than three hours ago
8	‾⟍	Steady or rising, then falling; or falling, then falling more quickly	

Code number	W	Past weather	
0		Clear or few clouds	
1		Partly cloudy (scattered) or variable sky	Not plotted
2		Cloudy (broken) or overcast	
3	⧸	Sandstorm or duststorm, or drifting or blowing snow	
4	≡	Fog, or smoke, or thick dust haze	
5	'	Drizzle	
6	•	Rain	
7	✳	Snow, or rain and snow mixed, or ice pellets (sleet)	
8	▽	Shower(s)	
9	⟨	Thunderstorm, with or without precipitation	

Figure C.8

CLIMATE DATA FOR SELECTED STATIONS

Stations		Jan	Feb	Mar	Apr	May	June	July	Aug	Sept	Oct	Nov	Dec	Year
1. Belem, Brazil	T	78.1	77.9	77.7	78.3	78.8	78.8	78.6	78.8	78.8	79.2	79.7	79.3	78.6
	P	12.52	16.02	17.17	15.04	10.43	6.50	6.34	4.57	4.72	4.13	3.54	7.76	108.74
2. Darwin, Australia	T	83.6	83.4	83.6	83.9	81.4	78.5	77.2	79.1	82.5	84.9	85.7	85.0	82.4
	P	16.18	12.37	11.18	3.08	.33	.09	.01	.02	.60	1.93	4.32	8.57	58.68
3. Darjeeling, India	T	40	42	50	56	58	60	62	61	59	55	48	42	53
	P	.8	1.1	2.0	4.1	7.8	24.2	31.7	26.0	18.3	5.3	.2	.2	121.7
4. Tripoli, Libya	T	53.4	55.9	59.4	64.6	69.1	74.8	78.1	79.3	77.9	72.3	64.8	56.3	67.1
	P	3.03	1.66	.97	.38	.20	.06	.02	.03	.39	1.42	2.66	3.76	14.50
5. Eucla, Australia	T	70.4	70.7	69.3	65.6	60.7	55.9	54.4	56.0	59.0	62.5	65.5	68.5	63.2
	P	0.73	0.78	1.18	1.09	1.13	1.10	0.89	1.05	0.51	0.73	0.70	0.66	10.55
6. Touggert, Algeria	T	50.5	54.6	61.1	68.8	76.9	86.7	92.0	90.1	83.6	71.7	60.1	51.5	70.6
	P	.24	.24	.35	.12	.16	.04	.04	.04	.08	.20	.43	.24	2.28
7. Norfolk, Virginia	T	42.1	41.8	48.2	56.8	66.5	74.4	78.6	77.3	71.8	61.8	51.3	42.9	59.4
	P	3.2	3.5	4.0	3.4	3.9	4.2	5.8	5.5	3.7	3.4	2.5	3.4	46.4
8. Dublin, Ireland	T	40	41	42	45	49	55	58	57	54	48	44	47	48
	P	2.2	1.9	1.9	1.9	2.1	2.0	2.6	3.1	2.0	2.6	2.9	2.5	27.7
9. Santa Ines, Chile	T	51	51	50	48	45	43	42	42	43	45	47	49	46
	P	6.80	5.76	6.67	7.12	7.52	7.29	7.70	6.58	5.42	6.31	6.42	6.73	80.32
10. Casablanca, Morocco	T	54.0	54.9	57.7	60.4	63.7	68.7	72.5	73.6	71.4	67.3	61.0	56.3	63.5
	P	1.97	1.89	1.93	1.54	.79	.24	.01	.03	.24	1.57	2.56	2.87	15.63
11. Capetown, Union of South Africa	T	70.7	71.1	68.9	64.6	59.5	56.7	55.2	56.7	58.5	62.4	66.4	68.7	63.1
	P	.67	.59	.87	1.93	3.70	4.29	3.70	3.27	2.28	1.57	1.02	.79	24.68
12. Wuhan, China	T	39	40	49	61	71	78	84	83	76	65	54	43	62
	P	2.1	1.1	2.8	4.8	5.0	7.0	8.6	4.6	2.2	3.9	1.1	.6	43.8
13. Addis Ababa, Ethiopia	T	60.8	62.8	64.6	63.9	64.8	61.9	59.7	59.4	60.3	60.3	59.5	59.5	61.3
	P	0.5	1.5	2.6	3.4	3.4	5.4	11.0	11.8	7.5	0.8	0.6	0.2	48.7
14. Chicago, Illinois	T	25.6	27.0	26.8	47.4	58.4	68.1	74.0	72.9	66.3	54.8	41.5	30.3	50.2
	P	2.1	2.1	2.6	2.9	3.6	3.3	3.4	3.0	3.1	2.6	2.4	2.1	33.2
15. Helsinki, Finland	T	22	21	25	35	46	56	62	61	53	43	35	28	41
	P	2.2	1.7	1.4	1.7	1.6	2.0	2.7	2.8	2.8	2.9	2.7	2.6	27.0
16. Norman, Canada	T	−19	−13	−2	19	41	54	59	54	41	24	−1	−15	20
	P	.4	.5	.6	.5	1.1	1.3	1.8	1.9	1.0	.8	.4	.4	10.7
17. Peking (Beijing), China	T	24	29	41	57	68	76	79	77	68	55	39	27	53
	P	.1	.2	.2	.6	1.4	3.0	9.4	6.3	2.6	.6	.3	.1	24.8
18. Pavlolar, Russia	T	0	1	13	36	55	66	71	66	54	37	19	5	35
	P	.4	.4	.6	.4	.7	1.7	1.7	1.9	.9	.8	.8	.6	10.0
19. Irkutsk, Russia	T	−5	1	17	35	48	59	65	60	48	33	13	1	31
	P	.6	.5	.4	.6	1.2	2.3	2.9	2.4	1.6	.7	.6	.8	14.6
20. Selagoncy, Russia	T	−42	−33	−14	11	33	54	59	51	37	13	−23	−38	9
	P	.3	.2	.3	.4	.9	1.4	2.0	2.4	.9	.9	.6	.5	10.9
21. Upernivik, Greenland	T	−8	−9	−6	6	25	35	41	41	33	25	14	1	17
	P	.4	.5	.7	.6	.6	.5	.9	1.1	1.1	1.1	1.1	.5	9
22. Ice Cap, Greenland	T	−41.8	−41.8	−31	−18.4	8.6	28.4	30.2	26.6	8.6	−11.2	−27.4	−36.4	−8.9
	P						NO DATA							

T = Temperature (° F)
P = Precipitation (in)

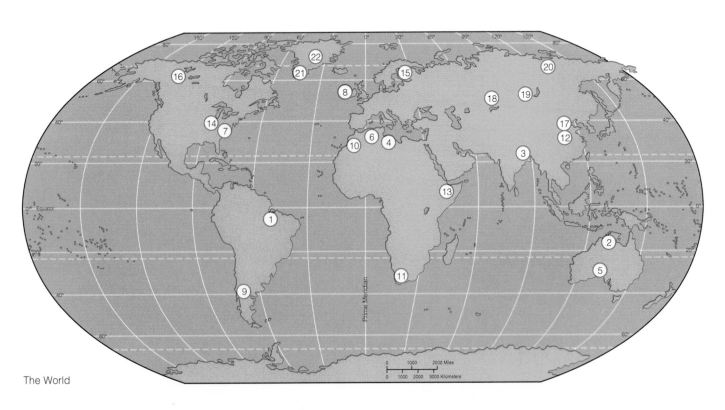

The World

Figure D.1 The numbers on the map show the location of the stations represented in appendix D. (Source: L.L. Moses, *Climatology Lab Manual.* Copyright © 1979 Allegheny Press.)

APPENDIX E

STAR CHARTS

Most of the heavens' brightest stars down to magnitude 4 are shown in figures E.1a-E.1c. The two polar charts (figs. E.1a and E.1b) include those portions of the celestial sphere that lie between 40° and 90° north and south declination. The equatorial-ecliptic chart (fig. E.1c) includes that portion of the celestial sphere between 40° north and 40° south declination. Right ascension is given on the horizontal axes of the equatorial-ecliptic chart and on the circumferences of the polar charts. The celestial equator horizontally bisects the equatorial-ecliptic chart, and the ecliptic circle makes an **S**-shaped curve about the celestial equator.

The positions and the magnitudes of the stars are indicated by a number from −1 to 4. All stars of magnitude 1 or greater are named on the charts. Some other stars of significance are also named (e.g., Polaris, Algol). Constellations are indicated by name and by tie-lines joining the brightest stars in each. The Milky Way is shown as an irregular, dark band crossing the charts.

If you live in the Northern Hemisphere, you should use the northern polar region chart (fig. E.1a) and the northern part of the equatorial-ecliptic chart (fig. E.1c). If you live in the Southern Hemisphere, you should use southern polar region chart (fig. E.1b) and the southern half of the equatorial-ecliptic chart. Locate a prominent constellation in the sky (e.g., the Big or Little Dipper in the north or the Crux or Triangulum Australe in the south). Then, using a flashlight, locate the same constellation on the appropriate chart. Rotate the chart until the constellation is oriented the same way as it is in the heavens, and then the chart will be properly oriented for locating and identifying other features in the heavens.

Figure E.1 Star charts of the (a) northern and (b) southern polar regions and of the (c) equatorial-Ecliptic belt. (Source: David Laing, *The Earth System: An Introduction to Earth Science,* © 1991 WCB, isbn 05972; Appendix V text on p. 537 and fig. V.1 on p. 538 depicting star charts.)

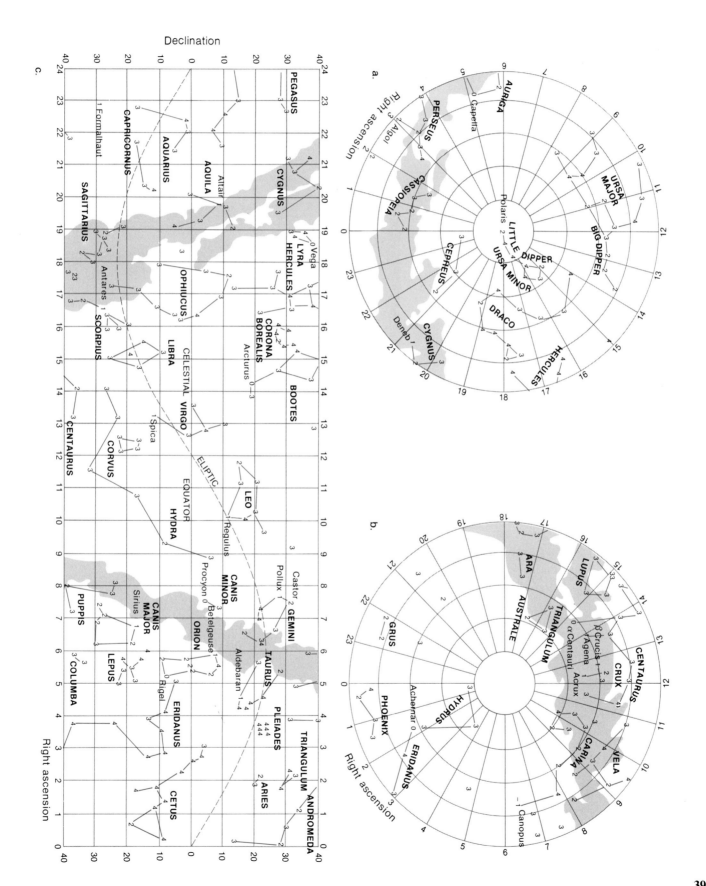

399

APPENDIX F

MINERAL IDENTIFICATION FLOWCHART

Mineral Identification Flowchart

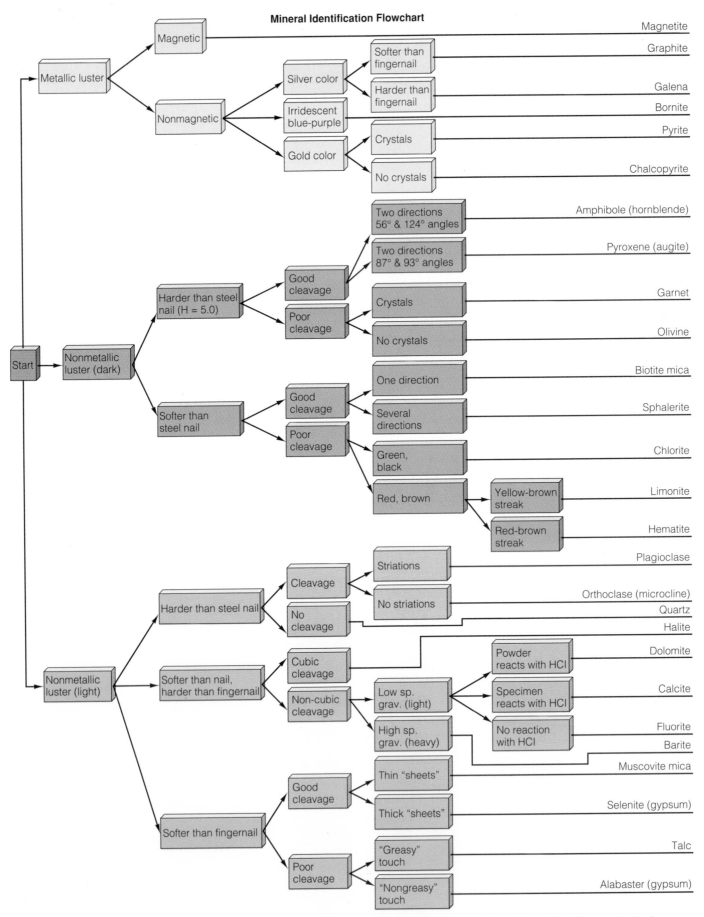

Figure F.1 (Source: From D.J. Thompson, *Basic Geology: Lab Manual.* Copyright© 1986 Allegheny Press. Used by permission.)

Appendix G

Rock Identification Flowchart

Mineral Identification Flowchart

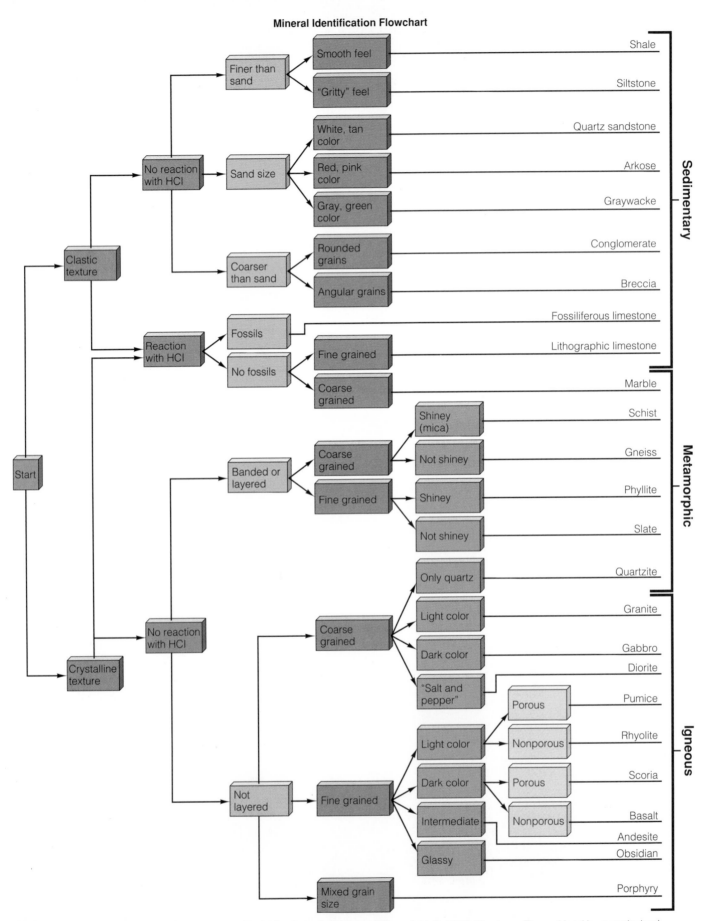

Figure G.1 (Adapted from D.J. Thompson, *Basic Geology: Lab Manual.* Copyright © 1986 Allegheny Press. Used by permission.)

APPENDIX H

THE 7TH APPROXIMATION SOIL CLASSIFICATION

At the beginning of this century, the U.S. Department of Agriculture (USDA) adopted a system of soil classification developed by Russian soil scientists. This scheme included three major levels of classification: from highest to lowest these are *orders, suborders,* and *great soil groups.* The orders are three in number and largely reflect the degree of maturity of the soil. **Zonal** soils are mature, well-developed soils; **intrazonal** soils are developed in places such as bogs, where local drainage conditions preclude normal soil development; and **azonal** soils are young and immature. The 9 to 12 recognized suborders are largely a reflection of climate and include categories such as desert, cold, temperate forest, and mountain. The great soil groups are soil varieties that are familiar to anyone with a rudimentary knowledge of soils. These include sierozems, chestnuts, podsols, prairie, and latosols. The great soil groups can be subdivided further into **soil families, soil series, soil types,** and **soil phases.**

Although the USDA system of classification underwent a number of modifications through the years, it was finally abandoned and replaced by a descriptive system that places emphasis on observable physical and chemical properties rather than environment or soil-forming process. This **Comprehensive Soil Classification System** (CSCS) uses soil names of Greek or Latin derivation and consists of six levels of classification (fig. H.1). In descending order, there are 10 **orders** (the names of which end in *sol*), 47 **suborders,** 185 **great soil groups,** 970 **subgroups,** approximately 4,500 **soil families,** and more than 10,000 **soil series.** Over the past 50 years, soil scientists in the United States have altered and modified this basic scheme (fig. H.2). This has been done by a series of **approximations,** each of which tries to relate the classification descriptions more closely to the soil as it is found in nature. For example, the 4th approximation was developed and published in 1955. The 7th approximation, published in 1960 and subsequently modified, is the latest version. Descriptions of the 10 soil orders of the CSCS follow.

1. **Entisols** (*ent,* recent) Soils with no distinct horizon development (azonal). Little evidence of the work of most soil-forming processes. Usually young, immature soils. Commonly developed on alluvial or eolian deposits. May form in any climatic regime. Erode severely when covering of natural vegetation is removed.

2. **Vertisols** (*vert,* turn) Characteristic of regions with distinct wet and dry seasons or those subjected to periodic drought. More than 35% swelling clay content. Soils expand during the wet season and shrink and crack during the dry season. Expansion often causes overturning of upper layers. Forms under grass cover. High in organic and base content.

3. **Inceptisols** (*incept,* begin) Immature soils with weakly developed horizons. No clay enrichment in B-horizon. Develop from the Arctic to the tropics, often on glacial deposits or volcanic ash. Brown soils because of an abundance of organic material. Native vegetation is forest.

4. **Aridisols** (*aridus,* dry) Desert soils with thin, moderately developed horizons. Low organic content gives A-horizon a light color. Subsoil (B-horizon) may contain accumulations of calcite, salt, or gypsum. Other high-mineral concentrations may lead to plant toxicity. Strongly alkaline. With irrigation these soils may be productive. These soils cover perhaps one-third of the world's land surface.

5. **Mollisols** (*molli,* soft) Thick, dark, organic-rich A-horizon and a clay-rich B-horizon (prairie soils). Rich in bases (Ca, Na, Mg). Weakly alkaline. Semiarid to subhumid climates. Grasslands. Also includes the chestnut and chernozem soils. Highly fertile with sufficient precipitation or irrigation. Dominant soils of the "grain belts."

6. **Spodosols** (*spodo,* wood ash) Ash-colored sandy A-horizon over a brownish, clay-rich B-horizon. Amorphous aluminum and iron oxides and humus. Strongly acid. Typically develop from sandy parent material in cool, humid climates and under forest vegetation. Relatively infertile. Includes some podsols.

7. **Alfisols** (*alfi* from *pedalfer*) Thin, brown to gray-brown A-horizon over a clay-rich B-horizon. More than 35% base (Ca, Na, Mg) saturation. Common under forest vegetation in the humid mid-latitudes. Moderate organic material, slightly acid. Agriculturally productive. Includes the gray-brown podsols.

8. **Ultisols** (*ulti,* ultimate) Red-yellow to red-brown color. Moist, clay-rich B-horizon. Less than 35% base saturation. Weakly acid. Wet-dry tropical to subtropical

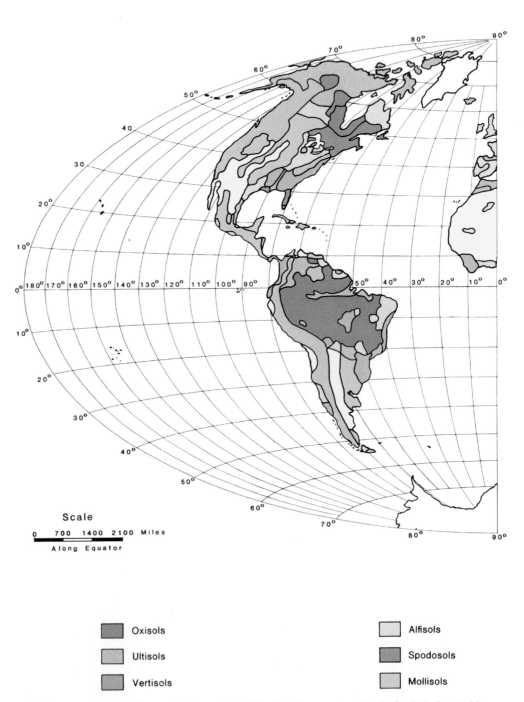

Figure H.1 CSCS soil map of the world. (Source: Arthur H. Doerr, *Fundamentals of Physical Geography,* 2d ed., © 1993; fig. 17.26 on pp. 294–95 depicting Comprehensive Soil Classification System map of the world, as shown attached.)

Aitoff's Equal Area Projection

	Aridosols		Histisols
	Entisols		Highlands
	Inceptisols		

climates. Develop by prolonged weathering of parent material. Includes the red-yellow podsols. Low fertility. Productive only with proper management.

9. **Oxisols** (*oxi,* oxide) Thick, intensely weathered soils with poor horizon development. Highly leached of bases. Weakly alkaline. Dominated by clay and hydrated oxides of iron and aluminum. Common in the tropics (rain forests). Includes the laterites (latosols). Infertile, dries to bricklike consistency when natural vegetation is stripped off.

10. **Histosols** (*histo,* tissue) Organic soils, including peat. Strongly acid. Forms in swampy areas such as the Florida Everglades and in cool, moist climates. Highly productive if drained. Fragile soils, easily eroded if disturbed.

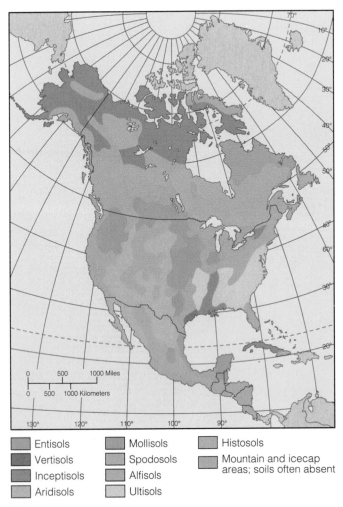

Figure H.2 CSCS soil map for North America.

APPENDIX I

TOPOGRAPHIC MAP SYMBOLS

Topographic Map Symbols

BOUNDARIES

National ....

State or territorial ....

County or equivalent ....

Civil township or equivalent ....

Incorporated-city or equivalent ....

Park, reservation, or monument ....

Small park ....

LAND SURVEY SYSTEMS

U.S. Public Land Survey System:

Township or range line ....

    Location doubtful ....

Section line ....

    Location doubtful ....

Found section corner; found closing corner ....

Witness corner; meander corner ....

Other land surveys:

Township or range line ....

Section line ....

Land grant or mining claim; monument ....

Fence line ....

ROADS AND RELATED FEATURES

Primary highway ....

Secondary highway ....

Light duty road ....

Unimproved road ....

Trail ....

Dual highway ....

Dual highway with median strip ....

Road under construction ....

Underpass; overpass ....

Bridge ....

Drawbridge ....

Tunnel ....

BUILDINGS AND RELATED FEATURES

Dwelling or place of employment: small; large ...

School; church ....

Barn, warehouse, etc.: small; large ....

House omission tint ....

Racetrack ....

Airport ....

Landing strip ....

Well (other than water); windmill ....

Water tank: small; large ....

Other tank: small; large ....

Covered reservoir ....

Gaging station ....

Landmark object ....

Campground; picnic area ....

Cemetery: small; large ....

RAILROADS AND RELATED FEATURES

Standard gauge single track; station ....

Standard gauge multiple track ....

Abandoned ....

Under construction ....

Narrow gauge single track ....

Narrow gauge multiple track ....

Railroad in street ....

Juxtaposition ....

Roundhouse and turntable ....

TRANSMISSION LINES AND PIPELINES

Power transmission line: pole; tower ....

Telephone or telegraph line ....

Aboveground oil or gas pipeline ....

Underground oil or gas pipeline ....

CONTOURS

Topographic:

Intermediate ....

Index ....

Supplementary ....

Depression ....

Cut; fill ....

Bathymetric:

Intermediate ....

Index ....

Primary ....

Index Primary ....

Supplementary ....

MINES AND CAVES

Quarry or open pit mine ....

Gravel, sand, clay, or borrow pit ....

Mine tunnel or cave entrance ....

Prospect; mine shaft ....

Mine dump ....

Tailings ....

SURFACE FEATURES

Levee ....

Sand or mud area, dunes, or shifting sand .....

Intricate surface area ....

Gravel beach or glacial moraine ....

Tailings pond ....

VEGETATION

Woods ....

Scrub ....

Orchard ....

Vineyard ....

Mangrove ....

COASTAL FEATURES

Foreshore flat ....

Rock or coral reef ....

Rock bare or awash ....

Group of rocks bare or awash ....

Exposed wreck ....

Depth curve; sounding ....

Breakwater, pier, jetty, or wharf ....

Seawall ....

BATHYMETRIC FEATURES

Area exposed at mean low tide; sounding datum ....

Channel ....

Offshore oil or gas: well; platform ....

Sunken rock ....

RIVERS, LAKES, AND CANALS

Intermittent stream ....

Intermittent river ....

Disappearing stream ....

Perennial stream ....

Perennial river ....

Small falls; small rapids ....

Large falls; large rapids ....

Masonry dam ....

Dam with lock ....

Dam carrying road ....

Intermittent lake or pond ....

Dry lake ....

Narrow wash ....

Wide wash ....

Canal, flume, or aqueduct with lock ....

Elevated aqueduct, flume, or conduit ....

Aqueduct tunnel ....

Water well; spring or seep ....

GLACIERS AND PERMANENT SNOWFIELDS

Contours and limits ....

Form lines ....

SUBMERGED AREAS AND BOGS

Marsh or swamp ....

Submerged marsh or swamp ....

Wooded marsh or swamp ....

Submerged wooded marsh or swamp ....

Rice field ....

Land subject to inundation ....

Figure I.1 (Source: U.S. Geological Survey.)

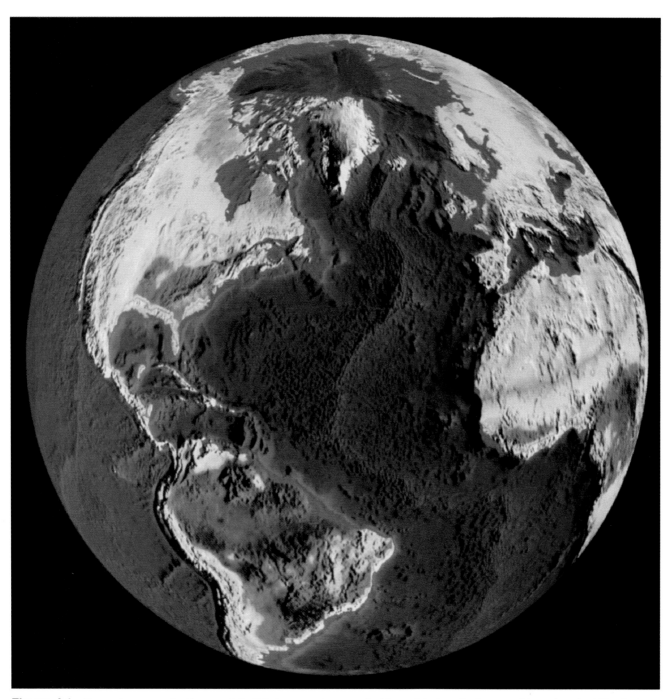

Figure J.1 Computer-generated image of Earth's topography compiled from data from both ground-based and orbiting instruments. Vegetation patterns are overlaid.

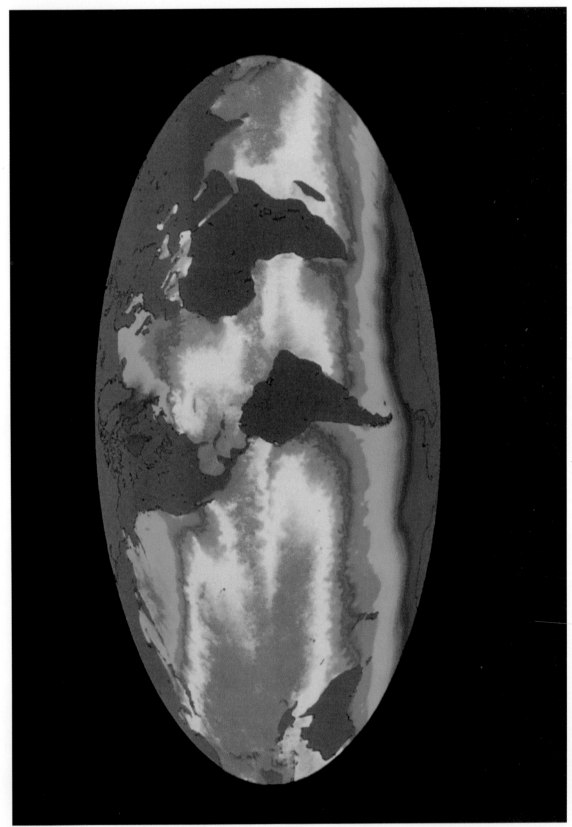

Figure J.2 Global sea surface temperatures obtained from infrared observation during July 1984 by the Advanced Very High Resolution Radiometer (AVHRR) on Board NOAA-7.

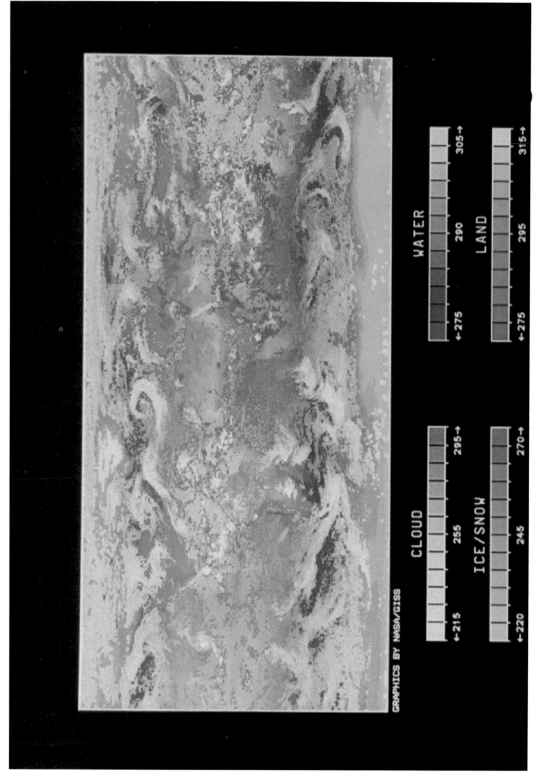

Figure J.3 World cloud cover pattern on October 15, 1983, assembled by NASA from weather satellite images.

413

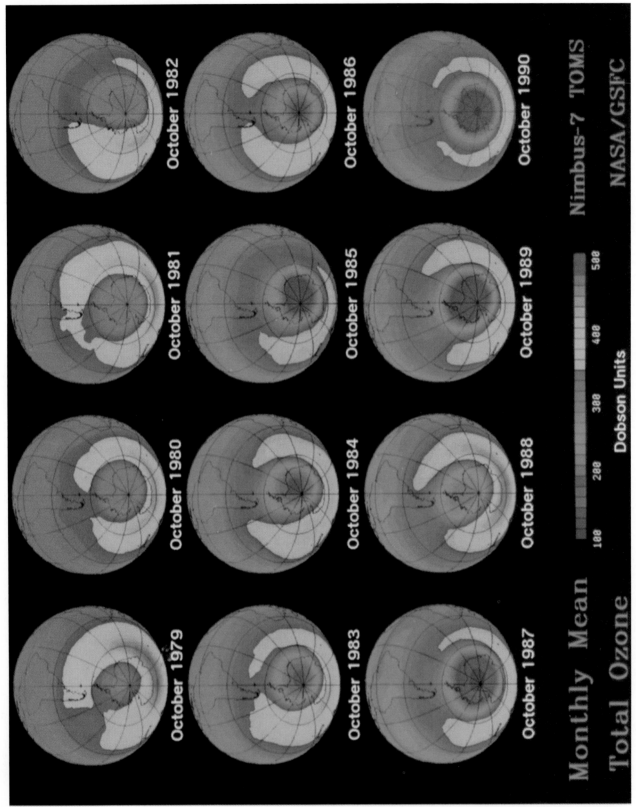

Figure J.4 Nimbus-7 TOMS images from October 1979 to October 1990 showing changing (decreasing) ozone levels over the south polar region.

Figure J.5 Viking Orbiter 1 mosaic of the surface of Mars showing the entire Valles Marineris, a canyon system more than 3,000 kilometers long and 8 kilometers deep.

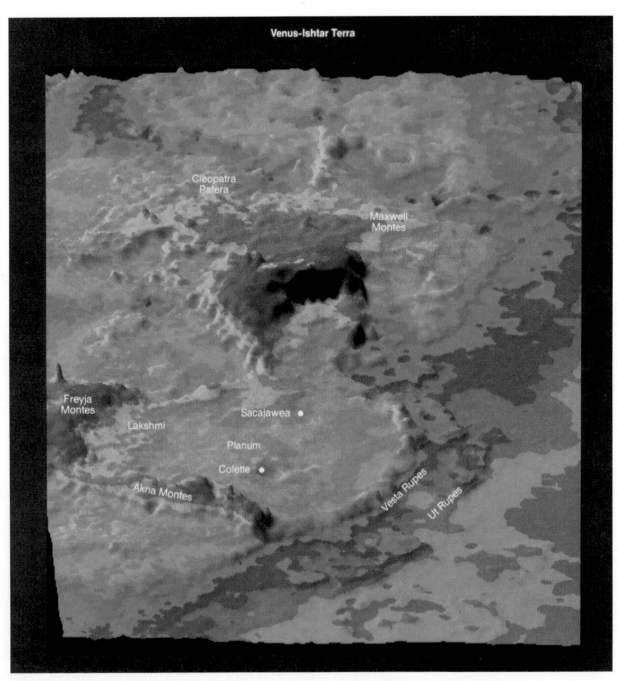

Figure J.6 Pioneer Venus image of a portion of the "continent" Ishtar on Venus. Maxwell Montes, thought to be an active volcano, stands higher than Mount Everest.

Figure J.7 Io, one of the moons of Jupiter, photographed by Voyager.

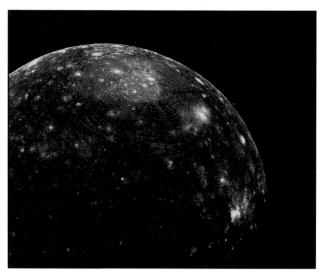

Figure J.9 Callisto, a moon of Jupiter, photographed by Voyager.

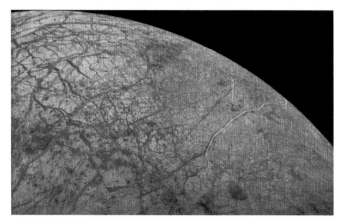

Figure J.8 Europa, a moon of Jupiter that is approximately equal in size to Earth's moon, photographed by Voyager. The lines are thought to be surface fractures.

Figure J.10 Ganymede, a moon of Jupiter, photographed by Voyager. Ganymede is the largest moon in the Solar System and is larger than the planet Mercury.

GLOSSARY

A

ablation Loss of ice from a glacier through melting, sublimation, and calving; usually a glacial ice loss that expresses itself as a thinning of the terminal end of a glacier.

abrasion Erosion by friction between glacial ice and rock or between bedrock and the rock embedded in passing glacial ice; also the wearing away of rock particles in transit within streams or in the wind.

abrasives Natural and manufactured materials used for cutting, sanding, grinding, or polishing.

absolute age dating Age in number of years for a geologic feature or event, usually determined by radiometric dating methods.

absolute humidity Ratio of water vapor to the volume of a parcel of air.

absolute zero Temperature at which there is no molecular motion; 0 K.

abundant metal Metal with an average abundance greater than 0.01% of Earth's crust.

abyssal plain Usually extensive ocean floor areas with relatively low relief.

abyssopelagic Water column above the abyssal plains, usually designated as the water below the bathypelagic and above the hadal zones.

acid mine drainage Waters that exit deep and strip coal mines; waters rich in sulfuric acid that originate from the weathering of pyrite in mined coal seams and coal refuse piles.

adiabatic cooling Internal cooling process that results when a parcel of air ascends and expands.

advection Wind; essentially, horizontally moving replacement air.

advection fog Ground cloud that develops over land or sea, usually in the winter or spring, that results from the cooling of warm air as it passes over cold land or water.

aftershock One of a series of tremors or temblors that invariably follow an earthquake over a period of days, weeks, or months.

A-horizon Soil zone below O-horizon; uppermost of the true soil layers; soil layer that is dark by enrichment with humus and deficient in soluble minerals lost to leaching; synonymous with the topsoil.

air mass Body of air that can be identified by its temperature, pressure, and moisture characteristics.

air pressure Quantifiable result of molecular collisions within atmospheric gases and between gases and planetary surfaces as a function of the overall weight of the gases; expressed as pounds per square inch, in millibars, or in inches.

albedo Measurable reflectivity of any surface.

alloy Any mixture of two or more metals, such as bronze, brass, and steel.

alpha decay Loss of a helium nucleus (two protons and two neutrons) from an atomic nucleus.

alpine glacier *See* Valley glacier.

amber Hardened tree sap.

ambient air Surrounding air.

anaerobe Earliest life form; also, anaerobic heterotroph; microscopic organism that exists in an anoxic environment.

anemometer Instrument that measures wind speed.

aneroid barometer Instrument that measures atmospheric pressure by means of an airless, corrugated container that expands and contracts to move a needle along a calibrated scale.

angiosperm Flowering plant that contains seeds enclosed in fruits.

angle of incidence Measurement in degrees from zenith that the sun's radiation strikes the surface of Earth.

anhydrite Evaporite mineral associated with very hot, evaporating conditions; $CaSO_4$.

Antarctic Bottom Water (AABW) Water that originates as cold water along Antarctica and sinks because of its increased density; a cold, moving water layer derived largely from the Southern Ocean.

anthracite Hard coal; metamorphosed bituminous coal.

anticline Arch-shaped or up-folded rock structure produced by compressive or mountain-building stress; arch-shaped rock structure with outward limbs that dip away from the axis, or crest.

anticyclone A downward and outward flow of air; a high-pressure system; a clockwise circulation of air in the Northern Hemisphere.

Antilles Current Surface current that joins the Florida Current to form the Gulf Stream.

antimatter Matter composed of antineutrinos, positrons, antiprotons.

aphanitic (fine) texture Textural term, most commonly appled to igneous rocks, in which the individual fine crystals are too small to be seen by the naked eye; associated with rapid cooling of lava; fine-grained.

aphelion Orbital position that is farthest from the sun; Earth's orbital position in January.

aphotic Dark water zone; ocean water zones below 1,000 meters (3,280 ft).

apogee Position in an elliptical orbit where the orbiting body is at its greatest distance from the object it orbits.

aquiclude Rock, soil, or sediment that prevents the passage of groundwater; impermeable rock, soil, or sediment.

aquifer Rock, soil, or sediment that allows relatively easy passage of groundwater; permeable rock.

arête Narrow ridge caused by glacial erosion; narrow rock ridge between adjacent alpine glaciers.

artesian aquifer Groundwater system in which water flows to (or toward) the ground surface under its own head through a well or a system of fractures.

artesian spring Flow that occurs when water under pressure in a confined aquifer reaches the ground surface.

artesian well Type of well in which water from a confined aquifer rises to the ground surface under its own head

asphalt (bitumen) Solid or highly viscous component of petroleum.

asteroid One of numerous rocky bodies that occupy an orbit between Mars and Jupiter.

asthenosphere Zone in the upper mantle characterized by low seismic velocities and composed of very hot, near-molten material that behaves plastically.

astronomical unit (AU) Unit based upon Earth's average distance from the sun (150 million km or 93 million mi) used for measuring distances within the Solar System.

astronomy Study of objects and processes beyond Earth.

atmosphere Sphere of gases that envelopes a celestial body.

atoll Semicircular coral reef island that forms around subsided, tropical oceanic volcano.

atom A naturally occurring or synthesized unit of matter composed of a nucleus and orbiting electrons; the smallest indivisible form of an element that maintains that element's chemical characteristics.

atomic number Total number of protons in an atom of an element.

aurora borealis (northern lights) Light produced as solar wind encounters nitrogen and oxygen in the upper atmosphere.

ℬ

backswamp Low-lying, marshy area created when floods in an alluvial plain deposit impermeable muds; marsh area of poor drainage and impermeable muds located between the natural levees of a stream and the valley walls.

backwash Downslope return water from waves breaking along a beach; the water that runs down the beach toward the surf zone.

banded iron formation (BIF) Low-grade ore (taconite) in which iron is concentrated in dark layers that contrast with light, iron-poor cherty layers, giving a banded appearance; also known as Lake Superior-type ore.

bar Elongate deposit of unconsolidated sediments that have been deposited by waves and near-shore currents; deposit of sand at the mouth of a distributary stream.

barrier island Elongate coastal island that acts as a barrier to waves and tides, protecting the mainland and separated from it by a shallow lagoon.

basalt Dark, mafic volcanic rock generally characterized by a fine-grained texture.

basaltic crust The oceanic crust; that portion of Earth's crust that is dominated by volcanic rocks of mafic composition.

basalt plateau Accumulation of horizontal layers of basalt erupted from multiple fissures during a relatively brief geologic span.

batholith Large, discordant, nontabular plutonic mass that has a surface exposure of at least 100 square kilometers (40 sq. mi) and that may extend for thousands or tens of thousands of square kilometers; Stone Mountain, Georgia, and Half-Dome in Yosemite are parts of exposed batholiths.

bathyal zone Seafloor along the continental slope, running from shelfbreak to continental rise.

bathymetry Depth and configuration of the ocean floor.

bathypelagic zone Uppermost aphotic zone in the ocean that extends from 1,000 meters (3,280 ft) to 4,000 meters (13,920 ft).

bauxite Principal aluminum ore composed of two or more aluminum oxides and hydroxides; a "fossil" soil or a residual deposit that includes diaspore and gibbsite found in the tropics or in areas with lateritic soils.

baymouth bar Spit that has closed off the open ocean entrance to a bay.

beach Subaerial and subaqueous depositional feature aggregated and shaped by waves, currents, and winds acting on various sediments; the color, composition, and texture of beach materials depends on the source of the materials and the wind, wave, tide, and current energy.

beach drifting Movement of beach materials in the longshore current and along the exposed beach face.

beach nourishment Application of sand to a beach to replenish that carried away by longshore transport and storm waves; also called beach replenishment.

Bernoulli theorem Principle that a constricted moving fluid gains momentum.

beta decay Loss of an electron from a nucleus as a neutron becomes a proton.

B-horizon Soil horizon below the A-horizon; known as the zone of accumulation or as the subsoil.

big bang Theory whereby the universe was created by an expansion of superhot plasma some 12 to 15 billion years ago.

bioclastic Sedimentary rock texture composed of broken organically produced fragments, such as shells, coral rubble, or decayed plant materials.

biogenic Derived from organic activity.

bioluminescent Illuminated by chemical activity within an organism.

bioturbation Overturning of sediments by burrowing organisms, such as worms, clams, and crabs.

bird's-foot delta Delta formed by many distributaries that has reminded some of the shape of a bird's foot.

bitumen A tarlike substance that is rich in carbon; opaque, noncrystalline substance found in pore spaces and in fluid inclusions in sedimentary rocks.

bituminous coal Soft coal associated with much of the Appalachian region and the interior coalfields. May be considered as a type of sedimentary rock.

black hole Object from which light cannot escape; the densest object in the universe, formed from the complete collapse of a star three or more times the mass of our sun.

blowout Uncontrolled (and dangerous) release of oil and natural gas from a well bore; a deflation basin, or bowl-shaped depression in sediment caused by wind erosion.

body waves Earthquake waves, the P-waves and the S-waves, that travel through Earth rather than around its surface.

bora Katabatic winds at the boundary of southern Europe and the Adriatic Sea.

brackish Water that is slightly to moderately saline.

brass Copper-zinc alloy.

breakwater Constructed barrier designed to break up the energy of the incoming waves and to create a quiet, protected harbor.

brine Saline water contained in the spaces of sedimentary rocks; water containing halite and other mineral salts in solution.

bronze Copper-tin-zinc alloy.

C

caldera Wide, deep depression much larger than a crater that is the remains of a violently erupting volcanic cone; large depression representing the missing upper portion of a volcanic cone that has been destroyed by a combination of disintegration, collapse, and ejection during a violent volcanic eruption.

caliche Hard calcium carbonate; ground surface crust that is deposited when water evaporates in an arid region.

calving The breaking off of ice masses from glaciers that extend to or into a body of water.

cavern Interconnected, extensive large subsurface voids caused by dissolution of limestone.

cay Small Bahamian island or reef near or just above high-tide water level.

Celsius Temperature scale using 0° as the initial freezing point of water and 100° as the initial boiling point of water.

cement Binding material holding together mineral particles in sedimentary rocks; cements are usually carbonate (calcite), silica (quartz), and iron compounds. Also, a product manufactured from limestone and shale that is used as the binding material in concrete.

centrifugal force Tendency for an object to move away from the axis of rotation or turning.

centripital force Tendency for an object to be pulled toward the axis of rotation.

chalcopyrite Most important copper ore.

chemical compound Combination of two or more atoms.

chemical weathering Chemical alteration of rock and mineral material in Earth's surface environment; synonymous with decomposition.

chinook wind Warm wind of late winter and early spring that increases evaporation and melts snow; from American Indian word for the "snow eater"; downslope wind.

chlorinity Total amount of chlorine in grams in a 1-kilogram sample of water.

C-horizon Soil transition zone between the weathered horizons above and the unweathered parent material below.

chromosphere Reddish rim of sun from which spicules, solar flares, or prominences appear to erupt; the "color" sphere.

cinder cone Volcanic cone, normally less than 450 meters (1,500 ft) high, consisting principally of pyroclastic debris and usually forming in clusters. Each forms over a short span of time (10 to 20 years).

circumpolar water Cold surface water of high latitudes that averages 0.5° C (32.9° F) at the surface with a salinity of 34.68‰.

cirque Semicircular depression gouged from rock by glacial ice and meltwater in mountainous regions; the starting point for an alpine glacier.

clastic Textural term applied to sedimentary rocks that are formed by the accumulation and consolidation of rock and mineral fragments.

clastic wedge Clastic sedimentary deposit of regional extent that has a generally triangular shape in cross section.

cleavage Splitting of a mineral along an internal plane of weakness in a crystal; in rocks the ability to split along parallel surfaces.

climate Condition of the atmosphere over a long period at a geographic locale.

clinker End product in the process of making cement.

coal Combustible sedimentary rock formed from the accumulated and lithified remains of plants; combustible sedimentary rock whose energy is rated in BTUs per pound.

coarse texture   *See* Phaneritic texture.

coke   Baked coal used in the manufacture of iron and steel; coal from which gases and liquids have been removed by heating.

cold current   Ocean surface current that flows from high latitude to lower latitude or longitudinally at high latitudes; slow, wide, and shallow current on eastern side of oceans.

cold dark matter   A type of matter that is indirectly detected by its gravitational effects on galaxy rotation; suspected dominant form of matter of the universe.

cold occlusion   A type of weather front that forms when trailing colder air undermines warm air.

column   Depositional feature formed in a cavern by the merging of a stalactite and a stalagmite.

comet   Wandering body of the Solar System composed of ice, dust, and rock material moving in a highly elliptical orbit.

compression   Squeezing force; a type of stress caused by two opposing masses.

concordant   Parallel to the surrounding rock beds; said of an igneous intrusive feature like a sill.

condensation   Process that turns water vapor into liquid water.

conduction   Transfer of heat energy between substances in contact.

cone of depression   Localized cone-shaped depression of the water table centered on a well and resulting from excessive pumping.

contact metamorphism   Local alteration of rock that surrounds an igneous intrusion; alteration of rock localized in a halo or aureole surrounding the intrusion.

continental crust   Thick unit of the lithosphere underlying the continental areas and composed primarily of rock with an average composition of granite.

continental glacier   Large mass of ice, snow, and rock debris that can reach thicknesses in excess of 3 kilometers (2 mi) and cover hundreds of thousands of square kilometers.

continental margin   Transition zone between a continent and deep ocean basin; that includes the continental shelf, continental slope, and continental rise.

continental rise   Sedimentary accumulation that is part of continental margin located between the continental slope and the abyssal plain.

continental shelf   Part of the continent that is submerged beneath shallow marginal seas.

continental slope   The portion of the continental margin between the shelf and the rise.

continental tropical (cT) air   A dry, hot air mass that originates over the American Southwest.

convection   Vertical heat transfer by a moving fluid.

convectional lifting   Thermally induced rising of air.

convection cell   Combined vertical and horizontal air movements that distribute fluids in a wheellike cell.

convective zone   An interior zone of the sun that lies between the radiative zone and the photosphere; a solar interior zone in which hotter gases rise through cooler gases.

convergence   The collision of crustal plates; also a net horizontal inflow of air under certain atmospheric conditions.

coral reef   Organic structure built by colonial organisms, principally corals, in warm shallow tropical water.

core   The central portion of a celestial body; the central portion of the sun, the solar furnace.

Coriolis force   A deviation from a straight path on a rotating body; a tendency away from a straight trajectory observed in moving fluids on a rotating body; a tendency on Earth for moving fluids to veer to the right in the Northern Hemisphere and to the left in the Southern Hemisphere.

corona   The outermost portion of the sun; the halo.

corpuscular radiation   Charged atomic particles that make up the solar wind.

country rock   *In situ* rock that surrounds and is intruded by an igneous rock body.

crater   The principal vent of a volcano that is usually circular in form and located at the summit of the cone.

creep   Slow, imperceptible movement of earth materials downslope at Earth's surface; a form of mass movement.

crossbed   Generally small-scale, thin sedimentary rock layers deposited at some angle to the horizontal by currents of water or air.

crust   Thin, outermost portion of Earth; outermost rocky units of Earth usually divided into two types: a thinner, oceanic (basaltic) and a thicker, continental (granitic) crust.

crystal   Outer, visible geometric form taken by a mineral as it grows and governed by its internal arangement of ions.

crystal lattice   Arrangement of ions in an orderly, repeated geometric pattern within a mineral.

crystalline   Composed of interlocking crystals; synonymous with nonclastic.

crystallization   Mineral growth from magma or water solutions; formation of crystals from molten mineral matter, molten metal, or water solutions.

cumuliform cloud   A vertically developed billowy cloud.

current   Movement of a fluid (e.g., water) from one place to another through a standing body of that fluid.

cyanobacteria   A primitive form of life; blue-green algae; possibly the first life forms.

cyclogenesis   The process that develops, sustains, and intensifies a cyclone.

cyclonic   An inward and upward flow of air; a low-pressure system; a counterclockwise movement of air in the Northern Hemisphere; a stormlike movement.

D

daughter atom   Atom derived from parent atom through radioactive decay.

declination   Magnetic direction or alignment with relation to the magnetic lines of force at Earth's surface; the angular distance to an object on the celestial sphere.

decomposition   *See* Chemical weathering.

Deep-Sea Drilling Project   Series of explorations of ocean floor that was carried out on the research vessel *Glomar Challenger*.

deep-water wave   Wave in water in which the depth is greater than one-half the wavelength.

deflation   Lifting and carrying away of sedimentary particles by the wind.

deflation basin (blowout)   Shallow topographic depression caused by wind erosion.

degenerate gas   Gas produced in white dwarfs by the separation of electrons from atomic nuclei.

delta   Alluvial deposit at the mouth of a river usually resembling a triangle or the Greek letter "delta"; deltas take their shapes under the influences of river discharge, sediment deposition, tidal flow, wave energy, and ocean currents.

dendritic stream pattern   Stream pattern, resembling the branching pattern of a tree, formed in a drainage basin by the random location and development of tributary streams.

density   Mass per unit volume of a material.

deposition   Settling under the influence of gravity of sedimentary particles in water, wind, or ice.

depositional center   Area of sediment accumulation that occurs on the continental shelf near the mouth of a river.

detrital food   Parts of decaying organisms and plant life that settle to the bottom of a water body.

deuterium   Isotope of hydrogen with a proton and a neutron in its nucleus.

dew point   Temperature at which an air mass becomes saturated with moisture.

diamond pipe   Pipe-shaped vertical igneous intrusion consisting of the diamond-bearing rock kimberlite (after deposits around Kimberly, South Africa).

diastrophism   Generic term for all crustal deformation including mountain building and ocean basin formation.

diatom   Algal life form with siliceous hard part (test); the most important oxygen producers in the sea, with population densities exceeding 200,000 per liter in the North Pacific.

diatomite (diatomaceous earth)   Sedimentary rock formed from the siliceous remains of shell-secreting fresh water and marine diatoms.

dike   Igneous intrusion that cuts across the orientation of existing rock; a tabular, discordant igneous intrusive body.

dimension stone   Stone that has been fashioned to the desired size and shape for the construction of dwellings and monuments.

dinosaurs   "Terrible lizard"; many varieties of Mesozoic reptiles that occupied land, sea, and air environments and that became extinct at the end of the Cretaceous.

dip   The angle that any planar surface makes from the true horizontal; especially used to describe sedimentary rock layers whose original horizontal orientation has been altered by tectonic forces.

dipole   Having two poles, such as the North and South Poles of Earth or of a bar magnetic.

discharge   Volume of water passing any point on a stream in a given time interval.

discordant   Any igneous mass that intrudes across the orientation of surrounding rock beds.

disintegration   *See* Physical weathering.

disphotic   The mesopelagic "twilight" zone in the ocean; the zone between depths of 200 and 1,000 meters (656 and 3,280 ft) in the open ocean.

dissolution   The chemically erosive action that occurs whenever water comes in contact with soluble materials; that action that produces caverns in limestone.

dissolved salt   Mineral material in solution.

distal turbidite   Deposit of fine sediment at the end of a turbidity flow.

distributary   Stream on a delta that distributes or channels the water from the main stream channel to the ocean.

divergence   The separation of crustal plates accompanied by the formation of a rift valley; a net outflow of air under certain atmospheric conditions.

Doppler effect   Measured changes in pitch as an object moves farther or closer to a point of reference.

downwelling water   Product of convergent currents driven by prevailing winds.

dry adiabatic rate   Rate at which unsaturated air rising through the atmosphere cools until it reaches its dew point; temperature change equivalent to 10° C per 1,000 meters (5.5° F per 1,000 ft).

ductile   Property that allows a metal to be drawn out into fine wire.

dune   Accumulation of windblown sand that is moving or that has moved but can be temporarily stabilized by a protective covering of vegetation.

dune complex   Site of moving and fixed sand dunes, with sand plains, ponds, and lakes.

E

earthquake   Trembling, shaking, or vibrating of the ground surface caused by faulting.

eddies   Large, relatively homogeneous water bodies that turn clockwise or counterclockwise as they spin off a pinched meander of the Gulf Stream; any swirling fluid.

ejecta   Matter thrown from a meteorite impact site or from a volcano.

Ekman Spiral   A change in direction of currents with increasing depth in the ocean caused by friction and the Coriolis force.

electrical absorption hygrometer   Measures humidity with a thin electrolytic film containing a water-absorbing salt.

electromagnetic radiation   Energy in a wave form that originates from every vibrating body in the universe and that can traverse a vacuum at the speed of light; includes light and radio waves.

electromagnetic spectrum   A graph of electromagnetic radiation based on an orderly progression of wavelengths.

electron   Negatively charged particle that "orbits" an atomic nucleus.

element   Matter with a specific chemical character; any one of 92 naturally occurring and 11 synthesized units of matter that serve as the building blocks of all substances.

emerald   Deep green variety of the mineral beryl.

emery   Natural mixture of corundum and magnetite used mostly on paper or cloth for polishing and finishing.

energy   Ability to do work or produce motion.

enhanced recovery techniques   Means, such as steam or water flooding, by which additional oil beyond that produced by either natural flow or pumping can be recovered from a reservoir rock.

eolian   Term that refers to any wind-related processes or wind-formed features.

ephemeral stream   Stream that flows only during and immediately after periods of heavy rainfall.

epicenter   Point on Earth's surface directly above the focus of the earthquake.

epicontinental sea   Shallow sea that covers continental areas during a transgression, usually to depths of 50 meters (165 ft) or less.

epipelagic zone   Upper 200 meters (656 ft) of the open ocean; also called the euphotic zone.

equatorial low   Atmospheric low-pressure system centered near the equator.

erosion   Wearing away of rock, soil, or sediment by moving fluids, ice, or gravity-driven mass wasting.

escarpment   Clifflike margin associated with a radical change in topography (such as a change from plain to plateau) or with earthquake fault lines.

estuarine circulation   Special kind of water circulation that occurs when fresh water discharging from rivers mixes with seawater; a flow of fresh water and brackish water over the salt water governed by density differences.

estuary   Bay, lagoon, or river mouth in which salt water and fresh water meet.

eukaryote   Advanced type of body cell characterized by the presence of a nucleus.

euphotic zone   A well-lighted, surface water layer in the ocean; *see* epipelagic zone.

eustatic change   A worldwide alteration in sea level.

evaporation   Process that turns a liquid into a gas; change from water to water vapor requiring the absorption of heat energy.

evaporite   Nonclastic sedimentary rock of either marine or nonmarine origin formed by the precipitation of salts from a water solution.

evapotranspiration   Loss of water vapor through plant and soil surfaces.

exfoliation   Predominantly physical weathering process whereby rock layers are stripped away, similar to the peeling of an onion; layerlike peeling of rock along fractures formed by the reduction in pressure (stress) as overlying rocks are removed by erosion.

exfoliation dome   Rounded rock mass, commonly granite, formed by exfoliation.

F

Fahrenheit   Temperature scale on which 32° is the initial freezing point and 212° is the boiling point of pure water.

fault   Break in a rock sequence or in Earth's crust or upper mantle along which displacement occurs.

faulting   Breaking and shifting of rocks in Earth's crust or upper mantle; immediate cause of most earthquakes.

fault scarp   Clifflike surface that results from faulting.

ferroalloy   Metal combined with iron to give it desirable properties, such as durability and corrosion resistance.

ferromagnesian mineral   Mineral rich in iron and magnesium, such as olivine, amphiboles, and pyroxenes.

fertilizer   Natural mineral supplements containing elements essential for plant growth.

fetch   Extent of open water across which the wind blows; an important factor in determining wave height.

fine texture   *See* Aphanitic texture.

fiord   Deep, steep-sided estuary created when a coastal glaciated valley fills with seawater.

firn   New, granular ice that constantly forms across a zone of accumulation of a glacier when fallen snow changes by melting and refreezing.

firn limit   Boundary separating the zone of dominant ice accumulation from the zone of dominant ablation on a glacier.

fissure   Elongate opening in Earth's crust through which volcanic material is ejected or extruded.

floodplain   Flat area between the stream channel and the valley walls that is periodically subject to flooding.

flood stage   Level of a stream overflowing its banks; period when a stream spreads out of its channel onto its floodplain.

focus   Actual location, or zone, at or below Earth's surface where faulting has occurred; also known as the hypocenter.

foehn   Wind that occurs in the Alps, similar to the chinook.

fog   Ground cloud formed by condensation in the air close to Earth's surface.

fold   Structural feature that expresses itself in a wave-shaped rock unit; wavelike structural rock feature that results from compressive stresses; some folds are symmetrical; others, asymmetrical.

foliated   Metamorphic rock texture in which minerals present are aligned parallel to each other under pressure, giving the rock a layered appearance.

foliation   Any planar feature (e.g., slaty cleavage, layering, banding) occurring within a metamorphic rock; one of the principal features in the subdivision and classification of metamorphic rocks.

footwall   Block of crust along a fault plane on which you could walk if the fault plane were open; side of a fault opposite the hanging wall.

foreshock   One of a series of tremors or temblors, usually so weak as to be undetectable except by seismographs, that precedes an earthquake.

fossil   Traces or remains of organisms preserved from the geologic past; any evidence of past life.

fossil fuel   Combustible material, such as oil, gas, or coal, that was derived from organic materials.

fracture   Random, uneven breakage of minerals other than along planes of cleavage.

Frasch process   Method of sulfur production; process employing air and hot water pumped under pressure down a borehole drilled into sulfur-bearing formations.

freshwater lens   Lens-shaped body of lighter, water at a river mouth that slowly mixes with the seawater.

friction   Force that impedes the relative movement of objects in contact.

front   Leading edge of a moving air mass.

frontal fog   Ground clouds associated with weather fronts where warm air masses come in contact with colder air.

frontal lifting   Dynamically induced lifting or overrunning produced at the boundary between two air masses of differing temperature and density whenever the warmer and lighter air mass rises.

frost   Thin ice layers produced whenever the dew point is reached below the freezing point of water.

frost heaving   Process of alternate freezing and thawing of water in pore spaces and fractures with the consequent downslope displacement of mineral grains, soil blocks, rocks, and constructed objects like sidewalks.

frost wedging   Type of physical weathering process; mechanical breakdown of rock due to the expansion of freezing water in cracks and crevasses.

fuller's earth   Variety of clay used as a soft polishing medium for metals.

fusion   Process by which atoms combine to make heavier elements; the process by which the sun produces its energy.

G

galaxy   System or aggregation of stars, dust clouds, and cold dark matter; aggregations of billions of stars.

gas injection   Enhanced recovery technique in which gas (natural or nitrogen) is forcibly injected into an oil-bearing rock formation to push and lower the viscosity of the oil so that it flows more freely to the well bore.

gem   Cut and polished gem mineral.

gemstone (gem mineral)   Natural substance that can be cut and polished into decorative minerals valued because they can be set in jewelry.

geologic agent   Any fluid substance that when in motion is capable of doing geologic work; any of several fluids, such as running water, groundwater, waves, currents, wind, and glaciers, that are capable of doing geologic work.

geologic structure   Rock feature that has been shaped by deforming forces.

geologic time scale   A time line of geologic and evolutionary events that is divided into eons, eras, periods, epochs, and ages.

geologic work   Erosion, transportation, and deposition of rock and mineral material by moving water, ice, or wind; the movement of earth material from one place to another.

geology   One of four earth science disciplines; the study of the solid Earth and its processes.

geostrophic wind   Constant horizontal ideal airflow parallel to isobars produced when the Coriolis force and the atmospheric pressure gradient balance each other.

glacial till   Sediment deposited by melting ice; heterogenous mixture of sediment deposited directly by melting glacial ice.

glacier   Thick mass (as a sheet or stream) of ice that moves when its weight responds to gravity and internal pressure.

glass   Unordered atoms in solid form.

globular cluster   Aggregation of millions of stars usually associated with galaxies.

gneissosity   Texture in metamorphic rocks that shows distinct banding of alternate dark and light minerals.

Gondwanaland   Southern supercontinent composed of Africa, South America, Australia, Antarctica, Madagascar, and India.

graben   Elongated, down-dropped block of Earth's crust that is bordered by two faults.

gradient   Drop in elevation as a stream flows a certain distance; atmospheric pressure change.

gradient wind   Wind produced with balanced forces around high- and low-pressure systems that flows along either a curved clockwise or counterclockwise path.

granite   Coarse-grained, light-colored intrusive igneous rock consisting principally of the minerals microcline, quartz, plagioclase feldspar, muscovite mica, biotite mica, and hornblende.

granitic crust   That portion of Earth's crust dominated by light-colored and relatively low-density crystalline rocks typified by granite; synonymous with continental crust.

greenhouse effect   Atmospheric heating caused by the addition of heat-absorbing gases, such as carbon dioxide.

greenstone belt   Zone of metamorphosed basaltic lava.

groin   Structure built perpendicular to the shoreline to absorb wave energy; artificial headland built to capture sand in its migration, to prevent beach erosion, and to protect a seawall.

ground moraine   Rock and soil debris left as a cover by a receding glacier.

groundwater   Water occupying pore spaces within the saturated zone of the subsurface soil, sediments, and rocks.

Gulf Stream   Largest and most intensified warm current in the Atlantic Ocean; fast western boundary current in the Atlantic that averages 5 kilometers (3 mi per hour) and carries more than 50 million cubic meters (65 million yd^3) of water per second.

guyot   Flat-topped, subsided oceanic volcano.

gymnosperm   Nonflowering, cone-bearing trees.

gyres   Large, clockwise and counterclockwise pelagic surface current patterns.

H

hadal zone   Deepest seafloor; water of the trenches.

hair hygrometer   Instrument that measures moisture in the air by means of a human hair that expands or lengthens upon absorbing moisture.

half-graben   Elongated, down-dropped portion of Earth's crust bordered on one side by faults and associated with crustal rifting.

half-life   Time required for one-half of radioactive parent atoms to decay into daughter atoms.

halocline   Water zone that shows a rapid change in salinity with depth.

hanging valley   Glacially eroded valley that lies above and tributary to another glacially eroded valley.

hanging wall   Block of crust along a fault plane that would be above your head if the fault plane were open and you could walk on it; block opposite the footwall.

hardness   Degree to which any mineral resists being scratched; comparative physical value of minerals in a scale devised by Friedrich Mohs on which the hardest mineral, diamond, ranks 10 and talc ranks 1.

hardpan   Concentrated zone of resistant soil components usually found in B-horizon.

headland   Most seaward extent of the land; a promontory.

heat   Quantifiable kinetic energy of particles of matter.

heat budget   Heat reception, storage and transfer over Earth.

heavy water   Deuterium oxide; moderator in nuclear reactors.

helium flash   Process whereby a star turns helium into carbon, occurring during later stages of stellar life cycle when star collapses upon itself.

Hertzsprung-Russell diagram   Graph that astronomers use to catorgorize stars on the basis of size, color, luminosity, magnitude, and surface temperatures.

heterosphere   Outer, less-dense and unmixed layer of Earth's atmosphere.

high tide   Largest water wave generated by the attractive gravitational forces of the moon and the sun and expressed as a high-water mark on shoreline features and structures.

homosphere   Lower, dense mixed portion of Earth's atmosphere.

horizon   Each soil layer or zone.

horn   Glacial erosional remnant that forms a relatively sharp mountain peak at the juncture of adjacent cirques.

horst   Elongated, upthrown block of crust bounded by systems of faults on each side.

hot spot   Stationary magma chamber in the upper mantle linked to the surface by a fracture system up which magma moves to form volcanic features; source of magma for the Hawaiian Island chain.

humus   Dead plant material disaggregated by chemical processes and microorganisms in the soil, resulting in a concentration of dark, fine-grained organic material.

hurricane   Earth's biggest and longest-lasting atmospheric storm; tropical cyclonic storm system that is tracked and named; tropical storm with winds in excess of 120 kilometers (74 mi) per hour.

hydration A wetting that occurs whenever water molecules bind to another element or compound; in hydration the hydrogen end of the water molecules bind to negative ions, whereas the oxygen end binds to the positive ions in another substance.

hydrocarbon compound One of thousands of compounds, including methane and ethane, composed chiefly of hydrogen and carbon, that in various combinations form petroleum.

hydrologic cycle Recurrent process that moves moisture into and out of the atmosphere and through the life forms, landforms, rocks, soils, and standing water bodies of Earth's surface.

hydrothermal deposit Metallic salts that are precipitated from the hot fluids associated with magma chambers and volcanic, undersea vents called black or white smokers.

hydrothermal solution Hot, watery fluids, rich in dissolved ions, that represent the final phase of the crystallization of a magma.

hygrometer Instrument that uses water absorption or accumulation to register humidity.

hygroscopic, or condensation, nuclei
Microscopic, airborne "dust" (pollen, organic debris, soil particles, volcanic ash, sea salts, etc.) that serves as a condensation surface.

hypocenter *See* Focus.

hypothesis An unproven conjecture used to explain certain observations.

hypsographic curve From Greek Hypos, "Height" and graphein, "to write," a graph that shows the percentage of Earth's surface that lies above or below sea level.

I

Iapetus Ocean A sea that separated North America from Europe and Africa prior to the formation of Pangaea.

iceberg Calved block of glacial ice that floats in the ocean.

ice sheet *See* Continental glacier.

ice stream *See* Valley glacier.

igneous rock Crystalline rock formed from molten mineral matter.

infiltration Soaking of water into and through the ground.

inner core Innermost zone of Earth occupying approximately one-fourth of its diameter and thought to be formed of a solid iron-nickel alloy kamacite.

insolation Acronym for incoming solar radiation.

intensity Measure of a storm's or an earthquake's strength in terms of structural damage and human perception; qualitative evaluation of an earthquake's strength that is, nevertheless, measured on a numerical scale, usually the Modified Mercalli scale.

intermittent stream Stream that flows during the wetter periods of the year when its bed intersects the water table and that dries up when the water table drops below the stream bed.

Intertropical Convergence Zone (ITCZ) A shifting atmospheric zone or boundary between the Northeast Trades and the Southeast Trades.

inversion An atmospheric condition in which warmer air overlies cooler air; an atmospheric

condition in which temperature rises with increasing altitude.

ion Atom with an electrical charge.

ionic substitution Process whereby ions of various elements of similar size and charge are able to substitute for one another at specific lattice sites within a mineral.

island arc Group of volcanic islands oriented in an arc pattern as seen in map view and associated with a deep-sea trench.

isobar Weather mapping symbol used to connect points of equal pressure; a line on a map used to indicate a uniform atmospheric pressure along its extent.

isotherm Mapping symbol used to connect points of equal temperature; a line on a weather map that indicates a uniform atmospheric temperature along its extent.

isotope Atom of a particular weight controlled by its number of neutrons; variant of an element dependent on the number of neutrons in the nucleus.

J

jet stream High-altitude, high-velocity winds.

jetty Structure built on one or both sides of an inlet, channel, or river mouth to keep the passage open to navigation; jetties divert the flow of the longshore currents to deeper water, thereby causing deposition of sediment away from the channel or river mouth.

joint Break in rock along which no displacement has occurred.

Jovian planets Four, mostly gaseous, giant outer planets of the Solar System: Jupiter, Saturn, Uranus, and Neptune.

K

kamacite Mixture of iron and nickel in the inner 1,600 kilometers (1,000 mi) of Earth's radius (inner core).

karst topography Composite of surface and subsurface features formed by groundwater dissolution of soluble rock, usually limestone; topographic and subsurface expression of dissolution in a limestone region.

katabatic wind Downslope wind.

kettle Depression caused by the melting of a block of ice left by a retreating continental glacier; small pond or lake in such a depression.

kimberlite Variety of peridotite usually associated with the diamonds of South Africa and elsewhere; both the fragmented nature of the kimberlite rock and the high-pressure origin of diamonds suggest that this rock was forced into the crust from some depth in the mantle.

kinetic energy The energy of motion; energy associated with and expended by moving bodies.

L

laccolith Concordant pluton made nontabular by a pronounced central thickening that arches up the overlying rock layers.

lahar Rapidly moving volcanic mudflow.

land breeze Nighttime seaward flowing wind developed as the land rapidly loses its heat while the water maintains its accumulated heat; relatively cool land produces a local atmospheric high pressure while the air over the water produces a local low pressure.

landform Individual physiographic feature affected by geological agents; landforms include mountains, plains, and plateaus.

lapse Decrease in temperature with increasing altitude.

latent heat of vaporization Stored energy of evaporation, which can be returned to the environment through condensation.

lateral fault Type of fault characterized by predominantly horizontal displacement of rock units.

lateral moraine Rock debris eroded from a valley wall that forms an elongate mound along the side of a valley glacier.

laterite Thick, reddish, strongly leached soil formed by intense chemical weathering in the humid tropics and subtropics.

latitude Distance from the equator as measured in degrees; most significant control on the amount of heat available for the atmosphere.

lava Molten rock material (magma) that reaches Earth's surface; extrusive molten matter.

leaching Dissolution and removal of mineral material from the soil by downward percolating water.

lee side Side opposite the windward side.

lifting Dominant mechanism for air-mass cooling and subsequent condensation.

light-year (lt-yr) Distance light travels in 1 year, or 9,467,280,000,000 kilometers (5,869,713,600,000 mi).

lignite Brownish colored soft coal with relatively low heat value.

lineation The subparallel to parallel alignment of elongated mineral crystals in a metamorphic rock.

lithosphere Earth's rocky outer layer, which includes the continents and ocean basins as well as the uppermost mantle and which is broken into a number of moving segments (plates); the more rigid, rocky material above the asthenosphere.

littoral zone Seafloor between high and low tides.

long profile Cross-sectional (side) view or diagram of a stream from its headwaters to its mouth; also called longitudinal profile.

longshore current Narrow flow of water parallel to the shoreline; narrow shoreline current generated by breaking waves.

longshore transport Movement of sediment by the longshore current.

lopolith Plutonic, usually gabbroic, structure similar to a laccolith but with a distinctive downward bow that results from the sagging of underlying rock.

Love wave Type of earthquake surface wave that manifests itself in a side-to-side motion of the ground.

low tide A level of the ocean's surface between the crests of high tide.

lunar eclipse A shadow cast by Earth on the moon; an obscuration of the moon that occurs whenever all or part of the moon falls within Earth's shadow.

luster Peculiar reflective surface of a mineral; appearance of a mineral's surface as it reflects light.

M

maar A flat-floored volcanic crater; a volcanic vent crater.

magma Liquid rock material with dissolved gases that lies below ground; molten mineral matter that can extrude as lava or that can crystallize to form plutonic igneous rocks and rock features.

magnetic reversal A change in orientation of Earth's north and south magnetic poles; a flipping of the magnetosphere.

magnetosphere Lines of magnetic force that encompass Earth and that are shaped by the positions of the north and south magnetic poles and by the solar wind.

magnitude Quantification of earthquake strength determined from seismic records; earthquake strength measurement based on amount of energy released and quantified in the Richter scale; scale of stellar brightness based either on appearance in the night sky (relative) or on probable appearance if a stellar object were 10 parsecs from Earth.

malleable Capable of being hammered into a desired shape; characteristic of most metals.

mantle Spherical rocky layer of Earth that lies between the crust and the outer core.

maria Dark areas on the moon thought by Galileo to be seas; basaltic flows at the probable sites of large meteoritic impacts on the moon.

marine invertebrate One of a large number of sea animals characterized by lack of an internal backbone.

maritime polar (mP) air Cool, moist air mass that develops over high-latitude oceans.

mass number Approximate weight of an atom defined as the sum of its protons (Z) and its neutrons (N).

mass wasting Downslope movement of earth material under the direct influence of gravity; gravitationally generated downslope movement of earth material facilitated by water or ice.

matter Anything that occupies space and has mass.

meander Bend in a stream.

medial moraine Elongate aggregation of ice-eroded rock that divides two alpine glaciers; combination of two lateral moraines.

mercurial barometer An instrument that indicates atmospheric pressure by the length of a column of mercury contained in a glass tube.

mesopelagic zone Pelagic ocean water that lies at depths between 200 and 1,000 meters (656 and 3,280 ft); the "twilight zone" of pelagic water; also called the disphotic zone.

metal Any element characterized by the physical properties of malleability, ductility, and conductivity and having the tendency to lose outer shell electrons to form positively charged ions.

metamorphic aureole Contact metamorphic rock in a "halo" surrounding an igneous intrusion; synonymous with metamorphic halo.

metamorphic rock Crystalline rock changed from its original state by heat, pressure, and the introduction of chemical fluids.

metamorphism Process involving heat, pressure, and chemical fluids that brings about solid state changes in rocks and minerals.

meteor Light trail of a meteorite burning in Earth's atmosphere.

meteoroid Material traversing the Solar System; meteoroids can be irons, stones, or stony-irons; some meteoroids contain carbon in dropletlike structures called chondrules (carbonaceous chondrites).

meteorologist Weather scientist; atmospheric physicist.

meteorology One of four earth science disciplines; predominantly observational and descriptive science of the atmosphere.

Milky Way Galaxy Aggregation of stars in various stages of development, dust, gases, radiation, and possibly cold dark matter to which our Solar System belongs; our own spiral (or possibly barred-spiral) galaxy, which has an overall diameter of 100,000 lt-yr and a central thickness of 10,000 lt-yr.

mineral Naturally occurring, inorganic element or compound with a definite internal arrangement of ions and a chemical composition that is fixed or that varies within narrow limits; a natural compound with definite crystalline, physical, and chemical properties.

model Mathematical or graphic representation of natural processes.

Modified Mercalli scale An earthquake intensity scale that indicates the power of an earthquake by the level of destruction it causes.

Moho (Mohorovicic Discontinuity, M-Discontinuity) Seismic boundary that divides the crust from the mantle.

molecule Smallest individual unit of a compound formed by two or more atoms.

moraine Landform produced by the accumulation of glacial till in varying shapes, thicknesses, and size; unsorted mass of rock and mineral material deposited by melting glacial ice. *See also* terminal moraine, ground moraine, recessional moraine, lateral moraine, and medial moraine.

mountain glacier *See* Valley glacier.

mountain system System of mountains (such as the Rockies and Appalachians) whose dimensions are hundreds to thousands of kilometers long and tens to hundreds of kilometers wide and which are usually formed by plate collisions.

mudflow Type of mass wasting especially characteristic of arid to semiarid regions where viscous mixtures of sediment and water flow downslope, following stream channels.

N

native element Mineral formed from ions of a single uncombined element, such as gold (Au),

graphite and diamond (C), sulfur (S), and copper (CU).

natural gas Dominantly methane (CH_4); gaseous component of petroleum; explosive gas found in coal mines.

natural leeve One of a pair of low, parallel ridges found on either side of a channel; a ridge along a stream channel formed when the stream leaves its banks during flooding and deposits its coarsest sediment immediately adjacent to the channel.

natural resource Anything taken from Earth to support life or a way of life.

neap tide Smallest tidal range that occurs approximately twice each month; a small tidal range caused by the alignment of the sun and the moon in a right angle to Earth that partially cancels their gravitational influence on the planet.

nebula Giant cloud of gas and dust that provides the material from which stars coalesce; giant cloud of gas produced by exploding star.

nebular hypothesis Account of the formation of the Solar System in which a large nebula condensed to form the sun and its attendant bodies.

nematath Series of volcanoes that mirror each other on both sides of a spreading ridge.

neritic zone Shallow water zone that extends from the littoral zone to the edge of the continental shelf.

neutrino Very small, virtually massless, high-speed particle associated with stellar explosions and solar activity.

neutron Uncharged particle in the atomic nucleus.

nodule Oval or spherical sedimentary structure that is often rich in metals, such as iron and manganese, and the nonmetal phosphorous.

nonclastic Without visible pieces (clasts); a type of sedimentary rock texture characterized by a lack of mineral or rock particles (clasts); the texture of chemical precipitates; formed by the precipitation of minerals from solutions.

nonfoliated Exhibiting no planar features; metamorphic texture that has no layers.

nonrenewable resource A substance that can be used by humans only once, such as coal or natural gas.

nontabular pluton Term used to describe a pluton that lacks a tabular geometric shape.

normal fault Fault in which the hanging wall drops relative to the footwall; result of tension.

North Atlantic Deep Water (NADW) Norwegian Sea water that sinks lower than the Mediterranean intermediate water.

O

oceanic crust The basaltic portion of Earth's crust; that portion of Earth's crust that is dominated by mafic minerals and dark igneous rocks; the crust that underlies the oceans.

oceanic ridge One part of an extensive, linear volcanic submarine mountain system associated with seafloor spreading.

oceanography One of four earth science disciplines; study of the oceans.

O-horizon Uppermost few centimeters of the

soil, composed of abundant decaying organic material mixed with some minerals.

oil Liquid component of petroleum consisting of hundreds of individual hydrocarbon compounds.

oil shale Shale permeated by a bituminous material known as kerogen.

ophiolite Obducted, generally metaliferous ocean crust that can be identified by a suite of petrological characteristics, including pillow basalts.

orbit The path of one celestial body around another.

ore mineral Mineral that can be mined and refined at a profit.

orographic lifting Mechanically induced process that moves air to higher altitudes, where cooling takes place.

outer core Mostly liquid iron sphere from the boundary with the inner core to the bottom of the mantle, approximately 3,200 kilometers (2,000 mi) out from Earth's center.

outwash Transported sediment picked up by glacial meltwater streams and carried down a valley away from the ice front.

overburden Rock and soil overlying an economic mineral deposit.

overthrust fault Low-angle reverse (thrust) fault along which older rocks are pushed over younger rocks.

overwash delta Fanlike deposit that forms in a lagoon whenever storm waves breach barrier islands.

oxbow lake Horseshoe-shaped, detached segment of a former stream channel, now filled with water.

oxidizing atmosphere Air that contains free oxygen (O_2).

ozone Triatomic oxygen (O_3), a colorless, odorless gas naturally produced in Earth's atmosphere and found concentrated in the stratosphere, where it blocks much of the UV part of the spectrum; a surface pollutant.

P

paleomagnetism Preservation in rock of the direction of Earth's magnetic field as it existed at the time the rock was formed; geomagnetism preserved whenever molten mineral matter cools below the Curie point and crystallizes to form igneous rock.

Pangaea United landmass named by Alfred Wegener for the supercontinent that formed during the later Paleozoic from all the present-day continents ("all land").

paradigm A significant model.

parallax Apparent positional change of an object that results from a change of perspective.

parent atom Atom of an unstable isotope that undergoes radioactive decay to form an isotope of a daughter element.

parent material Bedrock, glacial till, or any other earth material from which a soil has developed.

parsec (pc) Astronomical distance unit (3.26 lt-yr).

pause Atmospheric boundary in which little or no change in temperature occurs with increasing altitude.

pearl Hardened, spherical calcium carbonate secretions of oysters within their shells.

peat Woodlike fossil fuel with approximately 60% carbon content; first stage in the conversion of plant material to coal.

pedalfer Acidic soil with humus-rich A-horizon and clay-rich B-horizon that has been completely leached of calcite; characteristic soil of the humid eastern United States and Canada.

pedocal Calcitic, alkaline soil found in dry climates, where leaching is minimal; characteristic soil of the arid and semiarid western United States.

pegmatitic texture An igneous (plutonic) rock texture with very large crystals (grains).

perennial stream Stream whose bed intersects the water table and consequently maintains its flow all year.

perigee Position on an elliptical orbit where the orbiting body is closest to the object it orbits.

perihelion Orbital position of Earth when it is closest to the sun; Earth's orbital position in July.

permeability Characteristic of a substance that allows the free passage of a fluid; ability to transmit water or other fluids; extent to which pores spaces are interconnected.

petroleum Important fossil fuel formed exclusively in sedimentary rocks and consisting of a mixture of solid (bitumen or asphalt), liquid (oil), and gaseous (natural gas) hydrocarbon compounds.

phaneritic (coarse) texture Igneous rock texture in which the individual mineral crystals are large enough to be visible to the naked eye; texture produced by slow cooling of magma; synonymous with coarse-grained.

photomultiplier Instrument that measures the amount of light striking its surface.

photon Particle or package of light energy.

photosphere Visible portion of the sun with sun spots.

photosynthesis Process by which plants convert carbon dioxide, water, and sunlight into carbohydrates, while releasing oxygen.

physical weathering Reduction to smaller-size particles of rock and mineral material in Earth's surface environment; synonymous with mechanical weathering and disintegration.

pig iron Molton iron produced from the blast furnace and usually poured into molds referred to as piglets.

placer deposit Sedimentary accumulation of heavy mineral grains (e.g., gold, platinum, diamond, and tinstone) that were eroded from the primary vein deposits, transported some distance downstream or along a beach, and then deposited.

plains Extensive continental or marine (abyssal) surfaces characterized by low elevations and low relief.

planetesimal Term used to describe small bodies that accreted to form Earth and the other planets.

plasma Superheated form of matter.

plastic Tending to be slowly deformed without returning to original shape.

plate Lithospheric unit; one of approximately ten large and numerous small segments of Earth's outer shell (lithosphere) that are in motion relative to one another.

plate tectonics Theory that accounts for many observed geological phenomena that result from the movement of tectonic plates.

plateau An extensive elevated continental surface underlain by essentially horizontal rock layers often bounded by escarpments; plateaus may have small to great relief depending on the degree of uplifting and the extent of erosion.

pluton Mass of igneous rock that crystallized before reaching Earth's surface.

plutonic (intrusive) Generic term that includes all intrusive igneous rock bodies.

point bar Sand deposit on inside of a stream meander where stream velocity, and thus energy, is reduced.

polar high Atmospheric high-pressure system found above planetary poles.

porosity Percentage of open space within soil, sediment, or rock.

porphyritic texture Igneous rock texture characterized by two different crystal sizes, usually fine and coarse, that indicates a two-phase crystallization

potential energy Stored energy; energy that any body possesses by reason of its position or location.

precious gem Gem mineral of especially high economic value. The four usually considered as such are diamond, ruby, sapphire, and emerald.

pressure gradient Difference in barometric pressure between two adjacent sites.

primary wave Type of body wave that manifests itself in a push-pull (contraction-expansion) motion; first body wave to arrive after an earthquake; also called P-wave. Such a wave can be transmitted through solids, liquids, and gases.

principle of crosscutting Explanation for the time relationship between a dike and the rocks it intrudes; the rock intruded is always the older.

principle of lateral continuity An acknowledgment that sedimentary beds predominantly form as continuous units of similar rock materials deposited in a horizontal or near horizontal plane.

principle of superposition Explanation for the time ordering of sedimentary beds; younger rocks are always deposited on top of older rocks.

prodelta Fine-grained deposit of sediments lying beyond the delta front.

prominences Largest outward surface movements or jets of solar gas.

proton Positively charged particle that is part of the nucleus of an atom.

protoplanet Second stage in the formation of a planet created by planetesimals colliding and growing in size.

proximal turbidite Coarse-grained sedimentary deposit that occurs near the origin of a turbidity flow.

psychrometer Instrument that uses evaporation to measure humidity.

pycnocline Zone of rapid change in seawater density.

pyroclastic debris Generic term for all solid particles ejected from a volcano; any solid particle ranging in size from fine dust to large blocks that was ejected from a volcano.

R

radial velocity Speed with which an object moves away from an observer.

radiation A mode of transfer for heat energy that involves the passage of electromagnetic waves through space; any transmission of electromagnetic waves or atomic and subatomic particles.

radiation fog Ground cloud that occurs under clear skies whenever air is cooled because of loss of heat through radiation.

radiation zone Sphere within the sun that extends from the core to approximatley seven-tenths of the distance to the surface.

radioactive decay Process through which parent atoms ultimately change to daughter atoms by losing subatomic particles and gamma rays.

radiometric dating General technique for determining absolute ages of both organic and inorganic materials by examining ratios of parent to daughter elements.

rainfall line North-south line that divides the United States into two nearly equal parts with the area to the east of this line receiving on the average more rainfall than that to the west.

rarefaction Expansion of the ground or a spring with the passage of a P-wave.

rare metal A metal with an average crustal abundance of less than 0.01%.

Rayleigh wave Type of earthquake surface wave that causes the ground particles to move in elliptical orbits.

recessional moraine Ridge of till accumulated along the front of a retreating glacier; one of a series of linear ridges behind and subparallel to a terminal moraine, indicating either a pause in the glacier's retreat or a readvance of the ice.

red giant Type of star created when a main sequence star fuels itself by turning helium into carbon, resulting in an expansion of its outermost layers.

red shift Electromagnetic shift toward the longer wavelengths derived from the radial velocity of an object moving away from an observer.

reducing atmosphere Atmosphere containing little or no free oxygen.

refractory Said of a material that can be used in application that require high resistance to heat (such as the lining of a blast furnace); difficult to break down.

regional metamorphism A rock change that occurs over wide areas because of mountain building; especially indicative of great pressures generated deep in Earth's crust.

relative age dating Comparative age determination; technique for placing series of events in their proper historical sequence without regard for the actual dates of occurrence.

relative humidity Ratio (in percent) between the actual amount of water vapor present and the vapor carrying capacity of the air mass.

relief Difference in elevation between high and low points of topography.

renewable resource Material that can be generated within a relatively short span of time (e.g., crops, timber) or that can be reused (e.g., air, water).

reserve Measured, evaluated, and currently exploitable portion of a resource.

reservoir rock Porous and permeable rock layer within which petroleum accumulates.

residence time Duration of the presence of a substance in some location (atmosphere, ocean, etc.).

respiration Process in which animals use oxygen and release carbon dioxide.

reverse (thrust) fault A fault caused by compressive forces in which the hanging wall moves upward relative to the footwall.

revolution Orbiting; the trip around another object.

Richter scale Open-ended logarithmic scale with increments that mark increasing releases of energy during an earthquake; a measure of earthquake magnitudes.

rip current Backwash flow perpendicular to the shoreline and out to deep water.

ripple mark Small wavelike form that develops on the surface of dunes and sand bars and that may be preserved in sandstone.

rock Aggregate or physical mixture of mineral grains or crystals.

rock cycle Graphic method of portraying the interrelationships among the three basic rock types as well as their modes of disintegration and formation.

rockfall The vertical free fall of rock material; a variety of mass movement.

rockslide The movement of rock material down a steeply-inclined slip surface such as a fault plane or a sedimentary bedding plane; a variety of mass movement.

rotation Turning on an axis.

S

salinity Measurement of solid matter in grams contained in a kilogram (1,000 g, or 2.2 lb) of seawater after any carbonate has been converted to an oxide, any bromine and iodine have been replaced by chlorine, and any organic matter has been oxidized.

salinity profile A graph that plots ocean salinities through a water column.

salt dome Vertical column or pillar of salt that originated from a sedimentary salt layer at depth and that was squeezed upward plastically by the pressure of the overlying rocks.

salt marsh Low-lying, constantly wet area of silt, sand, and organic material at the boundary between land and sea, found mostly in temperate regions.

Sargasso Sea Large subtropic, Atlantic gyre bordered by a geostropic flow that helps intensify the Gulf Stream.

saturated solution Solution with no additional capacity for hydrating salt ions.

saturated (wet) adiabatic rate Variable rate a rising mass of air cools after its dew point has been reached.

schistosity Type of metamorphic foliation usually characterized by an abundance of subparallel mica crystals.

sea arch Feature cut through a headland by the joining of opposing sea caves.

sea breeze Landward breeze developed when air pressure is low over land and high over the water, occurring primarily during the afternoon and early evening.

sea cave Cave cut in a headland or cliff by wave action.

seafloor spreading Mechanism that forms new seafloor material as new crustal material wells up from the mantle and pushes older material aside, constantly renewing the igneous rock at the axis by replacement.

sea level Position of the ocean surface relative to the geodesic sphere.

seamount Submarine volcanic pile that does not reach the ocean surface to form an island.

sea stack Islandlike remnant of an eroded headland.

secondary wave Type of earthquake body wave that moves with a vertical shaking motion and that arrives after P-waves; also known as S-wave. Such a wave can only be transmitted through a solid (rigid) body.

sediment Loose rock, mineral particles, or organic debris formed by weathering, salts precipitated from solution, or the fragmentation of the remains of organisms.

sedimentary rock Any rock that forms by accumulation and consolidation of loose rock and mineral material and/or by the accumulation of fossils and salts.

seismic sea wave Long-period, high-velocity, open ocean surface wave created by earthquake, volcanic eruption, or landslide activity in or near the ocean; synonymous with tsunami; often mistakenly termed tidal wave.

seismic tomography Study of Earth's interior through seismic wave analysis by incorporating a number of seismic stations whose operators pool data to create a three-dimensional image.

seismograph Recording seismometer; an instrument that senses and records the passage of earthquake waves.

seismogram Graphic or photographic record of the passage of earthquake waves.

seismologist Earth scientist who specializes in the study of earthquakes and the study of seismic waves.

seismometer Instrument that is sensitive to the passage of earthquake waves.

semiprecious gem Any mineral used as a gem with the exception of the four precious gems (i.e., diamond, ruby, sapphire, and emerald).

shadow zone Area on Earth's surface that does not receive S-waves from an earthquake because they are damped out upon encountering Earth's liquid outer core (S-wave shadow zone); an area on Earth's surface that will not receive P-waves from an earthquake because of the refraction of the waves as they pass through the outer core (P-wave shadow zone).

shallow-water wave Wave in water that has a depth less than one-half the wave's length.

shelf break Edge of the continental shelf; boundary between shelf and slope; boundary between neritic and pelagic waters.

shield Broad low-lying expanse of Precambrian igneous and metamorphic rock that forms the core of a continent.

shield volcano Broad-based volcano with a low topographic profile though it may reach to very high elevations; largest volcanic structure composed principally of basaltic lava flows.

shoaling Becoming shallower.

sial Term derived from the chemical symbols for silicon and aluminum; term used synonymously with continental (granitic) crust.

side-looking sonar Imaging system that bounces energy off the ocean floor at an oblique angle to produce detailed images.

sidereal month Revolution of the moon compared against the stars as the reference point (27.3 solar days).

sigma-t (or T-S) diagram Graph of the combined effects of salinity and temperature on water that enables scientists to establish density characteristics for particular water masses.

silicates The largest of the mineral groups which includes most of the major rock-forming minerals and that, in turn, account for more than 90% of the volume of Earth's crust.

sill Tabular, concordant pluton usually basaltic (gabbroic) in composition.

sima Term derived from the chemical symbol for silicon and the word *mafic* and used synonymously with oceanic (basaltic) crust.

sinkhole Ground surface depression associated with the fall of a cavern roof.

slag Waste material formed in a blast furnace as a result of limestone (added as a flux) combining with impurities released from the coke and the iron ore.

slaty cleavage Variety of metamorphic rock foliation in which the parallel alignment of the minerals allows the rock to be easily split.

sling psychrometer Psychrometer that speeds up evaporation process through the use of two thermometers that can be spun rapidly at the end of a chain or cord to measure humidity.

slump Type of mass movement characterized by a rotational movement along a curved slip surface.

soil Portion of Earth's loose surface material that is capable of supporting plants with root systems.

soil profile Total vertical sequence of soil horizons.

solar eclipse Blockage of our view of the sun by the moon.

solar flare The largest of linear and usually arcing masses of gas ejected from the sun's surface in the vicinity of sunspots.

Solar System All the attendant bodies of a sun; the nine planets, their various moons, asteroids, comets, meteorites, and *Apollo* objects that are held in orbit around our sun.

solar wind Charged particles that exit the sun at high speed; corpuscular radiation.

sorting Sedimentary process that concentrates particles of similar sizes and shapes.

source region Point of origin for weather or sediments.

source rock Rock layer in which hydrocarbons formed.

spatter cone A relatively small, often symmetrical, conical structure built up from eruptions of lava.

specific gravity Density of a substance compared to the density of pure water at 4° C (39° F); the specific gravity of water is 1.0.

specific humidity Ratio between the weight of the water vapor and the weight of the air parcel containing it.

spectrophotometer Device used to separate light into different wavelengths to enable astronomers to study the spectrum of a star.

spicule Smallest and most common outward movements of solar gas.

spit Elongate subaerial deposit of unconsolidated sediments emplaced by the longshore current; an extension of the beach into the open and deeper water of a bay.

splay Fan-shaped deposit along river banks formed when the water level breeches natural levees.

spring Seepage of water that forms when the water table intersects the ground surface along steep slopes, a fracture system, or a sinkhole.

spring tide Largest tidal range that occurs approximately twice each month when Earth, the moon, and the sun are in a straight line so that their gravitational forces are enhanced.

squall line Moving zone of intense precipitation and storm activity associated with a cold front.

stable air An air parcel that has a tendency to return to its original position.

stable equilibrium Point at which ambient air and air parcel temperatures are equal.

stainless steel Alloy of iron that contains between 12% and 60% chromium.

stalactite Icicle-like mineral deposit projecting down from the roof of a cavern and formed by precipitation from slowly dripping groundwater.

stalagmite Moundlike or columnlike accumulation of mineral matter on the floor of a cavern that was deposited by slowly dripping groundwater.

stock Irregularly shaped, discordant pluton similar in composition to a batholith, but smaller (less than 100 square kilometers, or 40 square miles, of surface exposure).

stratiform cloud Layered cloud produced in stable air.

stratigraphic column Drawings or representations of layers or strata of sedimentary rocks depicting the depositional patterns.

stratopause Upper boundary of the stratosphere.

stratosphere Earth's atmospheric zone above the zone of weather; the atmospheric zone that contains the ozone layer.

stratovolcano Volcanic cone built up by a combination of andesitic (usually) lava flows and pyroclastic eruptions, often rising abruptly from the surrounding landscape to reach great elevations.

streak Characteristic chalklike mark some minerals make when they are rubbed on an unglazed ceramic plate known as a streak

plate; the color of a powdered specimen of a mineral.

stream Water moving in a confining channel downslope across the surface of the land.

stream head Point of origin (at higher elevations) of a stream.

stream mouth Downslope end of a stream; end of a stream that empties into the base level water body.

stream pattern Any of several possible areal patterns formed within a drainage basin by large streams and their systems of tributaries. Which pattern forms is governed primarily by topography and the structure of the underlying rocks.

strike-slip fault Type of lateral fault; a strike-slip fault derives its name from the dominant movement, which is along the line of strike (horizontal) rather than along the dip (vertical) of the fault system.

strip mining Common method of mining by which shallow mineral deposits are removed by excavating all soil and overlying rock layers (overburden) to expose the deposits.

strong nuclear force Atomic force that overcomes the natural repulsion of protons and binds the nucleus together.

strophic balance An equilibrium of centrifugal and centripetal forces.

subaerial Under air; above water.

subaqueous Under water.

subbituminous coal A brown coal; a coal with low heat value.

subduction Underthrusting of one lithospheric plate by another.

sublimation Phase change process by which a solid goes directly into the vapor state or vapor into a solid state.

sublittoral Seafloor immediately below the low-water (low-tide) mark; seafloor of continental shelf.

submarine canyon A gorge cut into the continental slope by the erosive action of turbidity flows.

subpolar low A low-pressure system that is established by ascending air between the polar and subtropic regions.

subsidence Dropping or settling of the ground surface relative to the surrounding area a type of mass movement.

subsoil Another name for B-horizon; the zone of accumulation.

subtropical high Seasonally fluctuating atmospheric high-pressure system of the subtropics bordered by the westerlies and the trade winds.

Sundance Sea Jurassic and Cretaceous period transgression that covered much of western United States.

sunspot Dark area in the photosphere of the sun with a temperature below 4,000 K.

supernova An explosive stellar event during which most of the mass of a large star is expelled at high speed; a stellar explosion triggered by a rapid collapse of a massive star; the brilliant light given off by an exploding star.

supralittoral Beach and dune area above the high-water mark.

surface wave Wavelike motion of the ground that expands from an earthquake epicenter; Love and Rayleigh waves.

surf zone Shallow water zone along the shoreline where the waves become unstable and break.

suspension Continuous movement of sedimentary particles above the ground surface or stream bed by moving air or water.

swash Final expression of a breaking wave as the thin sheet of water that flows farthest up a beach face.

syncline Structural downfold in which rock layers on either side of the axis are "inclined" toward or "with" *(syn)* one another; U-shaped structure.

synodic month Revolution of the moon considered against the sun as the reference point (29.5 solar days).

T

tabular pluton Igneous rock structure that resembles a table top, that is, relatively thin compared to its length and width.

taconite Lake Superior-type iron ore of low grade consisting of alternating hematite-bearing and chert-bearing layers.

tangential velocity Rate at which proper motion occurs, described in degrees and fractions of degrees of an arc.

tarn A lake in a cirque.

tar sands Sandstones impregnated with solid asphaltic hydrocarbons; nonliquid petroleum resources found in great abundance in Alberta, Canada.

temperature Quantifiable mean kinetic energy of molecules.

temperature range The daily or annual difference between high and low temperatures.

terminal moraine Linear ridge of till formed at the end or terminus of a glacier.

terrane Usually an oceanic plateau or an island group with a density between that of average continental and average oceanic crust, that may become sutured onto a continent.

terrestrial planets Four Earth-like planets closest to the Sun and similar in size and composition, including Earth, Venus, Mars, and Mercury.

terrigenous Derived from the continents or land.

Tethys Ancient tropical sea that divided Gondwanaland from Laurasia.

texture Size, shape, and arrangement of mineral grains or crystals in rock.

theory Explanation for phenomena that is based on fact.

thermocline Water zone of rapid temperature change with depth.

thermograph Recording thermometer.

thermohaline circulation Movement of ocean water driven by density differences that result from differences in temperature (thermo) and salinity (haline).

thermometer Instrument used to measure temperatures.

thrust fault *See* Reverse fault.

tidal bore Low wall or wave of advancing seawater that moves upstream during the onset of high tide.

tidal channel Stream channel across a tidal flat or an opening in a barrier bar through which the incoming and outgoing tides are funneled.

tidal creek A channel that runs through a salt marsh; a channel in a salt marsh through which ebbing and flooding waters flow.

tidal range The difference in meters or feet between the elevations of high and low tides.

tides Largest of all shallow wave systems; bulges of water that encompass whole oceans, making their wavelengths far in excess of the water depth.

tombolo Current-deposited, unconsolidated sedimentary structure that connects the mainland to an offshore island or sea stack.

topography General configuration of the land surface; lie of the land.

topsoil Nutrient rich, dark surface layer of soil; synonymous with A-horizon.

transcurrent fault Type of lateral fault along which there is no magmatic activity.

transform fault Type of lateral fault along which there is some magmatic activity and crustal spreading perpendicular to the fault.

transgression Progressive inundation of the land by a shallow sea.

transpiration The evaporation of water from plant surfaces.

transportation The movement of any earth materials on, above, or beneath Earth's surface.

trap Location (e.g., anticlines, faults, salt domes) where migrating petroleum becomes trapped by surrounding impermeable rock units .

trench Elongate and deep V-shaped valleylike seafloor feature associated with subduction zones.

tributary Stream that runs into another, larger stream.

tritium Isotope of hydrogen with a proton and two neutrons in the nucleus.

tropical storm Violent cyclonic movement with wind speeds reaching 119 kilometers (74 mi) per hour, cartographically expressed by several closed isobars.

tropopause Uppermost fluctuating boundary of the troposphere.

troposphere Zone of weather; lowermost sphere of Earth's atmosphere.

tsunami *See* Seismic sea wave.

T Tauri Second stage of a developing star during which the energy production of the star varies.

turbidity flow Swiftly moving (up to 35 km/hr, or 22 mph), sediment-charged, bottom-flowing current in a lake, sea, or ocean responsible for cutting submarine canyons.

U

ultimate base level Lowest level to which streams can erode their channels, usually taken as sea level because streams disperse upon flowing into the ocean.

ultraviolet (UV) radiation Electromagnetic radiation with a wavelength shorter than 0.4 μm; toxic radiation that causes sunburning and skin cancer; wavelength of radiation absorbed by the ozone layer.

uniformitarianism Principle that today's earth processes are not different in kind from those processes that have operated throughout earth history and that such history can be interpreted by studying the present.

unstable air Parcel of air that rises because it is warmer than its atmospheric environment.

upslope fog Ground clouds that occur at higher elevations with the adiabatic cooling of rising air.

upwelling Process in the oceans that brings cold, nutrient-rich water from depth up to the surface.

V

valley Linear depression in the land surface formed by stream erosion.

valley glacier An elongate mass of ice that flows downslope largely under the pull of gravity and which is confined to a former stream valley at high elevations; synonymous with ice stream, alpine glacier, and mountain glacier.

variable star Star that varies its magnitude over periods of 1 to 50 days.

vein Small, irregularly-shaped mineral body formed in the final stages of magmatic crystallization and often containing concentrations of economically valuable minerals.

vesicular (porous) texture Having small pores or cavities (vesicles) formed by the crystallization of lava around escaping gas bubbles; type of igneous rock texture.

viscosity Measure of any fluid's resistance to flowing; the opposite of fluidity.

volcanic (extrusive) Term descriptive of igneous activity that occurs on Earth's surface.

volcano Cone-shaped topographic feature composed of an accumulation of lava and pyroclastic material that has been ejected from a fissure in Earth's surface; extrusive igneous feature often classified as active, dormant, or extenct based on its history of eruption.

W

warm-core ring Gulf Stream loop that closes on the Sargasso Sea water from the south.

warm current Typically western boundary ocean surface current traveling away from or along the equator; fast, narrow, and deep surface current.

water Compound of hydrogen and oxygen that exists through the sharing of electrons.

water mass Large, three-dimensional body of water that has identifiable boundaries.

water table Boundary between aerated zone and saturated zone in the ground; upper limit of the saturated zone (groundwater).

wave Expression of energy in a medium; circular movement of water particles within a body of water in response to the passage of energy.

wave base Lowest level to which a standing body of water will be disturbed by the passage of a wave; a water depth approximately equal to one-half of the wavelength of a passing wave.

wave-cut platform Relatively flat shoreline area eroded by the action of the surf.

wave refraction Bending of a wave front a part of whose energy has been robbed by friction along a shoaling bottom.

wave train Successive series of wave crests and troughs.

weather Daily changes among gases of the atmosphere; temporary condition of the atmosphere at a given locality or region.

weather elements Quantifiable characteristics, such as temperature, humidity, and pressure, that define air masses.

weathering The chemical and physical alteration of rock and mineral material as a result of exposure to the gases of Earth's surface environment.

weather sphere Dense, lowest 16 to 24 kilometers (10-15 mi) of the atmosphere where weather occurs; troposphere.

white dwarf Last shining stage of a star's development where overall size is greatly reduced.

wind Advection; horizontal movement of air caused by pressure differences.

Z

zone of ablation Down-valley portion of the glacier where loss of old ice exceeds gain of new ice; zone below firn line.

zone of accumulation Portion of a glacier at the up-valley end where aggregation of new ice exceeds loss of old ice; zone above the firn line; in soil classification synonymous with B-horizon.

zone of aeration Uppermost ground layer that water traverses in its downward infiltration movement; soil zone in which the spaces between mineral grains are partly filled with air and partly occupied by the water.

zone of leaching *See* A-horizon.

zone of saturation Zone where the pore spaces in the soil, sediment, or rock are completely filled with water; the groundwater.

zooxanthellae Symbiotic algae that live with coral polyps in reef communities.

CREDITS

Line Art

Chapter 2

2.6, 2.8: From Charles C. Plummer and David McGeary, *Physical Geology,* 5th ed. Copyright © 1991 Wm. C. Brown Communications, Inc., Dubuque, Iowa. All Rights Reserved. Reprinted by permission. **2.9:** From Carla E. Montgomery, *Physical Geology,* 3d ed. Copyright © 1993 Wm. C. Brown Communications, Inc., Dubuque, Iowa. All Rights Reserved. Reprinted by permission. **Box Figure 2.I:** Adapted with permission from Lucy-Ann McFadden and Clark R. Chapman, "Interplanetary Fugitives" in *Astronomy,* August 1992, p. 31. © Kalmbach Publishing: Astronomy Magazine.

Chapter 3

3.6: From Robert N. Wallen, *Introduction to Physical Geography.* Copyright © 1992 Wm. C. Brown Communications, Inc., Dubuque, Iowa. All Rights Reserved. Reprinted by permission. **3.7D:** From Arthur Getis, Judith Getis, and Jerome Fellmann, *Introduction to Geography,* 3d ed. Copyright © 1991 Wm. C. Brown Communications, Inc., Dubuque, Iowa. All Rights Reserved. Reprinted by permission.

Chapter 4

4.6: From Robert N. Wallen, *Introduction to Physical Geography.* Copyright © 1992 Wm. C. Brown Communications, Inc., Dubuque, Iowa. All Rights Reserved. Reprinted by permission.

Chapter 5

5.1, 5.9A-B, 5.13A-C, 5.19, 5.26: From Arthur Getis, Judith Getis, and Jerome Fellmann, *Introduction to Geography,* 3d ed. Copyright © 1991 Wm. C. Brown Communications, Inc., Dubuque, Iowa. All Rights Reserved. Reprinted by permission. **5.3, 5.6, 5.7, 5.8, 5.12, 5.15, 5.24, 5.28:** From Robert N. Wallen, *Introduction to Physical Geography.* Copyright © 1992 Wm. C. Brown Communications, Inc., Dubuque, Iowa. All Rights Reserved. Reprinted by permission.

Chapter 6

6.9A-B, 6.16, 6.23, 6.28, 6.34: From Arthur Getis, Judith Getis, and Jerome Fellmann, *Introduction to Geography,* 3d ed. Copyright © 1991 Wm. C.

Brown Communications, Inc., Dubuque, Iowa. All Rights Reserved. Reprinted by permission.

Chapter 7

7.1A, 7.8, 7.17A-B: From Charles C. Plummer and David McGeary, *Physical Geology,* 5th ed. Copyright © 1991 Wm. C. Brown Communications, Inc., Dubuque, Iowa. All Rights Reserved. Reprinted by permission. **7.1B, 7.9:** From Carla E. Montgomery, *Physical Geology,* 3d ed. Copyright © 1993 Wm. C. Brown Communications, Inc., Dubuque, Iowa. All Rights Reserved. Reprinted by permission. **Box Figure 7.A:** After Sylvia Bender-Lamb, "Magma Energy Exploratory Well Long Valley Caldera," in *California Geology,* April 1991, p. 85. Used by permission.

Chapter 8

8.2, 8.30, 8.35A: From Arthur Getis, Judith Getis, and Jerome Fellmann, *Introduction to Geography,* 3d ed. Copyright © 1991 Wm. C. Brown Communications, Inc., Dubuque, Iowa. All Rights Reserved. Reprinted by permission. **8.11, 8.12, 8.13, 8.15, 8.22A, 8.23, 8.24A, 8.27, 8.29, 8.31, 8.32, 8.36A-B:** From Charles C. Plummer and David McGeary, *Physical Geology,* 5th ed. Copyright © 1991 Wm. C. Brown Communications, Inc., Dubuque, Iowa. All Rights Reserved. Reprinted by permission.

Chapter 9

9.2B, 9.19, 9.21: From Arthur Getis, Judith Getis, and Jerome Fellmann, *Introduction to Geography,* 3d ed. Copyright © 1991 Wm. C. Brown Communications, Inc., Dubuque, Iowa. All Rights Reserved. Reprinted by permission. **9.4, 9.5, 9.12:** From Carla E. Montgomery, *Physical Geology,* 3d ed. Copyright © 1993 Wm. C. Brown Communications, Inc., Dubuque, Iowa. All Rights Reserved. Reprinted by permission. **9.6 & 9.7:** From Barazangi and Dorman, *Bulletin of the Seismological Society of America,* 1989. Used with permission from the Seismological Society of America, El Cerrito, CA 94530. **9.8, 9.11, 9.13A, 9.14, 9.16, 9.23:** From Charles C. Plummer and David McGeary, *Physical Geology,* 5th ed. Copyright © 1991 Wm. C. Brown Communications, Inc., Dubuque, Iowa. All Rights Reserved. Reprinted by permission. **9.10:** From David Laing, *The Earth System.* Copyright © 1991

Wm. C. Brown Communications, Inc., Dubuque, Iowa. All Rights Reserved. Reprinted by permission.

Chapter 10

10.2A, 10.13, 10.16, 10.17, 10.24, 10.25, 10.27: From Charles C. Plummer and David McGeary, *Physical Geology,* 5th ed. Copyright © 1991 Wm. C. Brown Communications, Inc., Dubuque, Iowa. All Rights Reserved. Reprinted by permission. **10.3:** From Alyn C. and Alison B. Duxbury, *An Introduction to the World's Oceans,* 3d ed. Copyright © 1991 Wm. C. Brown Communications, Inc., Dubuque, Iowa. All Rights Reserved. Reprinted by permission. **10.4:** From A. Wegener, 1928, *The Origin of Continents and Oceans.* Reprinted and copyrighted 1968, Dover Publications. Used by permission. **10.5B, 10.6B:** From David Laing, *The Earth System.* Copyright © 1991 Wm. C. Brown Communications, Inc., Dubuque, Iowa. All Rights Reserved. Reprinted by permission. **10.18:** From Allan V. Cox, "Geomagnetic Reversals," *Science,* vol. 163, 1969, p. 240. Copyright 1969 by the AAAS. Used by permission. **10.21:** After P. Molnar, "Cenozoic Tectonics of Asia: Effects of a Continental Collision," *Science,* vol. 189(4201):419–26, 1975. Copyright 1975 by the AAAS. Used by permission. **Box Figure 10.B:** From Charles M. McCreery, D.A. Walker, and J. Talandier, "Hydroacoustics Detect Submarine Volcanism" in *Eos,* Feb. 23, 1993, pp. 85–86. Copyright © 1993 American Geophysical Union. Used by permission of the author. **10.28:** Adapted, with permission, from the *Annual Review of Earth and Planetary Science,* Vol. 11, © 1983 by Annual Reviews, Inc.

Chapter 11

11.4, 11.12, 11.21, 11.23, 11.26, 11.29, 11.31: From Alyn C. and Alison B. Duxbury, *An Introduction to the World's Oceans,* 3d ed. Copyright © 1991 Wm. C. Brown Communications, Inc., Dubuque, Iowa. All Rights Reserved. Reprinted by permission. **11.7:** Redrawn from G.L. Pickard and W.J. Emery, *Descriptive Physical Oceanography,* 4th ed., Copyright 1982, page 37, with kind permission from Pergamon Books Ltd., Headington Hill Hall, Oxford OX3 OBW, UK. **11.10:** Redrawn with the permission of Macmillan College Publishing Company from *Introductory Oceanography,* Fifth

Edition by Harold V. Thurman. Copyright © 1988 by Macmillan Publishing Company, Inc. **Box Figure 11.A:** From Hsu, Kenneth J., *The Mediterranean Was a Desert.* Copyright © 1983 by Princeton University Press. Reproduced by permission of Princeton University Press. **11.17, 11.18:** From Richard A. Davis, Jr., *Oceanography,* 2d ed. Copyright © 1991 Wm. C. Brown Communications, Inc., Dubuque, Iowa. All Rights Reserved. Reprinted by permission. **11.20:** From Arthur Getis, Judith Getis, and Jerome Fellmann, *Introduction to Geography,* 3d ed. Copyright © 1991 Wm. C. Brown Communications, Inc., Dubuque, Iowa. All Rights Reserved. Reprinted by permission.

Chapter 12

12.10: From R.A. Davis, Jr., ed., *Coastal Sedimentary Environments.* Copyright © 1985 Springer-Verlag, New York. Reprinted by permission. **12.11, 12.32, 12.34:** From Alyn C. and Alison B. Duxbury, *An Introduction to the World's Oceans,* 3d ed. Copyright © 1991 Wm. C. Brown Communications, Inc., Dubuque, Iowa. All Rights Reserved. Reprinted by permission. **12.14B:** From Richard A. Davis, Jr., *Oceanography,* 2d ed. Copyright © 1991 Wm. C. Brown Communications, Inc., Dubuque, Iowa. All Rights Reserved. Reprinted by permission. **12.25:** From James L. Sumich, *Introduction to the Biology of Life,* 4th ed. Copyright © 1988 Wm. C. Brown Communications, Inc., Dubuque, Iowa. All Rights Reserved. Reprinted by permission. **12.29:** From Carla E. Montgomery, *Physical Geology,* 3d ed. Copyright © 1993 Wm. C. Brown Communications, Inc., Dubuque, Iowa. All Rights Reserved. Reprinted by permission. **12.32:** From Alyn C. and Alison B. Duxbury, *An Introduction to the World's Oceans,* 3d ed. Copyright © 1991 Wm. C. Brown Communications, Inc., Dubuque, Iowa. All Rights Reserved. Reprinted by permission. **12.34:** From Alyn C. and Alison B. Duxbury, *An Introduction to the World's Oceans,* 3d ed. Copyright © 1991 Wm. C. Brown Communications, Inc., Dubuque, Iowa. All Rights Reserved. Reprinted by permission.

Chapter 13

13.1, 13.7, 13.9: From Carla E. Montgomery, *Physical Geology,* 3d ed. Copyright © 1993 Wm. C. Brown Communications, Inc., Dubuque, Iowa. All Rights Reserved. Reprinted by permission. **13.7:** From Carla E. Montgomery, *Physical Geology,* 3d ed. Copyright © 1993 Wm. C. Brown Communications, Inc., Dubuque, Iowa. All Rights Reserved. Reprinted by permission. **13.9:** From Carla E. Montgomery, *Physical Geology,* 3d ed. Copyright © 1993 Wm. C. Brown Communications, Inc., Dubuque, Iowa. All Rights Reserved. Reprinted by permission.

Chapter 14

14.26: From Carla E. Montgomery, *Physical Geology,* 3d ed. Copyright © 1993 Wm. C. Brown Communications, Inc., Dubuque, Iowa. All Rights Reserved. Reprinted by permission. **14.12:** From Arthur Getis, Judith Getis, and Jerome Fellmann, *Introduction to Geography,* 3d ed. Copyright

© 1991 Wm. C. Brown Communications, Inc., Dubuque, Iowa. All Rights Reserved. Reprinted by permission.

Computer Cartography
by Cartographics, Eau Claire, Wisconsin

Chapter 2
2.21B

Chapter 3
Box Figures 3.A-B, 3.9, 3.10

Chapter 5
Box Figures 5.A-C, 5.18, 5.30

Chapter 6
6.4A-B, 6.5A-B, 6.6B, 6.7, 6.8, 6.10, 6.12, 6.13, 6.14

Chapter 7
Box Figure 7.A

Chapter 8
Box Figures 8.D&F

Chapter 10
Box Figures 10.A&C, 10.1A-B, 10.7, 10.12, 10.19, 10.21, 10.22, 10.26

Chapter 12
Box Figure 12.A, 12.1, 12.13A-B, 12.16A, 12.19, 12.20, 12.27

Chapter 13
13.4A, 13.10, 13.17, 13.23

Chapter 14
14.7, 14.8, 14.9, 14.14, 14.20, 14.22, 14.24

Appendixes
D.1, H.2

Computer Illustrations
by Diphrent Strokes, Columbus, Ohio

Chapter 1
1.2, 1.3

Chapter 2
Box Figure 2.B, Box Figure 2.C, Box Figure 2.D, Box Figure 2.E, 2.2, 2.3A-B, 2.4A-B, 2.5, 2.10, 2.11, 2.14, 2.16, 2.17, 2.25B, 2.26

Chapter 3
3.2, 3.3, 3.7A-C

Chapter 4
Box Figure 4.A, Box Figure 4.C, 4.2, 4.5, 4.7, 4.8, 4.9A-C, 4.10, 4.11

Chapter 5
5.2, 5.4, 5.5, 5.10, 5.11, 5.15, 5.16A-B, 5.17, 5.20, 5.21, 5.22, 5.23, 5.25

Chapter 6
6.1, 6.3, 6.11, 6.18, 6.20, 6.22, 6.25, 6.29, 6.31, 6.33, 6.37, 6.38, 6.39, 6.40, 6.42, 6.43, 6.45

Chapter 7
7.3, 7.6, 7.11, 7.13A-D, 7.14, 7.18A, 7.20, 7.21, 7.24, 7.25, 7.26

Chapter 8
Box Figure 8.B, Box Figure 8.C, Box Figure 8.E, 8.3, 8.4, 8.5, 8.7A-B, 8.8, 8.10, 8.14A, 8.18, 8.19A, 8.20, 8.21, 8.22B, 8.25, 8.37, 8.39

Chapter 9
9.15, 9.17, 9.22

Chapter 10
10.11, 10.14, 10.18, 10.20, 10.23

Chapter 11
Box Figure 11.A, 11.2, 11.3, 11.6, 11.8, 11.9, 11.10, 11.13, 11.14, 11.27

Chapter 12
12.3, 12.12, 12.24, 12.33

Chapter 13
Box Figure 13.A, 13.3, 13.5, 13.11, 13.12, 13.14A, 13.16, 13.18, 13.21, 13.22

Chapter 14
Box Figure 14.A, 14.1A-B, 14.2A, 14.3, 14.4, 14.5, 14.10, 14.11, 14.18B, 14.19, 14.21, 14.25

Appendixes
C.1, C.2, C.3, C.4, C.5, C.6, C.7, C.8, F.1, G.1, H.2

Photographs

Chapter 1
Opener: © Greg Vaughn/Tom Stack & Associates; **1.1, 1.5:** NASA; **1.6:** Photo by N.H. Darton, U.S. Geological Survey; **1.7:** Courtesy of Otis Brown; **1.8:** © Steven Scher Collection/Black Star; **1.9:** © Paul X. Scott/Sygma; **1.10:** Photo by S.J. Williams, U.S. Geological Survey; **1.11:** © The Huntsville Times/Glen Baeske.

Chapter 2
Opener: NASA; **2.1:** NASA; **Box Figure 2.A:** California Institute of Technology; **2.12** (top two photos): National Optical Astronomy Observatory; (bottom two photos): California Institute of California; **2.13:** Yerkes Observatory photo; **2.15:** © The Anglo-Australian Telescope Board, photograph by David Marlin; **2.18:** NASA; **2.19:** NASA, Jet Propulsion Laboratory photo; **Box Figure 2.F, 2.20:** NASA; **2.21A:** NSSDC/Goddard Space Flight Center; **2.22, 2.23, 2.24, Box Figure 2.G:** NASA; **Box Figure 2.H:** Williams College-

Hopkins Observatory, photo by Karen A. Gloria;
2.25A: Yerkes Observatory photo.

Chapter 3

Opener: © Doug Sherman/Geofile; **3.1A:**
© Victoria Sayer-Ventura Star Free Press/Gamma-
Liaison; **3.1B:** AP/Wide World Photos; **3.4:** The
Observatories of the Carnegie Institution of
Washington; **3.5, 3.8:** NOAA; **3.12:** Erwin Raisz
© 1957, reprinted with permission of Raisz
Landform Maps, Melrose, MA.

Chapter 4

Opener: © Doug Sherman/Geofile; **4.1A,B, 4.3,
4.4; Box Figure 4.B:** NASA, photo by Owen B.
Toon.

Chapter 5

Opener: © Howard Bluestein/Photo Researchers,
Inc.; **5.14A:** Courtesy of Robert N. Wallen; **5.14B:**
© A.D. Coplay/Visuals Unlimited; **Box Figure
5.E:** © Steve Starr/Saba.

Chapter 6

Opener: © David Muench; **6.2:** © Kennan Ward;
6.17: © Richard Thom/Tom Stack & Associates;
6.19, 6.21: Courtesy of Dr. Charles Hogue, Curator
of Entomology, Los Angeles County Museum of
Natural History; **6.24:** © Brian Parker/Tom Stack
& Associates; **6.27:** Courtesy of Dr. Charles
Hogue, Curator of Entomology, Los Angeles
Museum of Natural History; **6.30:** Courtesy of
Robert N. Wallen; **6.32:** Photo by David
Swanland, courtesy of Save the Redwoods League;
6.35, 6.36: © John Cunningham/Visuals
Unlimited; **6.41:** © William Ferguson; **6.44:**
Courtesy of Dr. H.H. Wallen, Jr.

Chapter 7

Opener: © Rich Buzzelli/Tom Stack &
Associates; **7.2A, B, 7.4A:** © Doug
Sherman/Geofile; **7.4B:** © Wm. C. Brown
Communications, Inc., Robert Rutford and James
Zumberge/James Carter, photographer; **7.5:**
© Wm. C. Brown Communications, Inc./Doug
Sherman, photographer; **7.7A:** © Wm. C. Brown
Communications, Inc., Robert Rutford and James
Zumberge/James Carter, photographer; **7.7B:**
© Wm. C. Brown Communications, Inc./Doug

Sherman, photographer; **7.10A:** © Doug
Sherman/Geofile; **7.10B:** © Wm. C. Brown
Communications, Inc., Robert Rutford and James
Zumberge/James Carter, photographer; **7.12A:** Jim
Nieland, U.S. Forest Service; **7.12B:** Photo by Lyn
Topinka, U.S.G.S. David A. Johnston, Cascade
Volcano Observatory, Vancouver, WA; **7.15B:**
© Doug Sherman/Geofile; **7.16A, B:** U.S.
Geological Survey; **7.17C:** The Rahm
Collection, Western Washington University; **7.22A,
B:** © Doug Sherman/Geofile; **7.23A:** © Wm. C.
Brown Communications, Inc./Doug Sherman,
photographer; **7.27A, B:** © Wm. C. Brown
Communications, Inc., Robert Rutford and James
Zumberge/James Carter, photographer.

Chapter 8

Opener: © Gerald & Buff Corsi/Tom Stack &
Associates; **8.1, 8.9A:** U.S. Geological Survey;
Box Figure 8.A: NASA; **8.14B:** Courtesy of Frank
M. Hanna; **8.16:** NASA; **8.24B:** Courtesy of C.C.
Plummer; **8.24C:** The Rahm Collection, Western
Washington University; **8.36C, D:** Courtesy of S.
Jeffress Williams.

Chapter 9

9.1: © Rory Lysaght/Gamma Liaison; **9.2A:**
National Geophysical Data Center; **9.3:** U.S.
Geological Survey; **Box Figure 9.A:** Maryland
Geological Survey. Photo by John Thrasher.

Chapter 10

Opener: © David Parker/Photo Researchers, Inc.;
10.5A, 10.6A: Reprinted with permission of the
Carnegie Museum of Natural History; **10.9:**
Scripps Institution of Oceanography; **PLATES 1-
3:** *World Ocean Floor* map by Bruce Z. Heezen
and Marie Tharp, 1977. Copyright 1977 Marie
Tharp. Reproduced by permission of Marie Tharp,
1 Washington Ave., South Nyack, NY 10960;
10.28: Adapted with permission, from the *Annual
Review of Earth and Planetary Sciences,* Vol. 11,
© 1983 by Annual Reviews Inc.; **10.29:** Photo by
W.R. Normark, U.S. Geological Survey.

Chapter 11

Opener: © Denise Tackett/Tom Stack &
Associates; **11.11A:** Courtesy of Kathy Newell,
School of Oceanography, University of

Washington; **11.11B:** © Joyce
Photographics/Photo Researchers, Inc.; **11.24:**
Courtesy of Otis Brown and R. Evans; **11.25:**
Courtesy of Alyn C. and Alison B. Duxbury.

Chapter 12

Opener: © David Muench; **12.4:** From D.J.
Fornari, D.G. Gallo, M.H. Edwards, J.A. Madsen,
M.R. Perfit, and A.N. Shor, "Structure and
Topography of the Siqueiros Transform Fault
System: Evidence for the Development of Intra-
Transform Spreading Centers," *Marine
Geophysical Researches,* 11: 263–99, 1989.
© Kluwer Academic Publishers; **12.5:** © Alex S.
Maclean/Landslides; **12.8:** The National Park
Service; **12.9:** NASA; **12.15:** Courtesy Stephen P.
Leatherman; **12.16B:** U.S. Geological Survey;
12.21,12.22: NASA; **12.26:** © Manfred
Gottschalk/Tom Stack & Associates; **12.30:**
Courtesy of John Delaney, School of
Oceanography, University of Washington; **Box
Figure 12.B:** © Kelsey-Kennard Photographers,
Inc.; **12.31B:** Courtesy of Ken Adkins, School of
Oceanography, University of Washington.

Chapter 13

Opener: © Byron Augustine/Tom Stack &
Associates; **13.8:** Courtesy of Kennecott
Corporation; **13.14B:** © Doug Sherman/Geofile.

Chapter 14

14.6: Field Museum of Natural History,
neg.#75781, Chicago; **14.15:** Field Museum of
Natural History, neg.#GEO754OOC, Chicago;
14.16: © Albert Copley/Visuals Unlimited; **14.17:**
Field Museum of Natural History,
neg.#Geo80821C, Chicago; **14.18A:** Field Museum
of Natural History, neg.#GEO82665, Chicago;
14.23: © Mack Henley/Visuals Unlimited.

Appendix J

J.1: Photo by J. Frawley of Herring Bay
Geophysics, Dunkirk MD, NASA; **J.2:** NASA;
J.3: Photo by W.B. Rossow/NASA Goddard
Institute for Space Studies; **J.4-10:** NASA.

INDEX